**The Protection and
Conservation of Water Resources**

The Protection and Conservation of Water Resources

Hadrian F. Cook

Second Edition

WILEY Blackwell

Library of Congress Cataloging-in-Publication Data

Names: Cook, Hadrian F., author.
Title: The protection and conservation of water resources / by Hadrian F. Cook.
Description: Second edition. | Chichester, UK ; Hoboken, NJ : John Wiley & Sons, 2017. |
    Includes bibliographical references and index.
Identifiers: LCCN 2016039885| ISBN 9781119970040 (cloth) | ISBN 9781119334293 (epub) |
    ISBN 9781119334309 (Adobe PDF)
Subjects: LCSH: Water-supply–Great Britain–Management. | Water quality management–Great Britain.
Classification: LCC TD257 .C66 2017 | DDC 333.91/160941–dc23
LC record available at https://lccn.loc.gov/2016039885

A catalogue record for this book is available from the British Library.

This greatly revised second edition is dedicated to the former staff and students of Wye College University of London. Wye College was world-renowned, having worked for more than a century towards inter-disciplinary teaching, research and scholarship aimed at achieving better understanding of the rural environment worldwide. The Wye Campus was eventually closed in 2009 by Imperial College London. Wye College remains much lamented.

Contents

Notes on Author

Hadrian F. Cook is a lecturer and environmental consultant. Educated at the Universities of Sheffield, London and East Anglia, his academic interests include environmental history, environmental policy and floodplain management. He taught in schools and subsequently at Wye College University of London, at Imperial College and more recently on the staff of Kingston University. A Fellow of the Chartered Institution of Water and Environmental Management, he has also worked in conservation for the Harnham Water Meadows Trust at Salisbury and was a consultant to the UK research councils' Rural Economy and Land Use programme.

Preface

Allegedly a London cabbie asked the philosopher Bertrand Russell: 'Woz it all abaat then, Bertie'? The great thinker was stumped. Decades later, we too are invited to present ready solutions for the management of natural resources, but there is unlikely to be neither a Eureka, nor a 'Crick and Watson moment'. No technical fix and no economic model so far seems, by itself, to offer hope and we are invited to contemplate this. For current water resource management options present only a tantalising *à la carte* menu to policy makers and environmental commentators.

In the preface to the original book, I recalled the drought year of 1976 that I experienced as a student. I had transferred from the Faculty of Social Science to that of Pure Science and I recall twice being quietly told 'not to make too much of your prior affiliation with social science'. This well-intentioned advice was to prove outmoded in a career that, like many water professionals, found me with a foot in both camps – I had to make a return journey. More than ever, water is about politics and since the dry year of 1976 we have learned that hard subject barriers only hinder progress.

Then the brave new world of the 1990s in water governance started to feel creaky, and what is both interesting and galling for an author, no easy solution was presented. At privatisation of the water industry talks in England and Wales, neo-liberal values were in the ascendency, although critics talked of 'a privatisation too far'. Following the re-organisation in the wake of the 1989 Water Act, regulation was seen as not only a means of looking after the water environment but, because water is a natural monopoly, regulatory arrangements represented customer interests in the absence of true market competition. It was fear of market failure that drove changes. The big question was: how could governance arrangements generate a kind of pseudo-market that bought 'private sector efficiency' to erstwhile publically owned enterprises and also allow functionality as an environmental steward?

There would be fears arising out of privatisation of what hitherto had largely been a public service. Would it increase water bills? Yes. Might certain privatised entities fail? Kind of; Enron who owned Wessex water collapsed in 1998, but the water company was sold to YTL Corporation. Safeguards exist for privatised utilities. The troubled Yorkshire Water once flirted with 'going mutual' and Welsh Water (Dwr Cymru) has been running on a 'not-for-profit' basis to some effect; owned by Glas Cymru; it is a private company limited by guarantee. Foreign domination? If euro-sceptics fear this, the Thatcher and Major administrations seemed to care little for non-UK ownership of such a key industry. Re-structuring after 1989 saw fears over sovereignty trumped by ownership outside the UK in many instances. Matters European? We do not know the actuality if EU-derived regulation of the water environment following future changes in the UK's relationship with the EU. Investment? The industry was now free to invest through borrowing, but things like leakage and disputes over new reservoir construction continued and the real and complicated fears around groundwater quality and quantity have remained. Climate change? Scientific and political orthodoxy grew to a strong consensus – this process was not only real, but also largely human-induced and water resources were inevitably centre-stage. In the winter 2011/12 one borehole in the Wiltshire chalk apparently ran completely dry. Of course, one borehole does not a dry summer make, but then it rained and rained through the summer of 2012 and later there was serious flooding in early 2014. Have emergent governance arrangements made a difference? This is probably the hardest question. While we know a lot of technical information and are increasingly learning more about the human condition in environmental governance, limited achievements in reaching the Water Framework Directive goals by 2015 raised scarcely one cheer, so who has failed? Government? Regulators? Water undertakings? The great British public? Where is any modern National Water Strategy?

By the late 1990s a discourse that boiled down to 'water as a public service' versus management through privatisation and the profit motive seemed to morph into something far more intricate and subtle. While many problems could be solved technically given sufficient investment, there had to be limits to spending, and that large place referred to as 'the environment' required protection in order to provide environmental goods and services. And, for a political scientist, it contains groups of humans – the water industry itself, water consumers, flood victims, farmers, statutory regulators, conservation interests and so on – all graced with the name 'stakeholders'. These folk may well ask: Why cannot clever engineers make flooding a thing of the past? Is guaranteeing the consumer and endless, clean and cheap water supply a 'big ask', given all those scientists employed over decades? Why cannot farmers keep nitrates, phosphates, certain animal sources of pathogens and eroded soil out of natural waters? What should we do about transgressors of good practice and, given the strong legal basis for water governance, does compulsion actually work in the long-run? Why did not the huge investment opportunities created around 1990 greatly enhance regional and national water supply schemes? Actually, no new bulk transfer schemes have since emerged and it is natural to ask whether the 'national interest' can be served by private capital interests? Governments of various hues had the opportunity for investment before the 'credit crunch' of 2008. Finally, why has Scotland retained a water industry that is effective and operational within the public sector?

Recession or not, the main political parties became viewed as negligent in terms of environmental protection or promoting 'green living'. In general, there is a protracted disillusionment that strongly suggests unease with an entrenched neo-liberal ascendency. Might we enjoy a fairer world in terms of resource management and allocation? How is that to be achieved? But there are subtle changes and these would seem to relate to a change in politics towards single issues and a rise in public consultation or participation, however achieved. The left apparently abandoned ideas around 're-nationalisation' while centre-right, keen to reduce government borrowing and expenditure, once proposed expanding voluntary activity in the 'Big Society'.

If we next presume for the moment that we have sufficient technical information and 'present knowledge will do', then where are the changes in our thinking around water?

They are no longer only rooted only in natural science or engineering. There are economic considerations of course, but the impending revolution in water management relates more to the 'why, for whom and how?' than it does to presumptions about human need. Delegated public organisations were once created for, and charged with, the delivery of clean and continuous water supply. Is there a role for the 'third' or voluntary sector?

We may have seen strong elements of a 'rolling back' of the state, particularly in ownership of key industries, yet statutory regulatory bodies remain. The naked profit motive deemed the way forward at privatisation is criticised when dealing with natural resources. However, voluntary sector organisations are fairly described as 'value-driven' and 'not-for profit'. They are also an outcome of well over a century of historic alignment of voluntary action with human welfare. It will be argued that groups who view themselves as beleaguered (smaller farmers are a good example) increasingly have a say, and their concerns are being better heard, at least in catchment management, through a range of diverse organisations.

Just when you thought it was over, concerns over water shortage turned once more to excess of many kinds. The proverbial sediment is still settling from the floods of December 2015 that continued into the new year. These devastated many settlements, hitting houses, businesses, roads and bridges across the north of England and in parts of Scotland. The government went into overdrive trying to justify its flood policy and expenditure.

Flooding is very serious, although one can be excused for seeing the funny side - at least where politicians are concerned. Prior to the flooding across northern Britain of winter 2015/6, January 2014 saw upwards of 200% average monthly rainfall (1981-2010) across much of central and southern England to say nothing of the coastal battering that hit the south west and elsewhere. Preceded by previous serious events hitting Cumbria, York, Kingston-upon Hull, places in the Midlands and elsewhere since 2000, it seems that serious destructive flooding is more than ever an issue in Britain.

Perhaps recalling initiation practices at public school the elected Conservative member for Bridgwater and West Somerset (a descendent of Queen Victoria), Ian Liddell-Grainger MP, stated that:

> 'Chris Smith is a coward, a little git and I'll flush his head down the loo' (Mail Online, 6th February 2014).

Apparently this was on account of his handling of the flood situation on the Somerset Levels and Moors which, like flooding in 2015/6, truly horrified the country. It is reassuring that the noble Lord (Smith), then the Chairman of the statutory Environment Agency, made a more measured reply, genuinely agonising over the cost and prioritisation of flood defences.

Then as flood waters remained at unprecedented heights within living memory, the then Secretary for Communities and Local Government, Eric Pickles MP, unreservedly apologised for both government and Environment Agency (The Guardian 'Environment' 9th February 2014) causing one prominent academic in the field to claim he 'would be more use as a sandbag'. Among the actual points for discussion included costs of defences in general, and specifically dredging (or the alleged lack of it) within certain main Somerset rivers. Then, we all knew it was going to be alright when another descendant of Queen Victoria (Prince Charles) was filmed taking a boat through the flooded Levels.

The question of farmers and the floods is very interesting and within it we can see a dispute between agricultural systems. Clearly, small farming enterprises lost out badly and it was comforting to see not only a public outpouring of sympathy, but real practical help from the farming community across the country (Western Daily Press, 10th February 2014). Activities on the Somerset Levels and Moors are famously located in the livestock sector, and in recent decades farmers there are enrolled in agri-environmental schemes that provide for environmental benefits as well as agricultural produce. However, detractors from the agricultural sector were quick to blame their arable counterparts farming up-catchment, especially those who indulge in farming fodder maize (BBC News: Science and Environment 7th March 2014).

In June of 2014, when one could walk along the rivers of the realm admiring tide-marks from earlier in that year, a piece of Radio 4 announced that the Environment, Food and Rural Affairs Committee published its report into the winter floods of 2013-14. Launching its Report on Winter Floods, the Chair of the Environment, Food and Rural Affairs Committee, Anne McIntosh MP said:

> "We have repeatedly called on the Government to increase revenue funding so that necessary dredging and watercourse maintenance can be carried out to minimise flood risk, yet funding for maintenance remains at a bare minimum. Ministers must take action now to avoid a repeat of the devastation caused by the winter floods." (House of Commons Select Committee, 17th June 2014).

The Committee called for fully funded plans to address a backlog of dredging and watercourse maintenance. It is also important to maintain the growing numbers of artificial flood defences. There should be better funding arrangements from an Environment Agency and it should be flexible, including a degree of localism for:

> 'the devolution of maintenance activity to internal drainage boards and to local landowners, wherever possible. The Committee also urges the Government to address the confusion over maintenance responsibilities through a widespread education campaign.'

A group of MPs criticised the lack of investment in dredging (not only Mr. Liddell-Grainger). This may have mitigated some of the flooding to a small degree, as would more investment in physical flood defences, but economic cost is considered part of a strategy. A balance should be sought. Such debate may not sound as exciting (or ridiculous) as politicians trading insults, but professional interest lies in a need to learn lessons and optimise outcomes for all involved.

Sadly, people and property will always get wet. Lessons must be learned about rights and wrongs, angry MPs, dredging or farming practice. Self-evidently, these are political matters – the small 'p' being deliberate. There is a failure of faith in institutional action and likewise in confidence in scientific and technical knowledge or practice. Importantly, blame is generally sought with perpetrators identified. Taxes are paid to solve problems while political arguments, however crude, are generally framed at first around either excess or shortage of water. We may then ask if there are associated changes afoot set within such a discourse? These will inevitably reflect present changes in intellectual movements around 'civil society', perhaps a retreat from big government manifest in as 'statism', or indeed corporatism of big companies or institutions that are 'top-down' in nature, to be replaced by 'localism'.

Localism, once the preserve of the historic British liberals is manifest across much of the political spectrum. To coin a phrase oft used by that most successful of water NGOs, the rivers trusts, 'citizens are growing webbed feet'. In towns they are pulling the ubiquitous supermarket trollies out of rivers, in country they are organising to talk with all who 'work, rest and play' in rural catchments and in both instances farmers and water industry talk with young and old and often in a welcomed educational context. Yet the retreat from top-down must never be confused with an abdication of responsibilities by the state and its institutions. Most water professionals will agree there remains a need for good and effective regulation set within agreed boundaries, and this still costs public money!

The change has left any pure technological optimism behind. Terms like 'governance' replace old unpopular terms like 'government' and 'regulation'. For the Euro-fan there has been the Water Framework Directive (WFD), that, whatever its implementation problems, remains a kite-mark for river basin management. We must strive to meet its goals, even if member states often ask what these may actually be. Basic philosophical questions are asked as outcomes drive expectations and in its wake we ponder 'what is good ecological status?' More accurately is asked 'when was good ecological status?' Meanwhile and never hitherto heard of in fortress Britain is the requirement in article 14 for 'public participation'. After all, the spirit of such Water Acts as those passed in 1963, 1974, 1989 and so on was 'we know what is good for you'. While the ugly overworked term 'technocratic' was bandied about.

In recent decades, the revolution in water governance is political rather than technical or scientific and no practitioner has been left behind. As I undertook the accustomed conference and workshop round in connection with the UK Research Council's Rural Economy and Land Use Programme ('RELU' that was funded 2003-2014), staff of agencies were to be seen sporting badges sporting titles such as 'Public Participation Officer'. They spoke of engagement and were keen to talk with academics and with those whose livelihood was affected by regulation. That is a result.

I often start talks to professionals and students with a flip cartoon involving a be-suited and bowler hatted figure declaring himself 'from the English Rivers Agency (an invention)...and I am here to help you', this is a weak jest stolen from an old quip about the Inland Revenue (latterly HMRC). Association with farmers and others tells me that, at best, there is widespread scepticism around such approaches. The gap is exposed between WFD aspirations and actually doing something about such as diffuse pollution or flooding involving real farms and real communities. Yet efforts towards solving problems require not only investment in infrastructure or technical expertise but also 'human capital'. That is, individuals who visit and engage with communities as well as proffer advice.

We have all learned a new language, and that became the terminology of public participation, the voluntary sector and (to an extent) economics. We are invited to contemplate stocks of human capital, public engagement, knowledge transfer and various kinds of 'actors' and 'stakeholders'. Grown men and women, engineers, hydrologists and ecological scientists, twizzle the stem of wine glasses and mumble about 'opinion formers', 'capacity building' in diverse 'stakeholder communities'. From a top-down view (whereby there had to be losers in the game of environmental gain for a presumed greater good) we may be witnessing (or at least seeking) a new *uber-demokratie*. Now everybody is using a curious corporatist language in an effort to better manage water resources. With new emerging 'governance environments' we are entreated to think about markets, hierarchies, networks and (above all) communities. Regarding membership of the EU, we may ask when might Article 50 be invoked, what will be the UK's relationship with EU in future and what will be the impact on be the impact on water? There will nonetheless remain a race to reconcile the central state and local concerns, and all sectors: private, public and voluntary, are involved. This volume is humbly offered as the log within a continuing journey, rather than set of prescriptive answers.

Enjoy!

Salisbury, January 2017   *Hadrian F. Cook*

Preface to the First Edition

In many respects, Tuscany and Kent are very different. I smiled when an Italian student likened the landscapes of the two places, yet one Saturday during August 1995 I had to agree with him. Driving through Kent I observed that the harvest was well under way with golden fields of wheat being attacked by combine harvesters. Large bales of straw were left, like many cotton reels, among the uniform stubble, and in the intervening fields the grass was brown. By the end of the month leaves on trees were wilting and so were the understory plants in the broadleaved woods.

Almost 20 years earlier, 1976 was abnormally dry and will always remain in my mind. I graduated that year and recall the distraction of the exceptionally hot weather in Sheffield during finals, followed by a holiday in Wales, unusual for the fact that it did not rain. This was enough to turn a young geologist towards water resource issues in later life; but 1976 was not to prove unique.

Back to the 1990s, and in the late summer of 1995 a radio programme expressed concern for the East Anglian sugar beet crop, but the media preferred to concentrate on unease about the water industry's management of resources. Yorkshire Water was the first of a number of water companies who applied to the Department of the Environment for drought orders: reservoirs were reported to be at only one-third capacity. These were the first requests for' orders since privatisation in 1989. In mid-August, the Meteorological Office was able to make statements that this was the driest summer since 1727, and the hottest since records began. One estimate put the summer at a 1:200 year 'drought event'. Events naturally threw water supply issues into focus, and water companies in particular were criticised for not investing sufficiently in leakage control.

September proved to be above average for rainfall in areas of England (sadly the Lake District was not in this category) and there were floods in eastern Scotland. By late September, with rain lashing down outside, Yorkshire Water were defending their persistence in seeking drought orders. Meanwhile South West Water was criticised for water losses from their reservoirs. The 're-wet' was not to last: around the country reservoirs were not replenishing at a fast enough rate and conservation was deemed necessary 'well into autumn'. By April 1996 there was serious concern for groundwater reserves in Kent, a situation which persisted into the following year.

Whatever the cause of the shortages (and with rising demand, much can be attributed to inflated public expectation of the resource), there are problems to be addressed. Increasing bulk supply would not seem viable in most of the drier areas of Britain. Matters of water quality, of ecological conservation, a switch to demand management, economic issues of investment and pricing, political and ethical concerns about ownership and regulation, social justice, and use of water are all critical issues. 'Commodification' has led to a move from public service to tradable commodity and caused a massive break with the past. Water is viewed as a consumer item, and there are increasingly stringent regulations upon quality. There are signs of hope, however, when somewhere in excess of 97% of supplies meet stringent European Union standards.

The privatisation of water and sewerage utilities in England and Wales proved the most controversial privatisation of all. Attacked in the media for increased charges to the consumer, private water companies countered such criticism by talking about the high level of investment since 1989. Across the 10 privatised water and sewerage service companies, chief executives received salary packages totalling almost £1.7 million in 1994/95, while typically 25% of treated supply was lost through a leaky distribution system. Here is the stuff of controversy. Water companies are increasingly the target of takeovers, hostile or otherwise, and of mergers, real or threatened. With the water industry tradable, there are many potential problems.

Changes in environmental regulation since 1989 have proved uncontroversial, and indeed may already be a success story. Since 1 April 1996 there was a new Environment Agency for England and Wales and a Scottish Environment Protection Agency. Water shortage, water quality and environmental conservation continue to provide a real test for these – and indeed other – regulators to represent local, public and national interests.

Water managers have long stoically maintained that periods of low rainfall were a part of the natural order and not a portent of change. Then a meeting of the Inter-Governmental Panel on Climate Change in Madrid during November 1995 made an announcement to the effect that changes in climate were occurring, and these could most likely be attributed to anthropogenic activity, specifically the emission of 'green-house gases'. In the water world, changes in ownership and operational problems have overshadowed wide-ranging changes in regulatory institutions and water- and land-use policy. This book aims to rectify this oversight and place such new developments centre-stage, where they belong.

Acknowledgements

It is hard to recall all the individuals who have supported and inspired me, enabling the production of this volume that is humbly offered in an effort to capture the *zeitguist* in environmental management. Inclusion in the list below does not imply agreement with any opinions proffered. A fair list would include the staff and students of the former Wye College, University of London, colleagues, students and library service of Kingston University, the Harnham Water Meadows Trust, Salisbury, participants in the RELU programme, especially those encountered at conferences whose lapel badges have not all been retained in memory, and my immediate family: Clare, Greg, Ellie and Alasdair.

It would still be unfair if I did not include a few of my immediate associates who have helped in so many ways including reviewing my work and supplying information. They are: Dr David Benson, the late Dr Dylan Bright, Dr Laurence Couldrick, Dr Roger Cutting, Dr Stuart Downward, Dr Jon Hillman, Prof Kevin Hiscock, Mr Alex Inman, Prof Andy Jordan, Prof Keith Porter, Mrs Mary-Jane Porter, Mr Arlin Rickard, Prof Laurie Smith, Dr Kathy Stearne, Mr Jacob Tomkins OBE (who read and commented on the entire draft) and Dr Martinus Vink.

Introduction

A spectre is haunting Europe – the spectre is Integrated Water Resource Management (IWRM). In environmental management terms, the powers of old Europe are at last collaborating. The problem is, as far as politics are concerned, this important issue lags behind democratic enablement and action over climate change. More specifically, diffuse pollution remains a largely unsolved issue and serious flooding seems to re-visit Britain with a vengeance. Water resource managers would be excused for crying 'me too', after the proponents of climate change have had their say.

We have actually been victims of poor river basin management for longer than we have worried about climate change, for this only hiked up the environmental agenda during the 1980s and 1990s. For achievements based in river basin management date back to the hydraulic civilisations in Mesopotamia, Egypt, Sri Lanka, the Indus valley, China and various locations in the Americas. In Europe, spectacular water management is historically associated with the Romans (draining the Pontine Marshes and complex water supply arrangements involving aqueducts) or with Moorish Spain. Furthermore, it is interesting to observe medieval irrigation and drainage associated with monastic houses.

Whole books have been written about water in Britain, including London's water supply. Tales of the Great Stink from sewage in the Thames (which closed Parliament in 1858) and the daring-dos of Georgian, Victorian and later water engineers such as Joseph Bazelgette, C.H. Priestly or the John Rennies (father and son) who worked in so many areas of engineering, came to stress personalities. Also of continuing interest are magnificent water treatment works and reservoir construction in the British uplands. Those in the Elan valley and Lake Vernwy remain controversial in a devolving Wales. The whole of nineteenth century activity was underpinned by legislation, largely aimed at industrial pollution in the north of England, industrial areas of Scotland, and likewise in Wales. In the south, aside from a few industrial problems relating to light industry-was the problem was sewage. The Rivers Prevention of Pollution act 1876 and the vision and foresight of engineers had vision and foresight for not only practical matters, such as inter-basin transfers, but also administrative issues. For example, Frederick Topliss (1879) foresaw a need for Regional Water Authorities (realised 1974–1989) and his conjectural boundaries bare a recognisable likeness to those of the water utility companies today.

Catchment institutions really start with the mid-Victorian Lee and Thames Conservancies (Cook, 1999) but such organisations reflected more the concerns of elites for they were really concerned with salmonid fisheries, rather than with public health.

The Protection and Conservation of Water Resources, Second Edition. Hadrian F. Cook.
© 2017 John Wiley & Sons Ltd. Published 2017 by John Wiley & Sons Ltd.

Yet Victorian Britain was also a place of considerable advances, both legally and techni-cally. We are reminded of extensive sewage improvements in London sewers by Bazalgette (1858-1865) and imaginative water supply improvements in other British cities as well as pioneering legislation to protect rivers (Hassan, 1998). London's Metropolitan Water Boards, on the other hand, pioneered public water supply yet was an early exemplar of over-abstraction, of springs in Hertfordshire beyond the boundaries of London. John Sheail (1982) reports this change, noting the genesis of urban-rural conflict. In modern parlance, not only is there the expansion of that city's 'ecological footprint' beyond the metropolitan area, but there was, in the 1890s, serious signs of poor stakeholder engage-ment leaving resentment. Environmental history as a sub-discipline carries with it useful lessons for the 'sustainable development' debate.

In our own generation, we are indebted to the various editions of Malcolm Newson's book, *Land Water and Development* for raising many issues bridging technical and human problems. On the other side of the pond, New York City had commenced damming valleys for water supply reservoirs in the Catskills in the 1900s, a process that was to continue for some 60 years. This left a legacy of 'bitterness and resentment towards New York City', for it took at economically difficult times, fertile valley-bottom land from agriculture (Keith Porter, 2004, *pers comm*). Curiously, persuading the good folk of upstate New York to protect their catchment for water export to NYC became a major milestone not so much in the undoubted achievements of engineers, but also in community engagement. With associated economic support, it must be added.

Since the creations of all-embracing Regional Water Authorities in 1974 there have been three 'revolutions' in British water management, although not all reached all of the UK. The first was largely economic, that was the privatisation of the utility side (water supply and sewage) and its separation from regulatory bodies. By 1990, apart from ten privatised RWAs, there were a number of (already) private water supply companies in England and Wales and their number shrank to 16 with mergers and acquisitions. Scotland had a re-organisation into water boards, but the overall ownership remained public, as did those in Northern Ireland. For England and Wales, Ofwat represented consumer interests and enacted economic regulation in the absence of true market com-petition (water is a natural monopoly), the Drinking Water Inspectorate represented human health interests and the National Rivers Authority (incorporated in the Environment Agency in 1996) became the environmental regulator. The path was appar-ently set for cleaning up Britain's rivers and protecting groundwater resources.

The second 'revolution' at about the same time was scientific and engineering in nature. Attitudinal shifts within water management, through RWAs, privatised utilities or regula-tors, moved dramatically towards conservation. 'Soft engineering' and landscape values (for which we can partly thank Jeremy Purseglove's influential, if somewhat polemical book (1988 and 2015 editions) were all part of a means of managing water and landscape with conservation in mind. Things were moving beyond mere fish conservation towards a re-engineering that resurrected a presumed pre-industrial world. For all its philosophical problems, the drive towards conservation and re-creation of Britain's his-toric landscape was set for a generation, maybe more.

The third revolution was largely political, although inevitable social change (including attitudinal shifts) and economic changes (involving funding for voluntary sector initia-tives and new means of agricultural incentivisation) will produce a main theme of this book. This is the most nebulous of all three and it may take us into a firmly political direction, yet it displays a more uncertain outcome that either of the other two. This is the governance revolution. On the one hand, it incorporates 'top-down' regulation from the European Union and 'bottom-up' factors at the catchment and community level.

These changes have to live with technocratic institutional arrangements from the past, with statism, with regulated private undertaking and with conservation imperatives such as angling, flood alleviation, carbon sequestration, habitat enhancement, soil conservation, historic landscapes and more. It is of great significance that 'the public' engage with some more than others (flooding and water quality being two examples) although complaints about water charges may also be heard.

This clumsy grafting of top and bottom is egregiously political, yet it does not really refer to any of the UK's political institutions at levels below national government. This is partly historic; there is virtually no residual legislation relating even water as a public health issue for county, borough, unitary, district or parish councils. Yet we look to these to represent our interests. Like health, but not education, there is really no linkage with local government. Parish councils, on the other hand, may suffer from problems of engagement and seek a role, yet their commonest activity is in related areas of environmental management such as re-cycling. Certain organisations have sought to remedy the situation; Hampshire County Council had proposed a system of water advocates in in the former Hampshire Water Partnership. Otherwise, there is not much more to report if local authorities are to participate in the revolution.

Critics have been keen to point out the perceived shortcomings of Big Society. This is especially significant where there is a feared overlap of public provision with consequent losses of jobs and prospects for public employees. That aside, the role of the voluntary (or third) sector in UK has a long and distinguished history, being supported by governments of both left and rights since at least 1945 (Cook and Inman, 2012) because it can do certain things well, but substitution of public services is perhaps not one of them. As far as managing water resources, the sector was all but absent unless one counts angling bodies and those concerned with conserving water-based features such as canals and mills. It is from here that a voluntary sector initiative was not to replace, *but to innovate*. The Rivers Trust grew from what were basically angling pressure groups, but took on something of the campaigning zeal of earlier conservation bodies, including wildlife trusts, CPRE and the National Trust.

Such has been the success of rivers trusts since 2000 that both public agencies and private concerns are keen to engage with them. One example to be illustrated in this book will be the Westcountry Rivers Trust who, once it secured adequate funding has been working not only to improve the river environment for the Tamar catchment and beyond, but it becomes mediator in Payment for Ecosystem Services by South West Water to the farming community and plays significant roles in education and even 'bioregional planning'. It now has the resources to operate as what might be termed a 'competent authority' and carries with it considerable gravitas.

The gravitas that comes from trust in turn belongs to an organisation that, as an NGO, is 'value driven' and beyond meeting its own running costs is clearly 'not-for-profit'. In a world imbued with neo-liberal values, the voluntary sector is here promoted, yet the use of NGOs in pollution control is a sign of market failure, because classical economics tells us that pollution is one environmental outcome of a failure in environmental management.

In order to tackle the market failure implicit in the generation of diffuse water pollution, the profit motive is eschewed and an intemediary 'honest broker' is engaged. This need not be a public body, for while 'regulation may remain necessary', regulatory bodies are 'official' and may be less trusted than, say, NGOs while private concerns are profit motivated. Environmental politics becomes complicated.

The voluntary sector is valued by neo-liberal ideologues because it is free from state control and because, as they see it, it reduces state involvement in some responsibilities that require public expenditure. Funds have to be directed (or re-directed) as there is market failure – demonstrated by pollution and loss of habitat. The solution lies – we are told – in market mechanisms, for example, incentivisation is sometimes offered to farmers, this is a market mechanism. Otherwise we are left with motives that are not-for-profit, value driven, supported by individuals who are not themselves market-driven. Statism remains in the regulatory framework. Ownership, for the present, remains in private hands.

Perhaps ideological diatribes are not required for something as important as water. This book will review the panoply of scientific, technical, political, economic and social means of addressing the problems for water in Britain. Sadly, there can never be such a thing as pure objectivity, for normative approaches inevitably play well in most areas of conservation and management. And complexity: we may talk about 'wicked problems'. This intriguing definition is about matching technical uncertainty with corresponding societal uncertainty so that, to the faint-hearted, the problem looks intractable. Water resource issues have sometimes failed to be responsive to technical fixes, so that one is forced to look elsewhere. Diffuse pollution is regarded as a 'societal problem' because (for example) it arises from interactions of common property, open access resources (of water) with a widely distributed and essential human activity (such as producing food or manufacturing) – to say nothing of the treatment and disposal of wastewater. There remains the awful prospect that only limited strides have been made in solving problems of Britain's water environment and that we still need to address massive problems rooted in both societal and technical uncertainty.

To summaries this book, Chapter 1 deals with sustainability and water policy, outlining the issues presently at issue and also scope out the challenges. It poses the question: what is integrated water management? Chapter 2 reviews water availability in Britain: is there going to be enough? Chapter 3 explores the dynamic between institutions and legislative framework, its history and includes changes in abstraction licensing and asks if that really is enough in the age of stakeholder engagement? Chapter 4 introduces the catchment approach, Chapters 5 and 6 explore issues for sustaining bulk supply, noting the imperatives of climate change. It poses the question, are we doing enough in context? Chapter 7 explores the contemporary background to water quality issues, describing problems. Chapter 8 describes case studies of catchment problems, both urban and rural including the tools available. Chapter 9 describes solutions in land use change including technical fixes. Are these sustainable, if so how? Chapter 10 is concerned with emerging new governance arrangements and Chapter 11 leaps out of Europe looking as some successful examples around the world, asking how might positive lessons be learned, especially from nearby Europe, North America and Australasia.

1

Water, Policy and Procedure

There is a certain relief in change, even though it be from bad to worse; as I have found in traveling in a stagecoach, that it often a comfort to shift one's position and be bruised in a new place.

Tales of a Traveller, *Washington Irving (1824)*

1.1 Pressing Needs for Conservation and Protection?

Among the nations, the three constituent countries of Great Britain (England, Wales and Scotland) were early to industrialise and have been that way for around two and a half centuries. While this observation sets the scene for an account of the water resources of Britain, the last 30 or more years have seen dramatic changes away from the heavy industrial sector. Yet problems persist, particularly where 'technical fixes' have not provided solutions. Once it was assumed that regulatory measures, and especially 'end of pipe' pollution problems are solved (in theory) through consenting and licencing, yet diffuse pollution of waters persists from a range of contaminants and from a range of industrial and other activities. These result largely from the ways by which we conduct our economy and new solutions are sought. Not only is Britain definitively to manage its water resources on a catchment (or river basin) basis, but new political imperatives are emerging that require water management in part to become an extension of 'civil society;' this eclipses older ideas about 'technocratic management'.

This chapter outlines the present issues for sustainability and sustainable development in water resources, and it also scopes out the challenges. This is a tall order, for there is no agreed definition of sustainability or for any prescription of sustainable development. Such received wisdom on the subject is, however, helpful to a point for there is general agreement that three spheres of 'social' 'economic' and 'environmental' sustainability are involved. Figure 1.1 shows a common variant of the famous Venn diagram used in many accounts. Other commentators choose to re-name or expand this into other spheres, including the cultural and political. Certainly the latter is of great interest here, for changes in water governance are driven by political agenda and the political dimension can be seen as the driver for the others. While implicit in the social, to delve into the murky depths of cultural activities is also implied.

The Protection and Conservation of Water Resources, Second Edition. Hadrian F. Cook.
© 2017 John Wiley & Sons Ltd. Published 2017 by John Wiley & Sons Ltd.

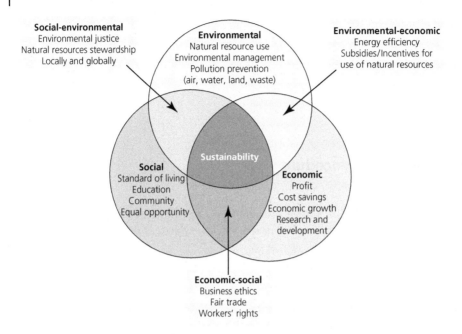

Figure 1.1 Spheres of 'sustainability' Rodriguez et al 2002.

Actually the Brundtland Report 'Our common Future' of 1987 (World Bank, n.d.) sought to define the notion of sustainable development, to use the oft-quoted definition:

> 'development that meets the needs of the present without compromising the ability of future generations to meet their own needs.'

This is a good stage-setter, but it gives an objective without ways and means. No doubt feeling a bit cynical, in some ways it has become that 'it is simply a new manifestation of an old, tired discourse.' Variously accused of being too western, too anthropocentric, even hiding the fact that the economic framework itself cannot hope to accommodate environmental considerations, perhaps a critique of the neo-liberal worldview that the answer lies within small adjustments to the market system, especially where the underlying presumption of economic growth remains. Is it, on account of being such a broad concept that it is open to wrong action, be that through good or poor motives? It can be therefore more catchphrase than a revolution in our thinking. We are sent back to thinking about equity, environmental justice and how the means of production are organised (Hove, 2004).

This book is about water in Britain, and it brings in a few comparisons from North America' from 'continental' Europe and from elsewhere. This constrains it to mature, developed economies. Furthermore, a volume that goes into the details of development theory is not its purpose. Mercifully, in water resource science and policy, there are boundaries that may be defined socially and politically (as stakeholder groups), as well as in a geological and topographic sense (Cook *et al.*, 2012). Even more helpful, if traditional aspects of standard setting and notions of 'carrying capacity' are included, presuming the calculations of these variables reflect real water quality and ecosystem issues, some of this work is done for us. For the World Conservation Union has described the idea of carrying capacity as 'improving the quality of human life while living within the carrying capacity of supporting ecosystems' (Gardiner, 1994).

Yet the real revolution in water management has been in the socio-economic-political sphere. Progress in water resource management as affects Britain seemed to falter from about 1990 onwards, and one manifestation was the UK research councils' Rural Economy and Land Use Programme (RELU) between 2004 and 2013 (RELU, 2015). Its objective was to take a long, sidewise look at UK land use issues by all interested parties, so that teams of social and natural scientists, policy makers and engineers, practitioners, theorists and all shades in between were involved. One illustration is the persistent issue of *diffuse pollution*, that is pollution (in this case of waters) that is not about attributing blame to an individual site or enterprise, *but to the way we do things*. Classic examples are nutrient pollution and pesticides from farming systems, loading of sediment to river systems from various land uses, phosphates and pathogens from sewage treatment and various hydrocarbons, salt and even heavy metal contamination in runoff from roads. In some way, we are all responsible and to go to the individual farmer or industrial concern is patently not only alienating, but it is difficult and expensive to enforce and would seem to contradict notions of 'natural justice'.

Mainland Great Britain is a small and overcrowded island and it has been stated that England, Wales and Scotland were early to industrialise. However, since around 1980 there has been a serious move away from the heavy industrial sector. Setting aside the momentous social changes of this, the often bitter legacy of neo-liberal 'Thatcherism', that changed communities often for the worse and made unemployment endemic in the economy, Britain was to suffer in the long-term for transition towards a service based economy. Industry continued, but in a muted form and with new products replacing some steel, coal products, ships and so on. Car production did continue, but it is under foreign ownership and operating under different constraints.

In parallel with industrialisation, an intensive agriculture has developed which largely concentrated in lowland areas. Official statistics for 2011 for the entire UK suggest that around 76% is in agricultural production (Defra, n.d.), this includes roads, yards, derelict land and associated buildings. Of the remainder, some 10% (and rising) is under forest and woodland and 14% under 'other', mainly urban and industrial, but also seminatural vegetation and recreational land use. Rivers, lakes, streams and canals cover some 2,580 km^2 and there is an unquantifiable volume of water beneath our feet as groundwater.

At the planetary level, the 'hydrosphere' is the arena in which hydrological processes occur and it is intimately related to geological, geochemical and biological planetary systems. As the above statistics suggest, for the land-based part of the hydrological cycle, to regard the hydrological cycle as wholly a 'natural' process belonging to some Arcadian, 'deep green' paradise world with little human intervention is of no use. Britain is certainly no exception.

With a population (2011 census) of around 61.3 million (53.0 million in England, 3.1 million in Wales and 5.2 million in Scotland), England and Wales had a population density of 371 persons per km^2 with several well-known large centres of population making it unevenly distributed and the figure is far lower for Scotland at 67 persons

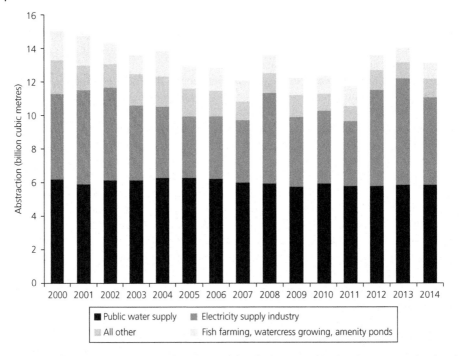

Figure 1.2 Estimated abstractions from non-tidal surface water and groundwater in England and Wales, 2000 to 2014 (Defra, 2016a) (*Source:* Environment Agency and Natural Resources Wales).

per km^2 with most living in just two large cities. Overall, England and Wales combined is among the most densely populated countries in the world. Pressures on water, as on other resources such as land and energy, are therefore considerable and potentially problematic.

Estimated water abstraction volume from non-tidal surface water and groundwater in England and Wales between 2000 and 2014 is shown in Figure 1.2 and fell steadily from an estimated 15 billion cubic metres in 2000 to 11.4 billion cubic metres in 2011, after which it increased again.

Overall categories continue to fall except spray irrigation, included in 'other' in Figure 1.2 (Defra, 2013a; 2016a).

The changes in abstraction levels between years include factors attributed to:

- Weather conditions, for example drier years could result in an increase in abstraction for agriculture and spray irrigation.
- Changes in the level of activity in different sectors.
- Improvements being made in the efficiency of water usage.
- Changes to abstraction licences.

The main reason for the overall decrease in abstraction between 2013 and 2014 is the fall in the level of abstractions used for electricity generation. Levels of abstraction for electricity generation fell by 16% from 6.4 billion cubic metres in 2013 to 5.3 million cubic metres in 2014. The abstractions for public water supply decreased slightly between 2013 and 2014 by 2% to 5.8 billion cubic metres in 2014.

For Scotland, the overall figure for abstraction is smaller, for 2012 it being about 737.7 million cubic metres from an overall far wetter climate (Scottish Government, n.d.a) and showing a strong downward trend of 13% between 2002/3 and 2009/10. Between 2010/11-2011/12, the volume of raw water abstracted also decreased but the figure is calculated using a different methodology. There was a 16% reduction in treated water produced between 2002/03 and 2009/10. Between 2002/03 and 2011/12, domestic water consumption increased by 7%, whilst non domestic consumption reduced by 18%. Of this around one third is for domestic supply. Leakage in the delivery system in Scotland in 2012 was around 27% of the total extracted with a long-term downward trend. Overall, however, water resources in Britain remain tight and continue to present problems for continuity of supply as well as water quality, something that will be dealt with later.

Perhaps to become too embroiled in statistics is to fail to set the scene properly. The curious reader can always visit many websites, be they governmental, regulatory, industrial, or linked to supra-national organisations such as the European Commission or United Nations organisations. The fascination many of us have for water is not only its barometer on how 'sustainable' our development may be, but more importantly how might this precious resource be allocated with equity, and how might appropriate governance structures be put into place.

Drivers of change have long been not so much domestic legislation, but EU directives, culminating in arguably the most dramatic and far-reaching Water Framework Directive or WFD (2000) of the European Union (EU). This bold step forward in integrating socio-economic and technical measures is correctly called the "Directive 2000/60/EC of the European Parliament and of the Council establishing a framework for the Community action in the field of water policy" (EC, 2014a,b). WFD calls for 'integrated river basin management for Europe' and incorporates notions of 'good status' for waterbodies (ecological, chemical and quantitative), achieving a 'good status' for all waters by a set date(s), base water management on river basins, streamlining legislation, finding "combined approach" of emission limit values and quality standards, getting the prices right and getting the citizen involved more closely.

In the author's experience, it was heralded by a warm welcome, even excitement by a whole range of water professionals. Only at the time of writing are notes of caution being sounded. However, it remains the yardstick and as far as water institutions are concerned, there has been a focus (or range of foci) that has caused a degree of re-orientation. No bad thing, some would say.

Perhaps as far as Britain is concerned, there are two key points. One is that before 2000, the EU set no regulation on quantity of water abstracted. Now it was a matter of supra-national legislation. The other being that the 'great British public' was to be involved in the decision making. How and why this development remained somewhat obscure is unknown, but it certainly sharpened up and advanced the political dimension in the watery debate. Meanwhile, other issues may have been regarded as being in hand. Getting the price right was part of an objective of water industry privatisation after 1989, itself a part of a 'neo-liberal' economic revolution that remains highly controversial to the present; the aim was to remove much of the cost of investment and regulation from the public purse. The concept of 'Good Status' is complicated and will be dealt with later, but would parallel moves towards environmental conservation and away from the dominance of 'hard-engineering' solutions.

Suffice to say, to anyone who grew up with notions of publically owned Regional Water Authorities between 1974 and 1989 as being stable institutions operating 'in the public good', remains bemused and intrigued by the changes in succeeding decades. Anything described as 'new' becomes suspect, for it too may soon become a curiosity of environmental governance history! We remain deeply concerned with changing regulatory environments.

1.2   A Conceptual Framework for Water Regulation

Water supply and usage is a matter of appropriating water from the land surface, sea and underground for some purpose while minimising the knock-on effects to both the resource and the wider environment including flooding (Defra (2014a). To summarise:

- *Water resources* are to do with resource balance within defined hydrographic units (such as a river basin or groundwater unit) and with sustaining bulk supply. In England and Wales, abstractions are controlled by licence controls and a parallel system has recently been developed for Scotland.
- *Water uses* include fisheries, recreation, conservation and navigation and they have to be identified as separate from instrumental abstraction for public or industrial supply. These uses will impact issues of both water resources and water quality.
- *Water quality* considerations should equally apply to chemical quality, (in terms of both direct toxicity as to other impacts of human activity including nutrient loading from agriculture of sewage effluent) as to biological, aesthetic and physical conditions that influence it. Water quality objectives are set within the envelope of absolute limits, now set by the EU that replaced governmentally set standards. They will also affect the uses to which water is put.
- *Flood defence* hitherto has involved public bodies taking charge in order to improve or maintain waterways, or to plan for flood management. There have been major re-evaluations since the flooding of 2000/1 across southern England and later serious occurrences further north.
- *Floodplain and wet habitat restoration* increasingly includes the relaxation of defences to save public money and to allow for the re-wilding of river corridors and coastal areas to create or renew floodplain functions and wetland habitats. Originally lead by Scottish public policy, this is now a major consideration across Britain.
- *Internalising costs*, starting with regulation the idea of funding of water resource management has been prevalent in England and Wales since the 1980s. While grant-in-aid for flood defence will always involve public expenditure, investment in the water industry *per se* is a matter of private investment.
- *Getting the citizen involved* is perhaps the single greatest development in water resource planning in 20 years or more. Moving beyond mere public consultation in planning to a point where communities have a say in water management involves public participation is revolutionary given the 'technocratic' past of the water industry. Here is a new need for total 'stakeholder' engagement.
- *The rise of new institutions* is needed to take public engagement forward. While there are always rumours of new statutory bodies, the *zeitguist* of the present day demands the growth of voluntary sector body involvement, including the rise of Rivers Trusts as new actors in river basin management and governance.

- *Planning of water resources and allocation of functions* is implicit in all the above; however, a competent authority that delivers 'regulation' is central for forward planning. This includes all of strategic planning of bulk supply options at a national scale (including bulk regional transfers) to water resource planning and related development within a catchment. Regulatory bodies are involved in the statutory planning system when, for example, urban development is involved. Integrated Water Resource Management manifest in River Basin Management Plans (formally 'Catchment Management Planning') should allow for the holistic appraisals of options and issues defined within river basins.

Britain had moved forward greatly with the adoption of Catchment Management Planning from 1990 onwards. The laudable aim was indeed to move water into the statutory planning process, something that had been evolving since at least the famous Town and Country Planning Act of 1947. Now the EU is underpinning this, and expanding its functions into new and democratic realms. The full history of this is covered in Chapter 3, however, here it is worth noting the various policies that have been put in place to deal with pollution, and are a part of the need to consolidate (or streamline) legislation.

Water quality has persistently continued to present problems for water managers and planners in Britain. While it is not so serious north of the border, it persists in England and Wales and predictably is worst in areas that are densely populated, or intensively farmed. One commonly heard distinction is between point pollution and diffuse pollution. Point pollution at least in theory is controlled by 'end of pipe measures' and can be licenced as 'consents to discharge' by a regulatory body, because the location, type of contaminant and potential perpetrator of pollution incidents is generally known – or knowable. There may be legal constraints on ordinary discharges, means of prosecuting for illegal discharges and the possibility of land-use controls in urban areas. Aside from criminal discharges that potentially lead to prosecution, point pollution can arise from operational errors and plant malfunctions. Unfortunately, the water industry itself is not exempt from this!

Diffuse (sometimes non-point) pollution remains highly problematic. Unlike urban areas, there are no real controls over rural land use. It is incentivisation through 'agri-environmental schemes' that are, and have been for some 30 years or more, the way forward. For example, Higher Level Stewardship (Defra, 2014b) incentivises farmers to manage heritage landscapes, that is 'to deliver significant environmental benefits in priority areas'. Arguably, benefits by way of reducing diffuse pollution to waters is a by-product, although reducing the farmed area under the plough, or in intensive livestock production, can only help reduce sediment, nitrogen and phosphorous loading. A more 'pollutant targeted' approach has been tried with controlling nitrogen inputs to land, since 1990. Now manifest as 'Nitrate Vulnerable Zones' (EA & Defra, 2014) these are compulsory (that is not incentivised) for: 'If your farm is in an NVZ, you must comply with the NVZ rules'. Driven by EU legislation, the Nitrates Directive (91/676/EEC) NVZs cover some 62% of the area of England, 3% in Wales and 14% in of Scotland (Scottish Environment Link, 2015). In a sense, this is a control on rural land use, although it is really management of land that is the objective.

Implicit in the above description is a question: What is politically and economically the best way to engage with communities and/or industries who might be potential polluters? The cost of 'regulation' may be met by a polluting industry as part of its profits, and in economic terms a means of 'externalising internalities' (Chapter 10). That said, pragmatic considerations are both political and economic. To do nothing until there is an 'incident', then issue first a fierce warning followed by a fine for non-compliance with a licence or regulation regarding water contamination does not easily engage the 'offending' stakeholders, nor is it especially cost-effective. The regulatory authorities are forever chasing (chemically or otherwise) potential miscreants and that costs technician and equipment time, to say nothing of ensuing legal costs.

Carrots (rather than sticks) are proving more than ever to be effective in terms of the politics of engagement and cost of regulation. This is the revolution in water governance. Farmers never were out to pollute for its own sake, far from it, but poor communications and isolation lead to misunderstandings; those misunderstandings lead to economic inefficiencies in terms of fertiliser use, animal welfare and so on. Rivers Trusts and others are proving new arenas, not only for the resolution of conflicts, but also acting long-term as Non-Governmental Organisation (NGO) intermediaries, sometimes termed 'honest broker' agencies for 'good stewardship' of land and water, and hence themselves providing leadership. Water companies, once vilified by creating poor ecological conditions by discharging poorly treated sewage, are in the business of payment for ecological services provided by land managers of one sort or another, and in doing so are protecting their own assets: the water environment itself. Before moving on, we can contemplate replacing the famed 'Polluter Pays Principle' by the 'Beneficiary Pays Principle.' Controversial in terms of blame allocation, but for the river basin manager or water consumer the real question is 'Does it work?'

1.3   The Historical Perspective

There is an agenda for 'green history' (Sheail, 2002), and water naturally plays a major part in that. In short, alongside social and natural sciences, there is a great need for historical enquiry in environmental studies, for *inter alia* it informs the sustainable development debate (Cook, 1999a). Furthermore, the breadth of historical enquiry (documentary, palaeo-environmental, archaeological etc.) is of itself integrative, making a parallel with the required 'holism' of the entire sustainable development debate in general and with catchment management in particular. In short, we must learn from past mistakes.

Human societies intervene in the hydrological cycle in order to exploit water resources. The usual ways are by constructing reservoirs to regulate river flow, by diverting surface flows and by sinking wells (later boreholes) into aquifers (rock formations that store water in economically useful amounts) to tap groundwater. Excess water is removed by improvements to drainage, often by the digging of ditches and by engineering works that widen and straighten channels. The objective has always been to meet local needs by smoothing out irregularities in the supply, on a seasonal basis.

The human dimension is further realised through legal systems. All major legal systems through time have adopted the principle that there can be no private ownership of flowing water (Kinnersley, 1988). The early 'hydraulic' civilisations of the Middle East and elsewhere show that water management in agriculture may go back millennia, typically in situations when canals were dug to divert water to fields growing crops.

The very survival of such civilisations depended upon water management at all scales. In Britain specifically, 'civilisation' came later in history, however the temperate climate enabled the 'Neolithic Revolution' that included the development of agriculture from about 6,000 years ago. Water management if only in the form of ditches dug to drain waterlogged soils date from prehistory (Cook, 1994). The following account traces the development of water resource exploitation from the perspective of surface waters (rivers, dams and flood alleviation), agriculture, groundwater, industry and urbanisation. Legislation is covered in Chapter 3.

1.3.1 River and Spring Diversion, Dams and Flood Alleviation

The development of water management is the starting point for appreciating the evolution of the water industry as it is today, a major player in Britain. Larger scale 'engineering' works date from the Roman occupation. Much was learned from the imperial city itself, for Rome was supplied via over 300 km of aqueducts (Binnie, 1995). Urbanisation brought piped water to Roman cities, and an aqueduct discovered in 1900 brought water to *Durnovaria*, Roman Dorchester from the river Frome in the late first century AD (Engineering Timelines n.d). Sheail (1988) traces the beginnings of direct river modification to the first century AD, the prime aims being improved land drainage, prevention of flooding and transportation. Modified areas included the Somerset levels and Fenland, where the Car Dyke in Lincolnshire may date from the time of the Roman occupation (Purseglove, 1988 p. 40). Also, during the Roman occupation, the first dams were constructed across small streams in connection with the fortifications on Hadrian's Wall and associate with mineral workings (Newson, 1992a). Advanced engineering skills were probably largely lost following the collapse of Roman administration in 410 AD and recovered in the medieval period. Water mills, likely introduced to Britain by the Romans, are known later from the Anglo-Saxon period (Stanier, 2000) and fish ponds, fish weirs, mill races and leats were recorded in the Domesday Book (1086).

After the Norman Conquest, there are well-reported water supply arrangements for the former Benedictine Christchurch Priory at Canterbury (the modern Cathedral), where from about 1160 AD, water was drawn from springs in the lower tertiary deposits outside the city through an aqueduct system of pipes to a water tower that survives today. The canons of St Gregory's were allowed to divert some of this water for their priory *en route* in exchange for a basket of apples each September (Hayes, 1977). Medieval pipes were typically made of elm (which was durable), although in 1236 the Corporation of London provided six-inch lead pipes to convey water from the River Tyburn to a conduit head at Westcheap. Cast-iron pipes were introduced in Edinburgh in 1755 but leaks prevented widespread use before the start of the nineteenth century (Binnie, 1995).

The historical dimension should also consider the development of what in the modern sense would be regarded as scientific ideas. Early ideas concerning the hydrological cycle could be intriguing, but somewhat off-track (Cook, 1999a). By the late seventeenth century, both English and French thinkers were approaching modern understandings of hydrology. Significant insights were provided by Rene Descartes, whereas Pierre Perrault developed the concept of the hydrological cycle, accounting for the disposition of rainfall by evaporation, transpiration, groundwater recharge and runoff (Nace, 1974).

Edme Mariotte established that precipitation was sufficient to account for river flow and Edmund Halley established that the water on land could be accounted for by precipitation alone and that oceanic evaporation could account for precipitation worldwide (Smith, 1972). Hydrological measurement thereby became a part of the empirical method, and during the eighteenth century, rain gauges were employed to establish the amount and distribution of rainfall in England. Indeed, hydraulics became a favourite subject of papers delivered to the Royal Society since its foundation in 1661 (Hall, 1989).

Pressure gradients enable water to flow, and where the topography is not suitable, water has to be pressurised either for direct supply, or more usually to fill a service reservoir (such as a 'water tower'). In 1581 an undershot paddle wheel system was installed beneath an arch under London Bridge, and this was used to pump river water into a tower and through a distribution system in the City (Hall, 1989, p. 14; Binnie, 1995). It was used to supply the lower part of the City, and was operational until1822. However, higher ground continued to require new gravity-fed arrangements. The eighteenth century saw more developments in pumping technology; lifting was first achieved via a windmill, then horse-driven pumps and finally a steam engine in 1768. By the end of the century, water supply developments were going hand-in-hand with canal technology, enabling a steam-pump and reservoir dominated industry to emerge in Victorian times.

Dam construction became widespread during the eighteenth century associated with water supply, to provide compensation water (a guaranteed supply of water released to keep a mill turning) and as canal reservoirs. By 1900, the era of large dam building had begun (Sheail, 1988). It was the growth of large centres of population, in response to industrialisation, which led to the widespread construction of domestic supply dams. Dam and reservoir construction accompanied by appropriate distribution systems is a means not only of smoothing supply by regulating the transfer of water from the upper to lower regions of catchments, but also of delivering water at the right place. An Act of Parliament in 1609 had enabled the construction of the 'New River', completed in 1613, to supply London. Water was brought from springs at Chadwell and Amwell in Hertfordshire, via a channel 1.2 m deep and 3 m wide with a typical fall of only 4 cm per kilometre, to the Round Pound at Clarkenwell, London, and thence to a cistern for distribution (Hall, 1989, p. 13). A century later, in 1709, the New River Company caused water to be raised to another small reservoir in Pentonville so that houses on higher ground could be supplied.

Reservoir construction on a significant scale meant the danger of river flows ceasing altogether during dry periods, and the major spur to construction was the canal system which itself had to be maintained with sufficient water. In order to maintain water in higher-level canals, while preventing rivers drying up altogether, inflow could be divided into two streams, one for reservoir inflow, the other encircling the new reservoir to the original channel. Sheail (1988) notes how much time was spent by Parliamentary Committees considering these 'compensation flows' compared with other aspects of reservoir construction, at a time when hydrological principles were poorly appreciated. In modern times, compensation flow remains an important design constraint, with flow discharged through a valve in the base of a dam. An organised system of river management emerged during the nineteenth century, and by 1900, both large dam building and inter-basin transfers of 'raw' waters had begun.

Victorian dam construction was common in order to supply growing urban areas and included Lake Vyrnwy in North Wales to supply Liverpool. Masonry dams were possible provided impermeable rock foundations were available, and Vyrnwy is an example. Of around 3000 dams in Britain today, almost half belong to the Victorian era, and most of these were of the 'embankment type', with a puddle-clay core. In this pre-soil mechanics era, experience dictated a slope of 1:3 on the water side, and 1:2 on the outside, irrespective of soil (Binnie, 1995). An example is the Upper Bardon Dam in Yorkshire. Such engineering knowledge was reached through experience, including dam failure and loss of life. Dam construction reached its zenith in the immediate post-Second-World-War period, but slackened off after 1960. Reservoir construction is no panacea for future water supply problems. Spin-offs such as water recreation have long been overshadowed by considerations of land-take from agriculture loss of natural habitats and loss of community where settlements are flooded.

The great age of dam construction can and never will be repeated. However, the dramatic alteration to landscape and environmental impact, as well as economic feasibility is often questioned in association with new proposal for impoundment. Proposals in water-stressed lowland southern England for lowland pumped storage reservoir construction at Broad Oak in Kent keeps re-emerging decade upon decade (CPRE, n.d.) being presently proposed by the water supply company South East Water. At the time of writing a second proposal, in southwest Oxfordshire by Thames Water (Oxfordshire County Council, 2013) is likewise proving controversial.

1.3.2 Agriculture

From the Neolithic period onwards, soil-water management systems have been employed in pursuit of food production. These enhance surface water drainage, control flooding, manipulate watertable levels, constrain livestock by 'wet fences' and possibly control soil erosion on steeper slopes (Cook, 1994; Cook and Williamson, 1999). Later developments allowed for the irrigation of pastures as watermeadows and catchmeadows. Modern soil-water management includes elements of these, although irrigation in production agriculture has overwhelmingly shifted to arable land.

Artificial features relating to water management, including features likely to be wells, are evident in the archaeological record (Cook, 1999a); other early examples of domestic water supply include clay-lined ponds from the Middle Bronze Age (Bradley, 1978, p. 50). These features permitted settlement and agriculture in areas with little or no surface water. Management of water evolved in an *ad hoc* manner to suit local environmental conditions, the needs of farming systems, early industry, settlements and navigation, in accordance with the social order and economic requirements of the time. Eventually safer and reliable methods of delivery of domestic and industrial supplies became possible.

Piecemeal land reclamation of wetlands continued after the Roman occupation of Fenland throughout the Medieval period. The presence of Norwich, once England's second city, would have provided an impetus for economic development, including drainage improvements in Broadland (Sylvester, 1999). Parallel developments in the Romney Marsh area of Kent and on the Somerset Levels and Moors and Fenlands meant that, in addition to land drainage for agricultural development, regional drainage was improved. Monasteries had the resources to effect large-scale drainage improvements in the Middle Ages, and it was largely due to their involvement that such regions experienced substantial reclamation for agriculture. Additionally, on Romney Marsh the Rhee Canal was constructed in the thirteenth century to enable access to several ports and keep the port of New Romney from silting up (Brooks, 1988). It is in the post-medieval period that drainage at a regional scale really reached its zenith, typified by but by no means uniquely represented in Fenland and spectacularly (from an engineering point of view) by the canalisation in the seventeenth century of the 'Bedford rivers' (Taylor, 1999) under the guidance of Sir Conelius Vermuyden, a Dutch land drainage expert brought in by Charles I, originally for drainage works at Hatfield Chase. Other areas impacted at this time included the Isle of Axeholm, wide areas of Lincolnshire, the Somerset Levels, along the Thames, and the Norfolk and Suffolk coasts, and coastal areas of Hampshire and Sussex experienced drainage works.

Early methods of underdrainage used such materials as brushwood or stones. In the nineteenth century these were replaced by tile drainage (Harvey, 1980, p. 71), to be replaced in the nineteenth century by earthenware pipes and tiles (Williamson, 1999). Since the 1930s and especially in the post-Second-World-War period there was a dramatic increase in the progress of underdrainage (Stearne and Cook, 2015), which became grant-aided by government, although this is no longer the case. In more recent decades, government grant-aid has been progressively replaced by payments to reinstate wetland environments in targeted areas under agri-environmental schemes such as the contemporary Higher Level Stewardship. This may be viewed as a historical irony to parallel the well-known payments to replace hedges that were once removed under subsidy!

Viewed overall, improvements to regional drainage were both large scale and effective, and included the deepening, straightening and widening of watercourses, and more efficient pumping, particularly common after 1945. Where gravity drainage was insufficient to rid an area of excess water, mechanical help was employed to provide a further spur to field drainage. Windmill technology, introduced into England in the twelfth century, was later to permit the mechanical lifting of water in the seventeenth and eighteenth centuries using scoopwheels (in effect a waterwheel operating in reverse to lift water), and later true pumps (Cook, 1999b). Where the land to be reclaimed was too low, or commonly where peat wastage and shrinkage occurred after reclamation, windmills attached to scoopwheels were employed to lift water up to arterial watercourses. Wind power could be assisted first by more efficient steam power in the nineteenth century and in the following century diesel and electric pumps became widespread in all areas requiring drainage.

Protection from flooding through improved flood defences, both from freshwaters and saline waters in coastal regions are also essential components of effective, regional-scale water management which today permits agricultural intensification. Policy documents such as 'Making Space for Water' (Defra, 2005) enjoin us to utilise inland and coastal management that potentially realise floodplain and coastal marsh functionality, although no detail is really given in respect of the operation (and hence social and environmental utility) of historic water management systems (e.g., Cook, 2010b).

Before the twentieth century, irrigation of arable and horticultural crops would be rare, excepting the 'sub-irrigation' practiced by maintaining relatively high ditch levels between fields in areas of wetland reclaimed for agriculture such as the Fenlands of eastern England. However, meadow irrigation was widely practised in England. It would have been particularly effective during the 'Little Ice Age' because it raised soil temperatures, protected the sward from frost, controlled weeds and flushed the soil with nutrients (Cook, 2007). Irrigations could bring about early grass growth in the late winter/early spring (the 'early bite' for animals) and boost hay production by re-wetting the topsoil later in the season. Topographically, 'floated watermeadows' could occupy either valley sides or alluvial floodplains.

Irrigation to reduce the soil water deficit is overwhelmingly a twentieth and twenty-first century practice, and largely undertaken since the 1960s on arable land. Post-Second-World-War irrigation has become a major draught on surface and groundwaters in certain regions of the UK. Overall, in England and Wales spray irrigation usually accounts for around one percent of non-tidal abstractions annually (EA, 2008; Defra 2015c), although regionally it is in East Anglia followed by the Midlands that together take the lion's share with the amount depending upon the year and locally the figure can be appreciably more. Water is also used in agriculture for other purposes such as washing livestock yards and in produce processing.

1.3.3 Groundwater Development

Groundwater development also has its history, and the occurrence of wells and clay-lined ponds is known from British pre-history. Inevitably, wells continued as a source of supply throughout history in both rural and urban locations with relatively shallow watertables, and where surface water sources were either remote or scarce. The availability of the resource prior to systematic exploitation in the nineteenth century would, nonetheless, be dependent on natural rates of discharge and recharge (Cook, 1999a,b). It is also probable that human interventions, through land-use changes and drainage improvements, affected groundwater availability from early times. For example, these may have caused a lowering of watertables in the chalk of southern England, requiring that wells had to be dug deeper in the fourth century (Sheail, 1988).

The quality of groundwater has generally been higher than surface waters on account of filtering of particulate materials and degradation of organic pollutants during transit underground, although solutes will travel freely through underground strata with the water. In modern times, groundwater sources require considerably less treatment than surface sources. In England, the chalk sources frequently only required chlorination.

During the nineteenth century, the advent of steam technology permitted the pumping of groundwater from larger wells and later from boreholes. Wells constructed at this time are typically 1-3 m wide and relatively shallow, although some reached to 125 m (Hall, 1989, p. 34). Such wells often were lined, and in suitable geology (typically chalk) had adits radiating at the bottom to facilitate the abstraction of water. Boreholes are sunk by drilling, or are percussion drilled. Evidently bores as deep as 300 m were achieved in the late nineteenth century (Whitaker, 1908, p. 4), and modern examples may be as deep as 500 m (Wilson, 1983, p. 87); generally, boreholes are less than a metre in diameter.

Groundwater abstracted in this way comes to supplement surface abstractions, and today, it is a resource in need of the most careful management in southern and eastern England. Sadly, groundwater has suffered from the 'out of sight, out of mind' problem, and, in 1922, in his presidential address to the Institution of Water Engineers, W.J.E. Binnie stated that:

> 'there is nothing to prevent a private individual sinking a well immediately adjacent [to a public source] by means of which the community may be deprived of its water supply.'

He continued to call for a licensing system to be introduced, a situation not realised until 1963 (Binnie, 1995). Nowadays, in England, around 30% of public supply comes from groundwater in Wales the figure is more like 3%. This source is especially important in the drier east of England (EA, n.d.a). For Scotland, around 5% of Scotland's public water supply comes from groundwater and sub-surface water largely is of good quality drawn from 96 boreholes and springs supplying around a third of a million people. Needless to say, it is invaluable in the production of whisky (Scotland's Environment, n.d), although pressure arises from agricultural pollution, water abstraction, causing water-table levels to drop, pollution from historic mining activities and from historic industrial activities. Nitrate Vulnerable Zone (NVZ) designations are for Moray/Aberdeenshire/Banff and Buchan, Strathmore/Fife, Lothian and Borders, and Lower Nithsdale, all in the east of the country (Scottish Government, 2015).

Pressures on the resource are serious and will be described later in some detail. To summarise, there are four key points to remember:

- Groundwater can be over-abstracted (reducing the resource availability over time).
- Groundwater can be polluted (contaminating supplies) because it is hidden and receives a range of contaminants with industrial and agricultural sources.
- Groundwater has an intimate relationship to surface waters, both flowing in rivers and streams, and standing as in wetlands. Over-abstraction has dire environmental consequences where it causes surface water to diminish or dry up.
- Aquifers may themselves be damaged or destroyed by engineering works and mineral abstraction (hence destroying the resource once and for all time).

1.3.4 Industry

Industry is a major user of water, and the changing patterns of water use in this sector are described in Chapter 9. From the eighteenth century onwards, industrial development occurred on a large scale, and all sectors used water in increasing volumes. For example, 'primary' industries used water at sites of mineral extraction or minewaters pumped from mines are frequently sources of contamination, including acidification. 'Secondary' industries consumed it during manufacturing that could involve polluting industries such as leather tanneries, dying and varnishing (Cook, 2008a) and also coking plants. Indirect effects caused by low pH precipitation has acidified some British waters since the mid-nineteenth century, culprits include the combustion of sulphurous coals and later the internal combustion engine. Transportation required water for the canal system and keeping certain (especially high-level canals) supplied could be a problem; and that water could be polluted through accidental spillage. One landmark piece of legislation was the the 'Rivers Pollution Prevention Act, 1876' prohibited the pollution of rivers by the discharge of sewage and other waste.

Problems therefore arise in respect of both quantity and quality of supply. We may identify three key and potentially negative aspects to water use at a location:

- Water is abstracted and returned to a water body with only physical changes.
- Water may be polluted in some way upon its return.
- Water may be abstracted and lost to the immediate environs.

Historically, watermills (known from the Roman period but probably only common in England since Anglo-Saxon times) 'return' water to a river with only a loss of kinetic energy caused by the drop required to drive the wheel; however, this did not prevent corn millers coming into conflict over use of low river flows (Newson, 1992a, p. 14) nor a requirement for nineteenth century millers to adopt steam technology for use at times of low flow (Cook, 2008a).

Today, water used for cooling in power stations returns river (or estuary) water with increased temperature. Although bulk supply is hardly affected, increased temperatures can damage fish stocks and reduce oxygenation of waters. Furthermore, antifouling chemicals used to prevent slime build-up may introduce toxic pollution, and drainage from coal stockyards may be acid. 'Heavy industrial' plants have a high potential for environmental contamination. The main offenders have been coal and bulk mineral extraction, power generation (including sulphurous emissions and gas-plant effluent), the products of blast furnace 'wet scrubbing' and steel and chemical manufacture including petrochemicals. Many cause a range of contamination such as pH reduction, suspended sediment or metalliferous contamination, deoxygenation and contamination from a variety of organic compounds including solvents and hydrocarbons. Waste from chemical or pharmaceutical production potentially contains a wide range of chemical substances. Light industry also has the potential for polluting waters, with products from the food processing industry, detergents, paper mills and dyestuffs often included amongst effluents (NRA, 1994a).

Then there is 'fracking' that is hydraulic fracturing of rocks to extract gas from hydrocarbon-rich rocks (sometimes referred to as 'shale gas'). Unlike, for example, drilling for water, oil or gas in a conventional manner, fracking has proven extremely controversial, as it involves pumping into wells under high pressure, a mix of sand, water and (unspecified) chemicals to extract the gas that can then be used in the 'dash for gas'. This revival in interest in gas for energy is hoped to provide less carbon dioxide emission per unit of energy extracted. It is hoped it will bridge the gap between large-scale fossil fuel usage worldwide and the adoption of non-carbon emitting nuclear power and renewable sources of energy. Alternatively, it is a dangerous geological process that involves investment that delays widespread adoption of such 'clean energy technologies' while posing geological problems including the generation of seismic activity, triggering local earthquakes, as well as groundwater pollution (BBC, 2015).

Contamination in rivers can be cleansed relatively rapidly due to low residence times in the channel. However, there is great concern in Britain over groundwaters at risk many of which have long residence times and can be contaminated from industrial effluent. An interesting twist to the problem of groundwater quality is the persistent problem of rising groundwaters, since the demise of many heavy industrial plants, and controls on licensing of abstractions in the second half of the twentieth century (Wilkinson and Brassington, 1991). Apart from problems of the flooding of underground tunnels and cellars and risk to foundations in such cities as Liverpool, Birmingham and London, there is the risk of a spread of underground contaminants. In former industrial areas, the intersection of watertables and contaminated land (seldom well documented) presents serious hazards to groundwater quality. Today, as a result of supra-national legislative provision from the EU, there are instruments including the Groundwater Directive 80/68/EEC and subsequent Water Framework Directive EEC 2000/60/EC aimed to protect first specific aspects of the water cycle, now calling for 'integrated river basin management' in Europe, to protect all aspects of waters and provide cleaner rivers and lakes, groundwater and coastal beaches.

1.3.5   Urbanisation and Supply: The Rise of an Industry

The rapid growth of industrialisation in the eighteenth and nineteenth centuries led to urbanisation, and a growing demand for continuous, potable water presented problems. The supply in London was intermittent until the late 1880s; in poorer districts supply was often from standpipes which operated typically three days per week limited to periods as short as one hour per day (Binnie, 1995). The eighteenth century saw the origins of certain water supply companies, mainly in London. By 1801, some 6% of the largest towns were supplied by statutory joint stock companies, a figure which rose to over half by 1851 (Hall, 1989, p. 18). In Brighton, a public water supply company dates from around 1825 (Headworth and Fox, 1986), when a small water company provided reliable, piped supplies in the Brighton area from a well sunk to the north of the town. A pumped supply was provided from a new well in 1834. Similarly, public supplies were provided elsewhere in Sussex for Worthing (1857), Chichester (1874), Bognor Regis (1877), Littlehampton (1888) and Sleaford (1896). In the nineteenth century, less leak-prone iron pipes began to replace the bored timber pipes used since Roman times (Courtney, 1994).

By 1901, most large towns had municipal waterworks on account of the inability of the private sector to keep up with demand, particularly in supplying large industrial centres with water services for poor populations. Examples include Bradford (Briggs, 1968, p. 156), where the Corporation took over the Bradford Waterworks Company in 1854. In 1872, an Act of Parliament transferred responsibilities from a number of small water undertakings in the Brighton and Hove areas to the Brighton Corporation, a process of takeovers which continued until 1918 (Headworth and Fox, 1986).

Serious human health problems only really arose in the context of the 'dirty end' of the emergent water industry, for innovations in water supply (reservoir construction and borehole construction) found easy, if costly, engineering solutions. Many statutory supply companies today trace their origins back to the second half of the nineteenth century. The Waterworks Clauses Act 1847 gave a standardised framework to water undertakings, where before individual Acts of Parliament had given rights over catchment and supply areas, and powers to install pipes and conduits and charge for water. Clearly, an industry emerged in parallel with others in manufacturing, agriculture and energy supply. The Victorians raised vast sums of capital by voluntary and municipal effort. This affected not only waterworks at the 'clean end', but also sewerage and sewage disposal.

Glasgow had utilised treatment through settling tanks and sand filters since 1808, yet filtration was only slowly added to the London supply after 1852 and abstraction was subsequently taken from the non-tidal waters above Teddington Weir (Binnie, 1995). The need for improved sewerage was highlighted in the 'Great Stink' of 1858, when London sewers became anaerobic. Furthermore, intake pipes in the Thames tapped polluted surface waters, and there is plenty of evidence of the poor condition of surface waters in industrial towns outside London (Briggs, 1968, p. 146). Subsequent modification of the main drains caused interception of major waterways to discharge downstream at Barking and Crossness (Binnie, 1995). By the end of the century settlement tanks and percolating filters were common in sewage treatment.

During the last quarter of the nineteenth century, some local authorities provided sewerage farms as well as waterworks (Briggs, 1968, p. 25), and until 1974 local authorities had the responsibility for sewage disposal. In reality, there were many problems here, including a sewage crisis manifest in the mid-nineteenth century when the struggle was on for effective means of treatment (Cook, 2008a). So problematic was this that Goddard (1996) can find a misplaced optimism in later nineteenth century, relating to sewage management. Waterborne disease, in the UK, sadly continued well into the twentieth century.

Sewage can contain human waste and other organic waste including animal carcasses, fats and oils. Organic wastes are likely to produce odours (especially under anaeorbic conditions), have a high biochemical oxygen demand (BOD) (resulting in low dissolved oxygen in receiving waters), and have high ammonia contents. They are also likely to contain pathogens (both bacteria and viruses) where not adequately treated. Furthermore, industrial processes can discharge to sewers. Cesspits frequently contaminated domestic wells, and when main sewers were first introduced in the nineteenth century, main watercourses became cesspools. Consequently, municipal sewerage schemes, often ambitious, were constructed, as in Birmingham where high mortality rates in some areas were linked to poor sanitation (Briggs, 1968, p. 223).

Nineteenth-century concern over waterborne diseases such as typhoid and cholera (of which there were epidemics in 1831/32 and 1848/49 killing almost 90 000 people) was linked to the rapidly urbanising population countrywide. One cluster of cholera cases was traced to a pump at Broadstreet in Soho, London, and led to the case for improved supply being accepted after 1853 (Binnie, 1995). Improvements in the distribution networks replaced *ad hoc* arrangements in supply, and most notably the adoption of low leakage pipes led to more sanitary arrangements for domestic supply.

Municipal water resource management schemes emerged alongside small private water companies, especially following the Waterworks Clauses Act of 1847 (Hall, 1989, pp. 17-18). The Act sought to standardise waterworks practice in terms of supply, pressure and quality, replacing the need for individual private Acts of Parliament for rights over catchment and supply areas, and powers to install pipes and conduits. The Act consolidated various preceding Acts passed with the objective of providing individual towns with private supply companies, declaring that the piped supply should be both pure and sufficient for domestic consumption. Bathing, washing and depositing rubbish into the undertaker's waters were forbidden (Elworthy, 1994, p. 47).

1.4   The Political Dimension

The Compact Oxford Dictionary (1996) defines policy as: 'course or principle of action adopted or proposed by a government, party, business etc.' otherwise the adjective 'prudent' and 'sagacious' may be furthermore involved! It is left to the reader to decide whether this is universally applicable to the UK and beyond.

'Policy' is something that is implicit for all kinds of institutions and hence affects, or is affected by, many stakeholder groups, actors and economic sectors. We may now talk of 'governance' implying a far more complicated picture than 'top-down' or 'command and control' and implying centralised decision-making (although this can occur at many levels). For modern political scientists to effect a revolution in our thinking around environmental management is in every way as dramatic as the legislation of the past two centuries or the achievements of Victorian water engineers, or the municipal pioneers and the scientists and technologists who came to vastly improve the water delivered to our taps. It will be demonstrated how power and decision making in water matters is being devolved to communities and to river basin groups, typified by the Rivers Trust movement, but for historical purposes we should start at the top.

EU Directives translated into UK law and policy formulations provide a basis for wider countryside planning and urban and industrial development including mining, quarrying and landfill. In short, water resource protection will be the single most important factor in shaping land use into the twenty-first century. Recently, this is particularly borne out because the call for Water Protection Zones in the uplands has the potential to cover vast tracts of land supplying upwards of 70% of drinking water for the UK from around 40% of its land areas (RSPB, n.d.), and as an 'ecosystem service provider' conservation of water supply at source goes hand-in-hand with many ecological and heritage objectives. Before the new millennium, it was disappointing to realise that water in the UK had been such a neglected resource, when compared with other areas of public policy. The literature abounds with examples from areas such as industry, agriculture, urban development and finance, making water policy look a poor relation, at least in historical terms. Arguably, it has only been since the Water Act 1989, and the sweeping changes subsequently introduced, that water emerged on the political agenda in a significant way. We hope that is changing and rapidly, not the least because of the Water Framework Directive (WFD).

Once waterborne diseases (such as cholera) were controlled by improving water supply during the nineteenth century, perceptions of hydrological 'problems' have been slow to emerge. This may be because hydrologists were slow to make their knowledge known to policy makers, or because the knowledge itself, or the political will to solve problems, lagged behind industrial, agricultural and urban developments. In Britain, land-use policy has been achieved only via market intervention, and hence land-use planning is rare outside urban areas (Newson, 1992a, ch. 8).

An example of historical slow response is the impact on upland afforestation. This severe land-use change was first identified as reducing water yield to reservoirs in the 1950s (Law, 1956), being followed by decades of research to consolidate these findings (Calder, 1990). Research in Scotland followed somewhat later. Issues of water quality (other than those associated with trade effluent and organic waste management) are even more recent. Concerns over nitrates in water were first aired in the early 1970s, and pesticides in surface waters later became an identifiable problem (Gomme *et al.*, 1991), subsequently these were detected in groundwater.

Water protection policy must therefore move on from urban (mainly point-pollution) controls within catchments to embrace rural (and predominantly diffuse) pollution. It must also consider differing timescales. Hydrological timescales normally operate over hours or days (longer when considering groundwater balances), increasing sediment loading to a drainage basin through urbanisation or arable farming switches to the 'geomorphological' timescale measured in years and centuries.

Policy making is, by its nature, interdisciplinary, incorporating not only hydrology and water resources but also hydraulic engineering, waste management, soil science, geomorphology, ecology and agricultural sciences and there is an increasing need for landscape/environmental history and archaeology. The social science compliment has to include not only economic considerations, but also politics and sociology considerations. One tangible requirement is in modern agri-environmental policy, when Higher Level

Stewardship (HLS) pays farmers and landowners to protect 'heritage' or ecological valuable landscapes (Defra, 2014b), for:

'HLS is being targeted in 110 areas across England. These target areas are where Natural England are seeking the most environmental benefits from HLS agreements for wildlife, landscape, the historic environment and resource protection. Outside these areas, we will consider all other applications depending upon the current national priorities and features present on the particular holding.'

Any regulatory body, if it is to reflect the modern ethos of integrated catchment management, has to reflect diversity in its institutional structure. The operation of the National Rivers Authority (1989-1996) was in exactly this position in England and Wales, and the EA even more so (Chapter 4) because it brings the additional 'media' of land and air to considerations of the water environment.

Before commencing a description of contemporary river basin governance it is useful to rehearse the needs for regulation in modern water institutions that arise for the following reasons:

- There is the need to balance the 'stakeholder' interests in water management. Waters are not only consumed by agriculture, industry and for domestic purposes, but are involved in conservation, landscape and recreation interests.
- As a 'temporary use resource', water availability can be finite within a specified time-frame. It may not automatically meet demand during dry periods, so the adverse results of mismatch are seen in environmental, supply and (potentially) health terms.
- Volume of supply does not always ensure that potential pollutants will be attenuated or diluted to a level where they apparently present no problem. This may be termed the 'carrying capacity' of the water environment. Policy and legal solutions, as well as 'technical fixes', are sought to ameliorate resulting problems.
- There are the ancient concerns of flood defence and land drainage, matters for the regulator on account of both the coordination effort required and capital costs involved in meeting the 'common good'. The latter are generally considerable.
- Although the banking crisis may suggest otherwise, there remains the politically (i.e., 'neo-liberal') motivated shift from state ownership of utilities (and indeed other industries) towards *state regulation of privatised undertakings*.
- This change therefore pulls away from a clear focus on water supply for human welfare that maintains sustainable abstraction and water quality towards the *profit motive*. This danger would seem to pertain in England but not in the remainder of UK.

While the notion of 'science into policy' is helpful because it stresses the scientific knowledge required for informed decision making, in practice it implies a one-way flow of ideas. A pluralistic approach is essential in solving water management problems. Policy, arising at the top from EU or UK legislation gives direction to a regulator, whereas the regulator also formulates policy, interprets legislation and (ideally) resolves conflicts. Being at the 'sharp end' the regulator operates in a preventative manner, and furthermore should be able to anticipate problems. Even this is an insufficient descriptor of development since the year 2000.

Water regulation in Britain has passed from 'technocratic' leadership to consultation-based management. This allows room for a democratic component whereby stakeholders' views are sought in the planning procedure. The needs of all 'users' of the water environment are ideally considered, including conservation and human and animal health interests. From this process, objectives for management are set. The question remains for whom are they set, how really are decisions made, and finally how is environmental and behavioural change effected on the ground?

It should also be noted that EU environmental policy based upon 'command and control' style legislation, typical of environmental quality legislation, has its critics, and this arises primarily due to the requirement for enforcement. The Union's Fifth Action Programme incorporated aspects of sustainable development, preventative and precautionary action and shared responsibility (Hillary, 1994) in an effort to move away from the (expensive) command approach. Water supply is not listed within the Programme (although industry and agriculture are), but Chapter 10 will demonstrate the preoccupation of the EU with water quality issues. Since then, WFD article 14 actively requires public participation.

Prior to 1989 the Regional Water Authorities (RWAs) had self-regulation under the principle of integrated catchment management. A well-known syndrome in environmental governance called 'poacher and gamekeeper' existed, with roles played by a single authority in resource regulation and operation of the utility sides. Regulation instantly became external to the utility side, and a new and stronger legal framework is in place. Because free markets in water supply are difficult to make work because of fixed infrastructure and the need for supply monopoly, neo-liberal ideology required a substitute for 'market forces' that, in theory would drive unit costs down and work to optimise economic efficiency. The state retreats, leaving only a regulatory framework and somehow market forces take over water supply and treatment, and solve all ills. If that cannot work we are in what is a 'market-failure' position where for example, if pollution is viewed as arising from the 'self-interest' of individuals involved in transactions, there arises a societal cost in its abolition for the welfare of individuals and the environment is in peril (Bannock *et al.* 2003). Water pollution (in this example) represents a societal cost or an 'externality' that is not internalised as a true cost of delivering water of sufficiently good quality (actually determined through regulation via EU Directives). Intervention comes via policy development and the debate about 'who pays', polluter, consumer or society at large will continue to rage (Benson *et al.*, 2013b). Even to get to the point of attributing blame (included in 'policing') requires a technical infrastructure, so that the cost of policing water quality and quantity issues incurs considerable cost.

The resulting shift in economic policy formulation after 1980 moved away from planning (often for state-owned industries or employing interventionist policies such as agricultural price support) towards a free-market approach which (in theory) prefers market mechanisms to decide how prices are set, and where investment is made. However, the protection of both public (consumer) and environmental interests is unavoidable, and hence is the *raison d'etre* for water policy development. In Chapter 3 it will be shown how regulation for both economic and environmental purposes has, in fact, led to an explosion of policy formulation affecting the operation of water utilities; merely to privatise the former RWAs as they stood might have been a disaster—on account of the poacher and gamekeeper principle.

Books dealing with policy theory *per se* (rather than practical application) have come from the social rather than natural environmental sciences, and have tended to concentrate upon policy analysis. That has been extremely useful but now there is a growing literature of what, for present purposes, we may refer to as 'applied political science'.

Such analyses makes suggestions for how communities, professionals and other interest groups not only engage with, but participate in problem solving commencing with engagement at the 'issue definition stage'.

Policy analysis is a process which considers both how policies are actually made (description) and how policies should be made (prescription). A consideration of policy theory is merited here because practically orientated water resource planners are then able to view the process in abstract. It is also helpful to reiterate the imperatives of public policy formulation following decades of 'neo-liberalism'.

In practice, natural monopoly regulation on private water and sewage undertakings is tight, and we will see how policies reflected in 'command and control' types of legislation could prove to be expensive and unpopular. Kinnersley (1994, p. 9) rightly identifies a need for new forms of coordination between water agencies, water users and government. He identifies causes as inertia among water undertakers and a realisation that good quality water supply is proving problematic.

A long-standing framework for policy formulation, based upon Hogwood and Gunn (1984, p. 2), is listed below:

1) Deciding to decide (issue search and agenda setting).
2) Deciding how to decide (issue filtration).
3) Issue definition.
4) Forecasting.
5) Setting objectives and priorities.
6) Options analysis.
7) Policy implementation, monitoring and control.
8) Evaluation and review.
9) Policy maintenance, succession, or termination.

An example of policy implementation in water quality is provided by the groundwater nitrate issue, discussed at length. Raised concentrations of nitrate in water have been perceived to be a problem to both human health (when ingested) and to natural ecosystems, although scientists and clinicians still argue over the extent of both. The greatest problem arises within unconfined aquifers over which there is intensive arable farming, although intensively managed pasture and the ploughing of grass also contribute. Furthermore, the reader is invited to compare that above with the more contemporary 'adaptive management cycle for catchment planning and process implementation' based in United States Environmental Protection Agency (USEPA) practice (Chapter 11).

The need for a pluralist approach is illustrated by the two kinds of explanation regarding how nitrate of agricultural origin comes to be in ground and surface waters in concentrations which are causing concern: one is scientific (based in soil and agricultural sciences), and the second lies in the need for food security and involves agricultural price support. This is policy-based and arguably provides the basis for the cessation of damaging land-use practices.

The North Atlantic blockade of the Second World War prevented the import of food into Britain. During the 1930s the British countryside had a high proportion of grassland, arguably a response to domestic agricultural depression. The ploughing of grassland over aquifers after 1940 released mineralised nitrate, and the subsequent use of nitrogenous fertiliser 'topped-up' the supply. Once it was thought that most of the nitrate in waters leached directly from inorganic sources; now it seems most is taken up by the crop, and it is the degradation of the increased plant biomass which releases mineralised nitrogen.

Agricultural price support was maintained after the war in order to maintain food security in Britain. On the Continent, the Common Market had gone down the same route and Britain became subject to the Common Agricultural Policy after her accession in 1973. By the mid-1980s, agricultural support policy linked to increasing use of fertiliser led to 'food mountains' and a nitrate problem in aquifers and rivers of the arable areas within parts of Europe. Lowland areas of Britain proved especially vulnerable.

By the late 1980s, the nitrate debate had reached fever pitch, and in 1990 some 192 sources exceeded the EC limit of 50 mg l^{-1} for nitrate (11.3 mg l^{-1} for NO_3–N) in drinking water. Farmers complained that they were victims of their own success, and criticism was the prize for doing what society asked of them. Fertiliser manufacturers were ready to blame the ploughing of grassland, or even felt exonerated due to the realisation that the effect of applying fertilisers was indirect (DoE, 1986). The RWAs were blamed for delivering polluted water to the consumer, while the environmental lobby blamed all three. It was the beginning of enforcement of the EU Drinking Water Directive of 1980 (80/778/EEC) which was to make the difference.

We may ask how the nitrate issue fits into the classic framework for policy making and implementation? And attempt to investigate this is given in Box 1.1. Hogwood and Gunn (1984, p. 10) admit their framework is not always followed in sequence, and there is much 'looping' and 'overlap' between stages. Although idealistic, this sequence helps

---

Box 1.1 The nitrate issue in terms of a framework for policy making and implementation

In terms of 'agenda setting' (1), the issue was identified as a potential problem once water analyses detected raised concentrations from the early 1970s. 'Issue filtration' (2) revolves around whether the existing institutions and procedures were sufficient to deal with the matter. The years before 1989 really represented a research stage, when the issue was explored (e.g., Foster *et al.*, 1986); the pre-privatisation water industry may have considered means of dealing with the problem (MAFF/STW/DoE, 1988), but little was actually done in connection with bulk water supplies, beyond source blending, to reduce concentrations.

'Issue definition' (3) involved objective analysis and the identification of probable causes, and, through various empirical and modelling exercises, quantification of the problem. Intensive farming practices were identified as the major cause, and therein was a probable cure. Predictions have been made as to likely future groundwater concentrations, and estimates made of likely affected acreage (Cook, 1991), these meeting the requirements of point (4), 'forecasting'. Setting objectives and priorities (5) involved meeting the EU limit of 50 mg l^{-1} NO_3 in drinking waters, and establishing that a programme of monitoring was desirable once the selected option was under way. However, the possible means of achieving these aims had meanwhile to be considered as 'options analysis' (6). These would seem to be treatment or blending of waters (MAFF/STW/DoE, 1988), or the prevention of pollution at source, the selected option in this case.

Policy implementation, monitoring and control (7) could then follow. Considering points 5, 6 and 7, the establishment of protection zones to avoid land management practices deemed inappropriate was selected. The establishment and operation of Nitrate Sensitive Areas (NSAs) over selected vulnerable groundwater areas and through which farmers were compensated for presumed loss on productivity, and Nitrate Vulnerable Zones (NVZs) aimed to protect all waters resulted, in the latter farmers are not compensated and protection measures as mandatory (Chapter 9). Evaluation in terms of farmer uptake of the NSA scheme and pollution monitoring and review (both 8) of the policy has been continual (Archer and Lord, 1993). This has led to policy maintenance (point 9) in the existing 10 areas, and the announcement in 1993 of new candidate NSAs. Since then, there are no more NSAs, but at the time of writing uncompensated NVZs comprise some 58% of England, having been reduced. Evidently, using an EU Directive, UK governmental thinking was to move towards compulsory restrictions as compensation payments over so much of the country would be economically unsustainable.

us to marshal thoughts of policy analysts around a specific problem. Policy and procedure (designation, establishment and evaluation) have to progress together to be successful and cost effective.

Britain has been (and is) engaged in a struggle with policy development to protect both waters and the wider environment. Synthesis of ideas of regulation and management is on the way, and we are able to learn from experience elsewhere in the world, especially other European countries and North America. These are explored in Chapter 10 and 11 of this book. However, the evolution of a policy to protect waters from nitrate pollution invokes top-down measures from central government in response to EU supra-national Directives, compulsion of a perceived 'polluter' and there is an implicit debate about compensation, violating the 'polluter pays principle' for the compensation is given for a loss of profit, a by-product of which is nitrate contamination of groundwater.

On the other hand, widespread NVZ measures protect both surface and groundwater from nitrates of agricultural origin and this requires compliance. This costs (public) money. Is there another way that is more participatory and smacks less of compulsion?

River basin groups in Britain, typified by the Rivers Trust movement, are becoming self-organising (Cook *et al.*, 2012). Hence there is a counter-action that has every potential to keep statutory authorities effective but at arms length from stakeholders as new institutions that are manifestations of 'civil society' take a lead in protecting water catchments used in public supply, in fishery protection, and more (Cook and Inman, 2012). Complicated new governance arrangements this may seem, but participatory democracy ought to be easier to sustain, be more efficacious (e.g., in sustaining in-stream ecology of rivers) and seek win-win solutions that improve compliance with statutory authorities, improve farm holding bottom lines and deliver environmental goods.

1.5 Summary

Early intervention in the hydrological cycle emerges as practices concerned with prehistoric water supply and field-scale soil-water management, probably in the Neolithic period. From Roman times, the beginnings of urban water supply, navigation and flood defences may be identified, and it is probable that some concept of legal regulation existed at the time. Early forms of industrialisation, a developing agriculture, and the need to provide both smaller settlements and towns with clean, abundant water led to pressure on the resource which may be charted *inter alia* by legal developments of both case law and statute law in England. As technology advances, we see the improvements in land drainage, flood defences and supply. These are linked to improved pumping technology and developments in hydraulic engineering such as efficient pipes for water delivery under a fairly constant pressure.

What emerges is a strong managerial theme, and the details of how this came about depended upon the political climate of the times. The Victorians saw water and sewerage services as a public good because of contemporary pressing issues of public health, and progress in legislation passed at this time is outlined in Chapter 3. With characteristic civic pride they either caused water services to be created through municipalities, or in many cases brought existing private supply companies into public ownership. The need for tight regulation became very clear in the 1960s, and in 1974, all- embracing RWAs were created within the public sector. RWAs embraced the principle of integrated catchment management, (WFD talks of Integrated Water Resource Management) which can mean all things to all people, but basically involves the total management of the hydrological cycle within a drainage basin (including wastewater management and flood control). As such, the need for all-embracing water management institutions became self-evident, and the RWAs dealt with issues of resource regulation, monitoring, pollution control and planning. Against this was the often-made point that the RWAs were both 'poacher and gamekeeper'. Self-regulation did not work in many instances (albeit the cause may have been as much public sector under-investment as institutional inefficiency) and this was witnessed by periodic assessments of water bodies and problems with drinking water supply.

By the 1980s, with political fashion strongly against public ownership, the water industry was, to the surprise of even many right-wingers, privatised through the 1989 Water Act. The remaining private water supply companies (only dealing with the 'clean' end, and not sewage disposal) had provided a link with a partially private sector past. In order to regulate the resource side (as opposed to the consumer interest side), the National Rivers Authority (NRA) and then the Environment Agency (EA) in England and Wales were created. In Wales in the latter case replaced by Cyfoeth Naturiol Cymru—or Natural Resources Wales (NRW)—in 2013. Parallel, though not identical, developments in regulation occurred in Scotland, although the utilities were never privatised.

Controversy over privatisation of utilities south of the Border has gone from raging indignation to a gentle simmer among environmental activists and political commentators of whatever stripe. Yet the industry functions well and meets demand; especially it can raise investment at will from private means in a way an effectively nationalised industry could not. However, jobs were lost, legislation had to be introduced to protect the vulnerable and private sector efficiency did not bring about even a stabilisation of water and sewerage service charges, let alone a reduction. Ironically, the Thames Tideway Tunnel project, promoted by Thames Water plc, is designed to end the release of untreated effluent into the River Thames and is underwritten by central government as it is too ambitious to fail (Pinsent Masons, 2012). From 2013, central government provided a £50 subsidy in domestic consumer bills aimed at addressing higher bills in the region compared with elsewhere (South West Water, 2012). Some might say these are significant examples of market failure post-privatisation of the water industry.

While geographic reference in terms of water services largely pertains (an inhabitant of Salisbury is served by 'Wessex Water', but energy services may be purchased by Scottish Power!), ownership of water companies is forever changing and, as observed at privatisation, there is no guarantee this if UK based, raising a small spectre of 'water security' to parallel that energy security. However, the sources remain British. There are chinks in the armour of a private water industry and controversy continues, especially questioning the profit motive (Tinson and Kenway, n.d.). Not only may professionals caught off guard still use words like 'public' and 'service' but Yorkshire Water was almost mutualised in 2000, Hyder (including the former Welsh Water) broke up with its water division becoming under the management of Glas Cymru (Welsh Water, 2013), a private company limited by guarantee (essentially not-for-profit) achieved this. The notion of an efficient but not-for-profit industry remains a possibility, something not imagined by the Thatcher government of the day. Remaining in public hands, the Scottish government passed legislation in 2005 allowing for competition in the water industry whereby business customers can potentially choose their water supplier, although Scottish Water continues to deliver water and remove wastewater (Central Market Agency Scotland, 2016).

Economists and politicians of left, right and centre will no doubt continue to debate the rights and wrongs of particular ownerships for the water industry, once seen as a public good operated through a public service. The most interesting political (with a small 'p' in this case) development of the present decade is the rise of NGO type river basin organisations, largely in the Rivers Trust movement. This parallels ideas from across the conventional political spectrum and seems acceptable to all. There have been a few years during which to assess the operation and cost-effectiveness of this 'democratisation' away from direct water environmental regulation and is explored further in Chapter 4.

2

Water Resource Availability in Britain

Water in Scotland? Nae problem, it just runs doon the mountainsides and intae the pipes!

Scottish water engineer

In Wales, if you can see the hills you know it is going to rain, if you cannot, this tells you it already is raining. I guess 1976 was an exception.

Welsh geographer

Thar jus' ain't no wa'ter in them hills like there was. Can't do nothin' with the meadows thus winter.

Might as well jus' stay in ther pub!

English drowner (of watermeadows), February 2011

2.1   Introduction

This chapter outlines issues of water volume supply. Water is described as a 'temporary use resource' because human societies interrupt and appropriate water from the natural hydrological cycle, only to return it later. Appropriation implies a hydrological surplus in the balance of inputs and outputs around the point of abstraction and this may be enhanced over time by storage and transfer schemes. Balancing water supply and demand involves consideration of groundwater resources, especially in the south and east of Britain. These resources are vital in meeting demand and maintaining surface water flow in many areas. This is not to underplay water shortages in areas highly dependent upon surface water sources, such as occurred in west Yorkshire and the South West during 1995 and the succeeding winter, but in these areas long-term water supply from precipitation is greater than in the drier south and east.

Significant human interference in the British landscape actually dates back about 6000 years, when clearance of the original 'wildwood' is thought to have commenced. Without human input there would probably be a reduced occurrence of low flows in rivers, while problems of wetland drying would be restricted to more cataclysmic natural events such as prolonged periods of low rainfall. Flooding of land areas, on the other hand, would be more frequent due to greater flows of water in river channels which periodically spilled over floodplains. Deliberate anthropogenic interference in the land-based hydrological

The Protection and Conservation of Water Resources, Second Edition. Hadrian F. Cook.
© 2017 John Wiley & Sons Ltd. Published 2017 by John Wiley & Sons Ltd.

cycle includes reservoir construction and retention dams (especially in the upper catchments), catchment and regional transfer schemes, flood embankments, sea walls, channelling and straightening of existing rivers and the lowering of watertables.

2.1.1 Flooding: The First Extreme

A flood is a surplus of surface water and is a natural event. Heavy or prolonged precipitation and/or snow and/or ice melt will cause a large input of quickflow into a stream channel, and its severity will depend upon the severity of quickflow-forming processes (Ward and Robinson, 1990, ch. 7). In practice, planners and the public require reliable estimates of serious floods using statistical methods (Stedinger, 2000), at their most useful where there is none, or insufficient direct river gauging. These include the amount and intensity of precipitations, and intensifying conditions within the catchment (controlled by channel network, channel length, catchment shape, surface storage capacity, antecedent soil water conditions, etc.). In the past, simple empirical equations were used to define the 'maximum flood' which could occur within a particular catchment (Wilson, 1983, p. 208). These generally took the form:

$$Q = CA^n$$

Where: Q is flood discharge (e.g., $m^3 \ s^{-1}$); C is a coefficient depending upon climate, catchment and units; A is the catchment area (e.g., km^2); and n is an index. However, this approach has a number of problems. The flood defined as 'maximum' is caused by an arbitrary large observed discharge, and the equation does not take account of antecedent soil moisture, rainfall, topography or other catchment variables. Frequency analysis of flood events based on time-series observations has now generally replaced such equations.

Floods need to be quantified in terms of flow rate and magnitude of runoff over time, and are modified by various catchment contributions including channel routing (Petts and Foster, 1985, p. 40). Flood routing involves finding economic solutions to floodwater management; today this is a process which includes not only engineering solutions but also conservation and land-use considerations employing a more 'holistic' approach. First, the flood behaviour has to be characterised in a quantitative way.

A graphical way of displaying this is the flood hydrograph (Figure 2.1), and for practical purposes this may be viewed as the response to a rainfall event (or other such catchment loading) of the catchment measures at a down-catchment gauging station. The arbitrary baseflow separation line is an attempt to separate 'background' discharge. Typically, this is from groundwater, but it equally could be compensation flow' from a dam in a regulated catchment where there has been a deliberate release of stored water from the runoff generated by the rainfall event. Precipitation intensity (as well as amount) is important in relation to the form of the storm hydrograph, including the magnitude of the peak and time to peak, although the latter is also affected by catchment land use. Intensities exceeding $250 \ mm \ h^{-1}$ can occur for short durations (say of 20 minutes), producing a rapid rise in the runoff peak (Petts and Foster, 1985, p. 42).

Reference to a 'one in so many year flood event', which defences are specified to withstand, represents a flow with a specified average return period. There is also the 'mean annual flood' ($Q_{2.33}$), which is a hydrological standard, defined as the annual maximum daily flow with a return period (on average) of 2.33 years. Although in gauged

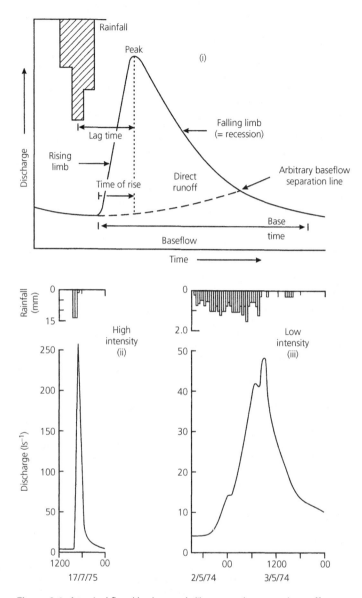

Figure 2.1 A typical flood hydrograph (i) nomenclature and runoff response to c. 28 mm of rainfall under high intensity, (ii) and low intensity, (iii) in the Yendacott catchment, east Devon. (redrawn from petts and foster, 1985, p. 40).

catchments this is derived from a simple ranking exercise of data derived from the gauging station, empirical equations have been derived for estimating the mean annual flood in ungauged catchments from mapped catchment variables.

The baseflow separation line in Figure 2.1 may be derived by various means, including drawing a line by eye, or better using appropriate computer software. However, its significance is that (in theory) it separates the flows concentrated in peaks from overland flows, generally a response to a particular set of rainfall and runoff circumstances, from the flow that is supplied from the soil and groundwater, including river banks. The Base Flow Index (BFI, the ratio of annual baseflow in a river to the total annual runoff) is the proportion of long-term flow that is thus derived and may be conceptualised as the hydrological pathway that will supply flow in a river over the long term.

Table 2.1 shows the BFIs and long-term flows for selected rivers (Cook, 2007) at selected gauging stations in the south of England. On the left are typical chalk rivers with high BFIs, on the right a range of other rivers with lower BFIs. The Tamar, Beult and Adur represent relatively impermeable catchments.

It is not only 'natural' parameters such as stream network or soil characteristics that have to be considered, but in a densely populated a country like Britain, human impacts such as urbanisation and impoundment are included in statistical estimation.

Such exercises are useful in elucidating the catchment variables affecting flood behaviour useful in ungauged catchment. A well-known example (NERC, 1975) is:

$$\bar{Q} = C\left[\text{AREA}^{0.94}\text{STMFRQ}^{0.27}\text{SOIL}^{1.23}\text{RSMD}^{1.03}\text{S1085}^{0.16}\left(1 + \text{LAKE}\right)^{-0.85} \right]$$

where $\bar{Q}$ is the mean annual discharge $(m^3\,s^{-1})$; AREA is catchment area (km^2); STMFRQ is stream frequency (junctions.km^{-2}); SOIL is a soil index derived from catchment soil maps, a measure of infiltration capacity and runoff potential (see section 2.2); RSMD is the net one-day rainfall of five years return period less the soil water deficit; LAKE is the fraction of the catchment draining through a lake or reservoir; S1085 is the stream slope $(m.km^{-1})$ measured between points at 10% and 85% of stream length; and C is a regional coefficient.

All of these variables, if increased, will increase the value of $\bar{Q}$, except LAKE, a measure of surface storage, which will reduce it by virtue of increasing the surface storage volume in a catchment and reducing the intensity of flooding e.g., $m^3\,s^{-1}$. There are other equations derived for urbanised catchments (the proportion of urban land will also increase the catchment value of $\bar{Q}$) and small catchments, generally less than $20\,km^2$ (Wilson, 1983, ch. 9) that is especially prominent in areas that have been urbanised. In the 'Essex, Lee and Thames areas', which are dominated by urbanised catchments in and around London, a different equation was derived:

$$\bar{Q} = 0.373\text{AREA}^{0.70}\text{STMFRQ}^{0.52}\left(1 + \text{URBAN}\right)^{2.5}$$

where URBAN is the urban fraction of the catchment. Note the strength of the exponent on the new term, an increase in the strength of the term STMFRQ (presumably reflecting the importance of engineered channels), and a corresponding absence of terms SOIL and RSMD which are characteristic of non-urban areas where soil and vegetation systems dominate catchment behaviour. In urban areas, covered and impermeable surfaces dramatically reduce the effect of soil drainage and soil water deficit. Subsequent studies for rural areas have suggested simpler forms than the first, six-term, equation may be employed (Wilson, 1983, p. 226), but the original form remains preferred.

As more information becomes available (Defra, 2009a), better modelling and statistical estimation techniques evolve and input parameters including scenarios of climate change have to be considered (Chapter 5). The possible outputs describing flood behaviour modelling become greater. Catchment descriptors can be developed that incorporate

Table 2.1 Base flow indices and long-term flows for selected rivers (Cook, 2007 after Marsh and Lees, 2003).

River Catchment Gauging station	County and Grid Reference	Mean Flow (cumecs)	Base Flow Index	Years of Record	River Catchment Gauging station	County and grid reference	Mean annual flow (cumecs)	Base Flow Index	Years of Record
Avon Amesbury	Wiltshire SU 151413	3.49	0.90	1965–2000	Tamar Crowford Bridge	Cornwall SX 290991	2.26	0.29	1972–2000
Nadder Wilton	Wiltshire SU 098308	2.90	0.82	1966–2000	Beult Stile Bridge	Kent TQ 758478	2.06	0.24	1958–2000
Frome East Stoke (total)	Dorset SY 866867	6.41	0.85	1965–2000	Adur West Hatterell Bridge	West Sussex TQ 178197	1.07	0.25	1961–2000
Itchen Riverside Park	Hampshire SU 445154	5.4	0.92	1982–1999	Great Stour Wye	Kent TR 049470	2.21	0.58	1962–2000
Test Broadlands	Hampshire SU 345189	11.01	0.94	1957–2000	Medway Chafford Weir	Kent TQ 517405	3.08	0.49	1960–2000

more variables; in theory, improving estimates for flood peaks in ungauged catchments. The Flood Estimation Handbook –FEH (Bayliss, 1999) replaced the Flood Studies Report, and while the latter is complicated due to the use of both statistical and hydrological process models able to make predictions of many catchment variables, it includes an estimation of median annual flood (QMED). QMED has an annual exceedance probability of 0.5, and a return period of two years.

For rural catchments, using catchment descriptors (Robson and Reed, 1999, p. 100–101):

$$QMED_{rural} = 1.72 AREA^{AE} (1000/SAAR)^{1.560} FARL^{2.642} (SPRHOST/100)^{1.211} 0.0198^{RESHOST}$$

Where:

AREA is the catchment area (km^2)
SAAR is the standard average annual rainfall based on measurements from 1961–1990 (mm)
FARL is an index of flood attenuation due to reservoirs and lakes
SPRHOST standard percentage runoff derived from HOST soils data (page 48)
SPRHOST = BFIHOST + 1.30(SPRHOST/100) – 0.987
RESHOST residual soils term linked to soil responsiveness
BFIHOST is baseflow index derived from HOST data
AE is the area exponent = 1–0.015 ln(AREA/0.5)

The r^2 compared with measured data is 0.916.

To stress the developing nature of flood estimation, a subsequent development for rural catchments, one subsequently developed equation is example developed has been:

$$QMED_{rural} = 8.3062 AREA^{0.8510} 0.1536^{(1000/SAAR)} FARL^{3.4451} 0.0460^{BFIHOST2}$$

where BFIHOST2 is the squared baseflow index derived from HOST soil data.

This model is analytically more simple because it uses only four catchment descriptors, where originally six were used in the QMED model reported in the FEH (Bayliss, 1999).

Using the 'Revitalised Flood Hydrograph (ReFH)' model (Kjeldsen, 2007) design flood hydrographs are generated for a specified initial soil water content, base flow from groundwater and a design rainfall event for a required return period. Soil water content rainfall must be specified on a seasonal basis, depending on the degree of urbanisation of the catchment under consideration. Summer conditions are specified for urbanised catchments and winter conditions for rural catchments.

These wholly or partially statistical approaches demonstrate that flooding may be characterised as extreme events, with peak flows which are close to being instantaneous. Other approaches use physical characteristics to model catchment behaviour in terms of flood behaviour. Flooding in Britain remains to the fore in public concern. Just to introduce the topic, over 5 million people in England and Wales live and work in properties that are at risk of flooding from rivers or the sea (Defra, 2014a) and flood defence is firmly the responsibility of the Environment Agency in England and Natural Resources Wales (NRW).

The Flood Risk Management (Scotland) Act 2009 causes a sharing of the roles and responsibilities of organisations involved in flood risk management across the public and communities, the Scottish Environmental Protection Agency, local authorities, Scottish Water and the Scottish Government who aim to take responsibility and coordinate actions to minimise the potential for reducing flood risk. Prior to that, responsibility had been shared between land owners and local authorities. Around 5% of the land area of Scotland is at risk of flooding from rivers or the sea and within this area 3.9% of all Scottish properties, both residential and business, are at risk (Scottish Government n.d. b). On the other side of the same coin are 'low flows', which are dealt with in the next section.

2.1.2 Low Flows: The Other Extreme

Low flows will naturally occur during periods of prolonged low rainfall; however, concern over low (or zero) flows is attributed to over-abstraction of both surface and groundwater. There are several ways to characterise high flows, and several methods of mathematical description of flow distributions exist, so methods of flood characterisation need not be over-complicated. However, descriptions of low flows present more problems.

An approach for estimating the mean flow is to derive a regression equation relating mean flow and catchment characteristics. This equation was based on 687 catchments with at least six years of data:

$$\text{MF} = 2.97 \times 10^{-7} \text{AREA}^{1.02} \text{SAAR}^{1.82} \text{PE}^{-0.284}$$

where:

MF = Mean Flow ($\text{m}^3 \text{ s}^{-1}$)
AREA = catchment area (km^2)
PE = Average annual potential evapotranspiration (mm)

Low flows can then be expressed as a percentage of the estimated mean flow (Gustard *et al.*, 1992). Alternatively, water balance approaches may be made for a catchment.

Low flows are far more prolonged than flood events, especially where groundwater fed. Several 'standards' exist and helpful summary reviews may be found in several textbooks (Petts and Foster, 1985, p. 39; Ward and Robinson, 1990, ch. 7). These include:

- The lowest flow ever experienced (e.g., in 1975–1976)
- The 95% exceedance flow, the flow which is exceeded on all but 18 days per year on average. This is used to determine surface water licensing policies in England and Wales and is calculated over a period of at least 12 years.
- The mean annual minimum 10-day flow determined for the 95 percentile, the statistic $Q95_{10}$, which is a measure of flow duration over a (specified) time period.

The 'mean annual minimum 10-day flow' or MAM_{10} used in the original Institute of Hydrology's Low flow studies report of 1980 (Institute of Hydrology, 1980).

- The mean annual minimum 7-day flow or 'dry weather flow' (DWF), which approximates to the driest week in the average summer exceeded 89 to 93% of the time, and may provide a more stable index.

It also comes as no surprise that empirical formulae have been derived which aim to directly estimate low flow discharges. One example quoted by Petts and Foster (1985, p. 39), deriving from work at the Institute of Hydrology, is as follows:

$$MAM_{10} = 11.2BFI + 0.0982SAAR - 6.81$$

where:

MAM_{10} is the minimum daily mean flow in a 10-day period; BFI is the 'baseflow index' (related to flow originating from any aquifers) and SAAR is the annual average rainfall (mm).

Low flows not only present problems in terms of resource availability, navigation (where applicable) or restrictions on the extent of the aquatic environment, but also cause problems for water quality. The ability of a stream to dilute pollutants and to maintain dissolved oxygen (DO) levels in waters is aggravated by warmer conditions in the summer, because less oxygen is dissolved in warmer waters. Where sewage or other organic pollution is present, there is a requirement for oxygen to enable the microbial breakdown of organic compounds. This is the biochemical oxygen demand (BOD), which will also work to deoxygenate waters. 'DO' and 'BOD' levels are specified in order to set water quality standards. Percentiles are also selected and applied to multiple sampling. For example, at the 90 percentile, at least 90% of samples taken from a water body in a period of time should conform to the standard in question.

2.1.3 Drought, Just Low Rainfall or Climate Change?

A glib, although reasonable, answer to a question demanding the causes of low flows might simply be 'drought'. Yet this response poses further questions; a satisfactory answer involves more than the statistical approach of 'standard setting' to define flow conditions, it virtually becomes a philosophical matter. Drought might be measured in terms of its impact on the natural world, typically reduction or absence of river flow or loss of wetland habitats, factors paramount in modern water management. This account will, however, concentrate upon the needs of human societies, because the concept of 'water resources' implies an anthropocentric view of the world.

In defining drought, it is important to lay aside any concepts of 'aridity'. This is a long-term climatic concept based upon average notions of a 'lack of moisture'. While it is inappropriate to describe Britain (which has sufficient water for domestic, agricultural and industrial use) as arid, the World Resources Institute classification of water availability has put Britain in the 'low' category based on availability per person (NRA, 1995d). Because drought is a more ephemeral (i.e., temporally limited) occurrence, it is manifest in abnormal reductions in water supplies (Agnew and Anderson, 1992, ch. 4). These authors list definitions derived from several workers around the world, many of the definitions being in quantitative terms and relate periodic rainfall to the long-term average. Hence, droughts are defined in terms of their component parts such as the meteorological, hydrological and agricultural or water supply aspects. For example, the British Meteorological Office definition states that absolute drought is at least 15 consecutive days with 0.25 mm or less precipitation per day, and partial drought 29 consecutive days when the mean does not exceed 0.25 mm per day.

Although the meteorological (rainfall) and hydrological (e.g., low flow) components of drought do lend themselves to being described by the use of relatively simple statistics, it is insufficient to leave the matter here. River flows defined as low are at best seen not as one-off events, but in the wider context of *flow regimes* appropriate to climate, hydrogeology, land use and, of course, *user expectation on the water.*

Interpreting river flows or reduced groundwater levels in a drought definition has to be more subjective, implying a large degree of user expectation.

Huscheke (1959) gives one definition of drought:

> [It is] 'a period of abnormally dry weather sufficiently prolonged for the lack of water to cause a serious hydrological imbalance (crop damage, water supply, shortage, etc.) in the affected area.'

Potentially the term 'drought' can be, and is, used for Britain. Agnew and Anderson (1992, ch. 4) also review concepts of 'agricultural drought'. Most concepts consider effects of reduced precipitation upon soil water deficit, crop performance and yield, and its impact upon animals, then widen it to economic and social aspects including adverse effects upon communities. The mental maps of the British imagination see droughts as 'over there' in more arid climes, a situation reported from Old Testament times and viewed in the modern Sahel through television screens. It is consequently felt appropriate to restrict the term to situations where people do find themselves in mortal danger.

Long-term records of rainfall, groundwater and surface water, which in some instances stretch back into the mid-nineteenth century, indicate that extreme events of low rainfall are a part of the natural climatic variation. Until then, it was regarded as difficult to ascertain whether a specified dry period is a portent of climate change. Since then, it has been used the herald a new era on climate change for Britain. Britain so far has enough water, even if maintenance of supply involves some hydrological juggling, and, consequently, writers on the subject of water resources in Britain often eschew use of the word 'drought'.

The term may be replaced by referring to *times of reduced rainfall* which occur periodically. In part, or whole, Britain experienced problems enough to cause concern over water supply in the years 1921 to 1923, 1932 to 1934, 1949, 1954, 1959, 1965, 1972/73, 1975 to 1976, 1984, 1988 to 1992, 1995 to 1997, 1997-1999, 2003-2006 and 2010-2012 fall into this category (Headworth and Fox, 1986; DoE, 1991; Courtney, 1994; Marsh *et al.*, 2007; Met Office, 2013). For example, the period 1989 to 1992 had rainfall 20% below average in southeastern England, but a recharge to the chalk aquifer was down 50%, because of reduced rainfall in winter (when most recharge occurs) in the years 1988/89, 1990/91 and 1992/93 (NRA, 1992b).

In hindsight, 1975-1976 was critical for many surface water users and, in many respects, was the most severe 'drought' in England and Wales (Scotland was barely affected) in 250 years. The mid-1970s drought saw rainfalls between October 1975 and March 1976 of only 62% of the 1916 to 1950 average; locally it was considerably less, as documented in Doornkamp *et al.* (1980). The impact on river flow was uneven: over 80% of mean in parts of western Scotland falling to around 40% in central England between October 1975 and September 1976. Groundwater was not so badly affected due to above-average recharge in 1975, but reservoir levels were seriously down. Other periods (not listed) have seen localised droughts affecting water resources. In resource terms the 1970s 'drought' was more damaging than the more recent one between 1988 and 1992. This was partly because by 1988 new reservoirs, such as at Grafham and

Rutland in the Anglian Region of the EA and at Kielder in Northumberland, had been completed and interlinked with wider supply networks (NRA, 1994b). In Kent, but the end of 1992, river flow was reduced below that of the mid-1970s because of the long-term fall in groundwater levels to the lowest since records began (NRA, 1992a), due primarily to successive years of low winter recharge. Even in the British context, we see how the impact of prolonged low rainfall has variable results.

Low rainfall returned to England and Wales during 1995. Interestingly, this coincided with a meeting in Madrid of the Inter-Governmental Panel on Climatic Change (IPCC) in November of that year, where, for the first time, climate change attributable to the emission of 'greenhouse gases' was beginning to be identified. Individual droughts can never be attributed to global warming, but the water industry had started to take such opinion seriously. In the 1990s the NRA was already, in any case, concerned enough to commission research (Arnell *et al.*, 1994).

The Secretary of State did express concern that water companies (especially Yorkshire Water, North West Water and South West Water) experienced problems during 1995 (NRA, 1995c). This was in the face of widespread issuing of hosepipe bans and drought orders, restricting the use of water for non-essential purposes, and despite there being good surface and groundwater reserves during February of that year. The dry spell commenced in March, and by August rivers were commonly at 50% of their expected seasonal flow. Indeed, the period April to August 1995 was, in some areas, drier than the corresponding period in 1976, and it followed the dry period of 1988 to 1992. Drought orders were presented to the Secretary of State and aimed to widen options for water resource development (including new sources). These were supported by the NRA, provided adequate steps had been taken to mitigate the environmental impact and manage demand. As consumption rose throughout the first half of the twentieth century, *quantity* of supply (as distinct from water *quality)* was deemed not to be a problem for agriculture, industry or domestic supplies. Historically, there has been little perception that water could be in short supply until the dry years of the mid-1970s but following the 'droughts' after 1990, further legislation was required (Water Act, 2003; Box 2.1).

The present position is one by which climate change is part of the planning procedure for the water industry. The industry sees itself not only as a 'victim' of climate change (in reality this translates into growing uncertainty about resource availability) but it is also a contributor, for it is energy-intensive and contributes upwards of 1% of national Green House Gas (GHG) emissions. Carbon is a part of its business planning, and its uses if anything, is rising; already we can see it is becoming difficult to differentiate GHGs, carbon use (and abuse) resource availability and policy. The kinds of aims across the industry are listed in the Box 2.2.

The Environment Agency will be able to encourage transfer of water resources between water companies and recover costs associated with drought orders and permits. This has impact across all sectors and includes small abstractors, canals, harbours and agriculture, including irrigators and industries mentioned including the water industry.

The private water industry is, in reality highly regulated (Chapter 3), with its interests are represented by Water UK (Water UK, 2016a). This organisation calls for a more 'joined-up approach' by government and regulators in terms of policy issues around incentives and mechanisms for carbon reduction, renewable generation and green electricity.

Box 2.1 Summary of the Water Act 2003

It was found, following droughts after 1990, that 'existing legislation did not enable adequate management of water resources and protection of the environment'. A government review led to the publication in 1999 of 'Taking Water Responsibly'. This signalled important changes, including a move towards time limited licences and the development of Catchment Abstraction Management Strategies (CAMS). The Water Act 2003 and makes significant changes to the licensing system and key pints are that are summarised:

1) All small abstractions, generally under 20 cubic metres per day (m^3 day^{-1}), will not need a licence
2) A licence will be needed for dewatering of mines, quarries and engineering works, water transfers into canals and internal drainage districts, use of water for trickle irrigation and abstractions in some areas which are currently exempt will now need a licence to make sure that they are managed appropriately and that any impact on the environment can be dealt with
3) Administration for making applications, transferring and renewing licences will be made simpler
4) There will be an increased focus on water conservation
5) Water companies will have new duties to conserve water and all public bodies will need to consider how to conserve water supplied to premises
6) The Government has new responsibilities for monitoring and reporting progress in this area and water companies will need to develop and publish water resources management and drought plans

---

Box 2.2 Means towards reducing greenhouse gas emissions (Water UK, n.d.)

- Reducing energy use (electricity and other fuels) through efficiency measures
- Water efficiency and leakage control
- Research and development into alternative low-carbon technologies
- Embedded renewable power generation including anaerobic digestion
- Purchase of green power and good-quality Combined Heat and Power
- Investment plans that include whole-life carbon impacts and costs
- Work with the supply chain to encourage low-carbon behaviour
- Reduced water wastage; saving energy and carbon in the industry, homes and businesses
- Exploitation of least-cost solutions for waste disposal
- Insulation against energy price volatility and lower water charges
- Improved coordination of regulatory policies
- Enhance sustainability of land management
- Improve carbon accounting
- Support carbon sequestration initiatives
- Support Sustainable Urban Drainage Systems (SUDS)

2.1.4 The Climate of Britain

The climate of Britain is temperate, that is to say there is generally sufficient water to support the needs of the population and water supply, and measures have evolved to smooth out the supply where there is a risk of it becoming deficient. Rainfall, however, is not evenly distributed across the country. The position of the islands on the western seaboard of Europe, and the preponderance of westerly and southwesterly airflows, bring moist oceanic air.

Precipitation may be characterised by:

- *The amount falling in a given time period* (typically monthly or annual) and this may be averaged.
- *The frequency of 'rainfall events' in time* (e.g., magnitudes of daily rainfall events [called return levels]) for events with return periods of 1 in 5, 10, 20, 30, 50 and 100 years
- *The intensity of rainfall* (typically, mm hr^{-1})
- *Spatial variation in the above* (related to geographical concepts such as topography or region)

There are many accounts detailing the amount and occurrence of precipitation. The term 'precipitation' includes all water deposited from the atmosphere, including not only rain, but snowfall, hail, sleet, deposition from mist, and so on. The problem, however, is that precipitation patterns are changing (Osborn and Maraun, 2008). Throughout the twentieth century, intensity of UK precipitation has increased during winter and to a lesser extent also during spring and autumn. There were more frequent spells of very wet weather and an increase in total precipitation, at least during the second half of the century.

Table 2.2 shows that rainfall is fairly evenly distributed throughout the year, although proportionately more rain falls in the summer for the east than the west, attributable to convectional rainfall events.

Rainfall totals vary widely. The mountainous western highlands of Scotland is one of the wettest places in Europe with annual average rainfall capable of exceeding 4500 mm (Figure 2.2). Other wet areas include the Lake District, Snowdonia and Dartmoor. Across Britain, the interplay of climate and topography gives higher rainfall in the west. Mean annual rainfall here may be in the order of two and nine times that in the east; values can be as low as 500 and 600 mm along the coast of southeastern England. Typical values for lowland England and eastern Scotland range from 550 to 1000 mm.

In wetter, cooler climates potential evapotranspiration (PE) calculations are lower. The range is typically between 400 and 550 mm per year (Ward and Robinson, 2000, Ch. 4) with lowest values over upland areas. However, in very warm and dry years in south and eastern England, this figure can exceed 600 mm (NRFA n.d.). It is the net radiation available for evaporation which is dominant in controlling the evaporative loss, and this radiation is reduced due to cloudiness in upland areas. Other factors, such as increased humidity in regions of increased rainfall, and increased advectional losses in windier coastal regions, also increase regional variations. The slightly higher potential transpiration figures, linked with often dramatically lower rainfall figures result in regional supply problems through adverse catchment water balances.

Some water company supply areas are currently below their 'target 'headroom' (Figure 2.3), this is the difference between available supplies and demand. Each water company calculates its target headroom to ensure it can reliably meet customer demand

Table 2.2 England 1981–2010 Averages. While this aerially averaged data are indicative and up to date, they mask considerable variability.

	Max Temp	Min Temp	Days of Air Frost	Sun-shine	Rainfall	Days of Rainfall ≥ 1 mm
Month	[°C]	[°C]	[days]	[hours]	[mm]	[days]
Jan	6.9	1.3	10.4	54.2	82.9	13.2
Feb	7.2	1.1	10.8	74.3	60.3	10.4
Mar	9.8	2.6	6.5	107.6	64.0	11.5
Apr	12.4	3.9	3.7	155.2	58.7	10.4
May	15.8	6.7	0.7	190.6	58.4	9.9
Jun	18.6	9.5	0.1	182.6	61.8	9.6
Jul	20.9	11.7	0.0	193.5	62.6	9.5
Aug	20.7	11.5	0.0	182.5	69.3	9.9
Sep	17.9	9.6	0.1	137.2	69.7	9.9
Oct	13.9	6.9	1.5	103.1	91.7	12.6
Nov	9.9	3.8	5.1	64.5	88.2	13.1
Dec	7.2	1.6	10.3	47.3	87.2	12.7
Year	13.5	5.9	49.1	1492.7	854.8	132.8

(*Source:* UK Met Office).

in a dry year, and this is reported to Ofwat. This means that water company customers in these areas are more likely to face supply restrictions in a dry year. The distribution is uneven, largely due to differences in population and resource availability, but some supply zones are next to others with a healthy surplus. This implies a need to transfer water surplus to requirements. As a generality, Scotland seldom faces water shortages although parts of the east are relatively dry (Figure 2.3).

Seasonality is another factor. Winter rainfall is important in refilling reservoirs and recharging groundwater, summer rainfall much less so because soils then tend to be drier and able to absorb rainfall to be transpired back to the atmosphere by vegetation. If lowland Britain is to experience wetter winters, then in principle that will be good for replenishing supply. On the other hand, hotter, drier summers would increase demand overall, especially in agriculture. Other hazards that may be affected are increased flood and soil erosion risk, higher intensity rainfall and possibly reduced infiltration capacities that will prevent efficient recharge of groundwater stocks following, and despite, increased precipitation. Surface water storage may be adversely affected by agrochemical pollutants and particulate material may be leached and disaggregated into rivers and reservoirs.

2.2 Soils and Geoloy

Water used in supply, for aquifer replenishment or runoff is available through the balance of climatic variables. Apart from evapotranspiration and land cover, soils and geology control the fate of water on a land surface. Groundwater volume vastly exceeds the volumes of fresh surface water (Hiscock, 2005) and there is a pressing need worldwide to protect groundwater sources from contamination.

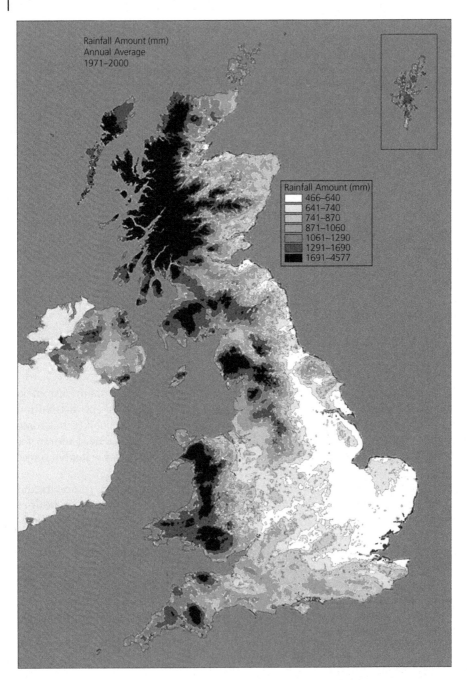

Figure 2.2 Annual Average Rainfall Amount (mm) for the UK 1971–2000 (*Source:* UK Met Office).

In Britain, older rocks are to be found in the north and west and younger in the south and east. It is these younger and overwhelmingly sedimentary rocks which provide the bulk of groundwater supplies. Older sediments tend to be harder, and hence more indurated, that is, better cemented, leaving a smaller pore space for water storage. In areas

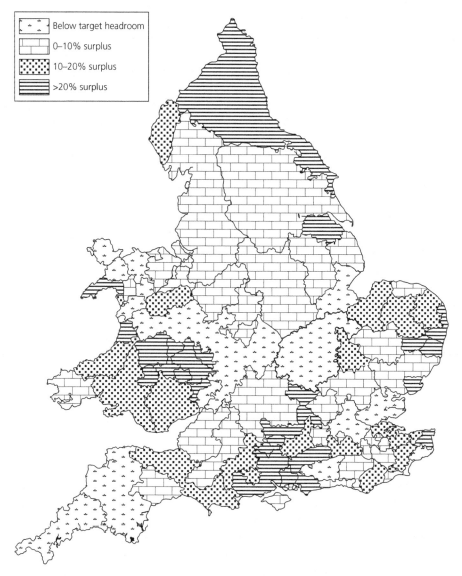

Legend:

- Below target headroom
- 0–10% surplus
- 10–20% surplus
- >20% surplus

Figure 2.3 Supply and demand balance for England and Wales by water company reported in 2008. In a dry year there is between 17 000 and 18 000 megalitres per day available so that areas below the 'target headroom' are likely to experience water use restrictions (EA, 2008). *Source:* Crown copyright.

of older rocks, surface flow predominates. In Chapter 1 it was established that groundwater abstraction in Scotland and Wales is small compared with England, and within England the contribution of groundwater to both supplies and streamflow is most significant in areas of the south, east and Midlands.

Figure 2.4 shows that the disposition of major aquifers in Britain are across the Midlands and south-east of England. Important aquifer formations alternate with relatively impermeable rocks such as shales, clays and mudstones. Rock formations may

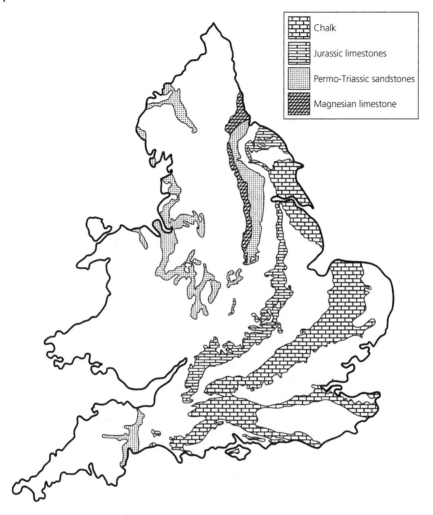

Chalk

Jurassic limestones

Permo-Triassic sandstones

Magnesian limestone

Figure 2.4 Principal aquifers in England and Wales. (Redrawn from Ward and Robinson, 1990, by permission of McGraw-Hill).

be separated into predominantly water-bearing *aquifers*, which store and transmit water, be it in useful amounts or not, and *aquicludes*, which are for most intents and purposes considered impermeable to water. *Aquitards* are beds which do allow transmission of water in significant amounts when considered over large areas and for long periods; in practice, most non-aquifers are aquitards to some degree. Formations such as the Chalk, Permo-Triassic sandstones and 'Lincolnshire' (Jurassic-age) Limestone provide essential supplies of water in parts of England and Wales.

Examples of confining (aquiclude) beds include the Gault, which confines the Folkestone Beds aquifer (part of the Lower Greensand formation) beneath, and the London Clay, which occurs stratigraphically above the Chalk. Recharge through supposed aquicludes has been noted in several places, such as in Dakota in the United States (Price, 1996, p. 73), making them aquitards! In recognition of the possibility of

---

Box 2.3 Use of the term 'groundwater'

Water for water beneath the ground, reserving the term 'groundwater' for the saturated zone only, presumably, because this represents the extractable portion of subsurface water. Here the term groundwater is preferred for all water stored or in transit through geological materials for the following reasons:

- The unsaturated zone is the route by which water recharges the watertable in most situations, and plays a role in pollutant transport and attenuation (Foster *et al.*, 1986).
- The fate of water in both zones is inseparably tied up and therefore merits consideration as a unified system.
- It follows etymologically that the word 'groundwater' is suitable, as it conveys the straightforward notion of 'water in the ground'. This logically includes soil water, that is, water in the soil and usually accessible to plant roots. In practice, and for reasons of scientific convenience, soil water is usually considered as a separate system.

---

transmitting water through confining beds, it is deemed important to include source protection measures over confined aquifers in groundwater Source Protection Zones (Chapter 10).

Groundwater is stored under two different kinds of physical condition: at or above atmospheric pressure and below atmospheric pressure. We must, therefore, consider two groundwater zones. These are, respectively, the *saturated zone*, where (in theory) all voids are filled and there is a positive hydrostatic head below the watertable, and the *unsaturated zone*. In the unsaturated zone, water is held in voids due to the capillary and adsorptive forces, at matric potentials below atmospheric pressure (Box 2.3).

Actually to talk of water being 'held' is a kind of shorthand; groundwater, like soil water, is most likely to be in constant motion at one timescale or another. The macroscopic flow in the unsaturated zone is generally vertically downwards (during drainage) and predominantly through fissures while water is typically held in finer pores. It is only when the top of the watertable rises through recharge that there can be a change to truly saturated conditions. Otherwise, transient conditions during heavy recharge through the unsaturated zone can lead to saturated conditions pertaining above the watertable for a limited period. Water movement can be in virtually any direction in the saturated zone beneath the watertable, depending on the local pressure gradients. Flow is predominantly through the fissures in fissured strata; however, in some aquifers, for example, the Permo-Triassic sandstones, inter-granular flow is of major importance. When the watertable is lowered, interstitial water may drain into the fissure system. This is the 'delayed yield' effect.

The 'soil' to an environmental scientist is the solid and semi-enduring interface between the atmosphere and the geology beneath. A product of solid geology or 'drift' geology (recent geological processes such as glaciation, wind deposition or alluvium or peat deposition) and usually termed 'parent materials' it is naturally highly variable. Other key factors in soil formation are climate, vegetation (natural or otherwise) and biotic factors, topographic disposition and of course, time, since the commencement of soil forming processes. Hydrologists talk of water falling on the surface of the land being 'intercepted' by vegetation cover, but partitioned at the soil surface, thereby invoking a whole sub-discipline of 'soil hydrology' within a wider context of 'soil science'. This partitioning means that water that does not merely rest at the surface to be evaporated back to the air from the soil surface itself or associated land cover, it either 'runs off' because the surface is wholly or partly impermeable, or it infiltrates into the soil profile itself. Once on a downward journey, it may be retained and used by plants or become part of a subsoil or aquifer system.

In practice, using information determined from the field, laboratory studies and soil mapping, classifications of soil water may be created that are useful in studies of runoff, low flow scenarios, flooding and soil erosion. One such example is the HOST (Hydrology of Soil Type) of the Centre for Ecology and Hydrology (Boorman *et al.*, 1995).

Applications of the classification have been developed that describe both the soil parameters required in low flow and flood estimation procedure response of catchments and also maps the characteristics. The original model was concerned with winter rainfall acceptance, to be used in conjunction with flood estimation. Parameters considered are soil water regime, depth to an impermeable layer, the permeability of the soil horizons and the slope of the land. HOST proper took this further to develop the model. The HOST classification is based on conceptual models that describe the dominant pathways of water movement through the soil and, where appropriate, the substrate beneath. Rain falling on the surface of some soils can drain freely, so that the dominant flow pathway is vertical. Alternatively, it may be diverted in the subsoil downslope as 'interflow'. If the underlying substrate is also permeable, the vertical pathway extends into the substrate, perhaps for some considerable depth. In all there are 29 categories derived from a matrix that considers on the one hand substrate hydrology (aquifer, mineral soil or peat soil) and hydrological pathway characteristics such as macro- and micro- porosity, by-pass flow, consolidation and permeability (Lilly, 2010).

2.3   British Aquifer Characteristics

The performance (and hence economic usefulness) of an aquifer depends upon water being stored, released and transmitted. This depends upon the relation between the fabric of the geological material and water. Aquifer characteristics such as transmissivity and the storage coefficient are determined by carrying out a field pumping test, whereby water is pumped at a constant rate from a well and levels in a nearby (observation) well are noted. From the rate of abstraction and changes in the cone of depression, calculations of aquifer characteristics are made. Real aquifers are extremely variable in their properties.

One simple measure of aquifer storage might be the porosity of the rock or sediment which acts as an aquifer, that is, the proportion of the aquifer filled with water in the saturated condition. Where the pores are small (such as in a Chalk aquifer where they are typically less than 1 μm), water will remain in much of the pore space even after the watertable has fallen, and there is no easy way of determining which of the pores are sufficiently small or isolated enough to retain water, or which will readily transmit water upon drainage of the matrix.

The porosity of chalk matrix is high, normally in excess of 30%, and often in excess of 40%. However, this porosity is not the prime cause of the rock's value as an aquifer. This relies upon the joints, cracks and bedding planes, termed 'fissures', which form between 0.5 and 5.0% of the volume; typically a figure of 1% is common (Headworth, 1994). These are wider than the pores, with widths typically between 50 and 5000 µm (Wellings and Cooper, 1983), and hence able to transmit water in useful quantities. They are typically spaced between 0.1 and 0.2 mm apart. In the field situation, as a watertable falls, or as pressure in the aquifer falls following pumping, water will be released to discharge. This is provided there is no influence from transpiring vegetation, where dewatering and upward movement may deplete the stored water.

There are four main measures of storage, release and movement of water in aquifer systems. They are storage coefficient, specific yield, storativity and transmissivity. The first three measures are described below; transmissivity is described a little later in the chapter.

- *Storage coefficient* is the volume of water which is released from a unit volume of aquifer per unit change of head. It is dimensionless, and certain authorities prefer to limit its use to unconfined aquifers.
- *Specific yield* is the volume of water released per unit surface area of aquifer per unit decline in watertable and applies therefore only to unconfined aquifers. For most purposes it is equivalent to the storage coefficient.
- *Storativity* is the preferred term for saturated, confined aquifers, that is, when no dewatering occurs. It is defined as the volume of water released from storage per unit surface area of the aquifer per unit decline in the component of hydraulic head normal to that surface.

In unconfined systems therefore, dewatering occurs as the watertable falls, hence values for specific yields tend to be high, often in the range of 0.01 to 0.03, whereas in confined aquifers storativities are lower, often between 0.00005 and 0.005. In confined systems, groundwaters remain under positive (above atmospheric) pressures, dewatering does not occur because flow out of a borehole is compensated for by lateral flow within the aquifer and no dewatering of the pores and fissures occurs. The smaller loss of water in confined aquifers, induced by falling hydraulic head, is attributable to changes in the porosity due to rearrangement of grains and expansion of water as the pressure is reduced.

Considering next movement of water through rocks, Darcy's law is the most important physical law governing the flow of liquids through porous solids. It can be expressed mathematically as:

$$v = K(dH/dl)$$

where v is the macroscopic velocity of water in the rock through a unit cross-sectional area of aquifer; K is the saturated hydraulic conductivity (permeability) of the aquifer; H is the hydraulic head; and l is the distance over which it applies. It follows that dH/dl is the gradient which causes the flow, termed the hydraulic gradient (dimensionless) (v and K typically have units of length per unit time, e.g., m s^{-1} or m day^{-1}).

The *transmissivity* (T) of an aquifer is a measure of how effectively water is moved through an aquifer to a well, and is the product of the saturated hydraulic conductivity (K) and the saturated thickness of the aquifer (b), that is, how 'deep' the saturated zone groundwater under consideration is above a supposed aquiclude. It can be expressed as:

$$T = Kb$$

(typically units of T are in m^3 day^{-1} m^{-1}, reducing by cancellation of units to m^2 day^{-1}.)

Table 2.3 Typical values of porosity and saturated hydraulic conductivity (K) for matrix and fissure flow.

Geological material	Typical matrix porosity (%)	Typical K_{matrix} m day^{-1}	Typical $K_{fissure}$ m day^{-1}
Clay	55	<0.0001	-
Silt	50	c. 0.1	-
Sand	45	1.0–50	-
Gravel	35	100–500	-
Limestone	0.1–10	10^{-10} (well cemented) to 5.0 (medium grained)	0.1
Sandstone	5–30	0.1 (well cemented) to 5.0 (medium grained)	1.0
Granite	10^{-4}–1	10^{-7} fresh) to 1.0 (weathered)	100

In order to illustrate these measurements, Table 2.3 shows some porosities for typical geological materials which form aquicludes and aquifers. High porosities and low K values for clay explain how this material effectively behaves as an aquiclude; although the overall porosity of clay is high, virtually all the water is in micropores and is held by adsorptive forces associated with the clay minerals. Significant transmission of water through clays only occurs via cracks. At the other end of the K-range are unconsolidated sediments, such as gravels, where water transmission via large intergranular pores is rapid.

The wide range of porosity quoted in the case of sandstone reflects more the effect of differing degrees of induration of the soft sediment during diagenesis, principally by cementation, than differences arising from differing pore volumes between primary particles. The figures quoted for K in granite reflect a strongly weathered, and hence fissured, example. On the whole, igneous and metamorphic rocks are poor aquifers.

Also given are K values for *fissure flow* ($K^{fissure}$), that is, conductivity considering only water movement through joints in the rock. *Matrix flow* is the conductivity of water through pores in the rock, K^{matrix}. K is very much higher for fissure flow in all cases. Overall values of K govern the flow of water through the aquifer to a well, borehole or natural spring. The matrix can control the retention properties of water and pollutants in groundwater systems.

Table 2.4 shows a range of typical characteristics for selected aquifers in Britain. Storativity, specific yield and transmissivity are defined in the text, mean annual replenishment is the annual mean recharge of the aquifer formation, and the 50% probable yield is the yield expected or exceeded from half the wells sunk in the aquifer. In bedded rocks, some horizons show more rapid flow to a well than others.

Preferential flow through such *flow horizons* frequently dominates groundwater behaviour in practical situations. For example, a 5 m layer of very high permeability with high storativity and transmissivity has been found to operate in the upper part of a chalk aquifer in Hampshire, where it dominates groundwater behaviour in the Itchen catchment (Headworth, 1994). Other complicating features include geological faults which provide preferential directions of flow for groundwaters. The Chalk dominates groundwater supplies, as might be suggested by the figure for replenishment in Table 2.4. It accounts for around 50% of all underground water supplies in England and Wales.

Table 2.4 Some aquifer characteristics (Monkhouse and Richards, 1982).

Aquifer unit	Storativity (*), specific yield (+)	Mean annual replenishment ** ml day^{-1}	50% probably yield** (ml day^{-1})	Transmissivity (range m^2 day^{-1})	Mean transmissivity m^2 day^{-1}	Environment Agency region/country
Upper/Middle Chalk	+0.01 to *0.0001	4608	3.20	10 368 to 1.0	1120	Anglian
Permo-Triassic sandstone	0.001 to 0.0001	1442	0.22	3542 to 16.4	225	North-east
Lincolnshire (Jurassic) Limestone	-	86	2.60	10 368 to 449	2850	Anglian
Magnesian Limestone	+0.01 to *0.0001	248	3.60	4406 to 19.9	830	North-east
Lower Greensand	+ <0.1	282	0.11	1209.6 to 11.2	c 250***	South-east
Carboniferous Limestone	-	2300	0.35	43.2 (single)	-	Scotland

Notes:

** These figures are typical across England and Wales

*** This is a typical figure supplied by the former Mid-Kent water from (now South East Water) pump testing

Although its transmissivity is highly variable, boreholes are usually located beneath dry valleys which display enhanced fissuring in southern England (IGS, 1970). In the Chalk beneath East Anglia, high transmissivity is similarly associated with topographic valleys which are in turn related to geological structure (Ineson, 1963).

Away from the influence of the surface, the matrix of chalk aquifers remains saturated in the unsaturated zone above the watertable; however, we would expect fissures often to be drained in this condition. Pollutants can diffuse their way into the saturated matrix from fissure flow as it occurs during recharge (Foster *et al.*, 1986) and this has important implications for the storage of pollutants in chalk. A second problem relating to the hydraulic properties of chalk is saline intrusion. This is due to the juxtaposition of chalk aquifers and the coast in areas as wide apart as East Yorkshire, East Anglia, the Thames estuary and along the south coast from Kent to Devon. Saline intrusion occurs where there has been over-abstraction, a particular problem when the aquifer is already fully exploited. Given the high population density along the south coast, this situation is common, and in vulnerable locations there is lateral invasion of saline waters. However, there is no simple interface of saline and non-saline groundwaters; seawater can intrude along discrete fissure lines (Headworth and Fox, 1986).

Fissuring is also the controlling factor in the aquifer characteristics of the Jurassic 'Lincolnshire Limestone' (where groundwater fluctuation is great) and the Permian 'Magnesian' Limestone. Both have reasonably high transmissivities and high values for 50% probable yield. The Carboniferous Limestone is underexploited as an aquifer on account of the lower probability of finding a fissure, or fissures, which will yield sufficient water. The figure for 50% probable yield masks enormous variation; it is possible to sink a bore and yield no water if no fissures are intercepted. Most of the aquifers in Table 2.4 are dominated by fissure flow, although pore flow through sandstone pores is also important in the Permo-Triassic sandstones and also dominates the Lower Greensand aquifers. The effect of intergranular porosity is seen in the higher specific yield of the Lower Greensand aquifers, and these display variable transmissivity values on account of lithological variation.

2.4   Groundwater and Surface Water Abstraction

Since the nineteenth century, active pumping of surface waters (from rivers, lakes and reservoirs) and groundwater has been possible, replacing more passive means such as lifting with a bucket or merely diverting surface waters into a pipe, leat or other supply channel. The result has been the possibility of both increasing abstraction at a point, and also over-abstraction, thereby increasing the possibility of long-term resource depletion, especially important when considering aquifer water balances. This situation potentially leads to the kinds of water shortage scenarios for eastern England shown earlier in Figure 2.3.

Table 2.5 shows the typical amount and percentage of total for groundwater abstracted in each region but this is defined by the water and sewage company that serves it wholly or partially (there are a number of water supply company areas) and corresponding to the former Regional Water Authorities as existed before 1990. Excluding extraction from tidal waters, these are shown in Table 2.4.

Table 2.5 Groundwater abstracted (all purposes) and percentage (UK Groundwater Forum n.d.).

Region (defined by Water Company)	Total annual groundwater abstraction M m^3 year^{-1}	Percentage of total from groundwater	Major aquifers
Anglian	325	47	Chalk, 'Crag', Jurassic limestones
Yorkshire	92	15	Chalk, Permo Triassic sandstones, limestones
Northumbrian	33	5	Magnesian limestone, 'Fell sandstone'
North West	142	15	Permo-Triassic sandstones
Severn Trent	40	51	Permo-Triassic sandstones
Welsh	16	4	Carboniferous limestone
South West	15	12	Permo-Triassic sandstones
Wessex	158	52	Chalk, Carboniferous limestone, Jurassic limestone
Thames	572	41	Chalk, Jurassic limestone, lower greensand
Southern	326	72	Chalk, greensand
Scotland	23	5	Includes Permo-Triassic and Devonian sandstones

The major aquifers are located in central, southern and eastern England. These are primarily the chalk that supplies around 60% of groundwater in the UK, the Permo-Triassic sandstones (around 25%), the remaining 15% comes from Jurassic limestone and minor aquifers including the 'Crag' of East Anglia, Magnesian and Carboniferous limestone, Carboniferous Fell sandstones, Devonian sandstones and 'greensand' aquifers.

If the total annual abstraction from groundwater for England and Wales is in the order of 1720 M m^3 year^{-1} and the total amount of non-tidal water abstracted from all sources in England and Wales in 2006/07 averaged almost 12,780 M m^3 year^{-1}, this means overall groundwater abstraction for England and Wales is in the order of 13%, much more than for Scotland, but masking considerable regional variation (Table 2.4). Over time this is around 10%; in reality, the overall groundwater figure into public supply specifically is more like 33% (EA, n.d. a).

Single global figures indicating the importance of aquifers as a source of water *supply* should be treated with caution. This is not only because, as Table 2.4 suggests, there is enormous variation from region to region (and even single regional values mask extremely high local reliance on this source), but because groundwater supports river flow and wetlands through seepage and spring discharge in so many areas. With climate change this is becoming forever more uncertain, and critical in maintaining wetland ecosystem services. These matters will be returned to later.

Management of groundwater resources may be viewed as a means to balance resources between seasons in Britain. Alternatively reservoirs may be viewed as an alternative to groundwater storage manipulation where the geology favours surface water. The balance of sources comes from surface waters including upland reservoirs, and streams and rivers. The highest proportions of the surface abstractions are in the Northumbrian, North West, Welsh and South West regions.

2.5 British Catchments

It comes as little surprise that the European Union's Water Framework Directive (2000/60/EU) (EC, 2014a) requires member states to manage water on the basis of river basins. Neither would it seem difficult to divide the land surface into areas over which water is captured and, with or without human intervention, is delivered via a drainage network to a point where it is either abstracted or further transferred to another part of the hydrological system. The term 'catchment' may be defined as the surface area from which a river collects surface runoff, and contains surface gradients down which water travels internally to the system to a point of discharge. In this book it is the preferred term, and it requires no further explanation, at least where surface waters are concerned. Another term describing the area from which water drains is 'river basin', which is better reserved for large river systems, and potentially multiple (and complicated) groundwater units. River basin is the term preferred in the WFD that can talk of 'River Basin Districts' (EC, 2014a) that is the basis for resource governance and planning. By this, a detailed account is required setting out how the objectives set for the river basin (ecological status, quantitative status, chemical status and protected area objectives) are to be reached within the timescale required. Two continental examples would be the 'Danube Basin' and the 'Rhine Basin' British river basins, such as the Thames, Severn and Tay are smaller in area but nonetheless form the basis for river basin management'.

Figure 2.5 shows the relationship between two groundwater and surface topographic divides where groundwater catchments and topographic divides do not coincide. Comparable situations are not uncommon in the extensive chalk aquifers of England where topographic catchments are frequently defined by redundant 'dry valley' systems, or where seepage zones are created by overlying aquicludes such as boulder clay. Most water falling over the chalk will, however, ultimately drain to groundwater. Another source of error in catchment definition derives from the places where anthropocentric interference is greatest and these are (lowland) urbanised areas. The installation of new channels, drains, culverts and gutter systems along roads and buildings will inevitably change catchment definition.

Hydrometric areas are either integral river catchments having one or more outlets to the sea or tidal estuary or they may include several contiguous river catchments having topographical similarity but with separate tidal outlets that are grouped for convenience. The Centre for Hydrology and Ecology bases numbers these 1 to 97 clockwise from northeast Scotland. There are further units for larger islands (such as the Isle of Wight, Anglesey and the Outer Hebrides) and for Northern Ireland that shares some catchments with the Republic of Ireland. Ireland operates a similar system (NRFA, 2014).

Mean annual runoff for England, Scotland and Wales is represented by the depth in mm of the average annual depth of rainfall available for river flow derived as the difference between precipitation and actual evapotranspiration. This figure is below 125 mm in parts of East Anglia (where below 25% of precipitation is actually contributing to the

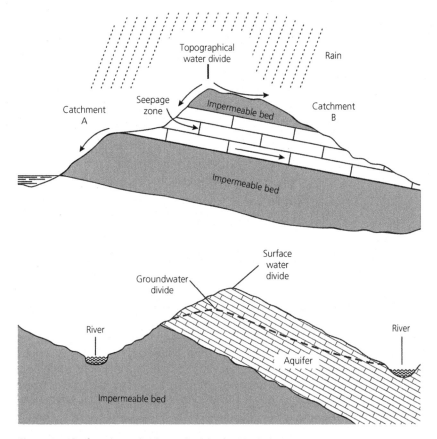

Figure 2.5 Surface water divides and subsurface hydrology.

river flow) but rises to around 2000 mm in parts of Wales and western Scotland (where as much as 75% of precipitation feeds river flow). Surface runoff is highly variable over time; for example, during the period October 1975 to September 1976 (which was abnormally dry in England and Wales), runoff was below 60% of the average across much of eastern and southern Britain (Ward and Robinson, 2000, p. 288). Similar extremes have been experienced since.

The catchment is thus recognised as the basis of comprehensive management under several major water Acts of Parliament (Chapter 3) and more recently by the European Union (Chapter 11). By way of illustration using historical data, Table 2.6 gives discharge information for 12 British rivers, together with their catchment areas, derived from DoE information and published in Petts (1988). All these rivers are regulated to some degree, through upstream impounding reservoirs or other surface or groundwater storage or augmentation. The drainage area refers to catchment area upstream of a gauging station, which is not necessarily the total geographical catchment, discharges are in $m^3 \, s^{-1}$ (or cumecs), and the ratio '95/mean' is the dimensionless consequence of dividing the 95 percentile flow (i.e., flow which is equalled or exceeded 95% of the time) by the mean. This is shown graphically in flow duration curves, such as those given in Figure 2.8, later in this chapter.

Table 2.6 Discharge (flow) characteristics for 12 British rivers (Petts, 1988).

River	Catchment area (km^2)	Mean gauged flow (m^3 s^{-1})	Max. daily flow (m^3 s^{-1})	95 percentile flow (m^3 s^{-1})	95/mean
Spey	2861	62.8	1089.22	19.16	0.305
Tay	4587	152.27	1222.56	46.98	0.309
Tees	818	16.54	317.15	0.86	0.052
Dee	1019	31.07	512.03	4.22	0.136
Severn	6990	96.38	529.53	22.46	0.233
Wye	4040	71.41	893.46	11.89	0.167
Trent	7586	80.99	815.53	27.08	0.334
Great Ouse	3030	14.32	311.49	0.99	0.069
Stour (Essex)	844	2.78	40.41	0.21	0.076
Thames	9950	67.54	1059.00	8.98	0.133
Medway	1256	11.62	269.30	1.39	0.120
Tamar	917	22.40	321.56	1.94	0.087

The Spey and Tay both display high 95 percentile flows, and both 95/mean ratios are over 0.3. This will reflect not only high rainfall in their upper catchments, but serious human interference on account of their being heavily regulated for hydroelectric power generation. Of the larger English or Welsh rivers, the Thames, Trent, Wye, Severn and Dee ratios are above 0.1, reflecting combined effects of Reservoir regulation, inter-basin transfers and effluent or groundwater augmentation. The Dee is a particularly highly regulated river with groundwater baseflow in its lower catchment. Otherwise, the Medway 95/mean ratio of 0.12 reflects both groundwater baseflow and reservoir influence in the Weald supporting the lower flows. Of the remaining rivers, the Great Ouse catchment has a predominance of clays, diminishing the effect of groundwater baseflow (incidentally, it is also prone to flooding), and low summer flows which are augmented by treated effluent. Both the Tamar and Tees have low ratios reflecting little or no groundwater components providing baseflow. Although the Essex Stour potentially has appreciable baseflow, being in a chalk aquifer, catchment groundwater is heavily exploited and abstraction has to be restricted to *maintain* baseflows; it is a heavily regulated river in a low rainfall region.

Casual observation shows that larger catchments tend to have larger mean channel flows. However, comparison of the Scottish River Spey with the Thames shows similar figures, yet the Thames catchment is around three and a half times larger. The cause is higher in-catchment precipitation for rivers which rise in the Grampian Highland region. A pronounced difference may be observed in mean flows between the smaller catchments of the Tamar (Cornwall-Devon border) and the Stour (Essex) which are of similar area, the latter in one of the driest parts of the country. Maximum mean daily flow gives a good indication of the high flow situation; the highest values are for the Scottish Rivers Spey and Tay, reflecting high precipitation and rapid runoff, and for the Thames, which has the largest catchment. Ineson and Downing (1964) estimated the groundwater catchment area of the River Itchen in Hampshire to be some 20% larger than that of the surface catchment for the portion of the catchment underlain by chalk, a situation confirmed by observation of the regional hydrogeological map (e.g., IGS/SWA, 1979). A more recent figure, derived from the Catchment Management Plan for the same part of the catchment (above the Highbridge and Allbrook gauging station) is around 45%, or 35% if the entire area of the catchment is included as far south as Southampton (NRA, 1992e). Low flows in the catchment have caused problems for the water quality, habitats and fish populations (EA, 2004). At the time of writing, the condition of the rivers Test and Itchen was found to be 'unfavourable' with a new partnership initiative being created to address some very difficult problems (Test and Itchen Catchment Partnership 2014).

The groundwater contribution to the main river channel above the Allbrook gauges is frequently from beyond the dry valley and winterbourne (seasonal) stream system which defines the surface catchment over most of the area, especially in the eastern part of the catchment. Since 1976, flow augmentation from groundwater in the upper catchment has been employed to support low summer and autumn flows, and this is further described in Chapter 5. The contrasting catchment sizes are shown in Figure 2.6.

The reliance of management on catchments (or river basins) arises from a historical reliance upon hydrological research applicable at that scale. Newson (1992a, Ch. 8; Newson 2008) states that rational management is not only likely to ask the question 'how', as do scientists, but is also very concerned with 'what if'. The development of catchment scale models with the advent of powerful numerical computation techniques enables scenarios to be explored, assisting in the proactive management of resources including 'what if' scenarios of surface and groundwater management.

Surface water abstraction from rivers, natural lakes and reservoirs is the largest source of water in many areas of Britain. Reservoirs are artificial means of impounding surface waters and may have a number of functions including storage, regulation of flow, power generation through hydroelectric schemes, and flood alleviation. Frequently, two or more functions are combined. The management of river catchments may be viewed in many instances as the control of water transfer from their upper to lower portions at sufficient rates to maintain flow and to meet demand.

River catchment management, in practice, may involve a kind of environmental juggling act. Classically, Schumm (1977) viewed a typical river basin in terms of transfer of water and sediment from the upper to the lower catchment with controls, natural or otherwise, affecting transfers within the system. Surface water abstraction must be managed so as to maintain an appropriate level of flow, yet, when demand is at its highest, flows will tend to be lowest, since maximum demand tends to occur during drier times of the year. Flows are often supported by groundwater and must also be sufficient to maintain aquatic and semi-aquatic ecosystems, carry away potentially harmful pollutants and permit navigation. Groundwater abstraction, in turn, must not dangerously reduce river flow. A number of measures can be used to help maintain river flow, particularly where demand from abstraction is high. In summary these generally are:

• Augmentation of flow from groundwater abstraction, either within a catchment where balances and hydrogeological conditions permit, or from elsewhere into the upper reaches of the streams.

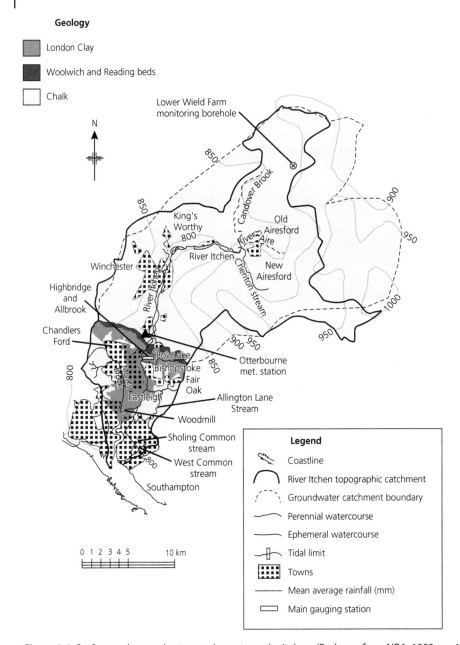

Geology

London Clay

Woolwich and Reading beds

Chalk

Legend

Coastline

River Itchen topographic catchment

Groundwater catchment boundary

Perennial watercourse

Ephemeral watercourse

Tidal limit

Towns

Mean average rainfall (mm)

Main gauging station

Figure 2.6 Surface and groundwater catchments on the Itchen. (Redrawn from NRA, 1992e, p. 12).

- Control of discharge by building a storage reservoir to impound water when flows are high; this is typically in the upper catchment to regulate flow lower down.
- Transfer of water across watersheds from another catchment into the upper reaches of the stream which needs augmenting. This frequently involves recharging reservoirs in the receiving catchment.

Box 2.4 River Catchment Conditions and Measurements

- *Catchment area.* The larger the catchment, the larger the volume of precipitation inter-cepted, other things being equal. 'Runoff' (as opposed to channel discharge) may be expressed in terms of discharge per unit area of the catchment ($m^3 s^{-1} km^{-1}$) and it may be observed that, other things being equal, peak runoff decreases with increased catchment size because the time taken to flow through the catchment to the gauging point ('time of concentration') increases with size. Aquifer systems within a catchment will also work to regulate runoff.
- *Catchment orientation.* Depending upon the prevailing wind direction, this will affect rainfall capture.
- *Catchment slopes.* The steeper the ground surface, the faster surface runoff will flow. In steep catchments, times to concentration will be shorter, and the peak discharges will be higher. Faster runoff will also reduce infiltration.
- *Baseflow index.* This is the proportion of the baseflow to the total flow component of a hydrograph. The index will increase with the baseflow component, and hence ground-water storage in an aquifer.
- *Soil water deficit.* This is the quantity of water lost through evapotranspiration from a soil profile since the 'field capacity' condition. (Field capacity refers to water retained against gravity following thorough wetting and free drainage.) It is important because there is higher rainfall infiltration into drier soils; wetter soils encourage runoff and hence increase flood risk.
- *Main stream length.* This is the length of stream in a catchment or to a gauged point.
- *Catchment shape.* Simply stated, longer, thinner catchments will tend to display longer times to peak runoff and have flatter peaks than more circular catchments. This is because there is a tendency for streams all to join the main channel close together in more circular catchments. Also, a storm is less likely to cover an entire elongated catchment.
- *Stream frequency.* This is a measure of stream occurrence in a catchment. *Stream density* is the length of channel per unit area of catchment; alternatively the number of stream junctions per unit area may be measured. In general, as stream frequency increases, both discharge and hydrograph peaks will tend to increase.
- *Lake and reservoir area.* This is the proportion of the catchment draining through lakes which usually receive a number of tributary streams. Increasing this variable will work to dampen hydrographs (i.e. reduce peak discharges).

River catchment management therefore involves timing the transfer of water from upper to lower catchment, taking into consideration water demand, maintenance of river flows and seasonality. It should not be forgotten that control of the sediment load by taking care to minimise soil erosion will prevent siltation of the channel and reser-voirs and hence assist in flood prevention in a catchment.

Box 2.4 describes some of the commonest variables and fixed measurements associ-ated with river catchments. Readers are invited to compare these with the variables listed in section 2.1.1.

Impermeable catchments are often termed 'flashy' because they lose little or no precipitation to groundwater, peaks in hydrographs are relatively high and the response to rainfall events is rapid.

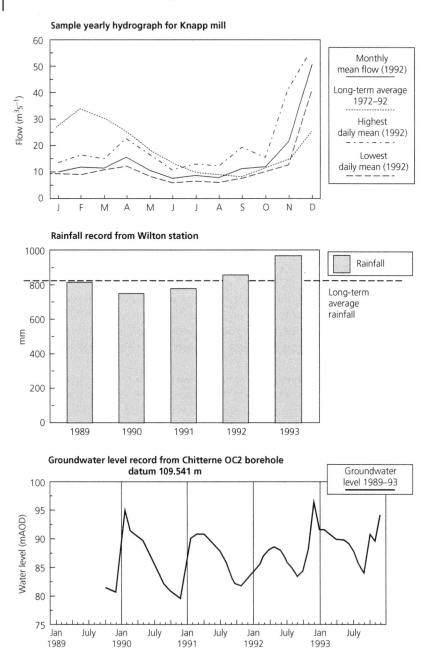

Figure 2.7 Sample Hydrometric Data for the Hampshire Avon Catchment. (Redrawn from NRA, 1994c, by permission of the Environment Agency).

Figure 2.7 shows some sample hydrometric data for the Hampshire Avon catchment, which, like the Itchen, is a predominantly chalk catchment in its upper reaches.

Aquifers such as the Chalk regulate rivers because they function as underground reservoirs, absorbing intermittent rainfall events and maintaining baseflow in the rivers.

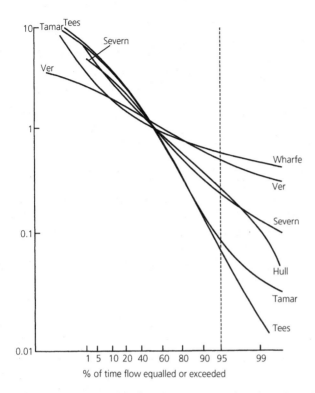

Figure 2.8 Dimensionless flow duration curves for selected British rivers. (Redrawn from Ward and Robinson, 2000, p. 259).

Hydrographs will display far fewer and lower peaks than for permeable catchments. Figure 2.7 illustrates this point; plots of monthly mean flow, highest and lowest daily means for 1992, and the long-term average all display a broad pattern which correlates with the seasonal groundwater levels. The long-term average curve agrees well with the year-by-year pattern of groundwater elevation, suggesting that river flow is strongly influenced by the 'baseflow' (i.e., groundwater) component.

Figure 2.8 shows dimensionless flow duration curves for selected British rivers. This kind of plot divides the daily discharge values by the average daily discharge for the time period under study to produce a dimensionless figure and presents the figures derived on the basis of exceedance. This approach allows rivers of differing discharges to be compared on a single graph. The figures thus obtained for flow are then ranked and plotted so as to show the percentage of time that a specified flow is equalled or exceeded. Steep flow duration curves (e.g., the Tamar and Tees) have quickflow dominated discharges, whereas gentler curves with flatter 'low-discharge ends' have appreciable groundwater storage in their catchments (Ward and Robinson, 2000, p. 259). Examples are the plots for the Ver and Wharfe, both of which have discharges supported by groundwater, the former from the Chalk, the latter from Carboniferous Limestone.

The statistic 95/mean (presented in Table 2.6) is readily found for each river on the graph using the 95 percentile level. The response of individual rivers to their catchment conditions is one part of the story the other depends upon the amount of available runoff. As shown by the empirical formulae at the start of this chapter, surface hydrological variables, soil factors and topography combine to determine the catchment responses to rainfall events. These responses are in terms of rate, variability and quantity of discharge.

2.6 Summary

Factors which affect the availability of the water resource are many, and relate to a strong human imprint on natural system systems. Fortunately for Britons, water is sufficiently available for most purposes for most of the time, including times of low rainfall. However, low rainfall spells since 1975 have highlighted not only resource management problems, but also the impact of water use on the wider environment.

Catchments, the unit of environmental management under consideration here, define areas which internally drain to a river or borehole provide the only rational unit for management, whether it is for considerations of quantity (water balance) or water quality. Care is required in defining catchment areas where groundwaters are involved; aquifer systems may convey water from outside a topographically defined catchment. It will be described how modern water protection measures require aspects of uncertainty to be built into groundwater catchment definitions. Nonetheless, catchment considerations in water protection are a part of planning in Britain and this should widen to consider integrated protection of land, water and air through the Environment Agency. Historic legal and management factors are next considered.

3

Institutions and Legislation for Resource Management

He [Professor Branestawm] wrote [a letter] to the Water Company to say that the water came out of the taps twisted, and was this all right? ...

... When they got back there was a letter from the Water Company to say they didn't mind the water coming out twisted, but if it came out in knots, to let them know.

The Incredible Adventures of Professor Branestawm
by Norman Hunter (1933)

3.1 Regulation: A Dirty Word?

Regulation has long been a driver in water governance in Britain, another factor being finance (Chapter 3.10). Inevitably, it requires a legislative framework, and water laws are known from most societies. Water has to be managed in the common good, assuring fair allocation yet protecting individual rights, provide for health and safety, overall water quality and all of these operating in the common good. To understand this complicated agenda and to appreciate the direction of change (both past and present) requires an historical perspective. Ultimately, legislation and management reflect the imperatives of a changing economy and population, as well as reflecting advances in scientific knowledge and governance. Britain is no exception.

There is, after all, a legal framework around virtually all economic activities. From the start it should be stated that contemporary thinking in water resource management still requires that there is regulation backed by legislative frameworks, even if the boundaries and responsibilities of both institutions and citizens change. In Chapter 10, it will be described how regulation may be moving to arm's length, while it is becoming desirable to 'democratise' water governance, that includes politics, rules, decision making and institutions. This has to be of direct economic benefit to stakeholders. By contrast, the word 'regulation' in the United States can be met with open hostility. Here, the term 'federal' can be read as 'state' in the UK, implying central government action that is potentially 'top-down'. In Australia and New Zealand, like nearby continental countries (France, Germany, Netherlands and Denmark to mention four) water governance clearly involves local government at some level.

It is argued that Britain has the potential to fall between several stools in terms of water governance. The long-term retreat of local government involvement, in favour of regionally based statutory agencies and also central government is all but total (Benson *et al.*, 2013a). This abdication not only bucks historic trends at home, but is at variance with much of the rest of the world—this chapter is about how we got to a seemingly anomalous position, while subsequent chapters will attempt to chart the way out including encouragement of public participation.

There is seldom any legal concept permitting ownership of flowing water and so natural forces and social frameworks may prevent politicians and institutions having unfettered control over catchments (Kinnersley, 1988). From a position of management via 'ancient custom' emerge statutes and, eventually, statutory bodies imposed by government to regulate water resources. Through emerging catchment management and the Water Framework Directive 2000 (WFD), modern practice aims to inject a strong element of democratic consultation, at least among interest groups or 'stakeholders', in order to moderate the impact of technocratic control.

3.2   English Law

Early English law had no clear distinction between public and private waters, although tidal waters including river beds belonged to the Crown, while non-tidal river beds belonged to the riparian landowner until half-way across their width (Kinnersley, 1988). There was an obligation on owners to maintain channels in good order, and a developing legislation controlling the water environment for the common good. There was no allowance for ownership of water in the sense of *proprietary interests,* and there has never been *absolute ownership.*

Water therefore flows through streams, rivers and estuaries without being owned by anyone. It can be *appropriated* by being contained in a vessel, tank or pipe, but otherwise the legal interest which is enjoyed in water falls short of being a right of property (Howarth, 1992). It is, consequently, the uses for which users are afforded legal protection. Common law legal interests are collectively known as *riparian rights and duties.* These protect abstraction for purposes such as agriculture, industry and domestic supply. They also aim to provide for fisheries and navigational rights. Landowners are therefore responsible for maintaining flows and not polluting waters.

Riparian rights and duties do not form a part of the criminal law; however, their infringement makes the plaintiff entitled to compensation. For example, if the riparian rights of a downstream landowner are infringed by irresponsible operations (perhaps causing pollution of a river) upstream, it is actionable under civil law. A right to abstract groundwater emerges from the case of *Chasemore* v. *Richards* (1859), while the riparian owner had rights to use the water under common law. In this case, the plaintiff sought action to prevent the Croydon Board of Health abstracting groundwater at a new waterworks to supply the growing town of Croydon, on account of reducing the flow of the river Wandle (today in south London) that powered his mill (Cook, 2008a; 2015). In his appeal to the House of Lords, it was held that the 'possessor of the soil' might sink a well, and pump from it to any extent however much it might affect his neighbours. Mr Chasemore lost, because it was decided there is no rights over 'percolating water' until it reaches the surface. This is clearly hydrogeological nonsense and the River Wandle Protection Act (1908) had to be passed somewhat later in order to protect the underground sources of the river.

Rights over waters are given to a riparian landowner to receive water in its natural quality and quantity, and an obligation placed on their owner for water to go from that owner without obstruction. Similarly, in rural areas a landowner had the right to abstract water from a well sunk and villagers had the right to draw water from a communal well (Elworthy, 1994, p. 46); both streams and underground waters are liable to cross many properties. The point that there remains no absolute ownership of water as a 'common resource' contrasts with concepts of absolute ownership of land which became entrenched during the post-Medieval period.

The criminal law contains a number of specific offences relating to water pollution, and thus provides the basis for the entire discharge consent system operated by some statutory body (Howarth, 1992). At different times there have been all-embracing Regional Water Authorities (RWAs) after 1974 responsible for resource regulation, planning and utilities, the regulatory National Rivers Authority (NRA), 1989-1996 and since 1996 the regulatory Environment Agency (EA). Prosecution through private criminal proceedings may be brought which are often less costly than civil proceedings. In recent times there is a move to transfer action to intermediary voluntary sector bodies in an effort to reduce legal actions (Cook *et al.*, 2011; Chapter 10).

Land drainage and flood protection have an ancient origin. On Romney Marsh (Kent and East Sussex), the 'Jurats', an organised body of men able to repair sea walls and control ditches, are known to have been effective in 1252. They raised money for their operations by charging a tax called the 'Scot', in practice an early water rate (Purseglove, 1988, p. 40). Similar institutions for flood defence existed elsewhere, such as on the Thames Marshes, where in 1390 a commission was established to inspect and repair flood defences and ditches. The first commission of sewers was established in Lincolnshire in 1258 following experience on Romney Marsh. The ensuing courts ensured that landowners maintained flood defences by maintaining dykes, walls and bridges in Fenland (Darby, 1956). The role of central government in these matters is reflected in the formal establishment of the Commissions in 1427 (Newson, 1992a, p. 15). In 1532, a Statute of Sewers was passed reinforcing legislation at a time when monastic lords were responsible for much drainage and reclamation. Monastic power was soon to be broken by the Reformation with resulting increased flood hazards in some areas, yet the courts survived into the twentieth century (Purseglove, 1988, p. 44). A balance had to be struck between the protection of the individual and the furtherance of the common good, and no balance could be struck without the cooperation of all the interests involved (Sheail, 1988). In the Romney Marsh area, as in so many similar reclaimed wetlands, the imperative of protecting from flooding enabled agricultural development to the present (Cook, 2010a).

Then there is the matter of pollution. By the early fourteenth century, English law had rules for sharing water which were capable of enforcement, but the resource needed protection from misuse. An Act of 1388, bearing the ungainly title of the 'Act for Punishing Nuisances which Cause Corruption of the Air near Cities and Great Towns', was aimed at avoiding foul smells from putrefied waters, and forbad the dumping of dung, filth, garbage and entrails in ditches, rivers and other water bodies. A General Sewers Act in 1531 aimed to 'cleanse and purge the trenches, sewers, and ditches', and a complex hybrid of common and statute law was in practice applied in water management, implying progress was to be made by precedent (Newson, 1992a, p. 17; Newson, 2008; NRA, 1994a). Like groundwater, there was poor scientific understanding. For example, there was no concept of waterborne disease until the nineteenth century, but early Acts represent milestones in pollution control legislation in situations of high population density. Specific Acts preventing the pollution of waters thus existed in towns from the fourteenth century, while the courts had powers to act against those causing nuisances to drinking waters under common law in rural areas.

The nineteenth century saw a rapid rise in the passing of statutes, and the main Acts of Parliament since 1830 which are of importance to water management and supply in England and Wales may be summarised as follows (DoE, 1971; Newson, 1992a, p. 17; Elworthy, 1994; Kinnersley, 1994; Binnie, 1995; Water Industry Act, 1999; Water Act, 2003; Flood and Water Management Act, 2010):

- Lighting and Watching Acts 1830/33
- Waterworks Clauses Act 1847
- Public Health Act 1848
- Metropolitan Waterworks Act 1852
- Local Government Act/Amendment Act 1858/61
- Salmon Fisheries Act 1861
- Rivers Pollution Prevention Act 1876
- Metropolis Water Act 1902
- Land Drainage Act 1918
- Land Drainage Act 1930
- Public Health Act 1936
- Water Act 1945
- Rivers Board Act 1948
- Rivers (Prevention of Pollution) Act 1951
- Clean Rivers (Estuaries and Tidal Waters) Act 1960
- Rivers (Prevention of Pollution) Act 1961
- Water Resources Act 1963
- Water Act 1973
- Control of Pollution Act 1974
- Water Act 1989
- Environmental Protection Act 1990
- Water Resources Act 1991
- Water Industry Act 1991
- Land Drainage Act 1991
- Environment Act 1995
- Water Industry Act 1999
- Water Act 2003
- The Flood and Water Management Act 2010
- The Water Act 2014

What is immediately apparent is the increasingly complex and specific nature of water legislation and the increasing frequency with which key acts have been passed. This growth reflects increased urbanisation, concern over public health and pollution from industry; later concern over water abstraction and environmental management. Yet it is not always easy to characterise a specific act as being concerned with public health, industrial pollution or (later) with the organisation and regulation of the industry, though switches of emphasis are clear. In 1830/33, the Lighting and Watching Acts prevented the washing of gasworks waste into streams. In 1863, successful civil litigation was brought against groundwater contamination from lead mining in the

Mendips which was polluting springs used in paper manufacturing (Gough, 1930); this was because water could be understood legally as flowing in underground channels and not in this case diffusing through porous rock. Skinner (1991) notes the less than clear definitions in UK law: neither common nor statute law recognise underground water flowing in defined channels as groundwater, but regard it as if it were a surface stream! The interconnectability (i.e., ability to move in three dimensions in porous strata) of groundwater is thereby missed. While the overwhelming bulk of nineteenth-century legislation is related to organic waste and disease control, most acts concerned directly with industrial effluents date from the twentieth century (NRA, 1994a).

The Waterworks Clauses Act 1847 provides a milestone in legislation designed to regularise and improve public water supply through public, municipal provision. It is an early example of legislation aimed at the protection and conservation, first of supply, and subsequently of the wider water environment. In 1852, the Metropolitan Waterworks Act enabled the introduction of filtration to the London supply which was only slowly adopted. The Public Health Acts of 1848 (amended 1861 and 1872) provided for the improvement of public sanitation, the protection of supply and the definition of urban and rural sanitary districts with responsibility resting with the local authorities. Several additional acts allowed for the establishment of by-laws to protect water supplies in specific urban areas (Sheail, 1992), as did the Metropolis Water Act 1902 for London, or the more localised Wandle Protection Act 1908 already mentioned. Here was then little further legislation in this field until 1945, when there was 'An Act to make provision for the conservation and use of water resources and for water supplies and for purposes connected therewith'. It was largely concerned with consolidation of sources of supply. An exception was the Public Health Act 1936, which consolidated the law on the nuisance of discharges of potentially dangerous or hazardous waste to public sewers.

In 1867, many shortcomings of the civil law were recognised by the Royal Commission on the Pollution of Rivers. Because the redress was for individual aggrieved parties to bring action for damages or injunctions (claiming nuisance or negligence), and because litigation was expensive and the outcome uncertain, progress on pollution was increasingly through the criminal law by which prosecution and punishment were possible, irrespective of individual rights (Howarth, 1992). The Rivers Pollution Prevention Act 1876 was designed to prevent discharges of sewage, mine waste and industrial wastes. It included the basis of modern discharge consents and was repealed in 1951. It aimed to prohibit the discharge of trade and sewage through the issuing of consents to discharge, although it did not apply to existing discharges. These were brought under control by the Rivers (Prevention of Pollution) Act 1961. The Clean Rivers (Estuaries and Tidal Waters) Act 1960 extended the powers of the River Boards (see below) to these areas. The Control of Pollution Act (COPA) in 1974 repealed much earlier pollution legislation and extended discharge controls to all tidal waters. It required public consultation in provisions for new discharges, and a public register of discharge consents and water quality monitoring data administered by the RWAs.

Part 11 of the Environmental Protection Act (EPA) 1990 superseded Part I of COPA in respect of the disposal of waste to land. Plants considered as 'heavy industrial' are regarded as having high potential for environmental pollution, and consequently they are targeted for 'Integrated Pollution Control' under the terms of the EPA. The main sectors involved are coal and bulk mineral extraction, power generation, gas-plant effluent, blast-furnace 'wet- scrubbing' products and steel and chemical manufacture, including petrochemicals. Statutory problems concerning surface water protection have largely been resolved over more than a century. For groundwaters, however, problems remained for much longer. While a scientific case may be made for considering all 'subsurface water' as groundwater, past legal definitions were seldom based upon scientific enlightenment.

Fortunately, as statute law concerning water resources becomes established, common law anomalies become less important, although there remains some confusion between definitions of groundwater for resource and pollution purposes. COPA prohibited the discharge from prescribed industrial processes and controls waste disposal to land where it may contaminate underground strata. Although the legal protection was subsequently enhanced by the EPA 1990 (NRA, 1992d), no legal definition of groundwater processes emerged. Skinner (1991) reports a common definition, following the Water Resources Act 1963, Section 2(2), and Water Act 1989, Sections 103(1) and 124(2)(b). Groundwaters are considered to be:

- waters contained in underground strata; or
- in a well, or borehole or similar work sunk into underground strata, including any adit or passage constructed in connection with the well, borehole or work for facilitating the collection of water in the well, borehole or work; or
- in any excavation into the underground strata where the level of water in the excavation depends wholly or mainly on water entering it from the strata.

The UK now has the opportunity to recognise water in the unsaturated zone as 'controlled'. Therefore, contaminated unsaturated zone water, which may percolate to the watertable, is also included. Although it is recognised that groundwater may be exposed at the surface and flow into excavations, groundwater flow in ordinary permeable strata is still not recognised.

The European Community Groundwater Directive (80/68/EEC) defined groundwater as:

> 'All water which is below the surface of the ground in the saturation zone and in direct contact with the ground or subsoil.'

Yet even EU legislation does not appear to recognise the unsaturated zone for what it is—a zone of both water and (potential) contaminant transmission which requires protection. Nor is there recognition of water exposed by excavation (Skinner, 1991). Consolidation legislation in the form of the Water Resources Act 1991 (NRA, 1992d) requires the 'competent authority' (since April 1996 the EA) to protect groundwater by controlling the majority of trade and all sewage effluent, and significantly, under Sections 92, 93 and 94, gives preventative powers for groundwater protection, a basis for the establishment of groundwater Source Protection Zones. It consolidates all provisions in respect of control of abstraction from groundwaters under the Water Resources Act 1963. Guidance given to planning authorities through Minerals Planning and Planning Policy Guidance notes emphasises the need to ensure adequate protection of groundwaters. The act furthermore is concerned with the enforcement of Codes of Good Agricultural Practice for the protection of controlled waters and also soil and air (Chapter 10).

Land drainage acts appear in 1918, 1930 and 1991. The 1918 act allowed the Board of Agriculture, county council or county borough council, to merge or form boundaries for an elective drainage body. The 1930 act became the 'benchmark' act that simplified and enhanced legislation passed since the sixteenth century and re-organise and maintain it on a catchment-wide basis (Sheail, 2002), including associated changes with the financial arrangements. In post-war Britain, grants were to become available for under-drainage that would cause conflicts between production agriculture and conservation (Cook, 2010a). The Land Drainage Act 1991 (amended 1994) enables Internal Drainage Boards to make changes that improve their efficiency (Land Drainage Act, 1991 and 1994). Furthermore, watercourses be should be maintained by its owner in such a condition that the free flow of water is not impeded; county council have powers of enforcement. The person responsible may also be prosecuted for nuisance under the Public Health Act 1936.

Internal Drainage Boards (IDBs) administer internal drainage districts; they are elected (rather than appointed), and have the function of controlling watertables and pumping water in areas where special drainage needs are required, such as reclaimed marshland. They are answerable to government and the EA, can receive some grant aid, and tend to represent the interests of landowners, especially the larger interests (Purseglove, 1988, Ch. 8). Local authorities are able to maintain the flow of watercourses under the Land Drainage Act 1991 and have power of entry to carry out drainage works. They must operate with reference to IDBs or RFDCs, now RFCCs. They also have responsibilities under the Public Health Act 1936 to deal with statutory nuisances of foul or choked and silted water bodies (NRA, 1992f).

The Water Industry Acts (WRA) 1991 and 1999 are the primary source of legislation governing the statutory water undertakings. The 1991 Act consolidated enactments which related to the supply of water, and the provision of sewerage services. There are general duties of water undertakers in Section 37 including the 'duty of every water undertaker to develop and maintain an efficient and economical system of water supply', making water supplies available to people who demand them, and maintaining and improving the distribution infrastructure. The Water Industry Act 1999 applied across England, Scotland and Wales. It clearly prohibited the pollution of from waste allowed to run to waste from any well, borehole or work water or prohibits abstraction. It made several important amendments to the Water Industry Act 1991. With present concern about water justice and the vulnerability of certain customers, it protected them from disconnection for non-payment of charges, streamlined financial arrangements for debtors and restricted the compulsory introduction of water meters (Water Industry Act 1991 and 1999). The wide-ranging Water Act (2014) is designed to make the water industry more innovative and responsive to customers. It should also 'increase the resilience of water supplies to natural hazards such as drought and floods', encourage metering and assist in the availability (and affordability) of insurance for those households at high flood risk including market reforms (UK Parliament, 2014).

3.3 The Rise of Water Regulation Institutions

Nineteenth-century development of the water industry was followed by a twentieth-century development of regulatory institutions. Water regulatory institutions arise because of a societal need to manage water resources in the common good and hence the requirement for a firm legislative background. Although long proposed as a means of managing catchments, the catchment is recognised as the basis of comprehensive management under the Water Resources Act 1963, the Water Act 1973, the Control of Pollution Act 1974 and the Water Act 1989, and importantly under the EU Water Framework Directive (2000). This makes the surface water situation clearer than for groundwaters that often do not conform to surface hydrological boundaries.

Integrated catchment management emerged after 1974 in the form of the RWAs, and the rise of catchment management planning after the creation of the NRA followed by the EA, brought hydrogeological considerations into contemporary water planning in England and Wales. The significance of the RWAs lies in the first concerted move in Britain towards integrated catchment management, decisively expressed in institutional terms. The following list outlines the developments leading to the establishment of the RWAs:

- *Mid-Victorian times.* River Authorities emerge in the Thames Conservancy (1857) and Lee Conservancy Board (1868)
- *1930.* Catchment Boards were established as a service for flood defence and land drainage in certain catchments following the Land Drainage Act 1930
- *1945.* Water Act creates framework for regional resource planning and establishes Central Advisory Water Committee
- *1952.* River Boards took over Catchment Board responsibilities, and gained new ones concerning river water quality, pollution control and fisheries, following the Rivers Board Act 1948 and the Rivers (Prevention of Pollution) Act 1951
- *1963.* River Authorities were established, with additional responsibilities of water resources following the Water Resources Act 1963

Since 1930, the responsibilities of regulatory bodies have increased. Indeed, modern responsibilities of the EA cover not only surface and groundwaters, but also air quality, waste regulation and ground contamination. Since the Second World War, another trend has been for the number of statutory water undertakings to reduce, starting with the Water Act of 1945 (DoE, 1971). This Act also authorised the establishment of a Central Advisory Water Committee to advise the Ministry of Health on conservation and use of water resources, and during the 1950s, this body drew attention to the lack of basic information for water resource planning (Smith, 1972). A dry summer during 1959 followed by floods during 1960 gave impetus to the 1962 'Proudman Report' of that Committee, which concluded that the management of water resources should be by multiple interests and should no longer be left to independent organisations. There were three main spinoffs: the legislation of 1963 and institutional responses in the establishment of the Hydrological Research Unit (1962) which became the Institute of Hydrology in 1968, and the creation of the Water Resources Board.

The Water Resources Act 1963 complemented developments in scientific hydrology and represented a real milestone in water resource planning. This was most notably by permitting the issuing of abstraction licences for surface and groundwater sources, with money obtained being used to finance conservation works (Smith, 1972). The licensing system came into effect in 1965. Licences of Right were only granted subject to providing to the reasonable satisfaction of the (then) River Authority that the quantities had been abstracted during the preceding relevant period (Section 34(3)). Clearly, new licences subsequently issued are not Licences of Right, but none the less are granted in perpetuity. Small abstractions for domestic purposes and farms were excluded, but licences for irrigation were required. However, 40 years later, these Licences of Right would need attention on account of a greater need for licencing control over water sources (see below).

The Water Resources Board (WRB) was involved in long-term planning including producing estimates that were useful into the 1960s. It enabled cooperation across regions to meet the needs of individual authority areas including the promotion of regulating reservoirs to increase river flows in periods of low flow, but it had no executive authority (Ofwat/Defra, 2006). WRB came into existence following the 1963 Act. It represented a central agency concerned with the conservation of water resources including strategic planning. However, the concept of 'Integrated Catchment Management' was still in its infancy and there was insufficient domestic legislation (Kinnersley, 1994, p. 46). It is interesting to note that this concept, under WFD, is expanded to talk of 'Integrated Water Resource Management' which should operate at a river basin level.

The Water Act 1963 had a requirement to preserve, control and develop water resources. There were 27 River Authorities replaced earlier River Boards (under the River Boards Act 1948), each with a role to conserve, redistribute and augment water resources in England and Wales on a regional basis. Each Authority comprised representatives from local and national government with between 23 and 45 members. Each River Authority had responsibility to conserve water resources that *was related to a river basin.* All groundwater and surface water abstractions were to be licensed by the River Authorities. Abstraction licences were awarded to existing abstractors and any organisation intending to abstract water would have to apply to the River Authority for a licence. A key development of the abstraction licensing regime was that River Authorities charged abstractors for the volumes licensed rather than the quantities actually taken. Abstraction charges took account of type of use and the effect each abstraction had on resource capacity. This required abstractors to concentrate on how much water they were likely to require in the future, rather than exaggerate reservations.

Funds raised from the abstraction licensing regime were used by the River Authorities to fund development of additional resource capacity.

In the years between the Water Resources Act 1963 and the Water Act 1973, water services were divided into three parts: *supply* (administered by statutory water undertakings), *sewage disposal* (administered by local authorities and joint sewerage boards) and *water resource management* (administered by the River Authorities, and the Thames and Lee conservancies). Within the description 'statutory water undertakings' there were 100 water boards, 50 local authority undertakings, 7 committees or authorities and 30 water companies. Sewerage authorities comprised 1364 local authorities, 27 joint sewerage authorities, the Greater London Council (GLC) and the City of London, and there were 27 River Authorities excluding the Lee and Thames conservancies (DoE, 1973). The Authorities submitted hydrometric schemes for resource assessment and had the role of predicting future demand (Smith, 1972), and this strategic planning role is a highly significant development.

The arrival of *multi-functional agencies* in the form of *Regional Water Authorities* (RWAs) following the 1973 legislation, was a response to this rather untidy and fragmented structure to the water industry. There were perceived obstacles to success in the areas of water quality, bulk supply and cost of supply, and there were legislative defects (DoE, 1971). Furthermore, the necessity to re-use large quantities of water was identified; hence there must be a single, comprehensive water management plan (including reclamation) for each river basin. From 1 April 1974 and continuing until privatisation in 1989, there were 10 RWAs in England and Wales which, together with 29 (then surviving) statutory (private) water companies, managed the entire hydrological cycle according to 'integrated catchment management'. This was to be a multi-functional, multi-disciplinary approach to water resource planning and management, and set-up to be predominantly owned in the public sector (DoE, 1973). At the time of privatisation, following the Water Act 1989, the 29 private water companies was subsequently reduced to 21 by mergers in 1994 (Burnell, 1994). Consolidation of water-only companies remains very much on the agenda.

Two further acts represent consolidation legislation. The supply of water requires three stages: abstraction of supply from natural sources, provision and delivery of supply mains, and supply of water, with attendant powers of control over misuse (Leeson, 1995, p. 116). The Water Resources Act 1991 sets out statutory responsibilities of the NRA (now the EA) and governs use of the resource. Common law rights qualify both surface and (probably) the ground abstractions in the rights of others downstream. The Water Resources Act 1991 reproduces the Water Resources Act 1963 as amended by the Water Act 1989.

The all-embracing Environment Agency (EA) was formed following the Environment Act 1995 and its significance is manifest in the Water Act of 2003. The EA will be able to encourage transfer of water resources between water companies and recover costs associated with drought orders and permits. It is concerned with the licencing system and these are changed in six key areas:

- all small abstractions, less than 20 cubic metres per day (m^3d^{-1}), will not need a licence;
- dewatering of mines, quarries and engineering works, water transfers into canals and internal drainage districts, use of water for trickle irrigation and abstractions in some areas which are currently exempt will now need a licence to make sure that they are managed appropriately and that any impact on the environment can be dealt with;
- administration for making applications, transferring and renewing licences will be made simpler. This will also reduce barriers to the trading of water rights;
- the status of licences has changed significantly, as all abstractors now have a responsibility not to let their abstraction cause damage to others. From 2012, the Environment Agency will be able to amend or take away someone's permanent licence without compensation if they are causing serious damage to the environment;
- there will be an increased focus on water conservation. Water companies will have new duties to conserve water and all public bodies will need to consider how to conserve water supplied to premises. The government has new responsibilities for monitoring and reporting progress in this area;
- water companies will need to develop and publish water resources management and drought plans.

EA and Defra consulted once more over reform of abstraction licencing (Wildlife and Countryside Link) in March 2013 with the results made public around the time of writing (Defra, 2016). The government response may be summarised:

- From the early 2020s, replacement abstraction permits will be issued with permitted volumes that at least reflect current business use and have a similar reliability to current licences. Hands-off flows and similar conditions will be standardised to simplify the system.
- At any time when flows are high, abstractors will be allowed to take water to store it with no seasonal permits.
- All abstractors directly affecting surface water will have conditions on their permits that enable flow based controls to protect the environment.
- Abstractors will be able to trade water in a quicker and easier way in catchments where there are potential benefits. Ofwat will work to consider trading mechanisms among potential users including balancing interests of large and small abstractions (including farmers).
- No permits will be time limited, providing a fairer approach. Information will be published so that abstractors and others can understand the environmental risks in their catchment; there is to be a stronger link with local communities.

These recommendations are proposing a lighter touch on abstraction licencing than might have been proposed, for there is a strong element of trusting abstractors to operate in a responsible fashion, while developing more market-based approaches to water allocation.

Then there is excess water. Despite the changes since 1970, flood defence remains an area of complexity. Flood protection from both sea and rivers involves numerous organisations. The EA has inherited a general supervisory role from the NRA under section 105 of the Water Resources Act of 1991. Under the act, work was carried out via Regional Flood Defence Committees or RFDCs (Defra, 2010a), now Regional Flood and Coastal Committees (RFCCs) (Lorenzoni *et al.*, 2015).

The Floods Directive 2007/60/EC (EC, 2015a) is a bedfellow of WFD, concerned with the assessment and management of flood risks since 26 November 2007. It requires:

> 'Member States to assess if all water courses and coast lines are at risk from flooding, to map the flood extent and assets and humans at risk in these areas and to take adequate and coordinated measures to reduce this flood risk. With this Directive also reinforces the rights of the public to access this information and to have a say in the planning process.'

It was a response to serious flooding across Europe between 1998 and 2009 which led to major damage and substantial loss of life, while the coming decades are likely to see a higher flood risk in Europe with greater economic damage.

The EA has direct responsibility for main-river, sea and tidal defences and also operates flood warning systems and for implementing the Floods Directive. It is required to arrange for all its flood defence functions (except certain financial ones), carried out by RFDCs, now RFCCs. Flood defence legally incorporates drainage (defined as including defence against water), including sea water; irrigation (other than spray irrigation); warping (introducing sediment to land through flood irrigation) and other similar practices involving the management of the level of water in a watercourse. There should also be the provision of flood warning systems. At the time of writing, there is a suggestion (and no more than that) from the UK Government for reform of flood management through the establishment of a new National Floods Commissioner for England (HOC, 2016), thereby stripping the EA of its responsibilities in flood management.

In order to carry out these functions, the EA through the RFDCs/RFCCs has various statutory powers including maintaining watercourses designated as main rivers, maintain or improve any sea or tidal defences, install and operate flood warning equipment, control actions by riparian owners and occupiers which might interfere with the free flow within watercourses and the supervision of internal drainage boards. RFDCs comprised staff of both the EA and appropriate local authorities (County, Metropolitan District or London Borough Councils). They could establish their own procedure subject to approval by the Minister, but their operations were disclosed and normally their meetings were held in public. Furthermore, they had the power to delegate to Local Flood Drainage Committees which covered specific geographical areas and whose composition was dominated by members appointed by appropriate local authorities. Flood management has since been concentrated in Regional Flood and Coastal Committees (RFCCs) (see Chapter 10).

RFCCs bring together members appointed by Lead Local Flood Authorities (LLFAs, these are county councils and unitary authorities) together with independent members with relevant experience for three purposes:

- To ensure there are coherent plans for identifying, communicating and managing flood and coastal erosion risks across catchments and shorelines.
- To promote efficient, targeted and risk-based investment in flood and coastal erosion risk management that optimises value for money and benefits for local communities.
- To provide a link between the Environment Agency, LLFAs, other risk management authorities, and other relevant bodies to engender mutual understanding of flood and coastal erosion risks in its area (ADA, 2106).

The Flood and Water Management Act (2010) is 'An Act to make provision about water, including provision about the management of risks in connection with flooding and coastal erosion'. It followed and largely implemented the recommendations of 'The Pitt Review' 2007 (Cabinet Office, 2008). In essence, following serious flooding in 2006 it was recommended there should be a wider brief for the Environment Agency and councils to strengthen their technical capability in order to take the lead on local flood risk management. It was reported that more can be done to protect communities through robust building and planning controls yet there had been excellent examples of emergency services and other organisations working well together, saving lives and protecting property. However, better planning and higher levels of protection for critical infrastructure were called for. There was a call for authorities to be more open about risk, while there is also much to learn from good experience abroad and people would benefit from better advice on how to protect their families and homes. Levels of awareness should be raised through education and publicity programmes, and recommendations made regarding how people can stay healthy and on speeding up the whole process of recovery, giving people the earliest possible chance to return to normality.

The Flood and Water Management Act of 2010 provides for better, more comprehensive management of flood risk for people, homes and businesses, helps safeguard community groups from unaffordable rises in surface water drainage charges, and protects water supplies to the consumer for as recent history had told, serious flooding can happen at any time. Kingston-upon Hull flooded in June 2007 and Tewkesbury in July of that year, while in November 2012 flooding occurred across the southwest of England, including some fatalities. This occurred again in early 2014, affecting the Thames valley and Somerset levels especially, that year being likely the warmest on record since 1772 (Parker *et al.*, 1992). Climate change projections suggest that extreme weather will happen more frequently in the future. This act aims to reduce the flood risk associated with extreme weather (Defra, 2014a).

The legal definitions of flood and coastal erosion are as follows:

1) "Flood" includes any case where land not normally covered by water becomes covered by water.
2) It does not matter for the purpose of subsection (1) whether a flood is caused by:
   a) heavy rainfall,
   b) a river overflowing or its banks being breached,
   c) a dam overflowing or being breached,
   d) tidal waters,
   e) groundwater, or
   f) anything else (including any combination of factors).
3) But "flood" does not include—
   a) a flood from any part of a sewerage system, unless wholly or partly caused by an increase in the volume of rainwater (including snow and other precipitation) entering or otherwise affecting the system, or
   b) a flood caused by a burst water main (within the meaning given by section 219 of the Water Industry Act 1991).
4) "Coastal erosion" means the erosion of the coast of any part of England or Wales.

It furthermore defines the concept of 'risk' in the context of flood risk management and coastal erosion risk management in terms of the operation of other acts such as the Water Resources Act of 1991 or the Land Drainage Act of 1991 or the Coast Protection Act of 1949. So that,

> 'The Environment Agency must develop, maintain, apply and monitor a strategy for flood and coastal erosion risk management in England and Wales (viz. a "national flood and coastal erosion risk management strategy").' The use of flood risk maps as well as agency co-operation and grants, to be administered by the EA and NRW.

The requirement of the EU for flood risk assessment is thereby met. Coastal defence from erosion (as opposed to flood) remains the responsibility of the EA and rests with local authorities which may receive grant aid for such works, but the Flood and Water management Act of 2010 required better agency cooperation and provides for the EA to local authorities across England and Wales are required to develop, maintain, apply and monitor a strategy for local flood risk management in their areas. The condition of both coastal and inland flood defences remains problematic, in essence major deficiencies identified in flood defence structures in 2001 were little improved by 2007 (National Audit Office, 2007). More recently, there has been concern over the level of local knowledge relating to flood defence and inevitably in a recession, sources of funding for flood defence.

3.4   The Scottish Dimension

In Scotland, the legal situation has hitherto been less complicated. Here, there is lower density of population and frequently greater effective rainfall than the rest of Britain. Consequently, there has not, until recently, been a licencing system for water abstraction. However, the Scottish Environment Protection Agency (SEPA) today requires 'some form of authorisation' for inland water abstractions, coastal and transitional water abstractions and construction of, and abstraction from, wells and boreholes (including pumping tests) (SEPA, n.d.b).

The situation north of the Border reflects the post-devolution situation, with a trend in Scottish governance to pull away from England and Wales while (at the time of writing) also remaining subject to EU Directives. Secondly, the water industry providing water supply and sewage services in Scotland is not, nor has ever, been privatised. The reasons for this will vary from that being an unpopular move in political terms, (especially following the Thatcherite debacle of the Poll Tax) to the relative severity of water supply and contamination issues being far less than those of lowland England particularly. Indeed, it may even come to pass that Scotland will export water southwards! The position in Scotland is therefore of a publicly owned water industry in the form of a statutory corporation called Scottish Water formed in 2002 (Scottish Government, 2014a) with a separate regulator (SEPA, n.d.c) that is analogous to EA and NRW and like EA, SEPA was formed in 1996 following the Environment Act of 1995.

The licensing system for abstractors, apart from agricultural abstractions, was pioneered in the (former) Forth Purification Board area. Subsequently, a decline in heavy industry is reducing pollution from that source (Kinnersley, 1994, p. 49). The right to abstract water from surface and underground sources depends upon Scottish common law and also upon specific statutes which govern abstractions for specific purposes. Between source and sea, the common law provides that a single owner has absolute proprietary rights over water within the watercourse. Otherwise, under multiple catchment ownership, the riparian owner is entitled, subject to the rights of other riparian proprietors, to make use of water flowing through his land provided that it is returned in quality, quantity and unaffected by force (Wright, 1995).

In early times Scottish common law catered for private rights to water. An act of the Scottish Parliament of 1621 secured supplies for Edinburgh and established that water can be taken a significant distance to consumers, that Parliament has power to establish specific uses, that public supply can affect existing (common law) rights and the public rights can work on land belonging to third parties (Macdonald, 1994). There were provisions for specific public water supplies by Acts of Parliament in 1833, 1855 (The Glasgow Corporation Water Works Act), 1893 and 1897.

Before 1800, there was little concern regarding the quality of Scottish waters (Hammerton, 1994) because population density was low and industrialisation in its infancy. During the first half of the nineteenth century, certain rivers seriously declined in quality, a situation found elsewhere in Western Europe, and River Commissions established in 1865 and 1868 undertook thorough investigations into the nature and causes of pollution of several rivers, both sewage and industrial. These investigations led to the Rivers Pollution Prevention Act 1876 which aimed to set the legislative framework. Water pumped from mine-workings, often acidified, was exempt from the Act, which was found to be flawed, primarily because local authorities (who were often the largest polluters because they discharged sewage) had no incentive to enforce the Act, and government inspectors were not provided for. Even transfer to county councils in 1887 made little difference because the latter mainly failed to use their powers: an early example of the 'poacher and gamekeeper' problem.

Only Lanarkshire County Council (backed by Medical Officers of Health and River Inspectors) worked to improve matters; elsewhere, river quality declined for 50 years after 1876. The County Council found that the legislation was insufficient to deal with its pollution problems. In 1936, the Advisory Committee on Rivers Pollution Prevention called for the establishment of River Boards; this was finally provided for in the Rivers (Prevention of Pollution) Act of 1951. Indeed, before 1939 there had been little attention paid to quantity or quality of supply, but wartime raised concern over security of supplies (Macdonald, 1994).

After 1944, water supply and sewerage provision in rural areas improved through a system of grants. Finally, there was a sound administrative basis for enforcing the law through catchment-based River Purification Boards (RPBs) (nine were created between 1953 and 1959), and in the north of Scotland 12 Purification Authorities were established, based upon county and borough councils. These authorities were also able to operate a discharge consenting system for sewage and effluent. Pollution inspectors, other scientific and administrative staff were appointed and water monitoring and river gauging commenced.

For comparison with the situation in England and Wales, it may be helpful to summarise certain main water-related Acts of Parliament applying to Scotland (Hammerton, 1994; Macdonald, 1994; Mackay, 1994; Wright, 1995):

- Rivers Pollution Prevention Act 1876
- Rural Water Supplies and Sewage Act 1944
- Water (Scotland) Act 1946
- Rivers (Prevention of Pollution)(Scotland) Act 1951
- Land Drainage (Scotland) Act 1958
- Flood Prevention (Scotland) Act 1961
- Spray Irrigation (Scotland) Act 1964
- Rivers (Prevention of Pollution)(Scotland) Act 1965
- Water (Scotland) Act 1967
- Local Government (Scotland) Act 1973
- Control of Pollution Act 1974
- Local Government (Scotland) Act 1975
- Water (Scotland) Act 1980
- Local Government and Planning (Scotland) Act 1982
- Water Act 1989
- Environmental Protection Act 1990
- Natural Heritage (Scotland) Act 1991
- Local Government (etc.) (Scotland) Act 1994
- Environment Act 1995
- Water Industry Act 1999
- Water Industry Act (Scotland) 2002
- Water Services etc. (Scotland) Act 2005
- Flood Risk Management (Scotland) Act 2009

Smith and Bennett (1994) recount the origins of flood defence legislation and show a lack of coherent, centralised information. Overall results of flood hazard are not easy to demonstrate, because of the statutory responsibilities being spread among the various bodies. The Land Drainage (Scotland) Act of 1958 provided for individual riparian owners to have primary responsibility for flood defence and this is most significant in rural areas, although regional councils have certain powers under the Flood Prevention (Scotland) Act of 1961 for non-agricultural land (mainly in the urban areas). The Local Government and Planning (Scotland) Act of 1982 was implemented in 1984 and gave the (by then) seven RPBs discretionary powers to install flood warning systems. Regional councils and riparian owners have responsibilities in terms of flood defence, but councils merely have permissive powers, not duties, in respect of land drainage. Howell (1994) supported a call for flood defence and land drainage to be brought under the umbrella of the Scottish Environment Protection Agency (SEPA) in order to simplify this situation.

The 1989 Water Act (which privatised water services in England and Wales) took the opportunity to improve provisions under the Control of Pollution Act (COPA) of 1974 (SOED, 1992) by enabling better matching with EU Directives on definitions and classifications of controlled waters, according to possible use, and enabled the Secretary of State to set Water Quality Objectives and improve pollution preventing powers. The RPAs were assisted to make charges and recover costs.

Wright (1995) notes the different arrangements for Scotland as summarised in the Water (Scotland) Act of 1980:

It shall be the duty of the Secretary of State [for Scotland]:

a) to promote the conservation of the water resources of Scotland and the provision by water authorities and water development boards of adequate water supplies; and
b) to secure the collection, preparation, publication and dissemination of information and statistics relating to such water resources and water supplies.

The 1980 Act makes provision for water authorities to apply to the Secretary of State either approving or authorising them to acquire water rights on a compulsory basis where public supply is involved. The Natural Heritage (Scotland) Act of 1991 established the Scottish Natural Heritage after the breakup of the Nature Conservancy Council (Environmental Protection Act of 1990), and merger with the Countryside Commission for Scotland. SNH was established in April 1992 and can recommend areas of outstanding natural heritage to the Secretary of State. It should also (Howell, 1994):

> 'have regard to the desirability of securing that anything done, by SNH or any other person, in relation to the natural heritage of Scotland, is done in a manner which is sustainable.'

Within the statutory planning process SNH is concerned with nature conservation, including environmental aspects of agriculture, forestry and fisheries, geological conservation and protection of physiographical features and, finally, the intangibles of natural beauty and amenity. It also covers water resources and, for the first time in Scottish legislation, the Act contains provisions for the control of abstractions for irrigation and commercial agriculture.

Pollution control has a complicated history in Scotland as elsewhere. Present arrangements (SEPA, n.d.g) are comprehensive and apply not only to surface waters but also aims to protect groundwater from discharges to land or direct to water bodies. Point source discharges include release of effluent or other matter to the water environment or land, via a pipe or outlet including sewage and trade effluent; surface water discharges from urban areas; and effluent from abandoned mine discharges. There is also regulation of diffuse pollution. Specific discharge listed are:

- sewage and organic effluents
- fish farms
- inorganic effluents
- thermal effluents
- surface water drainage
- waste sheep dip and waste pesticides
- storage and application of fertilisers
- keeping of livestock
- cultivation of land
- discharge of surface water run-off
- construction and maintenance of roads and tracks
- application of pesticide
- operation of sheep dipping facilities

Sustainable Urban Drainage Systems (SUDS) are promoted as part of control measures.

The Spray Irrigation (Scotland) Act of 1964 sought to control abstraction of water for spray irrigation because it had been shown that the dry weather flows in rivers in eastern Scotland were potentially at risk from over-abstraction (Hammerton, 1994), which is permissible under riparian rights. This was but a feeble echo of the licensing system introduced south of the Border after the Water Resources Act of 1963; licensing in Scotland had been virtually absent until the 1991 Act and progress slow.

The water industry remains in the public sector. Services were provided largely by the former Central Scotland Water Development Board under the Water (Scotland) Act of 1967 to areas of Central, Fife, Lothian, Tayside and Strathclyde Regional Councils (Burnell, 1994). It achieved this primarily through two major water supply schemes, Loch Turret and Loch Lomond (Wright, 1995). These are the basis upon which water statistics are collected and reported for demand and supply (SOED, 1994). The Local Government (Scotland) Act of 1994 provided for single-tier authorities from 1 April 1996. In this reorganisation, three public water authorities for Northern, Western and Eastern parts of the country provide water and sewerage services (Scottish Office, pers. comm., 1997). Water and sewerage have been local authority services provided by the nine regional and three island councils under the Local Government (Scotland) Act of 1973 prior to the formation of Scottish Water in 2002 following the Water Industry Act (Scotland) of 2002.

The Water Industry Act of 1999 applied to Scotland in respect of establishing a Water Commissioner. Later, the Commission regulated Scottish Water (much as Ofwat regulates the industry in England and Wales, see below) and aims to promote the interests of water and sewerage customers in Scotland' by making sure that they receive a high-quality service and value for money'. The Water Industry Act (Scotland) of 2002, as we have seen, established the modern water supply and sewage services industry while the Water

Services (Scotland) Act of 2005 replaces the Water Industry Commissioner (the Commissioner) with a body corporate, now called the Water Industry Commission ('the Commission'). The aim is to improve transparency, accountability and consistency of regulation in the water industry. There was a series of provisions regarding water and sewerage services. Specifically, these concerned prohibitions of common carriage where Scottish Water would use its water mains to carry water treated by a competitor and its sewers to carry wastewater from a competitor's customers to that competitor's treatment works in the water or sewerage systems, and prohibits retail competition for households. The carriage of water and sewage and set out definitions of "eligible premises" for the purposes of retail competition in respect of the "public water supply system" and "public sewerage system". The Act also makes provision for coal mine water pollution, providing a statutory basis for the remediation programme carried out by the Coal Authority to tackle and prevent pollution. Finally there are miscellaneous and general provision, including in relation to the offences created under the Act and the procedure for exercising the order and regulation making powers provided under the Act.

It was noted in Chapter 1 that the Scottish Government passed legislation in 2005 allowing for competition in the water industry in that was enacted 1 April 2008. Business customers (but not domestic) can potentially choose their water supplier, although Scottish Water continues to deliver water and remove wastewater. Competing supply companies carry out retail activities such as meter reading, billing and customer support. Supply companies bill their individual customers and Scottish Water bills the supply companies (Central Market Agency Scotland, 2016). Central Market Agency (CMA, a company limited by guarantee and owned by its members, administers the market for water and wastewater retail services in Scotland and self-describes as being 'at the hub of the new competitive arrangements. At the time of writing its members are: Scottish Water, Business Stream, Anglian Water Business, Wessex Water Ltd, Aimera, Severn Trent Select Ltd, United Utilities,Thames Water Commercial Services, Veolia Water Projects, Clear Business Water Ltd, Cobalt Water Ltd, Kelda Water Services (Retail) Ltd, Real Water (Edinburgh) Ltd, Commercial Water Solutions Ltd, Castle Water Ltd, Blue Business Water Ltd, NWG Business Limited, Source for Business Ltd and Everflow Ltd.

The Flood Risk Management (Scotland) Act of 2009 includes specific measures that will enable a framework for coordination and cooperation between all organisations involved in flood risk management, an assessment of flood risk and preparation of flood risk management plans, new responsibilities for SEPA, Scottish Water and local authorities in relation to flood risk management, and a revised, streamlined process for flood protection schemes. New methods to enable stakeholders and the public to contribute to managing flood risk, and there will be a single enforcement authority for the safe operation of Scotland's reservoirs (Scottish Government, 2012).

3.5   Privatisation in England and Wales

The election of a Conservative Government in 1979 that heralded the premiership of Margaret Thatcher (replaced in 1990 by John Major) represented a revolution in economic policy that changed the post-World War II consensus whereby the state not only intervened in the national economy, but also aspired to directly own key industries. These events are manifest today in that water and sewage services in England and Wales are not in the public sector, while Scottish Water is a legally constituted state corporation. Commentators describe the sale of public assets, de-regulation of markets (moving towards a supposed self-regulation) and other attendant features such as de-industrialisation and attacks on organised labour organisations as 'neo-liberalism' (Thorsen and Lie, 2007) can furthermore commonly be associated by conservative social values. However, privatisation was ideological at its roots. During the 1980s, the theory of 'Monetarism' that became linked with neo-liberal economic policy. Money supply was constrained in order to reduce inflation, and bring down the Public Sector Borrowing Requirement. Because such economic instruments were driven by the political fervour of the economic right (Minford, 1991) no publicly owned industry could hope to raise the required capital from government.

Such economic ideas replaced notions of widespread state ownership in such states as the former Soviet Union, or ideas of 'Keynsian' economic thought, otherwise commonplace in western countries. To Keynesians it is aggregate demand that shapes the economic policy, and that intervention by central government in spending (including 'quantitative easing') provided a cushion during times of economic crisis (Bannock, Baxter, and Davis, 2003) and reducing the impact of reduced demand in the economy, and of unemployment.

Over time, commentators have identified an underlying swing between public ownership (hence pressures to expand the public sector within a national economy) and the development of the private sector which is ideally 'market driven'. This swing is what has been described as an 'involvement cycle'. Proponents of this view (Vickers and Wright, 1989) identify periods of disenchantment with the market on the one hand and the state operation on the other, and the post-1945, economies in Western Europe tended towards regulation and public ownership, until the 1970s. When it is the turn of the state for disenchantment, this can lead to twin policies of deregulation and privatisation being followed. The expectation is that economic liberalism and market competition will constrain price rises and raise investment, involving the sale of public assets. Regulation concerns itself with controls of markets through controls on entry, exit, pricing, outputs, services supplied, markets served, consolidations and profitability in an industry (Swann, 1988). In a fully competitive environment, 'market forces' will give effective regulation, so economic, legislative and institutional intervention including state ownership, are reduced to a minimum.

In Britain, virtually all public industries have been privatised (sold to the market). Late in the Thatcher Government's term, the Water Act of 1989 separated out the *utility* (water supply and sewerage services) and *resource regulation* functions for the RWAs, thereby ending their classic 'goal conflict' of service provider and regulator. Following the act, all supply and sewage treatment/disposal followed the (then) 29 water-supply-only companies into the private sector, while public regulation became a way in which private monopolies were to be kept in check, and balances between environment, cost and consumer interest were to be sought.

Prior to the Act, there were impending problems which were other than regulatory; the all-embracing RWAs suffered from under-investment and had done so since the 1970s. There were two major problems: to raise the capital for maintenance and improvement of the ageing (often nineteenth century) supply and sewerage infrastructure, and to meet the increasingly stringent demands of EU directives on water quality. Because EU rules do not permit private companies to operate the regulatory rules contained in its Directives, the regulatory agencies had to remain in public ownership.

RWAs were not politically accountable and were 'technocratically' run, albeit in the public interest. There was no effective linkage with any elected body; prior to 1974 there had been elements of local/municipal authority control reflecting the Victorian roots of public utility provision. RWAs were furthermore presented as a manifestation of 'poacher and gamekeeper', for on the one hand they had responsibility for resource protection, and on the other hand, they were operating water treatment and sewage plants. There was no independence of regulation. Privatisation could give incentives to ascertain the needs of customers, the environment, on the other hand, required protection and will be dealt with in subsequent chapters.

The issue of *regulation* is thus separated from *ownership*. Following the RWA experience, few would argue that this is a bad policy development. Government privatisation policy was published prior to the sell-off (DoE, 1988) and is discussed by Newson (1992a, Ch. 7). The following summary is partly based upon these accounts:

- Privatisation would provide incentives to managers and other employees alike, and attract high quality personnel. This might involve wide share ownership in the utility among the workforce.
- Water was commodified (having been previously a service) ahead of privatisation; however, this was set against a background of under-investment and labour shedding after 1979 (the workforce in the water industry in England and Wales fell by about 24% between 1979 and 1989).
- Job losses continued after privatisation; Anglian Water, for example, shed around 900 jobs during 1994 alone. Privatisation would both raise capital from the markets and require a smaller workforce.
- New arrangements were needed for the protection of the water environment and economic regulation.
- Although there was regional/catchment-scale management, there was little attempt to link this into the planning process. Changes in regulation could change this. In operational terms, management issues might emerge which otherwise would be hidden in large state-owned technocratic institutions.
- By the mid-1980s, surveys were showing deterioration in water quality. New pollution sources blamed were spillages of slurry, silage liquor and diffuse sources from agriculture, and groundwater was also affected by nitrate. Changes in regulation should improve this while publicly quoted companies may look to their environmental image.
- RWAs faced no comparison with each other or other industries. After the changes, financial markets would be able to compare the performance of individual markets between companies and with other industries.
- Sewerage and sewage disposal infrastructure was not maintained or improved, and investment was badly needed.
- Because RWAs were multi-functional, they faced many problems. Solving these is a complicated matter of meeting the needs of real resource economics; meeting objectives of improved service quality, renewing assets, responding to growth, constraining price rises and pricing mechanisms may not help to limit demand in a classic supply and demand fashion. Events have subsequently shown that constraining price rises is not possible.

Responses to privatisation remain hard to characterise. Certain commentators have found find a growing service inequity, while others find enhanced consumer benefits

(Guy and Marvin, 1995). Insiders may celebrate privatisation in terms of the investment attracted (Frontier Economics, 2013) others may finds economic problems including rising prices for the consumer since privatisation (Tinson and Kenway n.d.). There is enormous variation, socially, geographically and environmentally, in the impact of privatisation. Furthermore, overall increases in price may be attributable to the drive to meet EU Directives, and otherwise water charges to the consumer might have fallen through efficiency gains due to regulation. Perhaps the point is that some 25 years later, privatisation in Britain as a 'workshop' of neo-liberal ideology remains a matter of controversy.

If a consumer is dissatisfied with the service from a garage, it is possible to go elsewhere. If the value for money provided by a pub is considered poor, maybe the same beer might be purchased elsewhere more cheaply in a particular town. The water industry, however, is a natural 'monopoly' (basically it is only possible to identify one supplier in the market) and it is extremely difficult to conceive of a means by which effective competition in water supply and sewerage services can be achieved. This is largely on account of the huge cost and space involved with its infrastructure. However, there are proposals including new entrants to the market:

> 'unbundling the current combined supply licence and creating a new upstream licence for companies wishing to introduce raw or treated water into an incumbent's network or remove and treat wastewater or treat and dispose of sludge from it. There should also be a network licence for those looking to provide infrastructure' (Defra, 2011a).

Such proposed reforms seem to be largely a call for arcane changes to the licensing system. One can imagine a situation analogous with the various rail companies running on the common rail network provided by Network Rail. While the UK-based reader can make up his or her mind as to whether rail privatisation has proved successful, he or she might also ponder whether in reality, the 'limits' of privatisation have in fact been reached! It has already been noted how there is legal provision for a 'competitor' under the Water Services (Scotland) Act of 2005, and competition around 'common carriage' is catered for in several pieces of legislation, including the Water Act of 2003. This Act of 2003 added 'a second-tier licensing scheme' to the Water Industry Act of 1991, with a stated objective of introducing more competition in the' retailing and production of water using incumbents' networks' (Reckon, 2007).

Historical, ethical and practical economic questions have been and will continue to be raised in the face of privatisation of a public service industry. Nineteenth-century improvers saw water and sewerage as services in the drive to improve public health, regardless of social class; and throughout the post-War period these services were seen as public utilities until the 1980s, when they had to be 'commodified', involving a change of perception before change could be effected. State ownership from 1980 was replaced by state regulation. Philosophically this undermines 'free-market' ideologies where statism is not favoured, but market failure is inevitable where there is no competition unless other measures are taken. If monopolistic practices are deemed anti-competitive (Bannock, Baxter, and Davis, 2003), then private monopolies outside a strong regulatory framework are a menace, on account of their potential to charge what they will for goods and services. The state has, since 1980, produced a range of 'regulators' for privatised monopolies.

The economic regulator (Ofwat, see below) regularly calls for more competition. Commencing with industrial users (for domestic consumers may prove more sensitive as there are more of them) consumers would be able to purchase water from a supplier of their choice. In the absence of a national water grid, or extensive means to affect bulk water transfers, the existing infrastructure would have to be used. High transfer costs arise due to pumping, and there is a leaky (and hence inefficient) supply network. Water may be purchased from an undertaking possibly hundreds of kilometres away, but the actual water consumed depends upon supply in the area of consumption!

There is envisaged a kind of domino effect, with water physically transferred from the area of purchase 'displacing' water in the receiving water supply areas. Further problems for regulators emerge in the need to establish fair play in 'common carriage' of water. Benefits to the domestic consumer, if any, will be slow to take effect in a utility which has seen consistent price increases since privatisation.

Having the flexibility to purchase water from undertakings around the country, given a fixed, common distribution system and high cost of transfer, is far from easy. There are, however, local situations where adjacent water supply undertakings provide differing options, for example, South Staffordshire Water and Severn Trent Water could compete to supply Wolverhampton, and there are areas in the south and east where smaller 'water-only companies' might effectively compete with the larger water service companies which arose at privatisation. Actually, the pattern here has been amalgamation, not competition.

3.6 Geographical Structure of the Water Industry

The 'Water Industry' is considered here to include the plant (infrastructure), human resources and institutional means whereby the hydrological cycle is deliberately interrupted to meet the needs of human society by diverting water in the ground and rivers (or other surface waters) and coastal/estuarine waters. The designation 'industry' commonly excludes agencies of regulation by resource management or economic instruments. The structure described is based upon Burnell (1994). In England and Wales, between 1 April 1974 and 31 August 1989, 10 publicly owned Regional Water Authorities (RWAs) controlled the entire hydrological cycle under the philosophy of integrated catchment management. In planning terms this was tidy, and the 29 private statutory water companies were issued with licences from the RWAs. After privatisation in 1989 there were 10 water service companies representing the former RWAs of the 1973 Water Act, serving some 38.5 million people with water supply, and 49.6 million with connected mains sewerage services. Their boundaries also largely corresponded with the original NRA boundaries which are catchment defined. The water service company industry boundaries thus most closely reflect those of the RWAs, unlike say those of the EA's regional structure that was reduced to the nearest political (i.e., local authority or similar) boundary. Given the historic complexity, only the modern map (dating from 2009) is shown in Figure 3.1.

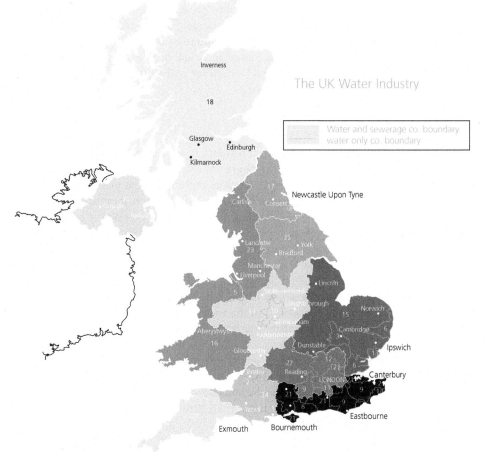

Water only companies

1- Sembcorp Bournemouth Water
2- Bristol Water
3- Cambridge Water
4- Cholderton and District Water
5- Dee Valley Water
6- Essex and Suffolk Water
7- Hartlepool Water (Anglian Water)
8- Portsmouth Water
9- South East Water
10- South Staffs Water
11- Sutton and East Surrey Water
12- Affinity Water
13- Affinity Water
14- Affinity Water

Water and sewerage companies

15- Anglian Water
16- Dwr Cymru (Welsh Water)
17- Northumbrian Water
18- Scottish Water
19- Severn Trent
20- South West Water
21- Southern Water
22- Thames Water
23- United Utilities
24- Wessex Water
25- Yorkshire Water
26- Northern Ireland water

Water UK, October 2012

Figure 3.1 The structure of the water industry in the United Kingdom (Saxon Water, n.d.).
Source: Water UK, 2012.

Numbers 15 to 17 and 19 to 25 represent the 10 former RWAs, separated from regulatory functions and today supply sewage and water supply services. Number 18 is the publically owned Scottish Water and Northern Ireland has its own arrangements (no. 26) as a government-owned company. Of particular interest are the 14 water supply only areas (representing 12 water-only supply companies and Affinity Water supplied 3 areas). These areas are derived from amalgamation of the 29 water supply only companies at the time of privatisation and were never in the public sector.

3.7   Regulation of an Industry

At the start of this chapter it was noted that regulation, while a 'necessary evil' is seldom popular with many stakeholders, and the degree of this unpopularity is also a function of political culture as well as a user's (or potential polluter's) relationship with a water body. The purpose here is merely to describe how it is at the time of writing, noting that this is no area where long-term changes are afoot that might have barely been realised at the time of privatisation, let alone in the days of the RWAs, a time of extreme 'top-down' control over the water environment in Britain.

It is no easy matter to describe the overall regulation of the water industry, nor was this the case in 1998 when the first edition of this book was published. In practice, much looks the same as the post-privatisation landscape was rapidly in place and in essence remains. The main regulatory functions in the water industry are: resource regulation, pricing and consumer affairs and drinking water quality. The agencies involved are, respectively in England and Wales, the EA/NRW, the Office of Water Services (Ofwat), the Drinking Water Inspectorate (DWI), and at arm's length, the EU.

Other agencies have interests relating to the industry, including legal, health, planning and government, and may also be considered as subsidiary regulators. These may be viewed as having influence rather than direct regulation. This is because they are statutory consultees in the planning process, they otherwise have a role in general policy formulation, or because they have a sphere of environmental regulation or management which may influence waters. The regulators of the industry are listed below (Rees and Williams, 1993; Burnell, 1994; Ofwat, 2015a; DWI, 2014; CCWater, 2016a):

- The European Union (EU) (WFD, drinking water quality, Environmental Directives, competition policy.
- UK Government (translating EU directives into law for England, Wales and Scotland).
- Subsequently the Scottish Government (responsible for water quality in Scotland) and
- The Welsh Assembly Government (Water Strategy for Wales).
- 'Ofwat' (the Water Services Regulation Authority) is the financial and the economic regulator of the water and sewerage sectors in England and Wales. Designed to ensure that companies provide household and business consumers with good service and value for money.
- Drinking Water Inspectorate (DWI) designed to provide independent reassurance that water supplies in England and Wales are safe and that drinking water quality is acceptable to consumers.

- Competition Commission (formerly Monopolies and Mergers Commission, (mergers and acquisitions of water companies, water supply licence modification).
- Office of Fair Trading, OFT (non-departmental, not-for-profit and independent of statutory regulators, is designed to ensure that markets work well for consumers through promoting and protecting consumer interests throughout the UK and ensuring that businesses are fair and competitive.
- Consumer Council for Water (the Consumer Council for Water [CCWater]) is a statutory consumer body for the water industry in England and Wales that is non-departmental, not-for-profit and independent of statutory regulators. It can bring complaints of domestic and business water consumers to Ofwat.
- Local government (public health, strategic planning, development control, flood protection, otherwise no direct control over water resource management).
- Department for Environment, Food and Rural Affairs, Defra (land drainage, reducing agricultural pollution, Codes of Good Agricultural Practice).
- The statutory Environment Agency was created, and from 1 April 1996 (in accordance with the Environment Act of 1995), it took over the functions of three agencies: the National Rivers Authority, Her Majesty's Inspectorate of Pollution and Waste Regulation Authorities (WRAs) such as the London Waste Regulation Authority. Subsequently formed for Wales, NRW dates from 2013.
- In Scotland, influence is wielded by the Scottish Natural Heritage, a statutory body, as well as voluntary sector bodies including Scottish Wildlife Trust, the Royal Society for the Protection of Birds (RSPB) Scotland and the National Trust for Scotland.
- Other agencies south of the Border indirectly involved as 'soft regulation' are both public and voluntary and include: the statutory Natural England, English Heritage and Forestry Authority and voluntary sector bodies such as the Rivers Trust, the RSPB, Blue Print for Water, the Campaign to Protect Rural England (CPRE), World Wildlife Fund (WWF) and Fish Legal (initially the Anglers Co-operative Association and later the Anglers' Conservation Association).

3.8   Environmental Regulation (I): The NRA

The inherent conflict of interest over water quality and abstraction rates on the one hand, and effluent discharges and legal requirement to maintain supply on the other, required a change to a situation where the utility side became separate from the regulatory. The NRA (1 September 1989-31 March 1996) was a non-(governmental) departmental body under the terms of the Water Act of 1989, with statutory responsibilities for the entire water environment.

The life of the NRA was expected to be finite and the advent of the Environment Agency was long anticipated, yet what was required was a radical departure in terms of regulation of the water environment. NRA was the only agency with responsibility for all matters relating to the aquatic environment, and separated from utility operation. During its lifetime there occurred radical and beneficial changes the water environment in England and Wales.

Before the formation of the EA, there were, inevitably, certain functional overlaps with Her Majesty's Inspectorate of Pollution (HMIP). The mission statement of the Authority (NRA, 1994e) read:

> 'The NRA will protect and improve the water environment. This will be achieved through effective management of water resources and by substantial reductions in pollution. The Authority aims to provide effective defence for people and property against flooding from rivers and sea. In discharging its duties it will operate openly and balance the interests of all who benefit from and use rivers, groundwaters, estuaries and coastal waters.'

Statutory requirements include inland surface waters, estuarine waters, groundwaters and coastal waters to three miles (these are controlled waters in terms of the Water Resources Act of 1991). There is also the matter of differing functions: flood defences (though not coastal protection which is and remains the responsibility of local authorities), navigation, water recreation, water quality, ecological considerations, resource management, open operation and balancing the needs of various stakeholders in the water environment. To summarise the responsibilities of the NRA (Cook, 1998, Ch 3) were:

- Regulating the quality of coastal waters and estuaries in accordance with EU standards
- Regulating the qualities of inland waters
- Regulating the quality of groundwaters
- Land drainage and control of land and coastal flooding
- Management of water resources
- Conservation and recreation involving water
- Freshwater fisheries
- Navigation (where the Canal and River Trust has no remit)
- Water resource planning and allocation

Publication of the 'Water Resources Strategy' (NRA, 1994f) was to be major development.

The Authority as a separate agency had head offices in Bristol and London, with (in 1994) eight regional offices. These regions were based upon the former RWAs and in 1993, Wessex and South West and Northumbria and Yorkshire regions were amalgamated. Most roles listed above, including strategic planning, were part of the remit of the former RWAs. However, functional separation of utilities from regulation, at a time when the EU, government and a host of other organisations and agencies became seriously concerned with issues of environmental quality, made the NRA a major player in the statutory drive towards sustainable development of water resources. The principles behind this, and the resulting planning framework of Catchment Management Planning, are described more fully in Chapter 4. The NRA saw principles of sustainable development (NRA, 1994f, p. 2) as:

- *Sustainable development* implies that there should be no long-term systematic deterioration in the water environment arising from water resource use and development. Balances struck should be cost-effective.
- *The precautionary principle* applies where significant environmental damage may occur, but knowledge on the matter is incomplete. Decisions made and measures implemented should err on the side of caution.

- *Demand management* aims to limit the total quantity of water taken from sources of supply. It states measures to control waste and consumption.

Such may be compared with the 'spheres of sustainable development' outlined in Chapter 1; for they are more specific, being strongly focussed on the water environment, and while they implicitly include 'social' values, these are hardly spelt out. Within these concepts lie a range of more specific policies including such as:

- Requiring water companies to achieve economically determined target levels of leakage and metering before new abstraction licences are granted for strategic developments.
- The promotion of water efficiency by industry, commerce, agriculture and domestic users.
- Where possible, redistribution of water resources in preference to the development of new sources.
- Favouring schemes which lead to an improvement in the water environment.
- Favouring schemes which meet the widest interests.
- Protecting and improving the quality of water resources.

These objectives establish the priorities for administering the legal basis of resource regulation, both reflecting the legislation during the mid-1990s and beyond and setting the trends for subsequent resource and legislative development. However, it will be demonstrated that such technical and economic measures hardly pave the way to modern ideas of modern inclusion, nor the role of other agencies such as those in the voluntary sector.

3.9   Environmental Regulation (II): The Environment Act 1995, EA and SEPA

Looking back over 20 years, a relatively stable environmental regulatory regime was established in the mid-1990s. Following the Environment Act of 1995, the EA for England, NRW and the Scottish Environment Protection Agency 'SEPA' for Scotland were formed on 1 April 1996. The Act is a highly significant piece of environmental legislation, and, to quote from the Preamble (19 July 1995):

> 'An Act to provide for the establishment of a body corporate to be known as the Environment Agency and a body corporate to be known as the Scottish Environmental Protection Agency; to provide for the transfer of functions, property, rights and liabilities to those bodies and for the conferring of other functions on them; to make provision with respect to contaminated land and abandoned mines; to make further provision in relation to National Parks; to make further provision for the control of pollution, the conservation of natural resources and the conservation or enhancement of the environment; to make provision for imposing obligations on certain persons in respect of certain products or materials; to make provision in relation to fisheries; to make provision for certain enactments to bind the Crown; to make provision with respect to the application of certain enactments in relation to the Isles of Scilly; and for connected purposes'

This passage conveys six key points. These are: transfer of functions from previously existing institutions, provision for contaminated lands, control of pollution in general, conservation of natural resources, conservation and enhancement of the environment (including fisheries), and control of certain substances. There is a strong echo here of the mission statement of the NRA to 'conserve and enhance the water environment'.

The NRA was combined with HMIP and local authority waste regulation to create the EA. Hitherto, HMIP, itself the outcome of earlier amalgamations, had as its focus industrial sources of pollution and there was a counterpart organisation in Scotland. The main legislative basis is the Environmental Protection Act 1990, the prime responsibility being the authorisation and oversight of industrial processes with the greatest pollution potential (Burnell, 1994), including overseeing waste disposal authorities. What is most significant about the legislative background to HMIP is the concept of 'Integrated Pollution Control' (IPC) under Part 1 of the 1990 Act. IPC aims to prevent the release of pollutants described as 'prescribed' (i.e., dangerous substances) to land, water and air, or else minimise releases and their effects (HMIP, 1994).

Operation of a prescribed process required authorisation. In theory, this was achieved through the 'Best Practical Environmental Option' (BPEO) and defined by the Royal Commission on Environmental Pollution as 'the option that provides the most benefit or least damage to the environment as a whole, at accepted cost, in the long-term as well as the short term'. The IPC approach was established because hazardous substances previously barred from release into one medium (say water) could be diverted to another (say air, through burning). A single regulatory system capable of targeting entire industries, processes or sectors was required. Trade and Effluent (Prescribed Processes and Substances) Regulations of 1989 defined certain 'Red List' toxic substances and certain processes, and required authorisation by HMIP of discharges from prescribed processes to sewers. Industries covered include fuel and power, waste disposal, minerals, chemical, metals, paper, tar and bitumen, uranium, coating, timber, animal and plant treatment (HMIP, 1994).

The Environment Act of 1995 (Part I) represents a major aspect of both innovation and consolidation. With respect to water and the (former) NRA, the Environment Act 1995, Part I, Chapter I, Section 2(1)a, is summarised below:

Water resources management (Part II of the Water Resources Act 1991).

In section 6 of the Environment Act 1995, it is the duty of the Agency to take all such action, in accordance with any directions given (this incorporates EU directives), to be necessary or expedient for the purpose (Environment Act 1995, Part I, Chapter I, Section 6(2)):

a) of conserving, redistributing or otherwise augmenting water resources in England and Wales; and
b) of securing the proper use of water resources in England and Wales.

These leave the general duty on water undertakers to maintain the water supply system while encouraging new methods of supply and distribution (such as bulk transfers).

Control of pollution of water resources (Part III of the Water Resources Act 1991).

It is furthermore heartening to read in section 6(1) of the Agency having duties, to such extent as it considers desirable, generally to promote:

a) the conservation and enhancement of the natural beauty and amenity of inland and coastal waters and land associated with those waters;
b) the conservation of flora and fauna which are dependent on an aquatic environment; and
c) the use of such waters and land for recreational purposes.

The last-mentioned involving consideration of the needs of the chronically sick and disabled. Other responsibilities include:

Flood defence (coastal and inland) and land drainage functions, land and works powers (Part IV of the Water Resources Act 1991, Land Drainage Act 1991 and Water Act 1989).

Flood defence committees are to carry out these functions.
Protection of sea and freshwater fisheries (various Acts).
Functions as a navigation authority, harbour authority, and conservancy authority transferred to NRA by the Water Act 1989 and Water Resources Act 1991.
Additionally there is a general function (Part I, Chapter I, Section 1) so that:

> 'The Agency's pollution control powers shall be exercisable for the purpose of preventing or minimising, or remedying or mitigating the effects of, pollution of the environment'

thereby incorporating the wider pollution legislation remit of HMIP through the Environmental Protection Act 1990 dealing with integrated pollution control for air, water and contaminated land. The Agency should also be notified by English Nature or the Countryside Council for Wales where land is designated a Site of Special Scientific Interest (Natural England and Defra, 2013; Natural Resources Wales, 2014), or otherwise of interest on account of its flora, fauna, geological or physiographical features. Similar arrangements apply to the National Parks and the Broads Authority. Recreation may also be promoted on waters at the discretion of the Agency.

The geographical structure of the Agency in England and Wales is not enshrined in the legislation. Of deep concern to water planners and managers was an option that boundaries would revert to local authority boundaries which pay scant regard to natural hydrographic boundaries. Preserving such boundaries is vital in securing the present development of, and future success in, Catchment Management Planning. Air pollution, by contrast, has no regard for any given boundary, and therefore virtually any administrative unit may be appropriate (minor air pollution problems such as bonfires, and also vehicle emissions, remain outside the responsibility of the EA/NRW). Waste regulation, having its origins in local government, would naturally favour the local government boundary option.

A radical solution might involve the use of hydrographic (i.e. existing EA and NRW catchment) boundaries and adjustment of local authority boundaries to fit. This would reflect the lead shown by New Zealand where re-organisation of regional government follows catchment boundaries. For the present, a typically British compromise has been struck. The Environment Agency Advisory Committee (EAAC(95)38) circular on geographical boundaries looked at several options, and of all possible combinations of administrative units and functions recommended an option with:

> 'Water catchment boundaries for water management functions at both regional and subregional level, and administrative boundaries for all other functions at both regional and sub-regional level.'

The key here is that the public face of the Agency was to appear as an integral whole, a unified set of boundaries across all its functions; internal coordination should, one hopes, cope with discrepancies between administrative and hydrometric boundaries which remain for internal water management and planning purposes. There is, however, still potential for confusion, between the 'publicly defined' boundaries, the catchment boundaries, and the administrative basis for air and waste disposal functions. However, with the principle of the catchment as the boundary for water purposes enshrined in the new Agency, there remains the possibility for radical shifts in local government boundaries at some later date.

Figure 3.2 shows the regional boundaries of the EA/NRW. These are largely amalgamations of administrative districts. The eight include Wales with much of central

Figure 3.2 Environment agency boundaries for England and Wales (Later Natural Resources Wales) Compared with regional boundaries (*Source:* EA). The area marked 'Welsh' became that of NRW.

Wales (the Upper Severn catchment) retaining the Severn Trent NRA boundaries for water purposes, but the public perception will be of one region for the principality.

Of the seven English regions all retain similar outlines to the NRA regions of 1994 with name changes producing Midlands and North East regions. These are only 'best fits' for administrative boundaries. For example, the Southern Region of the EA controls 2648 km of rivers and has three areas (Hampshire, Sussex and Kent), while the Southern NRA Region (maintained for internal water regulation purposes) controlled 2746 km of rivers and had six districts. Notwithstanding possible confusion over boundaries, the strong imprint of the NRA in the EA structure is to be welcomed. The national headquarters remain at Almondsbury, Bristol.

Subject to review, the basic functions of the EA in England and NRW are:

- regulate industrial processes with the greatest pollution potential to ensure that best available techniques (or technology) not entailing excessive cost (BATNEEC) are used to prevent or minimise pollution to the environment as a whole
- regulate the disposal of radioactive waste and (except on nuclear licensed sites) the keeping and use of radioactive material and accumulation of radioactive waste
- regulate the treating, keeping, movement and disposal of controlled waste to prevent pollution of the environment or harm to human health, in a manner which is proportionate to the threat posed
- preserve or improve the quality of rivers, estuaries and coastal waters through its powers to regulate, prevent, mitigate or remedy pollution to water
- take any necessary action to conserve, redistribute, augment and secure proper use of water resources
- exercise a general supervision over all matters relating to flood defence; it also has powers to take certain flood defence measures as approved by RFCCs maintain, improve and develop salmon, trout, freshwater and eel fisheries
- promote the conservation and enhancement of inland and coastal waters, and their use for recreation
- maintain or improve non-marine navigation
- regulate the remediation of contaminated land designated as special sites
- administer, in accordance with regulations on producer responsibility, registration of businesses and exemption schemes, and monitoring and enforcement of associated obligations.

The Environment Agency will also carry out work to obtain environmental information and promote an understanding of methods for environmental protection and management. Some of this work reflects statutory responsibilities, including to:

- assemble environmental data, from its own monitoring and other sources, so it can carry out its functions and form an overview of the general state of environmental pollution
- survey waste disposal needs and priorities, and advise the Secretary of State on his national waste strategy
- report on the state of contaminated land and, as necessary, produce site specific guidance to local authorities on dealing with contaminated land

- monitor pollution of freshwater, groundwater and the sea (up to three miles from the coast)
- publish information about the demand for water and available resources
- survey flood defences and flood risk areas
- notify the Ministry of outbreaks of notifiable fish disease
- follow developments in technology for preventing or reducing pollution
- assess, at the request of Ministers, environmental impacts of pollution or options for avoiding, limiting or cleaning up pollution
- carry out or promote research related to its activities (The Wastebook, n.d.).

Since April 2013, the EA operations in Wales have been transferred to NRW. This statutory body brings together the work of the Countryside Council for Wales, Environment Agency Wales and Forestry Commission Wales, as well as some functions of Welsh Government. The purpose is to 'ensure that the natural resources of Wales are sustainably maintained, enhanced and used, now and in the future' (Natural Resources Wales, 2014).

Water regulation in Scotland, until 1995, was predominantly through the River Purification Authorities (RPAs), which are responsible for pollution control. RPAs included the River Purification Boards (RPBs) on the mainland and certain local authorities on the islands. There were seven mainland-based RPBs under the Local Government (Scotland) Act of 1973 and three Island Councils which double up as RPAs in their areas (Burnell, 1994). They provided the basis upon which water quality assessments were undertaken and were responsible for dealing with pollution under COPA 1974. Today, SEPA comes under the same act, and SEPA will be a separate agency incorporating the notion of 'protection,' recognising the heavier emphasis on pollution control and lighter emphasis on flood defence and conservation which, after 1995, rests with other agencies.

SEPA's functions (SEPA, n.d.c) include helping the public understand and comply with environmental regulations while realising the many economic benefits of good environmental practice. This approach includes enforcement powers for tackling serious environmental threats. There is also expert advice and information on the environment:

Described in broad terms, our responsibilities include regulating:

- activities that may pollute water
- activities that may pollute air
- waste storage, transport, treatment and disposal
- the keeping and disposal of radioactive materials
- activities that may contaminate land.

Some other principal responsibilities include:

- monitoring, analysing and reporting on the state of Scotland's environment
- running Scotland's flood warning systems
- helping implement the National Waste Strategy
- controlling, with the Health and Safety Executive, the risk of major accidents at industrial sites
- operating the Scottish part of the Radioactive Incident Monitoring Network.

SEPA is concerned with not only successfully improving Scotland's environmental quality, but also has to factor in such variables as climate change.

While SEPA is primarily concerned with environmental pollution, responsibility for managing flood risk in Scotland is shared across the public and communities, SEPA, local authorities, Scottish Water and Scottish Government (SEPA, n.d.d).

3.10   Water Industry Finance, Economic Regulation and Governance

Like any industry, the water industry cannot achieve anything without suitable financial arrangements and this subject could merit a volume all of its own. Because water is a highly regulated industry (a response to it being largely a 'natural monopoly'), the industry is capital intensive and requiring long-term investments. There has been (in 2016) over £116 billion invested by companies in England and Wales since privatisation in 1989, with further investment coming from the public sector in Scotland. While the industry has to play its part in environmental quality as well as face climate change impacts, there are considerable recurrent costs to infrastructure as well as investment in new projects. For example, in 2012-2013, water companies spent £4.5 billion on investing in assets including to tackle leakage, flooding, drought and a further £5 billion on operating expenditure, including staff, services, power and chemicals. Overall for the UK, the industry plans to invest at least £5billion per year over five years, while there remain regulatory pressures to keep charges down, especially to domestic consumers, with price rises controlled by Ofwat.

Water companies are financed through a mixture of debt finance and equity finance (the most expensive source of funds), but the regulatory regime generally enables security for investors, including global investors. In tax terms, they are able to defer corporation tax from tax relief on capital allowances, helping to reduce bills to their customers. The sector as a whole is thought to contribute around £15 billion a year to the UK economy, and supports around 127,000 jobs directly or indirectly (Water UK, 2016b), while the Water Act of 2014 aims to make the industry more responsive to customers and further free-market reform (UK Parliament, 2014). Alternative voices claim the water industry (once a public service) has achieved much, but has also generated large profits that have attracted unaccountable investors driven by the profit motive, and there has been avoidance of taxation, the closing of pension schemes, reduced conditions of service and poor pay settlements (Tinson and Kenway, n.d.).

Water industry governance, according to Ofwat, requires 'effective leadership and governance of the water and sewerage sectors' with company boards operating to high standards, are accountable and transparent. It finds there is less progress against the holding company principles than the regulated company principles and there is concern for transparency (Ofwat, 2015a). Ofwat (the Water Services Regulation Authority since 1996) is the main economic regulator of the water industry for England and Wales and was created under the Water Act of 1989 as a statutory watchdog. It has all the characteristics of a 'neoliberal' agency for it not only deals with pricing and consumer affairs, but regularly calls for increasing competition. Ofwat's main duties (Ofwat, 2015b) are to:

- Further the consumer objective to protect the interests of consumers, wherever appropriate by promoting effective competition

- Secure that the functions of each undertaker (that is, water company) are properly carried out and that they are able to finance their functions, in particular by securing reasonable returns on their capital
- Secure that companies with water supply licences (those selling water to large business customers) properly carry out their functions
- Further the resilience objective to secure the long-term resilience of undertakers' (that is, water companies') water supply and wastewater systems and to secure they take steps to enable them, in the long term, to meet the need for water supplies and wastewater services

Further to these main duties concerned with avoiding market failure, other responsibilities of Ofwat include oversight of water undertaking efficiency, ensure fairness in fixing charges and service provison, promoting good regulatory practice and there is stated environmental concern around water undertaking activities.

Non-ministerial in nature, Ofwat today controls the economic and financing operation of water supply and sewerage disposal functions, and is independent of ministerial control. The current regulatory framework for water is set out in section 2 of the Water Industry Act of 1991 as amended by section 39 of the Water Act of 2003. Ofwat requires that (Politics, 2015) all water and sewerage companies submit to regulation by Ofwat as "water undertakers". It is concerned with economic efficiency and to improve and extend its network of water supply and sewage. It aims to ensure that regulations and standards of service are met by companies. Ofwat is empowered to issue "enforcement orders" against transgressors including fines. In extreme circumstances, it can seek High Court approval to appoint administrators to take over a water company, until the terms of the Order have been complied with.

Ofwat conducts five-yearly price reviews aimed not only to keep prices to a minimum, but also to permit water companies to make an adequate return on capital, thereby permitting investment into the water infrastructure. Price frameworks are set on a company-by-company basis, reflecting their business plans and projected revenues, but there is a maximum level above which prices may not rise. In 2009, price limits were set for water bills in each year between 2010 and 2015 then again for 2015-2020 (Ofwat, 2015b).

As the Water Services Regulation Authority, Ofwat requires that all water companies, and water and sewerage companies produce Strategic Business Plans, also known as Asset Management Plans (AMPs) for each five-year planning cycle. At the time of writing, each appointed water company submits a business plan at a price review. The business plan sets out:

- the company's overall strategy and the implications for price limits and average bills;
- its strategic objectives in terms of service performance, quality, environmental and other outputs;
- the activities necessary in the period to meet these objectives; and
- the scope for improvements in efficiency.

Prices for the consumer are set by the 'K-factor' (price limit) that is a percentage above the Retail Price Index for each year, although adjustment may be made for underpricing carried forward from year to year (Ofwat, n.d.). Between 2010 and 2011 and

2014 and 2015, this industry average ranged from −0.6% to +0.5, with considerable variation between companies. While this section is concerned with the operation of Ofwat, there are other essentially 'economic' regulatory bodies outlined in section 3.7.

Since 2005, a further development in consumer power has been the statutory Consumer Council for Water that self-describes as the independent representative of household and business water consumers in England and Wales (CCWater, 2016a). Sponsored by CCWater are the Customer Challenge Groups (CCGs) that are formed for most water companies and CCWater seeks to have CCGs set up in each company supply and (where appropriate) sewage disposal area. These are independent local groups of customer representatives and other stakeholders with a main aim of scrutinising and challenging water companies' business plans.

3.11 Drinking Water Inspectorate

The Drinking Water Inspectorate (DWI) was formed in January 1990 following the Water Act of 1989. Then a part of the Department of the Environment, today, it is reorganised within Defra. It is the job of DWI to ensure that standards for treatment from UK, EU and WHO standards are met (DWI, 2010). The Inspectorate (DWI, 2014):

- Provides independent scrutiny of water company activities for companies supplying drinking water to consumers in England and Wales;
- Works with other stakeholders for the improvement of drinking water quality and to secure drinking water safety;
- Commissions research to build a sound evidence base on drinking water quality;
- Publishes data on drinking water quality in England and Wales.

Its operation involves a rigorous sampling procedure appropriately spread throughout the water distribution system ensuring that accurate analytical methods are used, it investigate incidents which pollute drinking water, assesses chemicals and methods used to treat drinking water, instigate action to correct any problems, generally instigate research into water quality problems, advise the Secretary of State where prosecution of a water company is required, provides government with technical information on drinking water issues and in association with Ofwat, assess and respond to consumer complaints.

The legal standards in the UK are those which are set in Europe by the Drinking Water Directive of 1998 together with national standards set to maintain the high quality of water already achieved. The standards are strict and include wide safety margins. They cover:

- micro-organisms
- chemicals such as nitrate and pesticides
- metals such as lead and copper
- the way water looks and how it tastes (DWI, 2010).

Figures for drinking water quality compliance (2011) with the UK and European standards for England is 99.96%, Wales is 99.95% and Scotland is 99.89% (CCWater, 2016a; Scottish Government, 2014b).

3.12   Supra-national Legislation

Supra-national legislation comes to UK law in the form of standard setting, and policy framework setting. Since 1973, Britain has belonged to the European Economic Community (EEC), subsequently called the European Community and more recently the European Union (EU). Although aspects of public policy, especially in terms of agricultural support, had been common to much of post-Second-World-War western Europe, Britain now comes under EU Directives.

While pollution control legislation had not always been as successful as intended (prior to COPA 1974), Britain's pollution control has been long under the scrutiny of Brussels and rigorous standard setting was to find the country wanting. In setting frameworks for legislation and policy formulation, concepts comparable with 'BPEO' arise, such as the 'Precautionary Principle'. Having its origins as a discrete element of law in the Single European Act 1986 and representing a major initiative in environmental protection *per se*, it has been translated into water policy (NRA, 1994f) as:

> 'where significant environmental damage may occur, but knowledge on the matter is incomplete, decisions made and measures implemented should err on the side of caution'.

World Health Organisation (WHO) standards which, although not legally binding, carry universal authority influence both EU and domestic legislation. International pressure, especially in the arena of drinking water quality, meant not only that limits were set, but also that new substances, especially trace elements, came into the frame. This is in part a response to improved analytical techniques, and the revised list issued in 1993 includes bacteriological quality (coliform bacteria), metals (including certain heavy metals), organic compounds (such as aromatic hydrocarbons), nitrates, nitrites, trihalomethanes and pesticides. It is not appropriate to give long lists of substances and their numerical limits for drinking waters.

WFD represents consolidation legislation (Chapter 10) and is now intended to operate in concert with the Floods Directive. WFD rationalises EC water legislation, specifically seven "first wave" directives that were enacted prior to 2000, specifically those on surface water (and its two related directives on measurement methods and sampling frequencies and exchanges of information on fresh water quality); the fish water, shellfish water, and groundwater directives; and the directive on dangerous substances discharges, will be repealed (EC, 2015b). There remains the fear that UK environmental legislation may be weakened by a withdrawal, or even a weakening of bonds with the EU, but it is hard to make any clear predicition.

3.13   Summary

Legal aspects of water resource management are not only important because they define what is permissible, but also because they set in place institutions for *sharing and protecting water resources* and *regulating the industry and related aspects such as flood defence*. All-embracing water regulation institutions are modern, but some, notably those concerned with flood defence, are older. There remains no overall ownership of running waters, nor indeed of groundwater, which are the archetypal 'common property resource' prone to abuse. The statutory authorities concerned (such as EA and DWI) all operate in a legal and regulatory framework set by the EC and translated into member state law.

Because there was no concept of waterborne disease until the nineteenth century, there was no early scientific basis for appropriate public health legislation. That century saw a flurry of significant legislation following the discovery of the waterborne nature of certain disease. Subsequently, from the last quarter of the nineteenth century, legislation concerned with industrial pollution became important. The legislative position in defining groundwater flow does not reflect scientific knowledge, nor does the EU definition of groundwater protect water in the unsaturated zone.

In the nineteenth century, a right to abstract emerged. Rights over waters were given to a riparian landowner to receive water in its natural quality and quantity, and a responsibility to go from that owner without obstruction. In urban areas, the passing of Acts of Parliament to establish cleaner water supplies and better sewage disposal laid the legal foundation for the structure of the modern water industry.

What is apparent is the development of increasingly complex and specific legislation. Furthermore, the increasing frequency with which such key acts have been passed since 1830 increased into the post-Second-World-War Period. In other words, a rapid rise in urbanisation, industrial development and ensuing problems is reflected in the legislation. The trend since the Second World War has been for the number of statutory water undertakings ('utilities') to be reduced, starting with the Water Act of 1945. The River Boards Act of 1948 made provisions for the restructuring of the administrative aspects of river pollution control because much earlier legislation was ineffectual.

River Boards were established with responsibility for pollution control in the entire catchment area, including the right to sample any effluent discharge. Development then passed through the stage of all-embracing RWAs, which consolidated the idea of integrated water management on a catchment basis, to the NRA, independent of utility operation, and finally to the creation of agencies concerned with integrated pollution control of all media. The outcome was the EA which has been in existence since 1996. The EA is responsible for a range of pollution issues of land, water and air as well as being concerned with flooding and water abstraction. Scotland has its own, parallel arrangements through SEPA, although flood defence is more of a shared experience north of the Border.

It will be shown later how the planning process has come to embrace the water environment, and how the function of essentially technocratic agencies is becoming tempered through democratic consultation, the subject of Chapter 10. Supra-national legislation affecting Britain came first as water quality compliance required through a number of EU Directives, and these are evolving into consolidated 'Framework Directives' stands as a supra-national standard setter and regulator. The resulting WFD provides for both 'good status' for surface and groundwater bodies, and for the first time EU legislation has come to concern itself with the whole water environment, including volumetric abstraction. All contemporary water institutions are very much concerned with climatic change impact.

With the nuts and bolts of institutions in place and a long history of legislation, today represented in the passing of regular statutes, we may return to relatively simple questions such as is it fit for purpose? Will it work in the end? If not how might it be fixed? What might be the future of EU environmental regulations in a post-EU Britain? And last by not least, how can it all be rolled out to achieve true stakeholder engagement?

4

The Catchment Approach

Ways and Means

OK, so you want to conserve this marsh, then put it back how it was ... break down the sea walls and flood the bloody lot!

Kent farmer in a bad mood

4.1   Sustainable Development

Terminology can be confusing, but in Britain a 'catchment' is a subset of the larger 'river basin', the term used by Water Framework Directive (WFD) for large (discrete) drainage areas such as the Thames, Rhine or Danube. A watershed strictly means a topographic divide that separates water flowing in different directions (e.g., draining into different catchments). In North American usage, this is drainage 'divide'; the term 'watershed' being analogous to a catchment or river basin, a whole area of land draining to a definable point. Catchments and river basins are natural hydrographic units, and the most appropriate for the management of water resources.

For comparison, an environmental management framework that is neither a competing paradigm nor co-terminus with the river basin is the 'ecoregion approach'. In such a geographically defined area there are ecosystems that reflect variables of climate, soil, natural vegetation and landforms, factors that condition the nature of human activity, notably agriculture. Edwards and Dennis (2000) stress the need to relate catchments and ecoregions in order to accommodate considerations of changing land use in the control of diffuse pollution. Yet, it is the surface topographic divide that will define a catchment, something that may (in practice) merely define a sub-set of a larger ecoregion. While this is attractive to water resource managers it has to be remembered that groundwaters in aquifers beneath may not move in accordance with perceived rules of surface runoff (Chapter 2).

Inevitably catchment management is required to be 'sustainable', now a term universally acclaimed to be clichéd. The property of 'sustainability' might be described as a process—and a political process at that—for it cannot be in any way regarded as a single entity, not least because there is constant adjustment to policy and procedure that are normatively aimed to improve matters (Chapter 10). The political process is rooted in all of economics, politics and human behaviour as well as perceived (mis)use of natural resources. Management of water resources like anything else must maintain supply for human use without damaging the ability of the natural world to deliver ecosystem services, sometimes referred to as the 'stock of natural capital'.

The Protection and Conservation of Water Resources, Second Edition. Hadrian F. Cook.
© 2017 John Wiley & Sons Ltd. Published 2017 by John Wiley & Sons Ltd.

The Brundtland Commission of three decades ago famously declared:

'Sustainable development is development that meets the needs of the present without compromising the ability of future generations to meet their own needs'. It contains within it two key concepts (UN Documents, n.d.):

1) the concept of 'needs', in particular the essential needs of the world's poor, to which overriding priority should be given; and
2) the idea of limitations imposed by the state of technology and social organization on the environment's ability to meet present and future needs'.

The ideas of social equity are immediately apparent, and this was basically concerned with meeting the needs of the world's poor. As far as the environment is concerned, the second concept might in part be satisfied through good catchment management. Such thinking permeated the policies of water resource planners (NRA, 1994h) a publication which also contains the words 'environmentally sustainable water resources development' in the sub-title. For surface waters, Newson (1992a, p. 187), notes how difficult sustainability is to define, suggested that we should be working towards a practical definition is an essential technical and political venture in all fields at the close of the twentieth century.

Meanwhile, Gardiner (1994) quotes Agenda 21 of the UN Conference on Environment and Development held in Rio de Janeiro in 1992 in an effort to achieve 'sustainability in practice' for waters:

'Water resources must be planned and managed in an integrated and holistic way to prevent shortage of water, or pollution of water sources, from impending development. Satisfaction of basic human needs and preservation of ecosystems must be the priorities; after which water users should be charged accordingly'.

Sustainability may present serious problems for definition; sustainable development, on the other hand, permits the setting of targets with identifiable and measurable human and environmental outcomes. Nonetheless, establishing an adequate conceptual framework within which to explore notions of sustainable development is not easy. Goals should encompass the outcomes of socio-economic as well as natural or engineering science enquiry; indeed, certain authors prefer to see environmental issues in general as a process or relationships between human society and physical resources (Redclift, 1994).

This is helpful, because, like sustainability, 'development' is itself a dynamic, even if individual projects are, for convenience, packaged with definable goals. In this way, the knowledge base upon which decisions are made, and even the assumptions underpinning that base itself, has to keep pace with stated objectives of sustainable development.

One principle (Gardiner, 1994) might be: 'working with nature is more sustainable than trying to overcome nature'. Significantly, this calls for the end of notions of 'humanisation' (i.e., human domination) of nature. Yet, elaboration of the principles of catchment management may be explored, and issues of sustainability address questions concerning whether a system is able to maintain productivity (e.g., water quality or yields) in the face of environmental or economic pressures, and whether there is likely to be lasting damage to the resource base. To become implementable, sustainable development requires action plans which satisfy criteria in at least three dimensions: environmental, social and economic.

The vision for sustainable development was outlined at the 1992 United Nations Conference on Environment and Development (UNCED) or 'Earth Summit' and it still remains relevant today:

> "integration of environment and development concerns, and greater attention to them will lead to the fulfilment of basic needs, improved living standards for all, better protected and managed ecosystems and a safer, more prosperous future"

Rio +20 took place 20 years after its forerunner. In conclusion, it stated:

> 'We, the heads of State and Government and high level representatives, having met at Rio de Janeiro, Brazil, from 20-22 June 2012, with full participation of civil society, renew our commitment to sustainable development, and to ensure the promotion of economically, socially and environmentally sustainable future for our planet and for present and future generations.'(*The Guardian*, 19 June 2012). The document produced 'The Future we Want' was nonbinding.

Aside from massive protests and other political agenda being worked out alongside this important summit, the *Guardian* (23 June 2012) could state that:

> 'After more than a year of negotiations and a 10-day mega-conference involving 45,000 people, the wide-ranging outcome document—*The Future We Want*— was lambasted by environmentalists and anti-poverty campaigners for lacking the detail and ambition needed to address the challenges posed by a deteriorating environment, worsening inequality and a global population expected to rise from 7bn to 9bn by 2050.'

Castigated were self-interested career politicians and bureaucrats, likened to fiddling when Rome burns, with sustainable development providing a moral fig-leaf for politicians! If the objective of the Rio + 20 Conference was to secure renewed political commitment to sustainable development, it had failed.

By this point, the century had turned and a new generation turned to issues of sustainable development. It is as if the 'big picture' model was failing. This was all despite regular and significant reports from the IPCC about the dangers of climate change. Grappling with these problems, CIWEM produced a policy document that virtually opens with a statement of such failure:

> 'In June 2012, Rio hosted the greatest failure of collective leadership since the First World War. The Earth's living systems are collapsing because politicians see the environment as a brake on growth and not an opportunity to grow sustainably' (CIWEM, n.d.).

'Ouch!' is an understatement—and the passage continues:

> 'Consideration of environmental, social and economic components of sustainability is undermined by poor decision making, weak governance and institutional frameworks which, ultimately, allow too great an emphasis on economic growth to the detriment of environmental and resource conservation.'

One solution from CIWEM is to introduce testable concepts, for example:

> "To be sustainable an action must not lead, or contribute, to depletion of a finite resource or use of a resource exceeding its regeneration rate".

In this respect, there should be a fresh look at regulation, including support for businesses. There might be broader visions, clear recognition of environmental limits, an understanding of resilience in natural systems and better use of information and modelling. One approach, by way of seeking a renaissance, may be to re-construct whole governance arrangements from the bottom up. Scales operate vertically as well as geographically, communities associate with their river and a commensurate re-scaling of the economy may be required.

4.2 Frameworks for Sustainable Development and Water

Unlike minerals or fossil fuels, water has the characteristics of a renewable resource. Clearly, once notions of water balances were understood in recent centuries, then it becomes possible to 'live within our resources'. In a simple world, the only non-renewable water resources would be, for example, in the Middle East and North Africa where past climate change has created desertification, but there is still 'fossil water' dating from wetter times and it might be abstracted for human consumption, but without expectation of 'significant recharge' (National Academies of Science, 2007). Internationally, water resource availability is generally calculated on a country basis with distinction made between renewable and non-renewable resources:

- Renewable water resources are computed on the basis of the water cycle. They represent the long-term average annual flow of rivers (surface water) and groundwater reserves.
- Non-renewable water resources are groundwater bodies (deep aquifers) that have a negligible rate of recharge on the human time-scale and thus can be considered non-renewable.

There are problems. Elinor Ostrom (Chapter 10) was concerned with managing the commons and specifically with how communities succeed or fail at managing common pool (finite) resources (such as grazing land, forests and water resources as applicable), including how this is catered for in terms of 'polycentric, multi-level governance' (On The Commons, n.d.). In this context, water has to be viewed as a 'finite resource' because over-abstraction and pollution has the same effect as diminishing the stock of natural capital for a truly finite resource.

There is then the matter of sharing resources with neighbouring countries, and generally, this means issues of water resources and pollution for downstream users outside a particular jurisdiction, with the ever present problem of creating international friction. Then choices have to be made regarding allocation, not only between users, but economic sectors and maintaining sufficient water in both quality and quantity for environmental imperatives including groundwater recharge and channel flow (FAO, n.d.).

In-stream allocation of water can be established by computer models (Oklahoma Water Resources Board, 2105). Not only might minimum flows be established, but requirements for water availability and flow reliability at any location in the basin on an annual and seasonal basis (to assist licencing), estimate the amount of unappropriated water in the system, perform drought analyses to determine areas in the basin and water rights that are sensitive to water shortages, evaluate water policy decisions, including allocation of water, inter/intra basin transfers, interstate stream compacts, needs of climate change and climate change, etc. Such analyses assist in management and setting policy— which is important in implementing WFD.

Working towards a definition of sustainability in catchment water management might therefore involve the following considerations:

- Abstraction at all timescales should not compromise the ability of a catchment to maintain river flows, surface water bodies and groundwater storage.
- Land use, industrial production and sewerage management should neither compromise water quality for other users nor fail to maintain ecosystem quality.
- Development projects, be they for water resource, drainage or flood purposes, or if indeed arise from other origins (urbanisation, transport or industry), should sustain or enhance other catchment considerations (conservation/ecosystem diversity, amenity, heritage, navigation, flood control, water supply, etc.).
- Recycling of waters should be maximised and output of potentially polluting waste (which cannot be safely returned via the media of land, water or air) should be minimised.
- There should be governance arrangements that are capable of providing inclusion and fora that give a voice to all stakeholders.
- The needs of downstream users must be taken into account at all times.
- There should be adequate provision of legal, institutional, consultational, analytical and technical means to enforce policies, agendas and schemes set in the prosecution of the above aims.

Emerging themes are holism in management to meet multifunctional requirements, resource protection and the satisfaction of human and ecological needs before profit. The delivery of water and sewerage services must also operate within economic constraints. These kinds of considerations, and dilemmas, set the agenda for most kinds of management for sustainability.

Simmons (1995) identifies three approaches in goal identification:

- Curative, which is the modification of existing systems in order to move towards their sustainability.
- Pluralist, which is the combination of 'best practices' in different situations.
- Creative, which is the design of entirely novel systems to provide solutions to a deteriorating environment.

The curative approach is cautious and conservative but it allows lessons from the past to be implemented as policy. The pluralist approach allows for importation of ideas from elsewhere, and the creative approach will, in practice, arise from the synthesis of ideas in the light of notions of sustainability incorporating social and natural science as well as engineering. All are valid.

We are also able, at least, to make intelligent guesses regarding which components of a system are significant in a particular line of enquiry. Agriculture is responsible for much diffuse pollution and, at the system level, setting broad objectives for sustainable farming systems is possible (Cook and Lee, 1995). Sustainable agriculture seeks to effectively manage weeds, pests and diseases, to optimise biodiversity in farms, to reduce the environmental impact of land management, to maintain or improve both the quality and quality of produce, to stimulate rural communities and create fulfilling employment and to ensure that agriculture remains economically viable for the future. Similarly, in water management, we might seek effectively to control invasive organisms and pathogens in waters, optimise aquatic biodiversity, avoid over-exploitation of waters, manage flooding and water quality, maintain and improve the consumer product (potable water) and operate in an economically feasible manner. Here, is recognition of the economic, social and ecological dimensions.

Working within parameters set by nature decouples technical and other responses from automatically following the demands of unbridled development and market forces. Until the 1970s, there was a tendency to set parameters of water development by economic considerations only, leading to the well-known problems of low flows, over-abstraction of groundwaters, habitat loss and poorer water quality. New rubrics mean that we should operate within the ability of a system to function within ecosystem boundaries maintaining environmental quality and meeting human needs defined in terms of its carrying capacity. Basically, all criteria for sustainable development concern themselves with improving the quality of human life but also living within the carrying capacity of supporting ecosystems and not diminishing the stock of natural capital through poor management of natural resources, but recalling common access resources such as water is especially vulnerable to over-exploitation. Natural capital is the renewable and non-renewable resources, each of which can be affected in reversible and irreversible ways.

Since the 1990s, the trend in natural resource management has been very much away from definitions and great pronouncements, towards measurable outcomes. While there is financial costing built into all projects, other outcomes include chemical, ecological and physical aspect of water bodies. Such outcomes are enshrined in WFD 2000, for achieving Integrated Water Resource Management has to include measureable outcomes alongside the political, social and economic aspects of river basin governance (Chapter 3).

WFD sets concentrations for polluting substances and although achieving physical targets such as fiver flow targets has always been problematic, it is measurable. Such constrains abstraction to within 'environmentally acceptable' boundaries.

In setting prescriptions for sustainable development in catchments we have to recognise that conflicts will inevitably occur. Stakeholders often have differing objects, and social or economic needs may conflict with environmental objectives, implying that a democratic component is vital, or voices are heard and taken seriously. Governance arrangements should be able to resolve the imperatives of meeting a local need for water consumers that potentially undermine the interests of angling, navigation, water quality, conservation, industrial supply and so on—where flow in a river is seriously reduced. Neither are the goals of economic efficiency necessarily coincident with long-term environmental sensitivity.

In terms of conflict resolution, the emerging revised institutional arrangements that might be required will be covered in Chapter 10. The key (Figure 10.4) would seem to be adaptive management in catchment planning and process implementation that allows for feeding in of new ideas allied with not-for-profit organisations that deliver on the ground.

The low-cost route may meet the requirements of capital expenditure planning ('BATNEEC') in the short term, but may represent a plundering of the 'natural capital' for the future (Defra Business Link, 2011). Catchment planning has to be concerned with setting bounds on development activity; inevitably tensions arise. Setting bounds on catchment activities in order to achieve specified environmental and social objectives should be the driving force in decision making. Given finite human and financial resources, the decision-making process will be constrained by prioritising not only river stretches, groundwater areas, catchments or sub-catchments, but also issues. Then there is the matter of user expectation; stakeholders set constraints on environmental parameters. For example, coarse fisheries may dominate consideration in one location affecting consents and compliance applicable to industry, while elsewhere flood alleviation may affect land-use decisions.

There has long-stated interests in sustainable abstraction (NRA, 1995b), the EA currently can state:

> 'Water Resources is the term we use to refer to the quantity of water available for people and the environment. Abstraction is the removal of that water, permanently or temporarily, from rivers, lakes, canals, reservoirs or from underground strata. We need to make sure that abstraction is sustainable and does not damage the environment. We control how much, where and when water is abstracted through our licensing system. This system was introduced by the Water Resources Act 1963 and has been refined and changed as a result of the Water Resources Act 1991 and the Water Act 2003.... Our powers and duties enable us to regulate the use of water under existing licences and to decide whether to grant new ones. Where abstraction is damaging the environment we also have the power to amend or revoke existing licences' (EA, 2013b).

Actually, abstraction policy is constantly under review. The tone presented above is one of legislative teeth, more recently elements of self-regulation have been promoted (Chapter 3). In any case, UK is obligated under the supra-national WFD of 2000.

4.3   Environmental Assessment

The Council Directive of 27 June 1985 'on the assessment of the effects of certain public and private projects on the environment' (85/337/EEC, as amended), the Council of the European Communities required that a statement be made when any 'likely to have significant effects on the environment should be granted only after prior assessment of the likely significant environmental effects of these projects has been carried out'.

Subsequently Strategic Environmental Assessment (SEA) (EC, 2015d) became mandatory for EU states for plans or programmes which are prepared for agriculture, forestry, fisheries, energy, industry, transport, waste/water management, telecommunications, tourism, town and country planning or land use and which set the framework for future development consent of projects listed in the EIA (Environment Impact Assessment) Directive, or have been determined to require an assessment under the EC Habitats Directive.

EIA procedure requires a developer to provide full assessment of the impact of a particular project. The developer may request the competent authority (e.g., the Environment Agency) to state the scope of the EIA information provided. The developer should then provide information on the environmental impact, the 'environmental authorities' and the public must be informed and a consultation process followed. Finally, the competent authority decides, taking into consideration the results of consultations. The public is informed of the decision afterwards and can challenge the decision before the courts. The EIA Directive of 1985 has been periodically amended. These amendments require not only consideration of impacts such as pollution, human health issues, abstraction and impact of construction, but also CO_2 issues (such as capture and storage) and importantly public participation and environmental justice. Specifically, water issues addressed include surface and groundwater protection, avoidance of over-abstraction, large inter-basin transfers and impoundments (EC, 2015c).

The UK Government adopted the notion of Environmental Impact Assessment in 1988, and it is now a major part of the planning process.

EIA seeks to protect the environment through a local planning authority who must decide whether to grant planning permission for a project where there are likely significant effects on the environment. The public are given early and effective opportunities to participate in the decision-making procedures. There are five broad stages to the process, to summarise:

1) Screening—determining whether a proposed project falls within the remit of regulations and whether it is likely to have a significant effect on the environment.
2) Scoping—determining the extent of issues to be considered in the assessment and reported in the Environmental Statement. The applicant can ask the local planning authority for guidance.
3) Preparing an Environmental Statement (ES)—an applicant must compile the information reasonably required to assess the likely significant environmental effects of the development. Public authorities must make available any relevant environmental information they may possess.
4) Making a planning application and consultation—The ES together with the application for development to which it relates must be publicised.
5) Decision making—The ES, together with any other information relevant to the decision such as comments and representations, must be taken into account by the local planning authority and/or the Secretary of State in deciding whether or not to give consent. The public are informed of the decision and the main reasons for it (DCLG, 2014).

Informal environmental assessment within the EA for all flood defence, water resources and navigation projects, allowed for by the Water Resources Act of 1991 and which also requires that adverse effects are minimised, is referred to as Environmental Assessment (Brookes, 1994). Newson (1992a, Ch. 7) recounts the international progress of EIA in water planning since the 1970s, stressing not only that public participation is the key to success, but also the huge responsibility of engineering professionals in taking account of the precautionary measures thrown up by the process of EIA. The process is made especially complicated for water projects because of the multi-disciplinary angle required when employing the environmental science, planning process and engineering aspects in project development.

It is also interesting to compare the stages in EIA with the steps in broader areas of policy making outlined in Chapter 1. The two are complementary, and correctly executed EIA procedures will be of enormous benefit in wider policy analysis. Environmental assessments are inevitably project based, although there has to be a strong relationship to the overall planning process and to the wider process of policy formulation. The procedure was established in each of the eight NRA regions shortly before the birth of the EA in 1996 (NRA, 1994g):

> 'Water Resources Strategy' embodies the principle in the context of the strategic options for meeting water demand over the 30 years until about 2024, as a part of sustainable development. It is adamant that there will be neither over-abstraction nor excessive pollution while appreciating social and economic issues.

Strategic Environmental Assessment (SEA) became mandatory for government plans and programmes under the SEA Directive 2001/42/EC (Report from the Commission on the application and effectiveness of the Directive on Strategic Environmental Assessment [Directive 2001/42/EC]), COM/2009/469 (EC, 2015d). Each River Basin Management Plan—(RBMP) (Chapter 10) has to undergo an SEA, so they are applicable to water management:

> 'plans and programmes in the sense of the SEA Directive must be prepared or adopted by an authority (at national, regional or local level) and be required by legislative, regulatory or administrative provisions.'

Environmental assessments can take a long time to complete; however, preferred options can be set out at an early stage in order to facilitate the process of consultation and full investigation. The majority of EA projects are concerned with flood defences. The preparation of an environment statement therefore involves an enormous number of factors. First the nature of the project is identified (flood defence, irrigation, sewerage, impoundment, diversion, etc.) and the site selected. The SEA and EIA procedures are very similar, but there are some differences (EC, 2015d):

- The SEA requires the environmental authorities to be consulted at the screening stage;
- Scoping (i.e., the stage of the SEA process that determines the content and extent of the matters to be covered in the SEA report to be submitted to a competent authority) is obligatory under the SEA;
- The SEA requires an assessment of reasonable alternatives (under the EIA the developer chooses the alternatives to be studied);
- Under the SEA Member States must monitor the significant environmental effects of the implementation of plans and programmes in order to identify unforeseen adverse effects and undertake appropriate remedial action;
- The SEA obliges Member States to ensure that environmental reports are of a sufficient quality.

4.4   Evolution of Water Standard Setting

Emphasis in pollution control is broadening to include both diffuse and point pollution. Water quality, water quantity (especially flows), water body use and ecosystem requirements eventually all have to be considered.

To summarise, the approaches to setting standards for water quality are:

- Objective setting and periodic assessment are the basis for water quality schemes in British waters.
- Absolute standards may be set, which either should not be breached or may be used as some kind of comparator by which pollution is measured and action taken.
- Standards may be set deemed appropriate to local requirements. These relate to the use required of a water body with the standard referenced in absolute terms, but set locally.

The driving force in standard setting may be national or internally set guidelines (such as those set by the WHO), or EU Directives. Although voluntary compliance is highly desirable, there is legal backing for a standard setting to be enacted through Statutory Instruments.

Historically, consent setting for point sources derive from three sources: the EC Bathing Water Quality Directive (76/160/EEC), the Dangerous Substances (List I) Legislation arising from Directive 76/464/EEC (NRA, 1994a), and a revision of standards in the Surface water Abstraction Directive (75/440/EEC). The last-mentioned covers waters for human consumption supplied to the consumer.

Prior to the adoption of the WFD, water standards had evolved for both England and Wales and for Scotland. The situation for Scotland generally ran in parallel to those for England and Wales (Cook, 1998, Ch. 9). First of all it was 'determinands' (chemical or biological measures) that reflected largely the impacts of sewage discharge. These set standards for rivers, canals and estuaries that were based on dissolved oxygen % saturation, Biochemical Oxygen Demand (BOD—a measure of the biological demand for oxygen), and ammonia as 'ammoniacal nitrogen', a breakdown product of organic matter in waters. Some simple statistics were applied, and a grade from 'good' to 'bad was allocated. This was to evolve into the General Chemical Assessment (GQA) for rivers and canals that was to be accompanied by measures of biological quality assessment by species diversity and also predictions of species diversity that may result from pollution, drought or water quality improvement. GQA reporting ceased in 2009 (OMF, 2012). There was also proposed an aesthetic quality. One such scheme is the River Invertebrate Prediction and Classification System (RIVPACS), species groups recorded at a site were compared with those which would be expected *to be present in the absence of pollution and other harmful substances* and allowing for the different physical characteristics in different parts of the country. Various determinands were calculated for England and Wales and there were two summary statistics, known as ecological quality indices (EQI), calculated and then the biological quality was assigned to one of six bands based on a combination of these two statistics. In Scotland, a third EQI was also calculated by SEPA and the grading system based on a combination of all three statistics or EQIs (Defra, 2006b).

The advent of the WFD 2000 means that the systems that had evolved ceased to be used by 2009 (EA, 2009a). The next development recognised wider chemical parameters in a River Ecosystem (RE) Classification. Nutrients in river water remain a problem causing diffuse pollution EA. (2013a). The significance of which was the incorporation of pH, 'hardness' (as $mg\,l^{-1}$ of $CaCO_3$), dissolved copper and total zinc, recognising these toxic metals' presence in the environment. A stretch of river could then be classified as RE1 (good) to RE5 (bad) under the Surface Waters (River Ecosystem) (Classification) Regulations of 1994 (SI 1057).

The bringing into line of systems of water quality assessment, especially for rivers means there are serious questions to be asked for both spatial and temporal comparisons, specifically making it difficult for comparisons across the countries of the UK and beyond, and furthermore casts doubt on long-term trends in water quality and ecological changes.

The post-WFD water monitoring framework has emerged (FWR, 2010).

WFD requires that all surface waters and groundwaters within defined river basin districts must have reached at least 'good' status by 2015 (at the time of writing, details of the next deadline are under consideration), although this may need to be extended 'for technical or economic reasons'. It has done this for each river basin district by:

- Defining what is meant by 'good' status by setting environmental quality objectives for surface waters and groundwaters.
- Identifying in detail the characteristics of the river basin district, including the environmental impact of human activity.
- Assessing the present water quality in the river basin district.
- Undertaking an analysis of the significant water quality management issues.
- Identifying the pollution control measures required to achieve the environmental objectives.
- Consulting with interested parties about the pollution control measures, the costs involved and the benefits arising.
- Implementing the agreed control measures, monitoring the improvements in water quality and reviewing progress and revising water management plans to achieve the quality objectives.

This is wider ranging than merely identifying water quality standards and, while predictive methodologies such as RIVPACS have been used to inform WFD regulations, this Brave New World is far more holistic, as the above incorporates not just standards, but also implements water quality and related issues on a case-by-case basis incorporating consultation and economic measures. The precautionary principle stands and, in theory at least, the polluter still pays, however!

There remain water quality standards, and these specify maximum chemical concentrations for specific water pollutants and both groundwater and surface water bodies have to meet quality standards, specifically even if one determinand is breached, this means failure to meet 'good status', that is the combination of good ecological and good chemical status. A process referred to as 'intercalibration' aims to harmonise understanding and results of 'good ecological status across all member states (EC, 2015e). One example of new standards for ammonia is shown is shown in Table 4.1 compared with existing standards.

4.5 Trends in River Water Quality

Historically, the basis of water quality problems in Britain relate first to issues of sewage management (a result of urbanisation) and second to a plethora of pollution issues based in industrial development (including mining) and third are those from the use of agro-chemicals (Chapter 3). Hence areas that experience high population density

Table 4.1 UK Technical Advisory Group proposed (left column) and existing standards for ammonia in freshwaters (UKTAG, 2008).

Standards for ammonia					Existing standards The existing values are the thresholds used for the River Quality Objectives, RE1 and RE2, for England and Wales, and for Class A and B of the General Quality Assessment	
Total Ammonia (mg/l)					Total Ammonia (mg/l)	
(90-percentile)					(90-percentile)	
Type	High	Good	Moderate	Poor	High	Good
					0.25	0.6
Upland and low alkalinity	0.2	0.3	0.75	1.1		
Lowland and high alkalinity	0.3	0.6	1.1	2.5		

and/or industrialisation were inevitably prone to water quality problems. From the Great Stink of 1858, this resulted in practical measure and legislation, as well as long-term institutional evolution to address such problems. Inevitably the areas afflicted with poor water quality included the industrial belt of Scotland, the south of Wales, industrial cities in the north and midlands of England and, of course, the lower Thames basin.

In the debates that followed, where water quality increased, water undertakings, industry and regulators were keen to take the credit, while detractors talked of de-industrialisation (from about 1980) being the key factor in reducing point pollution. Diffuse pollution of waters was increasing blamed upon intensive agricultural production and fingers were pointed at farmers and the suppliers of agro-chemicals, rather than sewage undertakings, for diffuse pollution. Today, in theory at least, point pollution sources are controlled by consents.

It will be demonstrated that patterns of water quality and their changes over time are complicated and regional comparisons have to be treated with caution. However, between 1980 and 1990, calculated on the basis had been an overall decline in river water quality (NRA, 1991a), regional trends had shown that deterioration in terms of a consistent reduction in the percentage of 'good' and/or 'fair' and a corresponding increase of 'poor' and/or 'bad' stretches occurred in North West, South West, Thames and Yorkshire; the remainder either apparently improved or remained much the same.

For the period 1980 to 1985 there was a net downward change of 2% in river grading. Factors identified as causing difficulties for sampling and classification were the inclusion of non-routine samples from pollution incidents, a lack of consistency with respect to inclusion of samples suspected as being in error (e.g., those taken during flood events), differences in sampling frequency, pooling of data, over-ruling of grading by inexactly collected biological data, and issues relating to statistical interpretation.

Between 1985 and 1990 a downward figure of 4% was recorded but trends varied across the 10 former RWA areas. This was found to reflect shortcomings of the older scheme, especially differences in data collection between regions. Other reasons for a net decrease are identified as improved monitoring (casting doubt on the pre-1985 results), hot, dry summers in 1989 and 1990 increasing 'dry weather flows', and increased discharges from sewage works, industry and farms (DoE, 1992). The decline is quoted with some confidence, despite the probability of between 20 and 30% that any individual stretch may change class without any real change in water quality. This was attributed to thousands of determinations which, when averaged, smooth sampling errors. Although the average shows a steady decline, South West RWA displayed declines of 41 and 22% in the two periods 1980-1985 and 1985-1990, respectively. Such a change is difficult to explain in a region not normally noted for poor surface water quality. Hence the GQA scheme was born against an admission that water quality probably was declining (NRA, 1994h).

The summary of the report Water Quality Survey of Scotland 1990 (SOED, 1992), considering the chemical quality of rivers, lochs and canals, concludes that on the scale 1 to 5 and there was a net increase of 968 km in the length of Class 1 quality rivers between 1980 and 1990, and the length of Class 4 quality rivers has been reduced by 56% in the same period. The application of the River Invertebrate Prediction and Classification System (RIVPACS, see below) model in 1990 showed that more than 96% of Scottish rivers classified on the basis of biology are of either Class A or B. Already there was a dichotomy between water quality for England and Scotland.

Some 80% of river length in England was of 'good chemical quality' in 2009 (Defra, 2010d), when the GQA system ended and there appears to have been a convincing upward trend since 1990 and the challenge has been to bring reporting into line, eventually across Europe. All the EA regions seemed to show upward trends, with the overall worst in East Anglia but with the best overall improvement, in the southwest and the old industrial areas of the northeast, northwest and the midlands. Southern experienced profound problems 2000-2006 and Thames shows good improvement (OMF, 2012). This is discussed further in chapter 7.8. Unfortunately, these findings did not convincingly resonate with the criteria required under WFD (Chapter 10).

In Wales, the biological quality of surveyed sites 2007-2009 was largely unchanged, although there was a slight decrease from 88% in 2008 to 87% in 2009. In terms of chemical quality, 95% of river length was regarded 'good' in 2005-2009, and has maintained this level since 2005 (Defra, 2006b).

In Scotland, SEPA can report:

> 'The results also indicate that improvements in many aspects of water quality have been delivered through environmental regulation, cleaner technologies, improved sewage treatment and changes in agricultural practice. This is most notable for parameters such as biochemical oxygen demand, ammoniacal nitrogen, lead and sulfate, which have all generally declined in river waters. Other parameters showed a more complex pattern of regional and seasonal trends'.

Variation may be attributable to land use, to population density and levels of (past and present) industrialisation. For example, ammoniacal nitrogen (as the sum of dissolved ammonium and un-ionised ammonia) is associated with sewage effluent, agricultural fertilisers and other wastes. This determinand's concentration tends to be highest in the central Scotland and reflect the high population densities, although the River Don in Aberdeenshire also presents problems, for unlike elsewhere, ammoniacal nitrogen shows an increasing trend. (SEPA, n.d.a). In a European context, England is not doing that well in surface water quality (Defra, 2015b). There is a definite need to improve the quality of rivers, streams, lakes, estuaries, coastal waters and groundwater and only 27% of water-bodies in England are currently classified as being of 'good status' under standards set down by the EU Water Framework Directive (EC, 2014a). These concerns have long been echoed in the media. The EU Water Framework Directive calls for changes in the way surface water quality is monitored and reported and providing comparisons across countries in a 'Harmonised Monitoring Scheme' is challenging (Defra, 2010b). The period 2000-2009 has shown overall decreases in nitrate and phosphorus in rivers.

It is incumbent on EU member states that not only should standards not fall, but that they should have reached 'good status' in 2015, alongside inland water quality, although this might need to be extended (Defra, 2015b). A 1994 survey in England and Wales had identified 418 'bathing waters', finding that 345 (82.5%) complied, and an improving trend over time over the previous 20 years (NRA, 1995a). Specifically identified are bathing beaches and shallow marine environments supporting shellfish intended for human consumption. The Directive 2006/7/EC of the European Parliament and of the council of 15 February 2006, 'concerning the management of bathing water quality' repealed Directive 76/160/EEC, the previous bathing water directive. This is enacted in England, Scotland and Wales.

The highest rates of non-compliant (by microbiological criteria) or poor bathing waters were found in the United Kingdom (5.7%) along with Belgium (13.0%), the Netherlands (6.5%), Spain (3.8%) and Denmark (3.1%). Top was Cyprus with 100% compliance (EEA, 2012a). For estuaries and coastal waters, the situation around the coast of Britain improved 1992-2006 (Ospar Commission, 2009) and while the rate of improvement was impressive compared with some neighbours, evidently there is still some way to go to meet the new WFD standards for microbiological contamination.

4.6   Catchment Management Planning

River Basins are the rational environmental areas employed in Integrated Water Resource Management. This is (GWP, n.d.):

> 'Integrated Water Resources Management (IWRM) is a process which promotes the coordinated development and management of water, land and related resources in order to maximise economic and social welfare in an equitable manner without compromising the sustainability of vital ecosystems.'

Although worldwide, the catchment approach (as Catchment Management Planning, CMP) was a new concept in the UK when introduced in the 1990s (NRA, 1993a). CMPing presented somewhat anthropocentric objectives for water planning, but where quality is to be improved, stakeholder interests must be balanced, something reflected throughout WFD. In turn we must also talk robustly of ecological status.

With the arrival of the EA in 1996, waste regulation and issues of air quality are to be included with water. This produced a new generation of plans, the Local Environment Agency Plans ('LEAPs') which remained essentially catchment based (EA, 1997a). Comprehensive though these were, it was perhaps becoming un-wieldy for one plan to cover planning and protection for all the 'media' of soil, water and air. For water alone, convenience dictates that issues as diverse as pollution, abstraction licensing and flood defences are considered separately.

The WFD, however, requires management of all these aspects and more to be ultimately managed in respect of River Basin Districts. Continental-scale river basin management was promoted in the later nineteenth century. In the United States in the 1930s (USEPA, 2001, p. 10) the idea spread across the world. For example, the Tennessee Valley Authority has been discussed as a potential model ever since. The Tennessee Valley Authority (TVA) came into being in 1933 and there are now some 49 dam sites. It is undeniably a sound model providing basis for large-scale plans for the development of river basins.

The TVA has provided a model for integrated river basin management throughout the world, either through example or via technology transfer by ex-TVA executives (McDonald and Kay, 1988), yet the model has been replicated nowhere else in its entirety. Indeed, attempts elsewhere have been disappointing. The effect of the TVA has been to produce a massive electricity generating utility, achieved through regional development and public funding. The TVA influenced farmers (often through government agencies) to conserve soil, reclaim land from gullying, limit the use of fertilisers, diversify crops and encourage marketing through co-operatives. Otherwise, navigability has been improved, industrial development made possible by huge electricity generation (now supplemented by coal-fired and nuclear plants), forests have been created on marginal land, and rural poverty alleviated through welfare and educational opportunities available to TVA employees. Malarial mosquitoes are controlled by maintaining water levels from April to June, and flooding has been controlled where appropriate. TVA has drawn some harsh criticisms from conservationists; it was, after all, conceived in a 'pre-green' era. Other critics have a cultural problem with large-scale planning and public authorities which seem redolent of socialism, and are prepared to criticise its financial arrangements (Newson, 1992a).

We need perhaps to retreat from the whole river basin idea of TVA: for whatever reason it would not be appropriate in Britain, perhaps given the complexity of economic structures, its existing planning system and also the anti-statist *zeitgeist* developed since about 1980. Yet Britain had not been without earlier catchment-wide institutions. The English water engineer Frederick Topliss writing of the 1870s (Toplis, 1879), proposed that all issues of pollution, floods and storm water, water supply (quantity and quality) might be managed on what is essentially within a river basin management system. In Chapter 3 it was described how, anticipating the ideas of Toplis, catchment authorities emerge in the Thames Conservancy (1857) and Lee Conservancy Board (1868). Further regularision came in the establishment of Catchment Boards (1930), River Boards (1952) and in 1963, River Authorities were established.

The establishment of RWAs in England and Wales after 1974 caused these agencies to be responsible for the management of the whole hydrological cycle; however, there is no way that these were seen as an engine for economic development even if it was hoped that they were proactive in water planning and development. With the advent of WFD, the dream of Fredrick Toplis would be fully recognised.

Britain therefore has a solid pre-WFD track record in at least attempting to manage catchments, something that influenced WFD (2000). Holistic integrated environmental management requires a cross-functional approach involving several organisations with shared objectives. EA inherited, as a principal tool, the formulation of LEAPs based upon the earlier principles of Catchment Management Plans (CMPs). These aimed to set a vision for individual river catchments which takes account of both national policy and local community views; each one presents an action plan for the local authorities, landowners and other interests, reviewed annually for the next three to five years (NRA, 1995b).

CMPs were action plans which stated objectives for both water and associated land use, LEAPs then included issues of land and air protection and contamination including standard setting and water quality objectives. Importantly, they provide a forum for conflicts and enable the management of catchments proactively rather than reactively.

The EA, like its predecessors, has had to strive for democratic accountability, something that will be addressed in the next chapter, while the result of consultation is the drawing of Action Plans. Post-WFD, there is now a framework for the protection of waters and all their uses. Geographically Britain is divided into 12 River Basin Districts (RBD; Figure 4.1) and Scotland north of the Solway Tweed RBD constitutes one district (EA, 2010a).

A number of critiques may be raised around this classification. For one, 12 districts seems a low number, even for a small island state like mainland Britain. Second, hydrology does not respect national boundaries, for much of Wales belongs to the Severn RBD, the Scottish border looks a better fit but it is not perfect. Third, their size varies, large ones being the Scottish RBD, the smallest the Dee. Finally, each RBD may cover a range of geological conditions. Perhaps the most wide ranging is the southwest, which is mostly based on impervious Palaeozoic rocks including large granite intrusions that are exposed as moors from Dartmoor to the Isles of Scilly. Yet there are from Exeter eastwards a range of Mesozoic sediments that include aquifers such as Permo-Triassic and the extensive 'Wessex' chalk of Dorset and Wiltshire. These RBD areas form the basis for River Basin Management Plans and, helpfully are able to be grouped by catchment. For example there are nine for the South West region (EA, 2014a). On this basis is the reporting of ecological status, ecological potential and water quality (surface rivers, lakes, estuarine and coastal and also groundwater). River Basin Management Plans are to be prepared and renewed in six-year cycles and the first plans covered the period to 2015. The matter of public consultation and participation will be returned to in Chapter 10.

LEAPs were comprehensive and wide-ranging, but evidently the scope proved too much, for EA planning evolved towards them issuing a whole range of less generalised plans, most notably towards those concerned with abstraction -Catchment Abstraction Management Strategies or CAMs (EA, 2013b), or with flood defence—Catchment Flood Management Plans (CFMPs; EA, 2009b) as well as the holistic RBMPs EA (2014c).

4.7 The Actual Plans for River Basins

The list below gives an idea of the range of individual plans that are 'driven by government, water companies and the Environment Agency' that are concerned with achieving successful River Basin Management Plans (RBMPs). The EA Corporate Strategy 2010-2015 gives detailed insight relating to its policy: Creating a better place, outlining our vision for the environment. This includes a summary of outcomes designed for the use to measure success at managing the water environment. Within this are a number of specific strategies. (EA, 2013b) summarised below for England and Wales:

1) *Catchment Abstraction Management Strategies (CAMS).* The EA developed CAMS between 2001 and 2008 for all major catchments in England and Wales and these

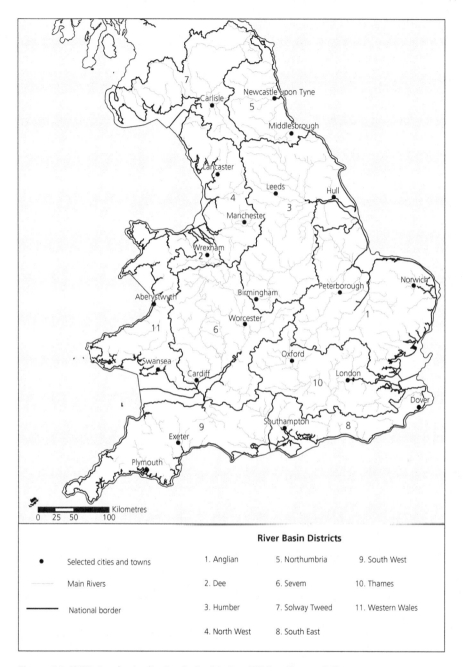

Figure 4.1 WFD river basin districts in England and Wales. (*Source:* EA).

were made available to the public for the first time. In general, abstractions over 20 cubic metres per day require an abstraction licence (under the Water Act of 2003). The granting of a licence he needs of the environment and existing abstractors and also whether the justification for the abstraction is reasonable. These procedures underpin the licensing decisions and contribute to delivering the objectives of the River Basin Management Planning. CAMS focuses on Resource Assessment and developing the Licensing Strategy, and the aim is sustainable abstraction.

2) *Water Resources Strategies.* The EA water resources strategies, Water for People and the Environment: Strategy for England and Wales and Strategy for Wales, is aimed at producing a consistent framework for water resources management across England and Wales and set out how they believe water resources should be managed beyond 2050. The strategy seeks to explain how EA will engage with many organisations to put in to practice actions that will:
   • raise awareness of the value of water;
   • manage water resources to better adapt to climate change;
   • make water use more effective;
   • promote incentives to reduce demand;
   • promote sustainable planning.

3) *Water Resources in England and Wales—current and future pressures report.* This report is designed to be updated in order to put the current and future pressures on water resources into context. It includes detailed information on:
   • how water availability changes from place to place;
   • water use, including information on the abstraction of water;
   • public water supply, leakage and household metering;
   • future pressures and trends, such as climate change and population growth.
   The report uses information from a variety of places including CAMS and Water Resource Management Plans (WRMPs) from water companies in order to assess the possible impacts of climate change.

4) *Water Resources Strategy Regional Action Plans (RAPs).* These take national actions (and principles) contained in the Corporate Strategy (see above) that translate into local activities. Water resources strategy Regional Action Plans (RAPs) report how these actions will be carried out at local levels, until up to 2015. RAPs also include water resources elements such as drought management. Initiatives such as CAMS and Restoring Sustainable Abstraction (RSA) will provide a mechanism for delivery of actions contained in the plans.

5) *River Basin Management Plans (RBMPs).* WFD requires EA to produce River Basin Management Plans for each of the 11 River Basin Districts in England and Wales. The first River Basin Management Plans were published in 2009, for example, that for the Thames (Defra and EA, 2009b) and they set the 'programme of measures' deemed necessary to ensure that inland and coastal waters achieve WFD 'good ecological status or potential' status (or an alternative objective). There should, furthermore, be no deterioration from their current status. For example, abstraction licensing is one mechanism in place to support River Basin Planning objectives. Other measures include the control of both diffuse and point source pollution, and the management of physical alterations to watercourses.

6) *National Environment Programme (NEP) and Asset Management Plans AMP.* NEP is a programme of investigations and actions for environmental improvement schemes designed to ensure water companies meet European Directives, national targets and their statutory environmental obligations. EA provides a list of investigations and solutions for the NEP. In creating the programme, there was consultation with the water industry and other organisations.

The NEP forms part of the final *Asset Management Plan (AMP)* that determines the overall level of investment made in water companies over a five-year period and based on prices set by the *Water Services Regulation Authority (known as Ofwat)*. Companies should incorporate these requirements into their proposed business plans.

7) *Restoring Sustainable Abstraction (RSA)*. Where abstractions are currently unsustainable, EA will investigate the causes and then implement measures to restore 'sustainable abstraction'. A change in abstraction licence (under the Water Act of 2003) could be involved, otherwise there are other actions needed that reduce environmental impact. Like the NEP there are three stages in this process: Investigation, Options Appraisal and Implementation.

8) *Environment Agency Drought Plans.* EA will produce 'drought plans' setting out plans for and management in drought situations. These range from high-level plans where there is a co-ordination in drought management activities throughout England and Wales, to local level plans where specific operational activities are addressed. These plans are reviewed annually and updated when appropriate. For example, EA has reviewed the 2012 drought and identified lessons learnt.

9) *Water Company Water Resources Business Plans (WRBPs)*. WRBPs are statutory obligations on water companies as business plans submitted to Ofwat who regulate the price customers pay for the supply of water and the treatment of wastewater. Ofwat reviews Water Company pricing in a five-yearly process known as the Periodic Review. Periodic Review, PR09 sets the price limits for the period 2010 to 2015. This is being updated. The EA will use the consultation period to check that WRBPs are consistent with statutory WRMPs.

10) *Water Company Water Resources Management Plans (WRMPs)*. WRMPs show how water companies plan to manage water supply and demand for water over a 25-year period. From 2009 they have published and consulted on their draft WRMPs that are regularly reviewed, and revised every five years. Ofwat use the Management Plans to assess the companies' supply-demand balance and the work they need to undertake as part of the Periodic Review. This information forms the basis for the Water Company Business Plans.

11) *Water Company Drought Plans.* Water companies prepare drought plans to show the actions they propose to take in order to manage water supplies during drought periods. They prepare and consult on them before submitting them to government.

Naturally, a useful but simplified dichotomy where societal water us is concerned remains whether to develop the supply side (develop new sources of raw water) or reduce demand (including re-use and recycling). Although this takes little account of environmental demand on water, these specific topics are further discussed in Chapter 6.

4.8 Catchment Flood Management Plans (CFMPs)

The above account is largely predicated on the presumption that most of the time, water supply and water quality is manageable and not only environmental, but also economic and business plans can be put in place. The winter of 2014 once more reminded us that unpredictability in water supply (in the form of excessive rainfall) presents what can only be described as disasters on a regional scale that impacts across the jurisdiction that is England and Wales. Catchment Flood Management Plans (CFMPs) consider all types of inland flooding, from rivers, ground water, surface water and tidal flooding, but not flooding directly from the sea as coastal flooding and this is covered in 'shoreline management plans'. They also take into account the likely impacts of climate change and impacts of land use change in a sustainable way. CFMPs will help the EA and its partners to plan and agree the most effective way to manage flood risk in the future (EA, 2009b).

Additional kinds of plan exist and have significance for waters. One such example is Water Level Management Plans—WLMPs (Water Management Alliance, n.d.). These plans provide a means by which the water level requirements for a range of activities in a particular area, including agriculture, flood defence and conservation, can be balanced and integrated (Chapter 8). Developed by Defra, these control water level required for specified activities (arable agriculture, grazing marsh, reedbeds, flood risk management and conservation etc.). They are particularly important when SSSIs are involved. They are of particular significance to Internal Drainage Boards, (IDBs) and WLMPs are produced by a consultation process.

4.9   River Basin Planning in Scotland

'Scotland is renowned worldwide for the environmental quality of our rivers, lochs and seas, which attract visitors and support our key industries. It is important for our continued economic success and well-being that we maintain this enviable reputation.' (Scottish Government and Natural Scotland, n.d.), SEPA, n.d.e).

Here is a direct difference in approach. While England and Wales seem to show an earnest desire to deal with what may be near-intractable problems and struggle to meet the strictures of WFD, the Scottish response is immediately to hit the buttons of a broader economic value. To Scots, the water environment is business, even a generator of foreign capital via tourism.

The Scottish RBD is essentially the whole country north of the Southern Uplands (Figure 4.2). The diversity of catchments here is, like the west country of England, alarmingly diverse when one comes to think of basic environmental management. While the agriculturally productive areas of the east, or Ayrshire in the west would be familiar to most European agriculturists, much is wild highland and rain-soaked islands where there is subsistence agriculture, including the well-known crofting system. In the meantime, the central valley is both densely populated, highly urbanised and contains what is left of the Scottish manufacturing industry.

With legislative provision made in regulations as the Water Environment (River Basin Management Planning: Further Provision) (Scotland) Regulations 2013 and came into force on 22 December 2013, we may see the process may appear behind that for England and Wales, but it looks to be considerably simpler. Perhaps the problems are not so severe with utilities in public ownership the temptations of profit motivation are removed. In any case, water quality in Scotland is generally better than in England and with the exception of the southeast, there are few problems in bulk supply.

Scotland, like England and Wales (next section) has a requirement to consult on water issues during the established ('statutory' planning process (the Water Environment (River Basin Management Planning: Further Provision) (Scotland) Regulations 2013).

Figure 4.2 The Scottish RBD. (*Source:* Scottish Government and Natural Scotland, n.d.).

These are: Transport Scotland, Scottish Water, Scottish Environment Protection Agency (SEPA), Scottish Natural Heritage (SNH) and Historic Scotland. In general, statutory consultees play three roles in the processing of planning applications:

- They will indicate what information they require from applicants in order to inform their assessment;

- They may provide relevant information to applicants; and ultimately
- They will comment on the planning application from the standpoint of their organisation

For example, Scottish Water will require, Flow and Pressure Test (Water), a Water Impact Assessment (WIA) and a Drainage Impact Assessment (DIA) in connection with any proposed development. In general, and in the light of WFD greater transparency is sought in water planning, and to quote from the regulations, directly focussed on the WFD. Regulations 2013 for Scotland Scottish Government and Natural Scotland (n.d.) are listed below:

'*Setting of environmental objectives*

3. (1) The objectives set under section 9(1)(a)(i) (environmental objectives) of the Act must, subject to the application of regulations 5 to 10—

a) for surface water—
   i) prevent deterioration of the status of each body of surface water;
   ii) protect, enhance and restore each body of surface water (other than an artificial or heavily modified body of surface water) with the aim of achieving good surface water status by 22nd December 2015;
   iii) protect and enhance each artificial or heavily modified body of surface water with the aim of achieving good ecological potential and good surface water chemical status by 22nd December 2015; and
   iv) aim to progressively reduce pollution from priority substances and aim to cease or phase out emissions, discharges and losses of priority hazardous substances;
b) for groundwater—
   i) prevent the deterioration of the status of each body of groundwater;
   ii) prevent or limit the input of pollutants into groundwater(1);
   iii) enhance and restore each body of groundwater, and ensure a balance between abstraction and recharge of groundwater, with the aim of achieving good groundwater status by 22nd December 2015; and
   iv) reverse any significant and sustained upward trend in the concentration of any pollutant resulting from the impact of human activity in order to progressively reduce pollution of groundwater; and
c) in addition, for each area and body of water falling within section 7(4) (protected areas) of the Act, achieve compliance with any standards and objectives required by or under any EU instrument described in that section by 22nd December 2015, unless otherwise specified in the EU legislation by virtue of which the area or body of water is protected.'

At the time of writing, details of the future deadline is under review. The approach seems to seek partnerships with various bodies, including these public bodies:

- Scottish Natural Heritage;
- Scottish Water;
- Forestry Commission Scotland;

- British Waterways Board;
- Fisheries Committee;
- local authorities;
- district salmon fisheries boards;
- national park authorities.

And also to set timetables, for example, it is planned to reduce the number of waterbodies affected by agriculture, sewage, acidification/abandoned mines, agricultural irrigation, hydropower, aquaculture, urban land use and urban flood protection, forestry and past engineering activities on abstraction, impoundment, pollution and physical alteration of watercourses in stages, to 2015, to 2021 and 2027. Targets for the latter date are especially stringent, although it is anticipated there will remain problems for acid mine waters.

Time will tell, Scotland, while facing fewer problems has confidence in her public institutions to deliver.

4.10   River Basin Plans and Planning in England and Wales

The explosion in water policies since the 1980s, represents an interloper into a statutory planning process, by then already some 40 years old. Model statements supporting the water environment have long a part of local authority development plans, the EA being a statutory consultee. Legal sanctions remain a backstop for recalcitrant individuals or organisations, but litigation is expensive and progress is best achieved through public engagement, consultation, agreement and education. Yet CIWEM can still call for 'a more holistic and sustainable approach is taken which is open and transparent, and encourages public and community involvement and debate' (CIWEM, 2012a). This includes more clarity as to which agency may be responsible for what in water planning terms (Cook *et al.*, 2012). Reading of such publications can leave one with a sense of insufficient 'joined-up-thinking' and this may be attributable to such as the diversity of organisations concerned with water or the real lack of understanding of the distribution of power versus responsibility.

Thames Water plc is one undertaking that has to confront such issues head on, being responsible for nothing less than water a sewage services in most of the London area, and a large proportion of up-catchment Thames basin (Thames Water, 2013). Hence, 'Every five years, water companies in England and Wales are required to produce a Water Resources Management Plan (WRMP) that sets out how they aim to maintain water supplies over a 25-year period'. Water planning horizons are inevitably far off for the water industry when compared with other aspects of planning, and whether it is the UK's 5-year planning cycle or the 6-year WFD cycle, there is a sense of perpetual modification, operating at least formally in the sense of 'adaptive management strategies' where one is struggling not only with urban development but also WFD implementation, biodiversity issues, and of course climate change. In this water demand forecasting is essential and self-evidently is linked to the ordinary planning process.

On the other side is how developers are obliged to address issues of the water environment and supply in applications for housing, business or industrial development.

Certainly there are a list on 'Statutory consultees' that should be approached regarding the impact on the water environment of a proposed development. These bodies are listed below, both statutory and non-statutory consultees (Communities and Local Government, 2009):

1) *Statutory consultees in river basin planning*
   British Waterways
   Commission for Architecture and the Built Environment
   Civil Aviation Authority
   Coal Authority
   Crown Estate Commissioners
   Department for Culture, Media and Sport
   Department of Energy and Climate Change
   Department for Environment, Food and Rural Affairs
   Department for Transport
   Environment Agency
   English Heritage
   Forestry Commission
   Garden History Society
   Health and Safety Executive
   Highways Agency
   Local and Regional Bodies (County Planning Authority, District Planning Authority, Greater London Authority, Local Highway Authority, Local Planning Authority, Parish and Town Councils, Regional Development Agencies and Regional Planning Bodies)
   Ministry of Defence
   Natural England
   National Air Control Transport Services and Operators of Officially Safeguarded Civil Aerodromes
   Rail Network Operators
   Sport England
   Theatres Trust
2) *Non-statutory consultees*
   Conservation Area Advisory Committees
   County Archaeological Officers
   Drainage Board
   Emergency Services and Multi-Agency Emergency Planning
   Health Authorities and Agencies
   HM Revenue and Customs
   Local Authority Environmental Health Officers
   Navigation Authorities
   Police Architectural Liaison Officers and Crime Prevention Design Advisers
   Schools and Colleges
   Waste Disposal Authorities
   Water and Sewerage Undertakers

Where a statutory consultee is concerned, planning law requires where consultation must take place between a local planning authority and certain organisations, prior to a decision being made on an application by a planning authority. These organisations have a duty to make an appropriate response. On the other hand, a non-statutory consultee may be approached where a local planning authority considers whether there are planning policy reasons to engage other consultees not designated in law but should have an interest in a proposed development (Communities and Local Government, 2014a). The problem with both categories is that while full and successful consultation may be undertaken, the planning authority is not obliged to take notice of the advice given, so that for example EA may be consulted and oppose a development of houses on a floodplain, its professional advice can be ignored in a planning decision. For:

> 'It is not possible for a statutory or non-statutory consultee to direct refusal and insist that a planning application is refused by the local planning authority.'

Although appeals to the Secretary of State may be successful if a case is made by a statutory consultee (Communities and Local Government, 2014b).

To elaborate, water issues may emerge in respect of other planning procedures. The Planning and Compensation Act of 1991 (Planning and Compensation Act of 1991) (NRA, 1994i) requires that decisions are made in accordance with such plans, 'unless material considerations indicate otherwise'. For non-metropolitan areas, the 1991 Act requires that the following plans be prepared:

- County Structure Plans, which provide a broad strategic planning framework and aim to ensure that development is realistic and consistent with national and regional policy.
- District or Local Plans, which set out detailed policies and specific proposals for the development and use of land. They should be in general conformity with the Structure Plan, making proposals for specific land allocations and setting out the policies for the control of development.
- Minerals Local Plans. These are concerned with the extraction of minerals such as floodplain sands and gravels, or other quarrying and mining activities, all of which affect either the hydrology or are potentially polluting activities.
- Waste Local Plans, prepared originally by the Waste Disposal Authority (now EA), are now integrated into the Minerals Local Plan. The Control of Pollution Act 1974 requires that disposal sites be licensed and do not endanger public health, cause water pollution or cause 'detriment to local amenity'.

These plans all require provision for the needs of utilities responsible for water supply and sewerage together with the other concerns of the EA. There is interaction with local authorities, and advice and influence proffered in the areas of water quality, water resources, flood defence, fisheries, conservation, recreation and navigation in the river corridor through statutory development plans. Where new authorities replace the older local authority structures, County Structure Plans and Local Plans will be replaced by Unitary Plans. This is not universally the pattern, and progress is slow.

One breakthrough of CMPing presented in plans published between 1991 and 1995, including catchments starting with the Medway NRA (1991c) and includes the upper Thames (NRA, 1995e) involved not only the comprehensive nature of what was on offer, but a commitment to the consultation of stakeholders. That quiet change in protocol would herald something of a revolution in environmental politics. If RWAs were criticised for being 'technocratic', implying 'we know best' and aloof, then a consultation process based at least in the statutory planning process would be welcome (Chapter 10). The term 'CMPing remains currency in Scotland (SEPA, 2003).

Apart from that, within the geographic framework described above, a series of procedures familiar to water professionals takes place, although what with the notions of 'good status' there is a stronger emphasis on outcomes. These have been limited to Statutory Water Quality Objectives, or at least represented an interplay between fixed standards and the needs of current water use. Concern about development was often concerned with the legacy of heavy industry, ongoing problems of agricultural pollution and pressure from urbanisation, especially increased sewage effluent, abstraction and flood risk management. Fisheries could dominate ecological considerations, but the needs of aquatic plants, birds and wetland conservation made inroads as appropriate. In1996, the newly formed EA inherited a strong legacy of water planning from its short-lived predecessor, the NRA. The next phase, described above, was to produce LEAPs, described above.

Article 14 of WFD 'to encourage the active involvement of interested parties', not only embodies the United Nations Economic Commission for Europe (UNECE) Convention on Access to Information, Public Participation in Decision-Making and Access to Justice in Environmental Matters or 'Aarhus Convention' 1998 (EC, 2015f), it produces legislative bases for the democratisation of environmental management. The 'revolution' was to go beyond mere public consultation into public participation. River Basin Management Plans are then the grandchildren of the CMPing process. Britain had to play catch-up in terms of environmental regulation, and at least after 2000 was in a good position to integrate the new and more exacting standards to RBMPing. Environmental managers and fellow travellers were tasked with becoming social scientists into the bargain!

Protecting water supply naturally falls on upland areas, not only because some 70% of water supplies in England and Wales, but also because these areas have relatively few breaches of water quality, although peat conservation can be a significant problem in certain areas including southwest England and the 'Dark Peak' area of Derbyshire. In this instance, measures to protect upland water sources run hand in hand with a need to minimise carbon exported and to preserve associated exosystems (such as mires and bogs) and preserve low stocking density grazing (ref needed).

4.11   What Does a River Basin Management Plan Look Like?

River basin management plans (RBMPs) are drawn up for the 10 river basin districts in England and Wales; the Dee and Western Wales river basin districts, managed by NRW. Solway Tweed is managed by SEPA (EA, 2014c).

Statutory groups involved in RBMPing are the Regional Fisheries, Ecology and Recreation Advisory Committee (members approved by Defra), the Regional Environmental Protection Advisory Committee (including local authority membership), a Catchment Flood Management Plan Steering Group (largely technical and conservation-based interests) and a Water Framework Directive and River Basin District Liaison Panel (Benson *et al.*, 2013a). Plans characterise the River Basin District (generally a large area), identify pressures and impacts upon water bodies within, provide an economic analysis of water use and summarise programmes of measures required for a particular RBD to achieve Water Framework Directive objectives. The plans will:

- Establish a strategic plan for the long-term management of the River Basin District;
- Set out objectives for water bodies and in broad terms what measures are planned to meet these objectives; and
- Act as the main reporting mechanism to the European Commission

The River Basin Management Plans have been developed with the River Basin District Liaison Panels and have six-year cycles. The process includes representation from stakeholders and economic sectors for their particular River Basin District. The Liaison Panel makes decisions regarding the River Basin Management Plans. There are also national stakeholder panels for England and Wales, made up of representatives from national organisations (Life, 2008).

Each country has to:

- prevent deterioration in the status of aquatic ecosystems, protect them and improve the ecological condition of waters,
- aim to achieve at least good status for all water bodies by 2015. Where this is not possible and subject to the criteria set out in the Directive, aim to achieve good status by 2021 or 2027,
- meet the requirements of Water Framework Directive Protected Areas,
- promote sustainable use of water as a natural resource,
- conserve habitats and species that depend directly on water,
- progressively reduce or phase out the release of individual pollutants or groups of pollutants that present a significant threat to the aquatic environment,
- progressively reduce the pollution of groundwater and prevent or limit the entry of pollutants,
- contribute to mitigating the effects of floods and droughts.

The plan describes the River Basin District and the pressures that the water environment faces including actions required, including those possible by 2015 and how the actions will make a difference to the local environment – the catchments, the estuaries and coasts, and the groundwater (Defra and EA, 2009a).

For example, the South West River Basin District Liaison Panel includes representatives of businesses, planning authorities, environmental organisations, consumers, navigation, fishing and recreation bodies, and central, regional and local government. The Environment Agency, working with local stakeholders, identifies actions needed to address the main pressures on the water environment. The Anglian region has a large emphasis on aquatic ecosystems, while Thames region is concerned with sewage discharges, diffuse pollution and physical modification of water courses. Both the North West and West Wales regions are concerned, among other things, with the legacy on industrialisation.

A criticism of such large Districts is the complexity of the topography, soils and hydrogeology within. The various annexes, in somewhat (electronically) weighty documents typically deal with the current state of waters, water body status, actions to deliver, protected area objectives, actions appraisal and justifying objectives, mechanisms for action, pressures and risks, adapting to climate change, 'aligning other key objectives to river basin management, economic analysis of water use, record of consultation and engagement, competent authorities and a glossary' (EA, 2014c).

Yet a Parliamentary Office of Science and Technology POST note (2008) can state:

'The EU Water Framework Directive (2000/60/EC) seeks to protect, improve and maintain the environmental condition of surface and ground waters. Under the directive, all inland, estuarial and coastal waters must aim to achieve "good ecological status" by 2015. More than 80% of water bodies in England and Wales currently fail to reach this status, something that had been the case since about 2008. This POST note outlines some of the challenges in implementing River Basin Management Plans (RBMPs) in the UK to meet Water Framework Directive (WFD) objectives' (POST, 2008). Largely to blame are land use practices, and specifically those that produce diffuse pollution from agriculture. Challenges remain.

4.12 Summary

The sustainable development debate, while embracing all natural resources and galvanised by climate change, inevitably embraces not only IWRM, but a new, all-encompassing form of catchment management including RBM through the agent of RBMPing, an outcome of EU framework legislation. Perhaps the major hope of Catchment Management Planning, now incorporated in RBMPing is that it will solve problems of diffuse pollution in affected catchments by encouraging sustainable development. Beneath a new planning process remains a set of tools, starting with Environmental Assessments that are concerned with anything from standard setting and environmental objectives to flood control and abstraction management.

Since 1990, catchment planning has evolved from being water based to incorporating air quality and waste management, and then once more divorced to embrace only surface and groundwater, yet packaged in a manner that concentrates on outcomes ('good status'), includes economic analyses and engages all stakeholder and the public (next chapter). With a good knowledge base at the start of the twenty-first century, and appropriate institutional measures, optimism is not so much misplaced as taking a long time to be realised. WFD demanded that surface waters reached 'good status' by 2015.

Diffuse pollution relates to a general failure in the planning process to prevent such pollution at source, and requires a multi-agency approach. Successes may be manifest where attempts to control the problem are being made by the implementation of Statutory Codes and initiatives such as Catchment Sensitive Farming (Natural England, 2016), where areas are identified for special attention on the basis of environmental vulnerability and where farmers are given financial incentives to minimise damaging activities.

Rural land-use issues often contrast with the rigorous treatment of near-channel issues and with the mechanisms for dealing with 'end of pipe' pollution and flood management. Examples are aquatic conservation, water quality, or where the EA exerts direct influence through the planning process such as in housing or industrial development. It is also conspicuous how soil information (as distinct from, say, geology) is often absent.

Abstraction controls are paramount as they affect not only the societal use of water but influence other uses such as flow in channels (affecting navigation and ecology) and concerns over conservation in rivers and wetlands. With many groundwater units under pressure, especially in eastern and southern England, the need to consider abstraction in a holistic fashion has never been greater.

Matters of flood protection are increasingly also placed centre-stage, and there is now a vehicle through which important decisions about hard engineering solutions can be maintained, enhanced, or relaxed in the interests of economy and habitat restoration. As a generality, flood defences take over from considerations of diffuse pollution in higher rainfall and impermeable catchments in western Britain. England, Wales and Scotland are rapidly becoming countries where environmental planning is, in theory, well advanced. With the principles in place, evolution will have to progress according to pragmatism and financial constraints. There is no doubt that the will is there from most quarters. The next chapter provides an analysis of social, political and economic factors.

5

Sustaining Bulk Supply

Consumption and Interference

YAHOO! IT'S SO HOT-THERE'S A WATER SHORTAGE! NO BATHS FOR US!-YES!
> *Dennis the Menace (to Gnasher) in* The Beano *no. 2823, August 24, 1996*

5.1 Introduction

The historic view has been that that Britain has more than enough water to satisfy current and future needs (NRA, 1995c). This view made presumptions about all of demand, climate stability and technological solutions that effectively evened out supply. It is misleading, for it certainly does not mean there is any great overall surplus. This chapter presents an overview of consumption and human interference in the hydrological cycle prior to discussion of problematic issues. At the root of water supply problems is the uneven supply and demand balance across England, Wales and Scotland, which generates many, and differing, demands on the resource. Paramount is the need to minimise adverse effects upon the environment as well as maintain quality of public supply.

5.1.1 England and Wales

About 99% of the population of England and Wales is served by the public water supply as mains water, the remainder comes under private supply (DWI, n.d.). Licences to abstract *in force in 2009* numbered around 20,000 (of which 16,000 were not time limited) and licences are required for sources that supply 20 cubic metres per day, or more. Licences are issued for both public and private purposes, including public water supply (the largest), agriculture including spray irrigation, electrical supply and other industrial use, fish farming and amenity ponds. The overall pattern of consumption is presented in chapter one but in more detail: of an estimated 11,399 million cubic metres in 2011, a fall in consumption is attributable to industry (other than power generation), agriculture and fish farming all falling by more than 50% since 2000. The only sector showing an increase in abstraction since 2000 is spray irrigation (included in the 'other' category in Figure 1.2) and in regional terms this figure is consistently highest for the Anglian Region, followed by Midlands (EA, 2008). Otherwise factors relating to the fall in water consumption may be due to weather conditions, for example wetter years could result in a decrease in abstraction for agriculture and spray

The Protection and Conservation of Water Resources, Second Edition. Hadrian F. Cook.

irrigation, changes in the level of activity in different sectors and improvements being made in the efficiency of water usage (Defra, 2013a). While the long-term reduction in consumption is welcome, there remains no reason for complacency for, as the trend to increase consumption of spray irrigation suggests, not only can some sectors reverse the trend, but there may well be a significant factor arising from climatic change.

Yet Britain, regarded in a global context is a 'low stress' country for resource availability (Maplecroft, 2012), although locally there are high stress areas in the south-east of England (WRI, 2003). For England alone between 2000-2001 and 2008-2009, average household water consumption in unmetered households rose from 149 to 150 litres per person per day, but that is not great. For metered households, water consumption decreased from 132 to 127 litres per person per day, a decrease of about 4% (Defra, 2010c). This is outstripped by many developed countries, for example the United States and Australia top the list (Data 360, n.d.). It also explains the pressure for household metering.

In population density, the UK at 247 persons per km^2 and is set at 32 out of 192, comparable with Germany, and far greater than the USA in estimates for 2012-14 (Worldatlas, n.d.). In England and Wales the average population density was 371 persons per km^2 in 2011; however in London this figure was 5,200 persons per km^2. If the London figures were excluded, the average population density for the rest of England and Wales was 321 people persons per km^2. There were 23.4 million households in England and Wales at census. The average household size was 2.4 people per household in 2011, just over half of the 4.3 residents per household in 1911. All regions saw population growth between 2001 and 2011, with the highest growth in London, the East of England and the East Midlands (Office for National Statistics, 2012). Following 1976, which had been a year of exceptional demand upon water resources, wider strategies than merely developing new sources have been considered. Demand forecasting is essential, while the 'precautionary principle' would require a margin of error to be built into estimates of water demand.

Summary of abstraction by EA region and NRW is shown in Table 5.1. The highest consuming domestic region is Thames, followed by the Midlands and Anglian, reflecting high population and size of region respectively. The largest arable agricultural regions reflect the importance of arable agriculture in each region and it is worth note that the electric industry, consuming cooling water, outstrips other industrial uses in most instances. By comparison, the Midlands and London have retained something of their industrial basis.

5.1.2 Scotland

Turning our attention to Scotland, Figure 5.1 summarises water use over eight years to 2009/2010. Like England and Wales, there is a definite downward overall trend, and like its neighbours the majority use is for domestic consumption (which of itself is rising) with the non-domestic consumption category that includes all industry and a small proportion for agriculture. The most striking figure is for leakage in the distribution that in 2009/2010 is estimated at a stunning 38%. There is agreement to reduce this. The average for England and Wales was between 14 and 27% for 2010/2011 (BBC, 2012) and like Scotland the trend is downward.

Scottish Water estimates that the average consumption of wholesome water per person in Scotland is around 150 litres per day, virtually the same global estimate as for England

Table 5.1 Estimated abstractions from non-tidal surface and groundwaters by purpose 2007.[1]

Units are	megalitres	per day							
	Public water supply	Spray irrigation	Agriculture (excluding spray irrigation)[2]	Electricity supply[3]	Other industry	Fish farming, etc.	Private water supply[4]	Other	Total
EA Regions									
England and Wales	**16,381**	**161**	**72**	**10,304**	**2,736**	**3,412**	**29**	**113**	**33,208**
England	**14,611**	**159**	**69**	**2,972**	**2,243**	**3,188**	**29**	**107**	**23,379**
NE[5]	2,154	11	5	418	142	465	5	5	3,204
NW	1,544	2	6	812	500	41	0	5	2,910
Midlands	2,466	33	5	1,003	1,043	12	6	5	4,574
Anglian	2,011	91	7	16	137	67	5	1	2,335
Thames	3,928	6	8	101	122	240	8	13	4,425
Southern	1,324	12	7	4	116	1,211	2	33	2,708
SW	1,184	3	31	619	184	1,152	4	45	3,221
Wales[6]	**1,770**	**2**	**3**	**7,332**	**493**	**224**	**0**	**6**	**9,829**

Notes:

[1] Some regions report licensed and actual abstractions for financial rather than calendar years. As figures represent an average for the whole year expressed in daily amounts, differences between amounts reported for financial and calendar years are small.

[2] Reduction in agricultural abstraction due to deregulation of licences as of 1 April 2005, however, this mainly affected the number of licences and not the estimated abstraction.

[3] Includes abstraction of hydropower.

[4] Private abstractions for domestic use by individual households.

[5] Several licences changed from surface water to tidal which are excluded from this table.

[6] The figures for Wales are for the Environment Agency (EA)/Natural Resources Wales (NRW), the boundary of which does not coincide with the boundary of Wales. Some abstraction licences for administrative Wales are included in the Midlands region.

(*Source:* Office of National Statistics, 2012)

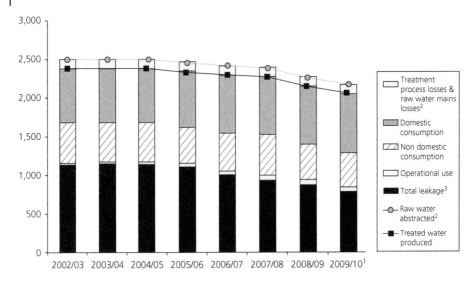

Figure 5.1 Public water supplies—water abstracted and supplied in Scotland: 2002/2003-2009/2010 in millions of litres per day. (*Source:* Scottish Government, 2010).

and Wales (Scottish Government, n.d.c). Abstraction from boreholes and springs reflects the geology, especially the distribution of aquifers in the Carboniferous sandstones and limestones, Permo-Triassic sandstones and Old Red Sandstone (Monkhouse and Richards, 1982), being of significance in the lowland regions. However, groundwater sources only accounted for around 5% for all Scotland (Scotland's Environment, n.d.).

5.1.3 Rising Demand?

Some time ago, in March 1994, a water resources development strategy was published for England and Wales (NRA, 1994g) that summarises the view of the former NRA on sustainable development of resources at the time. It followed several consultation and discussion documents and consultation with numerous other organisations. By region, the highest projections to 2021 were in the Anglian and South West regions, both at 27%, the lowest region was the North West where a drop was possible. The baseline date was 1991 reflecting a 30-year planning horizon, is usual for the water industry In Table 5.1, in 2007 estimated actual (public) water use was 16,381 Ml day^{-1}. The Water Resources Strategy (EA, 2009a). Since 2000, the highest public water supply abstraction in England and Wales was 17370 Ml/d in 2003 and lowest at 15799 Ml/d in 2009 (EA, 2015).

Since the first edition of this book, reflection suggests that quoting lists of statistics may look informative, but may equally end up as a reporting exercise for high-level number crunching. However to students of sustainability the presumptions in such 'scenario-building' are at least as informative. For example, the high, medium and low scenarios in the 1994 Water Resources Strategy are based in tangible presumptions. The high demand scenario was a linear extrapolation of rising demand in the 1980s, presuming growth in per capita consumption by a compound annual rate of 1% growth in metered and unmetered non-household consumption by a compound annual rate of

0.75% (all water supply companies), no increase in the proportion of domestic properties subject to metering above 1991 levels (all companies), and leakage levels per property remaining at 1991 levels. The medium and low scenarios presume reduced growth per capita, restricted growth in household consumption increases in metering in southern and central England and effective leakage controls. Trends for increasing demand for public supply have long had to be offset against an unpredictable future demand for water in the private sector (NRA, 1992 g).

Water Resources Strategies can employ scenarios based upon sustainable behaviour (by individuals with local governance that may encompass public ownership of key utilities), innovation (technology and knowledge led), uncontrolled demand (private company led with emergent economic disparities) and local resilience (adaptation of houses and communities through declining economic circumstances). The first two should lead to 'sustainability led governance', the latter to 'growth led governance'. Actually, little changes, for as long the water industry has had at its core the old tradition of expanding infrastructure (more reservoir capacity, more transfers, more boreholes) and conservation measures, generally centred on leakage management, but increasingly concerned with conservation, re-cycling and reduction of demand using domestic and industrial metering.

To answer the question at the top of this section:

> 'Our forecasts show that total demand for water is likely to continue to rise steadily over the next 10 years. By 2020, demand could be around five per cent (800 million litres of water per day) more than it is today' (EA, 2009a).

At the very least, this evokes the 'precautionary principle' for water resource planners.

5.2 Anthropogenic Influences and Climate, Problems and Solutions

5.2.1 Human Impact

Too little available water results in pressure on the resource, low flows and loss of aquatic habitats while too much can lead to flooding. Monitoring is central and the National River Flow Archive (NRFA) website regularly reports the data from recorded the Gauging Station Register, provides details of around 1500 UK gauging stations, and the Well Register provides information relating to around 160 index wells and boreholes (Centre for Hydrology and Ecology, n.d.a.). In addition to the official network of Meteorological Office stations measuring there are abundant other locations, generally private concerns many measuring rainfall alone. There exists for the EA and SEPA and through collaboration (including with the Centre for Hydrology and Ecology, British Geological Survey and the Meteorological Office), much information relating to hydrometric and climate data.

Since the 1970s, attention switched from flood research towards low flows and then back again. There have been dry spells in 1975-1976, and again in the early 1990s, while the concept of 'drought' actually defies accurate description on account of a large element of user expectation (Chapter 2). UK water undertakings have adopted the 1:50 driest year as defining a drought event but that is not universal. It is helpful to consider water shortage in terms of periods of 'low rainfall' leading to a 'hydrological drought'. The Met Office Rainfall and Evaporation Calculation System MORECS model provides a real-time assessment and predictions of evaporation, soil water deficit and effective rainfall for 40x40 km grid squares and can be adapted for different crops and topography, as well as being used in design for river flows (Met Office, n.d.a; Ward and Robinson, 2000, Ch. 4). Model construction is frequently updated.

The term 'drought' remained in EA usage, probably as convenient shorthand for planning purposes, but noting they are natural events and not restricted to any region. EA has drought plans in each region and water companies are obliged to produce drought plans (EA, 2011a; EA, 2012a). Indeed, policy formulation in dealing with 'drought situations' is high on the agenda because since 1989 there have been prolonged periods of reduced rainfall. There are recognised such imperatives as identifying the likely location and requirements of drought orders, working with abstractors to minimise drought impacts, operating its own schemes to minimise drought impact, advising government and promoting catchment-wide utility groups to optimise use of scarce resources.

Then, as noted in Chapter 2, flooding became important in policy terms, perhaps largely due to extreme flooding in 2000. The real problem for policy makers is that climate change appears to be producing more extremes.

Major anthropogenic activities affecting river systems may be summarised:

- Supra-catchment effects such as acid deposition and inter-basin transfers.
- Catchment land-use change such as afforestation, deforestation, urbanisation, agricultural development (arablisation or grassland establishment).
- Development of land drainage and flood protection schemes.
- Industrial, sewage treatment or transport-based developments.
- Instream impacts on water quality including organic, inorganic and thermal pollution and sediment loading.
- Instream impacts arising from abstraction, navigation, alteration of channel morphology or flow augmentation.
- Instream impacts arising from exploitation of native or introduction of alien species.
- Loss of wetland habitats arising from reduced surface flows, spring discharge or lowering of watertables due to abstraction of waters. Development of groundwater schemes that may compromise river flow or integrity of wetlands.
- Corridor engineering such as the removal of riparian vegetation, flow regulation.
- (through dams, channelisation, weirs, etc.), dredging and mining.
- Restoration or 're-wilding' of rivers, both channel morphology, floodplain and vegetation although this has a only small impact.

Arguably, the most fundamental means of river regulation has been the construction of surface reservoirs that control flooding, control intra-basin transfers or store water for purposes such as public supply, industrial use or irrigation. Other forms of regulation include the construction of flood defences, land and wetland drainage and losses to or augmentation from groundwater.

EA strategy sets out actions that will as far as good resource management is concerned, echo resources already under stress for basically these are comparable with the 1994 document (EA, 2010a):

- support housing and associated development where the environment can cope with the additional demands placed on it;

- allow a targeted approach where stress on water resources is greatest;
- ensure water is used efficiently in homes and buildings, and by industry and agriculture;
- provide greater incentives for water companies and individuals to manage demand;
- share existing water resources more effectively;
- further reduce leakage;
- ensure that reliable options for resource development are considered;
- allocate water resources more effectively in the future.

It is moot as to whether climate change is best viewed as an anthropocentric change, or an externally altered variable within a catchment, but for convenience it is convenient to deal with it under a different heading. But how did the EA respond?

5.2.2 Climate Change, the Water Industry and the Regulators

As far as water resources are concerned, the last 20 years or so have seen a revolution in attitudes to climate change in water resources (EA, 2010a). The 1994 'Water Resources Strategy' (NRA, 1994g) was based around policies of sustainable development, the precautionary principle and demand management (Chapter 3). Well into the late 1990s there was a seeming discernible lack of interest in or recognition of the potential for impact on Britain's water industry. While it is true that droughts were well known over decades (Chapter 2) and, especially the 1975/1976 example was well-described and its impact well-analysed and understood, the crucial link to scenarios of climate change was seldom made. May be the water industry saw nothing to worry about if infrastructure continued to be invested? Maybe it was because there were many who did not believe in climate change?' However, this comment implies nothing of the political overtones of present-day climate change deniers.

The EA has overall responsibility for water resources for England and Wales and has specific oversight for all of flooding, droughts, seal level rise and coastal erosion, for water resources. As a start, (EA, 2010a), Scotland and Wales have exploited up to 5% of actual abstraction as a proportion of effective rainfall, the North, Midlands and Southwest of England up to 10% and finally East Anglia and the Southeast jumps to between 20% and 45%. This range makes it among the most drought-prone areas in Europe, along with Cyprus, Spain and Italy.

Figure 5.2 shows there are significant pressures on the water environment, and while it is in East Anglia and the South East that are the most stressed areas (and these often correspond with stressed groundwater units) that these pressures are not just confined to the south east of England. A need for reform of the licencing system is manifest as the 'over-licenced' category that only has the potential for over abstraction.

While Figure 5.2 is a practical assessment, looking forward to times of presumed climate change presents some interesting ramifications, not only for water resources but for the well-being of Rivers. Figure 5.3 shows a scenario for changes in March river flow in 2050, Figure 5.4 shows September.

The modelled change shows there could be 10 to 15% increase (Black), 5 to 10% increase (Dark grey) or remain within the range 5% increase to 10% decrease (Light grey) compared with the base line calculated as a % of the 1961-1990 using the medium-high UKCIP02 scenario (UN, 2014).

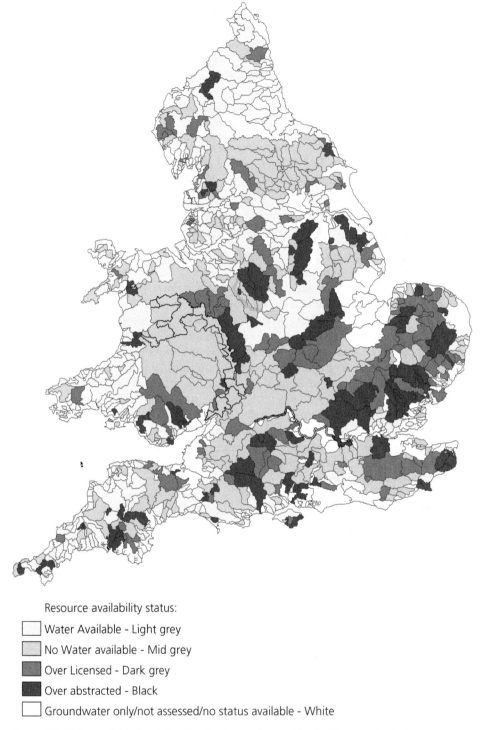

Resource availability status:

☐ Water Available - Light grey

▨ No Water available - Mid grey

▧ Over Licensed - Dark grey

■ Over abstracted - Black

☐ Groundwater only/not assessed/no status available - White

Figure 5.2 Water available for abstraction (surface water combined with groundwater). (*Source:* EA, 2010a).

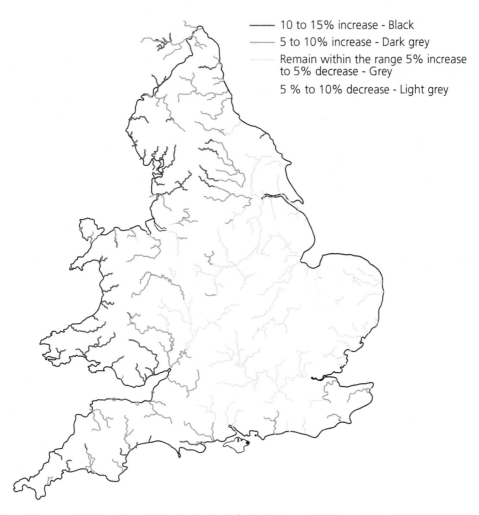

────── 10 to 15% increase - Black

────── 5 to 10% increase - Dark grey

────── Remain within the range 5% increase
to 5% decrease - Grey

5 % to 10% decrease - Light grey

Figure 5.3 A scenario for changes in March river flow by 2050. (*Source:* EA, 2009c).

5.3   Surface Waters, Reservoirs and River Regulation

Valleys or natural lakes may be dammed in order to increase the surface water storage capacity—geology permitting. Dam location relies upon low permeability or impermeable underlying strata, a supply of water of an appropriate quality for reservoir recharge and above all, suitable topography. Deeper valleys, such as the 'U' -shaped valleys of upland Britain, are preferable to the shallow valleys of the lowlands. Reduction of the surface to volume ratio in deeper valleys which are dammed reduces the opportunity for evaporation, hence increasing the storage efficiency of the water body. Small reservoirs concerned with supplying treated water to the consumer, and constructed to maintain an adequate pressure head, are termed 'service reservoirs', and are not given further consideration.

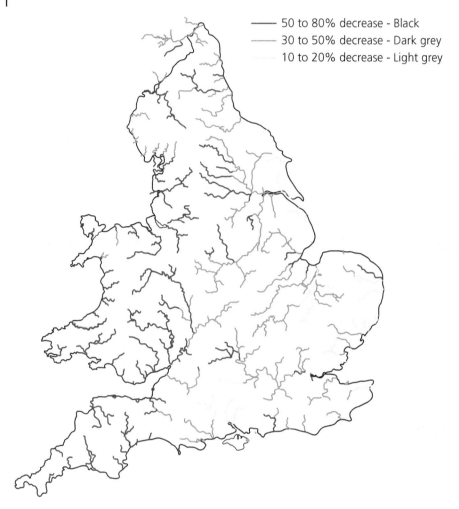

——— 50 to 80% decrease - Black
——— 30 to 50% decrease - Dark grey
——— 10 to 20% decrease - Light grey

Figure 5.4 A scenario for changes in September river flow by 2050. These model a 50 to 80% decrease (black), 30 to 50% decrease (dark grey) and 10 to 20% decrease (light grey). (*Source:* EA, 2009c).

A reservoir may:

- store water during a wet season or wet spell;
- regulate river flow enabling adequate flow during a dry season or dry spell;
- regulate river flow in order to control flooding, typically by controlling headwater discharges in the upper catchment;
- store water, smooth supply and create a head for hydroelectric power generation or other purposes;
- act as a settling pond for sediment loaded water.

Earlier reservoirs were constructed for bulk water supply, later ones tended to be more for the purpose of river regulation. Reservoirs vary in their operation according to purpose. For example, *pumped storage reservoirs* are favoured for lowland sites near river abstraction points (Kirby, 1984) and permit high flows in a river to be diverted and stored behind a dam where sediment can settle, bacteria are killed by light and the oxygen level may increase. Water quality thus improves while stored water awaits release for public supply. One example is Rutland Water in the Anglian Region of the EA and bunded bankside reservoirs are used in the London storage network to the same effect, mostly in the Stains and Windsor areas to the southwest of London.

River regulation reservoirs, on the other hand, store water from high flow and release it to support abstraction lower down the catchment. Reservoirs on the Severn, Dee and Tees are examples. For example, the Clywedog Reservoir in the Upper Severn catchment is drawn down in the winter to store floodwater which would otherwise produce problems lower down. It re-fills by 1 May to cover possible summer demand (Hall, 1989).

Dam construction in Britain may date from Roman times. However, worldwide Beaumont (1978) identifies three major periods of dam construction. Following a steady upward nineteenth-century trend, these are 1900 to 1914, the 'inter-war period' and a phenomenal burst of activity between 1945 and 1971 when 8180 major dams were constructed. The consequence is that around 15% of runoff in Europe is regulated by surface storage reservoirs.

Although during the twentieth century it is the United States which has been pre-eminent in construction, the pattern is consistent in Western Europe which also dominated the picture during the nineteenth century. Boon (1992) charts dam construction in Britain between 1801 and 1990 (Figure 5.5) and it is apparent not only

Figure 5.5 Progress of dam construction in Britain (1801–1990). (Redrawn from Boon, 1992, p. 14).

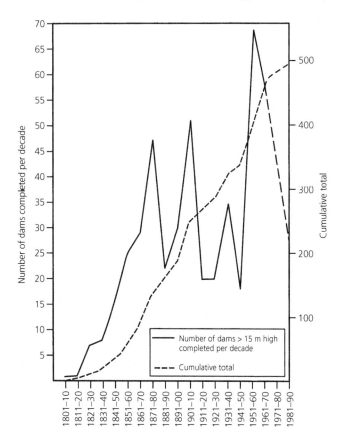

that there is a slacking off by the 1960s, but also there is an earlier peak, during the second half of the nineteenth century, the period of Victorian dam construction.

Over time the trend was to construct fewer, larger reservoirs; in Wales alone there are 14 reservoirs exceeding 1km^2 in area (Petts and Lewin, 1979). One such example is Lake Vyrnwy in north Wales on the Afon Vyrnwy, a tributary of the River Severn. In an area with a mean annual rainfall of 1905 mm, this reservoir had its dam constructed in 1888, covers some 453 ha and is capable of containing 57 000 Ml while supplying a reliable yield of 245 Ml day^{-1}. Another example is the Thirlmere Reservoir which supplies Manchester down an aqueduct 154 km long (Binnie, 1995).

Nottingham's supply includes water from three reservoirs in the Derwent Valley: the Howden commissioned in 1912, the Derwent in 1916 and the Ladybower in 1945 (Brown *et al.*, 1984). During the 1950s and 1960s, use of land for new reservoirs was unpopular, due to the taking of land from agriculture in the lowlands and loss of amenity in the uplands (Kinnersley, 1994, p. 99), although in time reservoirs may be seen as an amenity in themselves. Later projects include major undertakings such as the Kielder Reservoir (completed 1982) in Northumberland, and the smaller Wimbleball Reservoir in Somerset (completed 1978), which regulates the River Exe.

Criticism of dam construction in Britain is largely restricted to financial cost, land-take (and hence social cost) and impact on local ecology, especially loss of valuable terrestrial habitats. Benefits also include habitat gain and amenity as well as the reservoirs' intended purposes such as water supply, runoff regulation or hydroelectricity generation.

Impacts upon river flow merit further scrutiny. The minimum flow allowed out of a reservoir is termed 'compensation flow'. This concept may be seen in the context of (the conceptually problematic) *environmentally acceptable flow regime (EAFR)*, because of the complexity of river flows in space and time, and the complexity of the aquatic environment including fisheries as well as the needs of abstractors. In principle, minimum flows may be defined on a seasonal, rather than as a one-off, year-round basis; an example is discussed below in the context of the Darent. Flow augmentation is also achieved through effluent discharge. The context of developing Catchment Management Plans furthermore aims to balance all the demands upon a river (Chapter 9). Newson (1992a) lists the considerations necessary for compensation flow definition. These may be summarised as:

- the needs of existing licence holders to abstract water;
- requirements to dilute point-source effluents considering River Quality Objectives and public health;
- requirements (where appropriate) of power generation;
- requirements of adequate depth for navigation;
- riparian rights such as stock watering;
- needs of fisheries, including migration;
- ecological needs of plants and animals;
- amenity needs such as swimming and visual impact.

Flow beneath reservoirs would be regulated, and may be fixed, or caused to vary seasonally or maintained above a minimum level depending upon the design criteria, operation and catchment hydrology. Hydrographs downstream of a regulation point (such as a spillway) depend upon runoff characteristics down-catchment.

Figure 5.6 Impact on seasonal flows of a reservoir on the River Hodder, Forest of Bowland (after Petts and Lewin, 1979, p. 83).

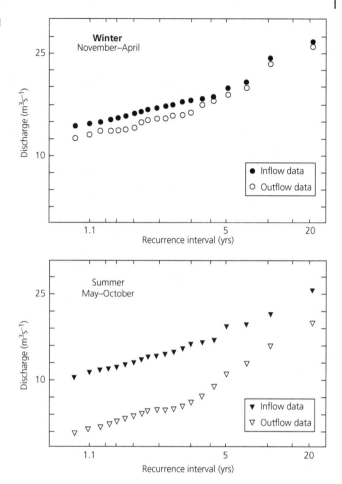

Many studies have examined the impact of reservoir construction on catchment, channel form, sediment load and wildlife. Figure 5.6 shows the impact on seasonal variations of inflow and outflow from reservoir impoundment on the River Hodder, Forest of Bowland. The Stocks Reservoir was constructed in 1933, and creates a 70% reduction in peak flows. This is a high reduction; by comparison a similar figure for Vyrnwy is 69%. Maximum regulation occurs during the summer months, and this may be shown by the large increase in recurrence interval for a given discharge in the summer inflow compared with outflow. This is a reflection of the reservoir filling during the summer and absorbing flood discharges compared with the winter when the reservoir is more often at spillweir capacity. The ratios of below-dam to pre-dam discharges for recurrence intervals at 1.5 ('bankful discharge'), 2.3 ('mean annual flood'), 5.0 and 10.0 years were 0.83, 0.86, 0.84 and 0.95, respectively. The increase in this ratio towards unity, and hence convergence of figures for larger (and rarer) flood events, has been observed in other catchments (Petts and Lewin, 1979), and is to be expected because of the greater ability of a reservoir to contain smaller (and more frequent) flood events.

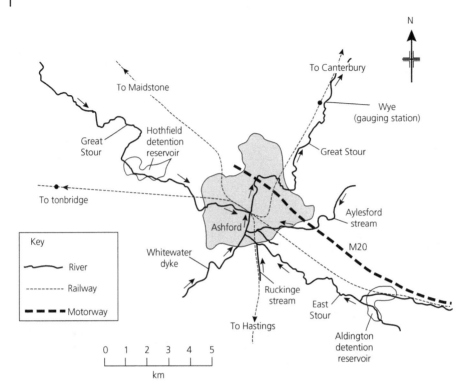

Figure 5.7 The location of the Ashford Flood Alleviation Schemes. (Redrawn from Gibbs and Lindley, 1990).

In lowland 'flashy' catchments (where runoff leads to rapid rises in the flood hydrograph), regulation of headwaters enables the amelioration of downstream flooding. For example, regulation of the headwaters of the Great and East Stour above Ashford, Kent, has improved flood protection measures where flood embankments and overflow relief channels in the town were considered inadequate. Protection is now against the one in 100-year return period flood (Gibbs and Lindley, 1990). Although there are additional, minor tributaries in this predominantly Weald Clay catchment, modelling predicted that 'detention reservoirs' behind retention dams on the main headwaters could be constructed in order to comfortably meet this requirement.

Figure 5.7 shows the location of the scheme. The retention dam is constructed economically using local Weald Clay, and with minimal visual intrusion. Their operation is entirely using a 'hydrobrake' controlling the outfall through the dam, which is a self-limiting vortex device. The lands behind these dams likely to be affected during impoundment are either conservation areas or are grazed. Account had to be taken of the need for compensation flow to a mill and a fish pass at Aldington, while flood protection and drainage had to be provided at Hothfield. The result has been a highly environmentally sensitive upstream flood alleviation scheme. However, it becomes debatable whether the improved options for floodplain development here, as elsewhere, are desirable, and some farmers now feel confident to plough to the edge of the river channel downstream of the reservoirs.

5.4   Aquifer Recharge and Climate Change

Commentators find a strong tendency to ignore groundwater in studies of climate change impact even at a global scale, although *predictions*, of groundwater recharge been forthcoming (Taylor *et al.*, 2012). Specifically in Britain it has so far proved difficult to detect resource balance changes that were clearly attributable to changes in climate, although the potential for recharge to be adversely affected has been recognised (Hiscock, 2005, Ch. 8; Herrera-Pantoja and Hiscock, 2008).

Because virtually all groundwater in the zone of active circulation beneath the unconfined portion is derived ultimately from precipitation, and because it is the unconfined aquifer which is pumped more often than not, groundwater reserves are susceptible to depletion under low rainfall. Recharge through the soil requires an actively draining profile, a condition seldom met during the spring and summer due to depletion of soil water by vegetation and crops. The maximum mean annual potential soil water deficit for 1961-1975 has been mapped for Britain (DoE, 1991, Ch. 3), and this shows values of between 150 and 250 mm over much of eastern England. Actual deficits beneath arable crops of the order of 100 to 160 mm are typical for East Anglia (Cook and Dent, 1990). Consequently, it is generally October, or even November, before a deficit manifest in August is eliminated and a soil will display free drainage beneath its profile. Opinions differ as to the efficacy of summer recharge in reaching groundwaters; recharge through preferential soil fissure flow in cracked dry soils and subsoils is plausible. However, as a rule of thumb, it is assumed that the soil water deficit accrued in the summer has to be overcome before soil drainage can occur.

The wettest June on record occurred in 2012 (Met Office, n.d.c), the total average UK rainfall was 145.3 mm, twice as much then long-term average for 1971-2000 average and wettest since 1910. Hence, the likelihood of eliminating soil water deficits across the UK would seem high. MORECS shows an unseasonally low deficit (Figure 5.8, black line, mid-2012). The very wet soil conditions meant that overall aquifer recharge (otherwise meagre in late spring and summer) was both substantial and sustained in 2012 and eliminated a cumulated groundwater deficit from previous years.

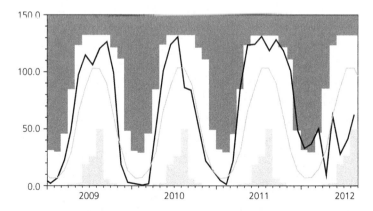

Figure 5.8 The black line shows variations in monthly soil moisture deficits (mm) for the English lowlands 2009-2012. The grey curve is the long-term average. The dark grey envelope above the curves and light grey envelope below the curves, envelopes indicate pre-2012 monthly maxima and minima respectively. Data source: MORECS (CEH, n.d.b).

For the three years previous to 2012, the elimination of the soil water deficit coupled with downward travel through the unsaturated zone is the more usual situation with aquifer recharge after October in southern and eastern England. This situation applies to unconfined or those only semi-confined aquifers, for example, where thick drift deposits such as boulder clay reduce infiltration. Confined aquifers present a more complicated situation, and long-term recharge will depend upon the balance between abstraction and horizontal flow from the recharge zone.

On average and over time, replenishment should exceed discharge and abstraction, a situation happily met in all but the driest years in eastern England. In some instances, continued abstraction from confined aquifers leads to long-term reduction in well hydrograph levels. An example in southeast Essex, where low transmissivities combined with over-abstraction from the chalk confined by London Clay has led to serious decline in well levels during the twentieth century, the so-called 'Braintree Depression' (IGS/Anglian Water Authority, 1981). This has remained a matter of concern, and hence investigation, into recent times (EA, 2013c).

Aquifer-wide and regional figures for recharge mask local variations which may be extremely critical in drier years, or where abstractions are high.

In order to illustrate year-to-year variations, in saturated zone water level a well hydrograph from the Wensum and Yare chalk groundwater unit for the Foulsham/Hindol Veston borehole in Norfolk is shown in Figure 5.9. The observation well is adjacent to the Wensum valley which contains highly permeable sands and gravels; although in its immediate vicinity there is semi-confining boulder clay over the chalk (IGS, 1976). Figure 5.9 shows well fluctuation over the period 1953 to 1993; units are metres above OD. Clearly visible are the effects of the drought of 1975/1976 and more so a period of low rainfall in 1988-1992 seen as falling levels.

Although seasonal patterns remain, the long-term decline in levels is apparent from the end of 1988, with recovery towards the annual mean occurring in the latter part of 1992. Low-flow responses were observed in flows of the River Wensum west of Norwich, which is to be regulated through well augmentation fields in its upper catchment. In Norfolk

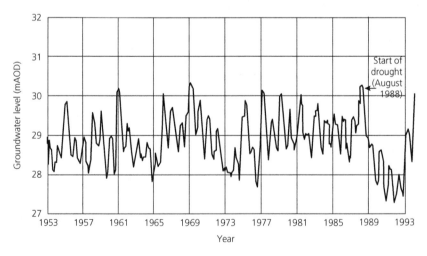

Figure 5.9 Well hydrographs from the Wensum and Yare chalk groundwater unit in Norfolk at Foulsham. (Redrawn from NRA, 1994c, by permission of the Environment Agency).

recharge was below 25% of the long-term average during this time. Long-term records indicate a 'drought' of the early 1900s as being of similar severity and duration in the area.

Over-abstraction remains an enduring problem. For example, Owen (1991) notes a problem in many valley heads where pumping groundwater diverts flow into a well, and away from supporting baseflow in perennial rivers, a situation highly dependent upon the local conditions such as vertical and horizontal flow conditions in the aquifer which may delay or enhance river discharge responses to pumping. Using the Colne catchment (Thames EA region) as an example, he notes how the upper 10 km of the reaches of its tributary, the Ver, became dry. These reaches were formally perennial or intermittent in their flow regime ('winterbournes'). This occurence may be directly related to increased abstraction from around 10 Ml day^{-1} in 1950 to near the licensed quantity of 48 Ml day^{-1} during the 1980s. Although groundwater concerns dominate low flow issues, other causes linked to surface water management also play a part.

The unfortunate combination, therefore, of over-abstraction and climatic change that reduces aquifer recharge is a comparable problem with that of river flow noted above. Essentially these respond the predicted increase in winter rainfall, but decrease in summer rainfall (Jenkins et al, 2009). Potential impact of climate change on groundwater by 2025 for the two most important English Aquifers. In all but one instance in Table 5.2 there is to be expected reduced recharge in groundwater.

In Scotland, a climate change plan has been produced by SEPA. This is very much concerned with energy and reducing carbon emission that are both issues closely related to climate change (SEPA, n.d.f). Average temperatures have risen more than 1°C since 1961 and average winter rainfall in some areas has increased by a massive 60%, but conversely summers can be as much as 45% drier. The seas surrounding Scotland have warmed by about 1°C in 20 years. As far as water quality and supply are concerned, both the expected periods of heavy and low rainfall can lead to deterioration of raw water quality and may impact on Scottish Water's ability to treat water, and there may be public health issues, and even water shortages, while flooding could lead to the deterioration of public and private drinking water sources. There is a need to develop means of linking climate change scenarios to water flows in both urban and rural catchments to assist future changes. The impact on natural and seminatural ecosystems will also need evaluation (Scottish Government, n.d.d).

Table 5.2 Potential impact of climate change on groundwater for England and Wales by 2025 (Medium scenarios).

Scenarios	Permo-triassic sandstone (West Midlands)	Chalk (Test-Itchen groundwater)
Long-term average winter recharge	9% reduction	3% reduction
Long-term average recharge	2% reduction	2% increase
Long-term average summer recharge	12% reduction	19% reduction
Middle range river flows (Q50) for rivers mainly fed by groundwater	13 to 21% reduction	6% reduction
Dry period river flows (Q95) for rivers mainly fed by groundwater	10 to 25% reduction	5 to 6% reduction

(*Source:* EA, 2009c).

5.5   Low Flows in Rivers and Possible Solutions

Natural periods of reduced rainfall that may be linked with climate change will inevitably lead to flow reduction, but modern concern regarding low flows frequently relates to over-abstraction. Heavily regulated rivers have relatively high 95 percentile flows because maintaining flows during dry spells is an objective of regulation. We might also expect to see regional effects of rainfall and groundwater support where these make appreciable differences between catchments. Selection of the 95 percentile flow in rivers may seem an arbitrary standard, but it is found to be suitable for ecological purposes (NRA, 1994e), something that remains to the present as the default 'hands off flow' or HOF (EA, 2013b, Chapter 8).

Chapter 2 showed how the condition of 'low flow' may be characterised by a range of statistical parameters. Reduced surface runoff occurs during dry spells, such as 1975-1976 (Ward and Robinson, 2000, Ch. 7). As a generality, this period of low rainfall is regarded as being largely a surface water drought, other dry periods, especially where low winter precipitation compromised aquifer recharge, experience pressure on groundwater resources. Such distinction has to be made with caution because for example in eastern England especially, periods of serious groundwater depression impact upon both river flow and wetlands.

As an alternative to process-based models of catchment response to climatic variables, statistical models are used. Historical river flow data can be reconstructed from known long-term averages of actual evaporation and from rainfall data. In this way, statistical catchment models of river flow can be reconstructed as far back as the mid-nineteenth century (Jones *et al.*, 1984a). This approach has to be taken because reliable gauged data only date back, in most cases, a few decades. Before the mid-nineteenth century, reliance upon historical records of flood and drought has to be made, unless river flow is reconstructed from extrapolations of other palaeoclimatic data such as tree ring analysis (Jones *et al.*, 1984b). Although assumptions are inevitably made with respect to climatic stability in the long term, and indeed catchment variables such as land use and cover which may affect catchment response, such reconstructions give valuable insights into probable long-term river flows.

Scenarios of climatic change impact on river flows has been a topic of research that informs policy consideration for some time. Currently accepted scenarios have been summarised in Figure 5.4 for the year 2050 paint a picture that caused concern (EA, 2009c; UN, 2014). Before the Madrid conference of the Inter-Governmental Panel on Climate Change (IPCC) in 1995, there was no suggestion that any climatic trends, drier or otherwise, had emerged from the ordinary climatic variability, also termed 'noise'. Dramatic change, however, would be of great concern due to the delicate seasonal balancing of resources in some areas with changing precipitation patterns; curiously, recent years have seen an increase in frequency of dry winters which aggravate the low flow problem.

This problem was first described in general terms followed by the NRA 'Water Resources Strategy' (NRA, 1993b; 1994g) which considered the problem in detail. In 1990, 40 priority locations (i.e., specific stream lengths) were identified as experiencing low flows, of which 20 were identified as needing the most urgent attention. The number has subsequently been reduced. They include two streams in the Colne catchment, the Misbourne and the Ver, in the priority category. In the regulator's view

(NRA, 1993c), low flows are due to 'excessive authorised abstractions rather than due to drought, although of course, drought will exacerbate such problems'. The blame is firmly laid at the feet of the Water Resources Act 1963, when Licences of Right were issued in perpetuity in 1965 (NRA, 1994g).

Once it was believed there were ready solutions. Unfortunately the problems remain not only with us, but should be seen in terms of a range of other problems such as pollution, general ecology and issues around channel morphology. In short, it is part of a suite of issues supposed to be addressed in the WFD. While certain rivers have improved overall and the EA lists the Wandle, Thames, Stour (Worcestershire), Taff, Darent, Dee, Nar, Stour (Dorset) and Mersey Basin (NCE, 2011). Specifically over-abstraction causes problems, for example the upper Kennet, upper Lee, Mimram, Beane and Itchen are included among those suffering from abstraction problems (WWF, 2010); there remains particular concern for chalk streams.

Certain instances of rivers suffering from over-abstraction have been addressed in the early part of the present century. Solutions proffered typically revolve around changing the abstraction regime and localised measures such as lining river beds where water leaks in to adjacent aquifers. Legal and regulatory measures through revoking *abstraction licences* are not straightforward, and may involve costs. In Chapter 3 it was noted that the Water Act of 2003 enables the alteration of licences to abstract.

5.6 Impact of Over-Abstraction on Wetlands

Although the environmental effects of over-abstraction are most obviously seen in terms of reduced river flow, the loss of wetland habitat causes major—and increasing concern. To the impact of agricultural and other forms of land reclamation must be added the threat posed to (surviving) wetland ecosystems by drought and over-abstraction. It is not only the biodiversity, the potential of wetlands to improve water quality, the possibility of water retention but also the ability of active wetlands to sequester carbon. The half-century 1934 to 1984 saw the loss of 90% of surviving East Anglian Fenlands, leaving around $10\,km^2$ unaffected (POST, 1993), and these are often threatened by water shortage. Freshwater wetlands may be taken to include fen (base rich peatlands), bog (acid peatlands) riparian wetlands (e.g., riverside valley fens) and marshes.

These support a range of habitat types and vegetation, including open water habitats, *Sphagnum* bogs, wet heath, wet grasslands, reedbeds and 'carr'. It is water chemistry (especially pH, nutrient status and calcium content), together with the hydrological regime (e.g., depth and frequency of flooding, and residence time of water in the system), which largely conditions the species composition. Specific examples are discussed in Chapter 8, with the main areas of concern outlined in Table 5.3.

The Joint Nature Conservation Committee (JNCC) finds lowland wetlands under threat from Drainage, water abstraction and water pollution, peat digging, air pollution, inappropriate site management, and fragmentation. Overall, Britain still has some of the finest wetland sites in Europe, and increasingly these are being protected under the Ramsar Convention which places an international obligation to protect and conserve them. On Redgrave and Lopham Fens, Suffolk and Wicken Fen, restoration is proving successful, as elsewhere in the peat Fenland region (Chapter 8). Others continue to experience problems, and Ramsar can conclude that many wetlands across the UK

Table 5.3 The effects of water shortage on lowland wetlands.

Effects	Site examples
Serious drying out of wet heath	The Flashes, Frensham, Surrey
Drying out of calcareous heath	Cock Marsh, Maidenhead, Berks
Threat of salt-water intrusion into drying fen	Horsey Mere, Norfolk
Drying out of dune slacks	Formby, Lancashire
Loss of plant and animal species diversity through lowering of groundwater through abstraction	Redgrave and Lopham Fens, Suffolk
Lowering watertable in fen and reduced flow in ditches, both required to support habitats	Wicken Fen, Cambs

(*Source:* POST, 1993).

remain under threat despite being designated for their importance. For example, coastal wetlands face inundation resulting from sea level rise and the North Norfolk Coast Ramsar site has important areas of saltmarsh, reedbed and freshwater mere which are breeding areas for bittern, avocet and marsh harrier, as well as wintering grounds for waders and wildfowl. Grazing marshes are being damaged by poor water management, for example in the Pevensey Levels (East Sussex), low water levels have resulted in the habitat drying out, preventing wading birds from feeding their young (Edie Net, 1999).

5.7   Impact of Land Drainage on Hydrology

Interaction between land drainage schemes (such as installing underdrainage) and any associated channelisation are relevant because, at the field scale, underdrainage may commence enhanced progress of water to the sea. On the way they will also affect downstream flood behaviour; specifically any change in land use that increases runoff has the potential to adversely affect flood behaviour. It is significant that the speeding up of catchment discharge, a process change in rural areas which have undergone engineering 'improvements' in pursuit of flood prevention or resulting from land drainage, is even more marked in urban areas (Section 5.10). This section is concerned with rural land drainage.

In effect, whole regions may experience drainage, largely to improve options for agriculture. Classic examples include the Fens (both silt and peat areas, the Somerset Levels and Moors (Cook, 1994). In 2013-2014 this area was the recipient of serious flooding perhaps in part aggravated by up-catchment arable operations (see introduction) and on Romney and Walland Marshes. Concerted land drainage may have ancient origins but accelerated since the Second World War, enacting dramatic landscape, hydrological and agricultural transformations (Cook, 2010a). Secondary treatments at the field scale involve subsoiling and the creation of unlined drains or 'moles', which facilitate the movement of water away from the upper horizons of the soil towards the main underdrains. The heavier the soils (and hence the lower their permeability), the more effective are these treatments.

Land drainage, catchment-scale drainage and flood management, therefore, tend to be rural issues whereas supply and sewage effluent problems are generally urban. However, land drainage and flood alleviation directly affect fewer people than do, say, issues of water supply or sewage disposal because they occur in rural catchments. The objectives of agricultural drainage are to improve crop yield, typically by between 10 and 15% (Rose and Armstrong, 1992). This is not an enormous benefit, considering the volume of past EU overproduction, and the losses of environmental goods not taken into the calculation (Purseglove, 1988, p. 232). Indeed, up to 50% of the cost of underdrainage between 1940 and 1986 was met by government grant aid (Green, 1979; Morris, 1989), after which it declined. Economics such as these were always going to be brought into question.

Rose and Armstrong (1992) report that field underdrainage lowers the watertable, but does not greatly affect the water balance. The route that water takes from the field surface to the outfall ditch or river is affected; for example, in clay soils underdrainage diverts runoff from the surface to the subsoil drains, attenuating and delaying its progress.

Figure 5.10 illustrates the mechanism of underdrainage at the field scale. On the undrained plots, during the field capacity period, the undrained land remains saturated with high watertables preventing infiltration. On drained plots, watertables are lower so that rainfall infiltrates. The result of underdrainage is therefore increased subsurface flow (the objective) causing the frequently observed reduction in surface runoff and broader, flatter hydrographs.

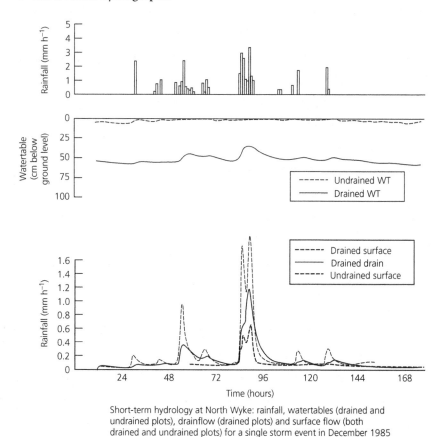

Short-term hydrology at North Wyke: rainfall, watertables (drained and undrained plots), drainflow (drained plots) and surface flow (both drained and undrained plots) for a single storm event in December 1985

Figure 5.10 The impact of underdrainage. (Redrawn from Rose and Armstrong, 1992).

At the catchment or regional scale, changes in mean annual flood frequency have to be considered. In regional drainage plans, and in small catchments, as the proportion of under-drained land rises, the hydrograph peaks may also rise again (Green, 1979). This could be due to a switch from overwhelmingly surface flow to dominant subsoil flow instead of a combination of the two, thus 'desynchronising' flood behaviour. Hence the effects of land drainage on river peaks and discharge have to be related to variables such as size of drainage scheme, location in catchment and size of catchment (Hill, 1976)—no hard and fast rules control catchment behaviour and impact of land drainage at any scale.

Regionally, drainage improvements to the Somerset Levels between the 1930s and 1970s lengthened the grazing season by four months from between May and September to between March and November, otherwise flooding remains (WWA, 1979). With watertable restoration, agri-environmental schemes provide for landscape conservation and water level management plans (Chapter 8), what was once a drive to control flooding and increase agricultural production is now integrated into conservation and landscape considerations.

The extent of regional drainage and associated underdrainage in eastern lowlands, as in the Romney Marsh area, could be phenomenal. Between 1940 and 1972 underdrainage affected over 15 percent of the total area of Essex, and this represents over 30% of that county's arable land. During the 1970s, several parts of eastern England had one or another form of underdrainage installation and improvement running at over 2% of the total area per annum (Green, 1979). Since the turn of the century, however, there have been controversial policies put in place to not only halt, but also reverse this trend of increasing drainage efficiency. Restoration of large areas of the Fens and the use of wetlands and washlands in 'Making Space for Water' has been officially sanctioned as government policy, including the provision of funding (Defra, 2005). Ironically, it now sounds controversial in the light of events on the example on the Somerset Levels (Introduction). ·

5.8 Hydrological Effects of Urbanisation

Britain will continue to urbanise. In affected areas flood issues are at the forefront, not only because of the changes to the surface cover (scientific and engineering criteria) but also because of the imperative to safeguard urban dwellers and their property. The multifaceted nature of flooding in urban areas has been discussed at length by several authors. For example, Parker and Penning-Rowsell (1980, Ch. 6) list its effects. These are essentially divided into two kinds: *intangible* (anxiety resulting from flood events real or anticipated, waterborne disease, disruption of communications), which are important but to which no accurate monetary value may be ascribed, and *tangible*. In this case, direct contact with floodwater and sediment or other waterborne matter causes quantifiable (i.e. indemnifiable) losses or disruption to economic life (e.g., the effects upon roads, electronic communication and utilities). This is well-established but however urban flood impact is described, serious flooding across Britain in recent years has re-focussed attention on urban flooding.

In Chapters 2 and 3, the development of Sustainable Urban Drainage Systems ('SUDS') as options for flood mitigation was alluded to. A subset of suites of options that may be termed 'Water Sensitive Urban Designs 'delivers greater harmony between water, the environment and communities' (Ciria, 2016). SuDS measures include drainage systems that mimic the natural conditions (including permeable surfaces, filter strips, swales, wetlands and detention areas). While design and implementation continue apace, the economic cost of clean-ups or of investment to mitigate problems attributed to climate change are a matter of concern. In 2007, urban flooding due to drainage systems being overwhelmed by rainfall is estimated to cost £270 million a year in England and Wales; 80,000 homes are at risk (POST, 2007).

In the urban situation, changes are not enacted for reasons of catchment water yield or flood behaviour, these are *by-products* of land cover changes. Urbanisation represents maximum human interference at the very point of the hydrological cycle where water (in its three phases) may be *exchanged* between land and atmosphere. In terms of magnitude of interference, comparable changes are associated with afforestation which reduces catchment water yield, or conversely deforestation which may increase the flood hazard and attendant problems such as soil erosion (Chapter 8).

In summary, urban surfaces change the following:

Interception. Interception loss may be close to 100% where evaporation from building surfaces or storage precludes drainage. In most areas the effect of vegetation is dramatically reduced.

Infiltration is drastically reduced where surfaces are roofed, concreted or tarred and surface runoff encouraged in its place. An exception is for 'soakaways'.

Runoff is channelled into gutters, drains, channels and sewers in a manner designed to speed water away from areas where it may be a problem, be it at a domestic scale or larger scale, and this may cause flooding to occur. Hydrograph responses are therefore different.

Surface depression storage may be reduced compared with most rural land uses.

Catchment boundaries may also change by the installation of engineered features in lowland urban areas. These temporally store or speed the passage of water out of an area through storm drains, culverts and flood relief channels.

Water quality issues are inevitably different from those in rural catchments. They reflect factors such as industrial activity, sewage management, and sediment and pollutant yield in the catchment. Where infiltration occurs urban waters can have profound effects on groundwater quality.

Figure 5.11 shows the effects of three stages of urbanisation in the Canon's Brook by Harlow New Town (after Hollis, 1974). Clearly the discharge peaks of the unit hydrographs are becoming greater and earlier in time as urbanisation progresses. However, detailed changes are often more subtle than this; smaller, frequent floods may increase during urbanisation, while extreme events are barely increased at all. This is because large events are generated by high-magnitude storms over fully saturated catchments (Richards, 1982, p. 260), while smaller events (in urban situations) will always generate more runoff than a comparable situation in rural areas which will not necessarily be over saturated soils. Channel cross-sections will seek to accommodate the increased urban flood peaks; width and depth may also both increase.

Packman (1979) reviews the problem of modelling changes in flood frequency and magnitude in urbanising catchments, finding that there is a net increase in runoff, reduction in lag-time to hydrograph peak, and a flattening of the 'flood frequency growth curve'. This latter property suggests that increased urbanisation will increase the magnitude of a flood (measured in terms of peak discharge value only) for a specified return period.

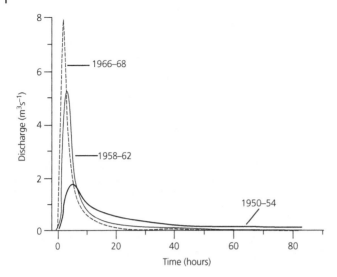

Figure 5.11 The effects of three stages of urbanisation in the Canon's Brook by Harlow New Town. (Redrawn from Hollis, 1974).

It is not always possible to obtain detailed hydrological data on catchments before and after urbanisation, hence the need for statistical and processed-based modelling procedures. Urbanised areas are prone to modify the local climate. They present extremely rough surfaces to air conveyed over them and vertical movement is further aided by convection due to the 'heat island' effect resulting from the clustering of buildings. This combined effect of forced and thermal convection may be associated with large interception losses (from impervious urban surfaces) and increased condensation around nucleii (present in increased numbers due to air pollution). The net effect is to increase precipitation over urban areas compared with the surrounding rural regions. Such an effect has been proposed for London (Atkinson, 1979), and, if significant, would be a further complicating factor in urban hydrology.

It is furthermore important to differentiate between on-site changes to waters and those downstream, that is, with an impact remote from the activity supposed to have been the cause (Chandler *et al.*, 1976). Most notable is the probable negative effects of waste disposal on groundwater quality, and the positive effects, both on-site and downstream, of land cover change and quickflow.

Any catchment changes which seek to speed up runoff potentially add to the 'flashiness' downstream. Parker and Penning-Rowsell (1980, Ch. 6) cite one example where 30 new houses at Llandudno Junction were flooded to a depth of 1 m during October 1976 (ironically immediately after the mid-1970s drought) due to installation of an undersized culvert. The urban situation is, for better or worse, controlled due to engineering knowledge and control through the planning process with the EA as a statutory consultee. In rural areas, the situation is not so tightly controlled.

Since then, there has been much interest in accommodating scenarios of climate change and assessments of likely impacts on flooding; however most responsive examples (steep-sided, small or urban catchments) would be expected to be most vulnerable to such changes (Defra and EA, 2004).

5.9   Flood Control and River Engineering

Floods occur when runoff exceeds drainage capacity, for example where the rate of precipitation exceeds the infiltration capacity of the soil. The consequence may be that channel or storm sewer capacity is exceeded (Ward and Robinson, 2000, Ch 6).

5.9.1   Engineered options

Historically there has been a strong effort to improve the efficiency of surface channels, and inevitably this has simply involved making channels discharge water more rapidly to the sea, or at least to where it becomes someone else's problem. The means employed are the capital-intensive and (often) environmentally insensitive 'hard engineering' solutions such as deepening, widening, straightening, re-profiling and clearing surface channels, building embankments, retention dams and so on. Process modelling, however, may help by enabling the de-synchronisation of flood behaviour so as to reduce catchment peak discharge.

Rivers have been managed for centuries but it is largely since the latter part of the nineteenth century that interference on a scale commensurate with mechanisation emerges. Furthermore, river engineering evolved through trial and error in a fashion directly comparable with reservoir construction (cf. Binnie, 1995) and in recent decades the experience of the Victorian (and earlier) engineers have been supplemented by advances in fluid mechanics, soil mechanics and materials science. A certain confidence emerged, rapidly followed by the triumph of 'hard engineering' solutions.

A polarity of views emerges that differentiates between the *river* (with all its cultural, geographical, botanical and wildlife connotations) and a *drain* (merely a way of ridding a catchment of excess water). Replacing a natural channel with a smoother engineered section increases 'efficiency' in removing excess water which might otherwise cause flooding. In doing so, by modifying or removing sediment, meander and vegetation features, aesthetic and ecological features are sacrificed in the interest of channel performance. This may, however, be reversed in order artificially to restore natural channel and floodplain features where flooding can be accommodated.

A summary of hard (also called 'traditional') engineering methods has been produced by Hey (1994), and these may be contrasted with 'environmental options'. These are recounted below with an outline of possible impacts upon channels:

- *Resectioning* is the dredging and/or widening of main channels to increase in-bank discharge capacity. In urban areas, rectangular-section channels are constructed in order to minimise space. Effects on upland rivers can lead to headcut erosion on alluvial reaches and deposition in the engineered reach as the river attempts to re-establish its width, depth, slope and erosion pattern downstream. Harder engineering solutions involving bank revetments and check weirs in dredged sections and bedload sediment traps further upstream are attempts to control the transport regime of the river and maintain the design channel. Periodic dredging may be all that is required in lowland areas.
- *Realignment,* often associated with resectioning has the effect of straightening the channel and increasing the gradient. Effects in upland rivers are similar to those of resectioning; additionally the river will attempt to meander in the engineered reach.

Check weirs and head sills in the bed can maintain the gradient and prevent headcut erosion. In lowlands, bank revetments can help prevent scour, but the problems are not so severe.

- *Flood bank constructions*, levees or embankments, are often constructed parallel to the river. The closer to the channel, the higher they have to be in order to accommodate the same discharge. Effects in uplands can cause erosion of the engineered reach, and deposition further down. There are few comparable problems in lowland rivers.

Hard engineering solutions, generally pilloried for ecological destruction such as loss of habitat and isolating the channel from the hydrological environment of its floodplain, may be criticised ultimately on the economic and practical sustainability of the solutions offered in many areas where near-total flood control is required. Hey (1994) recounts how reaches of rivers have become destabilised following the implementation of channel straightening. He quotes the case of the lower Mississippi, where serious headward erosion within main channels and tributaries caused serious sedimentation (and hence flooding) problems in the modified reaches. Destabilising the sediment transport capability and regime of a river leads to channel instability on account of the introduction of new initialising conditions along the river's long-profile. Elsewhere, such as in eastern England, low bedload transport rates may not present a frequent (hence costly) need for dredging, but a variety of 'river training' schemes still destroy instream habitats. Furthermore, river regulation may reduce peak flows and hence destabilise the ability of the channel to transport coarser sediment. Reservoirs often cut the supply of finer sediment leading to a coarsening of the bedload, sometimes of aggradation and even loss of natural floodplain functions below the dam. Given time, the spaces between the coarse bedload material may fill with fines, rendering it impermeable and unsuitable for salmon or trout spawning.

5.9.2 Environmental Options

These are characterised by considerations of the broader environment, including conservation implications, channel stability sustainability of the solution chosen, and economic constraints (Hey, 1994). There are four kinds of approach which may be summarised as follows:

- *Relief channels* may be constructed to divert flow (above a given river stage) away from the main channel. Effects potentially include a reduced carrying capacity for the main channel with deposition of coarser material. Instability is possible at both the intake point and the re-entry point of the relief channel. Because lowland rivers have much material transported in suspension, there are fewer problems here.
- *Partial dredging* of a limited section of the river, or even alternately dredging each side, increases cross-sectional areas, while leaving the unaffected area as a source of flora and fauna for recolonisation. Effects in upland rivers include a rapid re-filling of the dredged portion with fresh sediment; there are few problems with periodic (5- to 10-year) dredging of lowland rivers.
- *Distant flood banks* (where space is not limiting) are preferable to high, narrow channelised banks, and may be located at the edge of the meander belt. Provided that there is sufficient space, this is the best flood alleviation option.
- *Two-stage channels* are created by excavating the upper section of the floodplain adjacent to the river; higher flows are contained within the berm while the low

flow channel is preserved. It is best suited to lowland situations where careful design is required so that the alluvial watertable is not too close to the surface to prevent grazing.

Natural recovery of engineered channels will depend upon available energy ('stream power') and sediment supply, and takes place (unassisted) over decades. In terms of reinstating meanders, a period of 125 years was quoted for a gravel-bed river in Brookes (1992). Meander adjustment takes place during peak flows which, like the ability to transport sediment, are reduced in frequency by river regulation. Reinstatement of the river environment (including urban, hard engineered channels) involves artificial reintroduction of water-edge plants, re-engineering channels by deliberately scooping out pools, depositing sediment for riffle-bars and even re-cutting meanders (Purseglove, 1988, Ch. 7). Making Space for Water (Defra, 2005) as has been noted integrates ideas of wetland creation and rural land use that is tolerant of flooding, reviewing SUDS and seeking cost-effective solutions, including 'soft options' in coastal erosion and flood defences. Nonetheless, the tardiness of adopting certain measures has subsequently been open to question (Environment UK, 2012).

5.9.3 Scottish Rivers: A Case Study of Regulation

There is an erroneous and common perception that rural Scotland contains largely untamed wilderness, however Scottish rivers are heavily regulated (Gilvear, 1994) through major dams, diversion dams (allowing inter-catchment transfers) and around hundreds of km of tunnels, pipes and aqueducts all modifying the catchment hydrologies, predominantly in upland areas. The main cause is hydroelectric schemes. Regulation for water supply is by means of supply reservoirs, and by river abstraction. It is true that the upper catchments are generally unregulated, and that the high rainfall combined with impermeable geology leads to inherently flashy catchments. These tend to be small in area, and with summer flows under 15% of the mean flow and floods exceeding 300 times the dry weather flow. Such effects are naturally damped by lochs and peatbogs within catchments.

However, once rivers reach agricultural and urban areas there is a constant problem of flooding. Flood alleviation in Scotland is accomplished primarily by flood warning schemes, channelisation and flood embankments (Gilvear, 1994); many of these were constructed in the late eighteenth and nineteenth centuries without thought for downstream effects. Such river training will inevitably affect the frequency of flood-plain inundation. Subsequently Gilvear *et al.* (2002) found grounds for hope through improvements in many aspects of the natural heritage status of Scottish rivers (including salmonid fisheries). However, the degradation of many river reaches has also occurred through increased water abstraction, diffuse pollution, unsympathetic river engineering and the overgrazing of riverbanks.

A further complication is that of climatic change and climate trends in Scotland show that since 1961, heavy rainfall events have increased significantly in winter, particularly in the northern and western regions where winter rainfall has increased by almost 60%. There is concern about the damaging floods in recent years: Perth 1993, Strathclyde 1994, Edinburgh 2000, Elgin 1997 and 2002, Glasgow 2002, and Hawick 2005 (Scottish Government, 2008). While temperature trends are predicted to increase in most regions and most seasons, rainfall amount and intensity and days of heavy rain may increase

(SNIFFER, 2008). Smith and Bennett (1994) note a trend in increase in the mean annual flow of as much as 40% in the Rivers Dee, Findhorn, Tay, Clyde, Nith and Teviot in the years 1970 to 1989. This correlated with a rise of mean annual rainfall for Scotland from 1270 mm in the 1970s to 1554 mm in the 1980s. The trend is especially important during winters and springs which are becoming wetter. Predictions of increased flood risk have been noted for some time (Scottish Executive, 2002). Gilvear *et al.* (1994) describe a high number of flood defence breaches on the Rivers Tay and Earn during a 100-year return period flood in January 1993. Although in part the breached portions could be attributed to location (e.g., the outside of river bends, overlying former river courses, or embankments parallel to floodplain flow such as those along tributaries), there were others attributed to poor or inadequate construction.

Conventional, and costly, 'hard engineering' solutions may be feasible in urban areas. If increased discharge or even the cost of maintaining existing defence becomes prohibitive, then solutions include the construction of spillways in areas of agricultural land to permit localised flooding prior to a major overtopping. This option requires 'floodplain zoning' in terms of land-use planning, and might be a natural outcome of agri-environmental planning, including permanent set-aside and the option for wetland habitat improvement.

5.10 Summary

The degree of hydrological interference through the agencies of vegetation change, artificial drainage, flood defence, groundwater and surface abstraction and impoundment is considerable, yet still there are problems of shortage and excess of water in Britain. Modern strategies of water management involve balancing resources to cover for periods of shortage (including 'low flows') and protecting against flooding. Land drainage and urbanisation have been achieved at considerable loss to the natural environment, be it arable intensification or urbanisation. Into this complexity comes the confusion of climate change, now central to all processes of water resource management including surface and groundwater abstraction, low flow prediction and flood estimation. Meanwhile, the costs of hard or 'traditional' engineering solutions to flood protection and channel engineering are increasingly considered unacceptable on various counts, notably aesthetic, ecological and economic; their sustainability is in serious question, while dredging of main rivers may be making some kind of comeback. The focus of modern channel and floodplain engineering is to strike a balance which generally involves *utilising the natural floodplain and channel processes.*

Choice of solutions will depend ultimately upon local conditions, environmental and human requirements, and funds. Fortunately, hydrological modelling has emerged as a powerful tool in decision making and has proven its worth in both balancing scarce resources (especially conjunctive use of surface and groundwaters to maintain abstraction and sustain river flows) and in flood characterisation and alleviation studies during a period of consensus around climate change, for problems of excess, like shortage of water, cannot be separated from that of supply for societal use. How might we address future management? WFD has emerged as a legal and policy driver and many branches of Government have issued documents attempting to tackle many of the mounting problems. Like everything else it will be a delicate balancing act of economics, ecology, politics and the elusive 'sustainability'. There remain no easy solutions.

6

Sustaining Bulk Supply

Possible Solutions

> *Mr Bidder – President [of the Institution of Civil Engineers] observed, that although the Paper treated of an unimportant river in the neighbourhood of London [the Wandle], it opened out questions of great social and scientific interest; the water supply, water rights, the pollution of rivers and the remedies that ought to be applied, consistently with the improvements now taking place in social science, and with the drainage of the towns and the villages in the vicinity.*

> *Discussion following presentation of Frederick Braithwaite's 1861 paper On the Rise and Fall of the River Wandle; its Springs, Tributaries and Pollution. Minutes and Proceedings of the Institution of Civil Engineers.*

6.1 Introduction: Supply or Demand Management?

The above quotation, displaying great concern for the state of the river Wandle (today in south London, but in Braithwaite's day conveying a certain rural ambience) is both interesting and depressing; written in mid-Victorian times it could be applicable more than 150 years after George Parker Bidder's presidency of the Institution of Civil Engineers (Cook, 2015). This chapter looks at policy for the future use of water resources in England and Wales. Scotland may have local problems (the over-abstraction for hydroelectric schemes is an example) but generally there is no need for development to meet bulk supply.

In England and Wales, about 10% of freshwater resources are abstracted (this figure excludes abstraction to support power production that is often returned directly to the environment). Water resources are considered to be 'under stress' or over stretched if this figure is more than 20%, so that when viewed regionally, Southeast and Eastern England can be classified as an area 'under stress from water abstraction' with more than 22% of freshwater resources abstracted. Water resources are under greater stress only in drier countries of southern Europe (EA, 2008). More water is available in the north and west and this begs questions for regional water resource management and inter-river basin transfer. Figure 6.1 shows the water transfer schemes for England and Wales (SW refers to surface water abstraction, GW to groundwater abstraction). In reality, west to east transfers are well developed when compared to north to south.

Solutions for regional water balance problems may be characterised as either resource or demand management. Water undertakings will combine both approaches to a greater

The Protection and Conservation of Water Resources, Second Edition. Hadrian F. Cook.
© 2017 John Wiley & Sons Ltd. Published 2017 by John Wiley & Sons Ltd.

Legend

- ■ Reservoir
- ······ Regulated reaches of river
- ◯ Reservoir group boundary
- ⟶ Water transfer

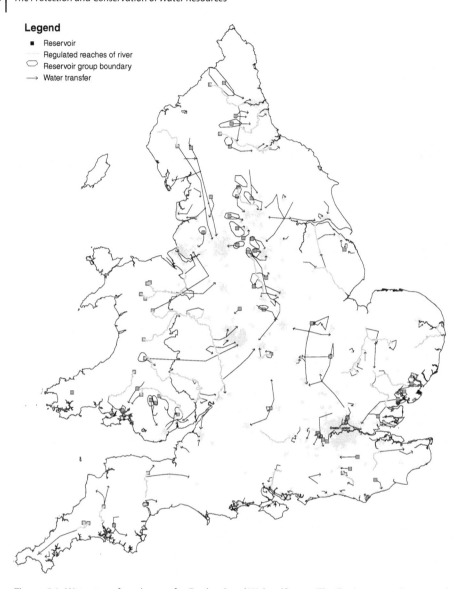

Figure 6.1 Water transfer schemes for England and Wales. (*Source:* The Environment Agency, adapted from a map by Thomas Mackay and NRA information).

or lesser extent. *Resource management* (also termed *supply management)* increasingly depends upon two main options: better use of existing resources (with hydrological modelling playing a major role in balancing resources), and bulk transfers of raw water between catchments (Figure 6.1). Privatisation threw into focus the poor state of much of the infrastructure of the water industry which has led to both appraisal and research into demand-side management options. *Demand management* has become a tool for water use managers in comparatively recent times; the main options are leakage control, metering, effluent recycling, restrictions in times of low rainfall, effecting behavioural change, and pricing. *Water efficiency* then differs from water conservation in emphasis, for it focuses on reducing water waste rather than restricting use (Waterwise, 2016).

Since 1 February 1996 there have been new duties on water undertakers to 'promote the efficient use of water by its customers' under a new Section 93(a) of the Water Industry Act 1991 (Duxbury and Morton, 1994, p. 182). Actually, little has changed, for the Chartered institution of Water and Environmental Management (CIWEM) can summarise the position:

> 'Commitment to the twin-track approach that assesses demand management and new resource options on an equal long run economic basis, taking full cost and benefit account of environmental and social effects' (CIWEM, 2012a).

This is not sitting on any fence, but is a realisation of the complexity of water governance, sometimes categorised as 'wicked problems' (Chapter 10). For reasons including environmental impact, geological constraints, land-take and public opposition, options for constructing new reservoirs are limited in Britain, and the case for their construction, or the extensions of existing impoundments, remains controversial among stakeholders, to say the least. The solution of water resource issues using dam construction would have been favoured before the 1970s, but there are currently few plans. The frequently discussed Broad Oak Reservoir Scheme near Canterbury has yet to be commenced, the Bewl water reservoir expanded and there has long been a suggestion for new reservoir near Abingdon. The call for new public supply reservoirs as part of a wider strategy remains very much on the agenda of water resource planning (Medway Council, 2007).

6.1.1 Groundwater Exploitation

It has already been stated that groundwater development can seldom be separated from considerations of local surface flows, although past practices of riverside abstraction from groundwater tended to separate the two, often with scant regard for maintaining adjacent river flows (Headworth, 1994). With the continual sinking of wells, adits and boreholes throughout the later nineteenth century and continuing through the twentieth, and increased pumping in order to meet rising demand (most notably since 1945), problems of shortage have been manifest. Since the Water Resources Act 1963 (and subsequent legislation), the water industry is not permitted to 'mine' groundwater (i.e., abstract beyond recharge capabilities in the long term) without considering local environmental impacts.

Objectives in groundwater management may be summarised as:

- Management to meet the needs of the entire water environment including striking a balance between abstraction rates and uses of surface waters.
- Maintenance or improvement of adequate river base flows both for downstream abstractions and for in-river needs such as effluent dilution, navigation, fisheries and aquatic ecosystems, in seasonal context.

- Maintenance of groundwater levels to preserve wetland sites, support springs and meet ecological objectives.
- Protection of groundwater quality, thereby protecting surface waters thus fed, conserving the resource and meeting EU standards.
- Meeting the volumes required for public supply.
- Meeting the needs of other 'raw' (untreated) water customers such as industrial or agricultural users.
- Meeting the needs of other protected rights, particularly for domestic wells and other purposes abstracting less than 20 cubic metres per day (these are unlicensed).

In meeting objectives of effective management, three general approaches to groundwater management are available, and are used either independently or together. Water in England and Wales is based on 'CAMS' (Catchment Abstraction Management Strategy) units that may be based in surface or groundwater, or combined water sources.

Develop new sources. Extending the 'frontier' of groundwater exploitation as an option relies upon reaching hitherto unexploited sources and these are increasingly rare, notably since the 1963 legislation. High-yielding aquifers seldom have a surplus (and operating within a margin of error, the 'precautionary principle'), others may not be easy to exploit. For example, although there is abundant Carboniferous Limestone throughout Britain, water undertakings are reluctant to develop this fissure-flow dominated aquifer on account of its unpredictable fissure patterns. Furthermore, certain waters from the aquifer are of poor quality and require expensive treatment.

By contrast to the Carboniferous Limestone, the Chalk is a widely used aquifer. In the Anglian region Chalk dominates groundwater resources, Norfolk has untapped groundwater, Suffolk less, with Essex and Cambridgeshire having little room for manoeuvre unless groundwater schemes are devised whereby careful management options are employed. In the Midlands EA region there are a number of groundwater units licensed beyond their recharge ability EA (2008).

Change abstraction patterns. Exploitation can be arranged so as to prevent deleterious effects such as saline intrusion, or loss of flow in rivers which cross permeable strata. 'Conjunctive use' of surface and groundwaters aims to run them in an optimal way, recognising their different hydrological characteristics. Implicitly, it involves the integrated (and combined) management of aquifers and surface water systems to optimise production and to allocate water between competing demands. For example, a reservoir may fill early in winter, and groundwater use means only that surplus flows are lost to the sea. Conjunctive use here may involve reducing groundwater abstraction (and hence benefiting long-term storage) while the reservoir is drawn down. The latter may replenish later in the season in any case. Conversely, where groundwater is a more robust source, the groundwater component is increased to avoid surface source failure, or unacceptably low flows.

Groundwater in British aquifers tend to be 10-12°C (UK Groundwater Forum, 2011a), unless there is a reason to be higher such as localised thermal effects. Groundwater baseflow therefore has the potential for reducing diurnal and seasonal variations in river water temperatures: reducing those in summer, and increasing those in autumn through to spring. Abstraction should be at a sufficient distance from the river channel to minimise adverse interaction between surface and groundwater levels. The chemical quality of groundwater is important, including 'hardness' and nitrate concentrations,

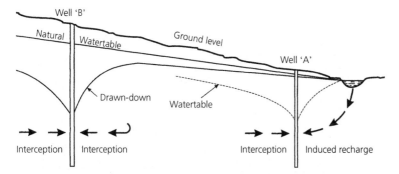

Figure 6.2 Relations between well pumping and groundwater fed rivers in continuous aquifers. (Redrawn from Owen, 1991).

but lower dissolved oxygen contents in groundwaters may present problems including pollutant attenuation and for fisheries. In the Water Resources Act 1991 Section 20 there are powers for the EA and undertakers to work together in management Schemes and under the Water Act of 2003 licences can be revoked without compensation. However, this is easier said than done; benefiting low summer flows can, by depleting groundwater levels, reduce (otherwise higher) winter baseflows.

Augment groundwater from surface waters. The difference between this and the previous option is that recharge is in part by artificial augmentation *in excess of that available from effective rainfall.* Methods employed in artificial recharge of groundwaters are generally by active pumping into injection wells, or allowing water to infiltrate via lagoons or surface watercourses. Experimentation is required in order to explore the feasibility of proposed schemes, and tests do not always prove successful. Maintenance of water quality is essential; chemical or biological pollutants may contaminate the aquifer while suspended sediment in recharge waters may block pores and fissures required for infiltration. Storing waters in the ground is comparable with surface reservoirs but the timescales vary. Critical periods (full to empty) seldom exceed 18 months for surface reservoirs; for groundwater storage it may be many years (Bland and Evans, 1984).

Figure 6.2 illustrates the relation between well pumping and groundwater-fed rivers in continuous aquifers (Owen, 1991). Depletion of river flow consists of two components, interception of groundwater which would otherwise reach the river by flow to a well (wells 'A' and 'B'), and induced recharge which is water actually removed from the river in response to groundwater gradients generated by pumping (well 'A'). Draw down at a well will cause affected springs to dry up and, in a continuous aquifer, to relocate at a lower topographic level.

In lowland Britain, most channels are alluvial and much of the alluvial material is fine, typically silts and clays, and unfissured. The aquifer beneath may be highly fissured and hence have a high vertical saturated hydraulic conductivity compared with the alluvium. There is thus a resistance to water flow between the channel and groundwater, a situation realised through hydrological modelling and taken advantage of in surface water augmentation schemes.

To combat low flows in dry periods, across England and Wales there are today some 50 river augmentation scheme, a doubling of those planned or commissioned by 1991, (Owen *et al.*, 1991), mostly tapping chalk and Permo-Triassic aquifers. Augmentation generally has environmental protection as a stated purpose when it is important to increase flow in dry periods, once a trigger point is reached. It also a policy option for Scotland (Scottish Government and Natural Scotland, n.d.).

6.1.2 Groundwater and River Support in the Anglian Region

The EA Anglian Region is worthy of closer scrutiny, for there has long been concern over managing water abstraction on account of low rainfall. Average annual rainfall is low in the region, the regional annual average figure being around 600 mm, some 70% of the national average, and rainfall is lowest in coastal regions (EA, 2012a). Abstraction from rivers creates problems in maintaining flows, and although groundwater recharge is also limited by low rainfall, surplus of winter precipitation recharges the aquifers, mainly the chalk. Forecast of total demand by the water industry includes considerations of climate change, increasing household, consumer expectations employment patterns and innovations. Investment is required to meet supply, although climate change itself may not be a significant driver. There is an anticipated increase in demand forecast to 2035, incorporating the 'headroom' margin between supply and demand that is regularly reviewed in making forecasts (Anglian Water, 2010).

Infrastructural investment in the Anglian water resources area is well established and Figure 6.1 shows overall water resource development in the region of the east of England. Groundwater schemes are supporting flow in many rivers of the east of the region and are able to support surface water abstractions in the lower catchments such as the Great Ouse system (that enters the Wash north of Denver) and the major towns of Essex where demand is balanced using reservoirs. Not only is river flow supported, but also public supply needs, to support river flows (to compensate for surface water abstractions) and to maintain wetland habitats, an important consideration in areas with falling watertables.

Part of this, the Lodes-Granta groundwater scheme, is designed to improve the water environment to the north and east of Cambridge as well as providing water for public supply. In an area over 600 km², surface flows (including those in the high-level carriers called the 'Lodes') in Fenland (to the north and west of Cambridge; Figure 6.1) are to be augmented through limited aquifer development so as to support abstraction and set Environmentally Acceptable Flow Regime (EAFR) and protect wetland habitats such as Wickham Fen (Tester, 1994).

Under suitable situations artesian overflow replenishes surface waters. In Lincolnshire around 30 'wild bores' or abandoned artesian boreholes (Barton and Perkins, 1994) discharge into surface watercourses including the River Glen. Uncapped the bores are uncontrolled, largely unwelcomed, and represent a loss of good quality groundwater. Most of the wells have now been regulated.

6.1.3 Leakage Abstraction in the Brighton Block

The chalk of the South Downs has long been considered in terms of a series of aquifer 'blocks', that is escarpments with the scarp face facing north and separated by tidal rivers (see Figure 6.3; south coast below London). Running from east to west the blocks are referred to as the Eastbourne, Seaford, Brighton, Worthing and Chichester blocks (Headworth, 1994, 2004) schemes of Southern Water 1970-1990.

Since the 1950s a 'leakage abstraction management' scheme has been associated with the Brighton Block. The aim has been to restrict salinization of the aquifer through maintaining seaward flow in the aquifer throughout the year. The Brighton Block is defined by the outcrop of the Gault clay aquiclude, located stratigraphically below the aquifer formations to the north (the chalk and upper Greensand are apparently in hydraulic continuity), the River Adur to the west, the Ouse to the east and the sea to the south (IGS/SWA, 1978). The overall pattern of groundwater flow is towards the margins of the block, away from the east-west crest of the Downs which reaches over 120 m above OD, and a north-south spur projecting into the northern outskirts of Brighton. Near the coast, however, groundwater gradients are low with the general level of the watertable around OD level. Saline water intrusion (this process is worse along fissure zones) presents a serious risk to groundwater quality. Chlorides have persisted in exceeding 40 mg l^{-1} Cl at shallow samplings along the coast and that increase with depth continue to cause concern presenting a threat to water supply (Entec, 2008; IGS/SWA, 1978; Jones and Robins, 1999, p. 72).

A new (and innovative) abstraction policy commenced in 1957 aimed to combat increasing salinities, a problem present since the nineteenth century (Headworth, 1994, 2004). It involved the use of several sources operated by Brighton Corporation and abstraction patterns were changed so as to conserve inland aquifer storage and place greater emphasis on intercepting outflows from the block to be put into public supply (Headworth and Fox, 1986).

Figure 6.3 shows the operation of the scheme. Aquifer management policy centres on making the maximum use of boreholes located along the coast during winter months, allowing aquifer recharge and reduced demand in order to capture outflows towards the sea. During the spring and summer, following groundwater recharge, abstraction from the coastal region is reduced, and more is pumped from inland sources where

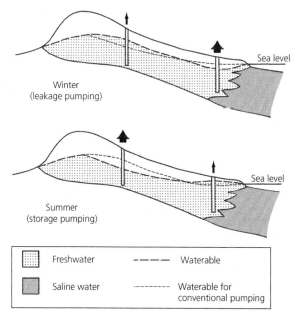

Figure 6.3 Operation of the leakage abstraction groundwater management scheme. (*Source:* Owen *et al.*, 1991, p. 26. Redrawn and reproduced by permission of Oxford University Press).

levels have recovered. Since the 1950s, control of groundwater levels by means of careful monitoring of resources, abstraction control and predictive modelling (especially near to the coast which is sensitive to seasonal tidal variations and long-term salinities) has controlled the risk of saline intrusion.

Since the 1970s, numerical models have been used in aquifer management (Headworth, 1994). This multi-faceted management scheme, which may yet require desalination and recharge from sterilised and fully treated effluent, has so far been able to meet demand following drier winters, despite the fact that the volume of water going into supply fell from around 190 Ml day^{-1} in early 1989 to around 150 in mid-1992 (Miles, 1993). Finally, instantaneous controls are applied to minimise the chloride levels in water supplied. The scheme remains in place, although there has been a shift towards operational efficiency from groundwater storage conservation (Headworth, 2004).

6.1.4 Supporting Flow in the Southern Chalk Rivers

Local river augmentation using groundwater in the Southern Region of the EA supports flow in the River Itchen started as desk studies in the 1970s (Headworth, 2004). Figure 2.6 shows the catchment. Originally, the Itchen was regulated by its tributary, the Candover. Operation of the Candover Scheme relied upon groundwater abstraction in the upper catchment being transmitted via pipelines to supplement low summer and autumn flows (IGS/SWA, 1979) for in this important watercress growing catchment, watercress production around Arlesford is supported by flow from artesian bores, suggesting good availability of groundwater in the upper Itchen groundwater catchment.

In Chapter 2, it was noted how the Itchen catchment was about 20% larger than the surface-defined catchment; hence, there is an ability to augment surface flows provided that abstraction points are sufficiently remote from the supported river channel. The function was to permit abstraction from the lower catchment close to the tidal limit and retain flow in the upper catchment, and to this end the scheme is regarded as successful. Augmentation occurred during the low rainfall periods and pump testing of this successful scheme continues, especially about the ability to supply during drier periods (EA, 2011b).

The River Alre tributaries have been considered for augmentation in order to further support flow and abstraction in the Itchen. Hydrogeological investigations carried out in the chalk aquifer (NRA, 1991d) showed how average transmissivities in the Candover catchment were of the order of 2000 m^2 day^{-1} whereas those in the Alre catchment were typically three to four times greater. Storativities were low, typically 0.003 to 0.01. Such high transmissivities are not good for river augmentation schemes, because they cause a rapid expansion of the cone of depression resulting from pumping towards the river with a resulting reduction in net gain (as illustrated in Figure 6.2). On the other hand, high storativity is desirable as it is a measure of water released upon watertable draw-down. Long-term pumping tests tended to bear this out, and if the river augmentation scheme is to be commissioned in the Alre catchment, licences to abstract would have to be limited.

The Darent, at one time, was regarded as the worst affected 'low flow' river in England and Wales. In the period approximately 1890 to 1910, abstraction from the chalk approximately doubled due to the activities of the Metropolitan Water Board, later through and industrial activity and increased 'export' of resources to developing urban areas; by the 1960s the situation was becoming critical (Cook, 1999a). The river Darent in Kent provides habitats for many important invertebrates, fauna and fish, however it suffers from over-abstraction through Licences of Right. There has been not only unacceptably low flows, but sometime flow ceased altogether in dry years, for example 1976, 1989-1992, 1990 and 1996, particularly in the mid-catchment that is prone to dry up completely (Herbertson, 1994).

This river derives its flow from the lower greensand and chalk aquifers and the development of numerical integrated catchment models, which link the behaviour of groundwater and river flow to recharge and abstraction patterns, enables the pre-abstraction behaviour to be modelled. The links between groundwater behaviour and river flows are thereby established, and provide a model environment in which to test alternative options for the relocation of abstractions. In the Darent catchment, five water companies were authorised to abstract 155 Ml day^{-1} and private industry 18 Ml day^{-1}. Although aquifer recharge exceeds actual abstraction by 15 to 20%, and abstraction is limited to 70% of that licensed by voluntary agreement, flow is frequently below the EAFR. These figures take account of ecological needs such as conservation of aquatic species. The needs of fish and invertebrate populations led to the setting of a mean EAFR at 97 Ml day^{-1} at Hawley. A 95 percentile flow was set at 23 Ml day^{-1} and a minimum target flow at 26 Ml day^{-1} on account of the requirements of brown trout. Flow velocities in excess of 0.5 m s^{-1} with depths in excess of 0.3 m are required; the requirements of the flora might vary widely between minimum flows of 29 and 173 Ml day^{-1}.

Figure 6.4 shows a good example of early application of integrated numerical catchment modelling. The output shows flow modelling for a 1:20 year drought without abstraction, with the target flow under these specified conditions, and scenarios of 70 and 100% abstractions. Evidently, abstraction has reduced flows to zero in stretches of the river.

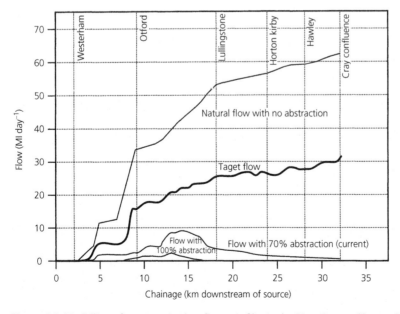

Figure 6.4 Modelling of comparative low flows profiles in the River Darent. (*Source:* Re-drawn from Herbertson, 1994).

The implementation and maintenance of the Darent Action Plan, required development of groundwater management by the EA, Thames Water plc and other organisations. Modelling had shown that simply reducing abstraction rates along the valley would not be sufficient to meet the EAFR. In addition to restricting abstraction in the mid-catchment chalk boreholes, and up-catchment lower Greensand bores, cost-effective solutions may involve flow augmentation by artificial springs (augmentation boreholes) set along the riverside, and conjunctive use of surface waters by importation of water from former chalk quarries at Northfleet (outside the catchment) and from the London Ring Main (River Thames water) in winter to enhance recovery of groundwater levels.

To solve low-flow and ecological problems, the last 20 years have seen reduced licensed abstraction, the construction of three boreholes to increase flow in dry periods and re-profiled channels to maximise the benefit of reduced licensed abstraction for flow-sensitive species. One example being reduced Thames Water plc (the successor to the metropolitan Water Board) reducing its abstraction from 1996. The Darent is now in Good Ecological Status under the Water Framework Directive. Due to remaining risk of deterioration, the final stage of the Darent Action Plan will involve reducing abstraction further.

6.1.5 Shropshire Groundwater Scheme

While the chalk, almost ubiquitous in the south and east of England, would seem the focus of groundwater management, there are other instructive schemes. Within the Midlands Region of the EA there was a large projected demand on aquifers of the Shrewsbury area and abstraction from both the Severn and Trent rivers. Such demand puts pressure on the resource development associated with the River Severn which is required to support abstractions (NRA, 1993d). This scheme caused flow of the Severn to be augmented from the Permo-Triassic sandstone formations underlying much of North Shropshire (NRA, 1994j). The scheme commenced with feasibility studies in 1971, gaining formal approval from the Secretary of State in 1981.

The Shropshire Ground Water Scheme had been implemented so far in four out of eight phases in six areas (UK Groundwater Forum, 2011b; EA n.d.b) and is licenced for a total of $330,000 \, m^3 \, day^{-1}$. The Scheme is designed to be used, on average, once every three years to meet peak dry weather demands for water. Water is pumped from groundwater reserves stored within the Permo-Triassic Sandstone formations of north Shropshire and released in conjunction with surface water reservoir releases to balance the demands of abstractors while safe guarding the ecological needs of the river environment. Example summers when it has been called upon have been 1984, 1989, 1995, 1996 and 2006 when river flow was used to augment river flow.

Interesting features include its inclusion with releases from the Clywedog Reservoir in the headwaters of the Severn during periods of prolonged drought. Boreholes are constructed in the sandstones for both observation and monitoring, and underground pipelines transfer water to outfalls on the rivers. At times of low flow, water from abstraction boreholes is discharged through pipelines to the Severn and its tributaries. At full development, it is anticipated that the scheme will transfer water to the river in this way during one year in three, for periods not exceeding 100 days. The scheme will be carefully monitored in order to prevent adverse effects upon existing abstraction, surface water bodies and downstream abstractions from the Severn.

6.1.6 Groundwater Augmentation Schemes

In order to assist the balance of groundwater (in a specified catchment) for abstraction with recharge and river flow, it is sometimes feasible to recharge aquifers artificially in excess of their natural recharge, because the storage available may be large. Total abstraction should be less than average recharge and may be reduced by an amount required for spring flow. Seasonality is important; where groundwater levels are high enough to maintain summer/spring flows, the subsequent seasonal rise in levels causes wastefully large winter/spring flows (Bland and Evans, 1984).

Despite considerable interest and experimentation, operational groundwater recharge schemes are comparatively rare in Britain, but this is changing. A working example of conjunctive use, including groundwater augmentation, is the Stour Augmentation Groundwater Scheme, one of the former Anglian EA region schemes and shown in regional context (see Figure 6.1). In the area to the north-east of London and south east of Cambridge, the Essex rivers Stour, Colne and Blackwater that drain to the North Sea are themselves supported by bulk, raw water transfers from the Ely Ouse to the north of Cambridge. On account of a large storage capacity and restricted natural recharge of the chalk (because of the exceptionally low effective rainfall in south-eastern Essex), the River Stour can be used to augment the aquifer in places where the river cuts through the confining Boulder Clay. By periodic abstraction from the chalk, far enough away from the river to allow re-circulation, the river can be augmented in times of drought, and will subsequently recharge, through its bed, over subsequent periods. By supporting the Stour, ultimately the yields of the Abberton and Hanningfield reservoirs will be enhanced.

Groundwater may have performed its up-catchment functions, it may not be pure and yet may be potentially lost to the sea, as with the Brighton Block. Such water under suitable conditions can be abstracted and treated. A technique is 'aquifer storage and recovery' (ASR), developed in the United States and it may be suitable for use in south and east England (ASR, n.d.). ASR is a method which relies upon storing surplus winter groundwater below ground in aquifers that can be poor in terms of hydraulic properties or quality, the water only to be abstracted as required. With care, contamination need not occur and it potentially balances average and peak demands, thus deferring the need for new resources to meet peaks. The main problem normally associated with recharge schemes is clogging due to suspended solids in recharge water; however, problems associated with bacterial growth, gas entrainment, and unfavourable chemical interactions have also been described from around the world. Proactive groundwater management of this kind is not always possible and the success of such schemes depends upon many variables. This is illustrated below with reference to experiments at Hardham, Sussex, between 1966 and 1981. In the area, synclinally disposed Folkestone Sands of Cretaceous age comprise an elongate but closed basin 9 km by 3 km. It is underlain by impermeable strata, while confined in its centre by overlying Gault Clay. The aquifer has a maximum storage of 158 Ml. At first, artificial recharge was by gravity from the River Arun via recharge lagoons (IGS/SWA, 1978) and a dual-purpose (recharge/abstraction), single-main borehole (O'Shea, 1984). Groundwater levels in the Folkestone Sands fell rapidly in the late 1960s and settlement occurred at the surface; consequently two lagoons were dug into the sands and gravels overlying the aquifer to assist recharge. Soon recharge rates of 2.5 Ml day^{-1} could be sustained (Headworth, 1994), and subsequently it became one of the most thoroughly researched sites for artificial recharge in Britain with an induced recharge of 15 Ml day^{-1} above a natural recharge of 9.9 Ml day^{-1} from rivers which flow across the basin (Owen *et al.*, 1991). The aquifer, being composed largely of quartz grains, was fairly inert and chemically compatible recharge waters were used.

However, ASR overall is proving slow for adoption in the UK. Experiments into recharge of the Folkestone beds relied upon good quality, fully treated and chlorinated water for recharge, relatively low suspended solids, inhibited bacterial growth and a downhole recharge valve to prevent gas entrainment. Untreated and unsettled river water soon prevented aquifer recharge through clogging and biofouling and attention switched to borehole recharge (Headworth, 1994). A new borehole was drilled through the confining Gault Clay to a depth of 63 m into the Folkestone sands which permitted a recharge rate in excess of 4.1 Ml day^{-1} over a month. The aquifer had an average transmissivity of 600 m^2 day^{-1} with a storativity of between 3×10^{-2} and 3×10^{-1} (O'Shea, 1984). With potable quality recharge water it was concluded that recharge was feasible and de-gassing was not required. Subsequently aquifer models were produced.

The slow pace of development owed much to the problem of local ground settlement, something most noticeable on the peaty alluvium rather than Gault Clay. There was further assessment of this problem in 1994. It had long been hoped that development of conjunctive use of the river and aquifer would provide an enhanced supply of water for the Crawley area. The Hardham basin represents a potential major groundwater source for Southern Water. Aquifer recharge through boreholes is preferable to lagoon recharge using river water, although this may be possible locally. The recharge aspect has not been developed to the operational stage, but there remains great interest in the idea.

Groundwater augmentation is established in north London, where eventually there will be a drought yield of over 170 Ml day^{-1}. In 1994 it was calculated that the London Supply Area of Thames Water Utilities Ltd had a resource deficit of 159 Ml day^{-1} (O'Shea, 1984, 1995). The scheme described is located on the northern limb of the London Basin syncline, and involves developing the chalk/basal sands (lower Tertiary) aquifer which is locally confined by London clay. The chalk is some 250 m thick, the Thanet beds only 14 m. There is an established artificial recharge scheme in the Lee Valley, adjacent to the reservoirs, using artificial recharge to meet strategic drought demands and comprising five larger diameter older wells and nine newer boreholes.

However, groundwater development in the Enfield-Haringey area will be by development of recharge/abstraction boreholes adjacent to the New River aqueduct and will number a further 23 upon completion of the scheme. Figure 6.5 shows the general layout of the new scheme; interestingly it involves an early feature of the water supply of London, the New River completed in 1613. Recharge is to be from the normal distribution network using fully treated drinking water, with the aqueduct transmitting untreated groundwater during operational use.

Pump testing revealed 15 new bores to yield between 1.9 and 6.6 Ml day^{-1}, while transmissivities vary between 75 and 2200 m^2 day^{-1}: a considerable variation. Operational tests and groundwater modelling will eventually produce an optional management model defining optimum recharge and abstraction strategies. The upper 30 to 40 m of the chalk aquifer were found productive. In average years flows in the Thames and Lee combined with the Lee Valley reservoir storage will meet demand and there will be a sufficient surplus for borehole recharge. During drought situations, when surface

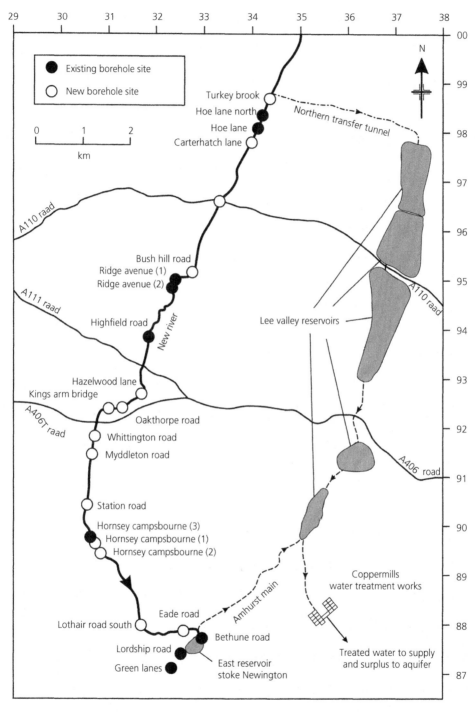

Figure 6.5 Groundwater development in the Enfield-Haringey area. (*Source:* Re-drawn from O'Shea, 1984, 1995).

storage and flows fall, there will be a switch to abstraction mode from recharge. Then flow from the New River will discharge into the East Reservoir at Stoke Newington in the south, and then via the Amhurst Main into the Lockwood Reservoir for treatment and supply. Meanwhile, to the north flow from north of Turkey Brook will be transferred to the Northern Lee Valley reservoirs, via a transfer tunnel. This northern route is on account of the maximum capacity of the Amhurst Main being limited to 90 Ml day^{-1}.

6.2   Catchment Transfers

Bulk, untreated or 'raw' water transfer between one river and another worldwide is one way of meeting demand and manipulating supply by augmenting water in one catchment from another. Inter-basin transfer schemes are known from the United States, the former USSR, Canada and Australia. Newson (1992b) aptly describes the situation in Canada as being a result of water being 'in the wrong places'. Here, 9% of global runoff serves around 35 million people, yet most available water is in the scarcely populated north. This is a familiar problem where there are demographic and water supply imbalances.

In Britain too there is imbalance. The north and west are wettest, yet the bulk of the population lives in lowland areas in the South, Midlands, East and in Scotland. Before 2000 it was thought demand would rise, fortunately this has not been the case although there can be sectors that buck the trend of falling demand across Britain such as domestic demand in Scotland and irrigation in England and Wales (Chapter 1) may increase from time to time. However, complacency is never called for, and there may be many reasons, including growing demand under scenarios for climate change that suggest geographic or temporal issues with maintaining supply and ecological conservation that cause issues of infrastructure to remain a matter for discussion at least. Although it may prove feasible for small transfers, say around 50 Ml day^{-1}, large-scale transfers involving the existing canal system have not been favoured for some time, due to likely environmental impacts (NRA, 1994g, p. 51).

Conceptually we may imagine the scales to be:

- Transfer from upper to lower catchment controlled by dams or augmentation from groundwater. This is transfer *within* a single catchment.
- Transfers between one river and the next, termed *intra-regional* transfer.
- Those between regions, termed *inter-regional* transfers.

Frequently, reservoirs are employed in order to balance resources at critical points during transfers (Figure 6.1). Reservoir deployment, enlargement and construction played a major part in the 'Water Resources Strategy' of 1994. Options were redeploying water from Lake Vyrnwy, the enlargement of the Craig Goch Reservoir in Wales, the construction of a South West Oxfordshire Reservoir (in connection with the Severn-Thames transfer), and there remains the possibility of a new reservoir construction in East Anglia, either at Great Bradley near Newmarket or in Fenland close to the site of transfer from the Ely Ouse to Essex.

Here may be observed a change in view, for the EA remain cautious around large-scale bulk transfers; the private water companies such as Thames Water practice this and hence remain interested (Vaguely interesting, 2012).

Water UK, the water industry organisation, without aggressively promoting bulk transfers is clear that is it an option, subject to constraints such as environmental desirability and economic feasibility. The White Paper on managing water resources (2012) is pro 'strategic' (i.e., local) movement of water, for example from one water company to another (UK Parliament, 2012) but is presently opposed to large-scale transfer, for example from the Kielder reservoir in Northumberland southwards, or increasing bulk transfers out of Wales into England, something that may also be politically sensitive as well. The Kielder Reservoir, for example, completed in 1982, was supposed to meet demand in a region dominated by heavy industry, support flow in the rivers Tyne, Tees and Wear, and provide hydroelectric power. However, there remains the distinct possibility that this, the largest man-made lake in Europe and supplying a potential of 900 Ml day^{-1}, with a surplus of up to 500 Ml day^{-1}, could be used to supply water throughout the east of England as well as regulate rivers (NRA, 1992 g).

If further connections are made between the Tees and Swale with regulation of the Yorkshire Ouse, and from there water is transferred to the River Witham in the Anglian Region, this could be realised. Ultimately, water falling in Northumberland could help to meet the supply in Essex if transfer schemes already envisaged are put into play. However, the 'Water Resources Strategy' sees long-distance transfer north to south as less economically viable than west-east dominated transfers. It remains possible that marginal demands in parts of Yorkshire might be met in this way.

However, it is factors such as the cost of large-scale engineering projects, of environmental impact and (where transfers are not gravity assisted) the cost and carbon impact caused by pumping large volumes of water. To quote:

> 'Defra's decision to focus on maximising "strategic" interconnections allowing water to be transferred over relatively short distances is a sensible starting point and we look forward to the Environment Agency's forthcoming overview of interconnection options. Defra should remain open to considering whether large-scale infrastructure may be an appropriate and cost-effective solution in some circumstances. Defra and the Environment Agency should in particular look to exploit interconnection opportunities presented by other large infrastructure projects, thus minimising environmental impact and economic costs.'

The Position Statement of the Environment Agency on large-scale transfers (July 2011) is directly addressed at the water companies, who have a statutory duty to prepare, consult, publish and maintain Water Resources Management Plans (WRMP) under the Water Industry Act (1991) sections 37 A to D, as amended by section 62 of the Water Act of 2003 (EA, 2011c) is cautious. The Agency asks that all water companies consider all reasonable options in planning process in developing final WRMPs that are applicable to 2035, preferring short distance transfers and better sharing between companies. Large-scale transfers are not excluded and these have an historic basis, but are not presently included as an option. There is concern over the environmental impact and cost of large-scale transfers from the north of England or Wales and differences in water quality may have impact on ecology and on flow regimes. Such will only be progressed if they can be 'demonstrated to be an economically and environmentally viable option to meeting a forecast water supply deficit'. It is even feared that this may compromise the 'good ecological potential' required by the Water Framework Directive.

The most suitable set of options is selected as a result of this assessment process based on costs, yield and environmental effects. Water companies must furthermore plan for droughts, and restrictions on water use in severe droughts are part of water companies' plans, as required by water resources legislation. EA assess the wider needs of all water users and the environment to ensure sustainable water supply for agriculture, power generation and industry as well as public supply. EA produces a water resources planning guideline, setting out detailed guidance and best practice methods for producing WRMPs. The map (Figure 6.1) shows the major river, reservoir and groundwater abstractions, water transfers and regulated rivers in England and Wales. Whilst these schemes are primarily for public supply, some of them also provide water for other abstractors and to maintain the environment.

From the point of view of an independent scholar, the jury remains 'out' on bulk transfers. One guess is that the future will eventually require a limited development of the idea when localised transfers linked with other measures such as demand management and changes in licencing structures cause concern.

6.3   Other Options

Certain alternative options may appear futuristic but remain interesting because they reflect past confidence in technological solutions to water supply problems. They have mostly been rejected for reasons of cost, operational difficulty or environmental impact. Many have been dubbed 'esoteric' (NRA, 1994e), many have the absence of localised water abstraction as a common feature.

Desalination of seawater may be most likely of the alternative means of water procurement and it could deliver huge quantities of freshwater at coastal locations (inland if saline groundwaters are used). Distillation requires large amounts of energy, and in arid countries a potential energy source is solar energy, while in certain Middle Eastern countries oil provides a cheap energy source. Alternatively there are methods available using the waste heat from power stations, such as derives from exhaust turbine steam and off-peak generation. The multi-stage flash process (MSF), (NRA, 1994g) operates by feeding cold salt water into the system via a long pipe passing through several chambers where it acts as a condenser and gains heat from the condensation. It is subsequently heated in a steam-fed heat exchanger and enters the first chamber where part of the salt-free water 'flashes' to vapour which condenses on the colder feeder pipes as freshwater, and the process is repeated back through the chambers operating at progressively lower pressures, back finally to the chamber which first imported the cold saline water in a pipe. 'Reverse osmosis' (RO) is an alternative which enables seawater to demineralise at lower energy costs than distillation plants. Water moves naturally from a weaker salt solution to a stronger solution through a semi-permeable membrane, until equilibrium concentrations are reached on either side. If a pressure differential is applied, with higher pressure on the side with the concentrated solution, water will move towards the weaker solution side.

Desalinisation is still an option mentioned by the EA (EA, 2013d), and there are attendant carbon emissions, so that other options are preferred. Nonetheless, Thames Water runs the Beckton desalination plant, working on the RO principle, the first of its kind in UK and capable of supplying up to one million consumers in times of drought taking water from the Thames estuary. If this technology improves, desalination may play a more significant role in balancing supply and demand in England and Wales.

A National Water Grid would move *treated* water as the National Grid for electricity supplies power. The idea, been mooted since the 1940s, relies upon engineering developments that would permit the linking of regions with an excess of water by means of large pipelines and aqueducts (NRA, 1992f). Local grids for treated water connecting adjacent areas do exist, and the EA policy is to encourage these (EA, 2011c).

The London Ring Main commenced operation by Thames Water in 1994 (Thames Water, n.d.) and comprises 80 km tunnel, an average diameter of 2.5 m, providing up to 300 Ml/d of water per day delivered to consumers via three major treatment works. It is effectively a 'ring of water' around London, allowing water to flow in either direction. It is gravity operated, reducing energy consumption.

Tidal/estuarine storage Estuaries represent a point of interface between freshwater and seawater, and to impound landward freshwater free from tidal influence a barrier must be constructed. Britain is well supplied with potential sites, yet no barriers have actually been built, for like tidal barrages for energy generation, these are contentious. It is unlikely any will ever be built for freshwater storage purposes and there would be implications for conservation, fish migration, boat passage, flooding, sedimentation, landward groundwater levels and salinisation, and water quality issues, mostly arising from the impoundment of industrial and sewage pollution.

Importation of water under the sea from France has been considered and there is a small surplus in the Pas de Calais groundwater. With economic developments associated with the Channel Tunnel (there is insufficient space here for a water main), it remains unlikely that there can be water exchange in the same way that there is for electricity. Other methods of bulk importation suggested have been filling tankers, towing water-filled bags termed 'flexible tankers' (NRA, 1994l), and towing icebergs. It is possible that global warming might increase the breaking up of Arctic icebergs, thus facilitating the process. Importation of bulk water into Britain is currently not considered a viable option.

Rainmaking has shown promising results in southwestern Australia where silver iodide smoke is released from an aircraft onto cloud tops causing the condensation of water droplets provided that there is atmospheric moisture present. The substance used has to have a crystalline structure similar to that of water crystals (Open University, 1984) which form clouds in middle to high latitudes. Contemporary literature does not mention this as a serious option for England and Wales.

6.4 Demand Management

Demand management is an established side of the twin track approach to meeting demand outlined in section 6.1 and it may be defined as' the management of the total quantity of water taken from sources of supply using measures to control waste and consumption, (NRA, 1994g). This also reflects the EA's statutory duties under section 19 of the Water Resources Act 1991 to conserve, redistribute and otherwise augment water resources and to secure their proper use. Meanwhile, a serious warning shot has been fired in terms of changes to licencing arrangements under the Water Act 2003 (Chapter 3). Otherwise stated, new resources are only to be provided when the alternatives have been utilised and environmental considerations should be given proper weight in decisions.

In terms of water utility regulation we talk of demand management. In Chapter 3 it was noted how water supply is a relatively inelastic commodity which does not respond greatly to price signals, and hence is prone to market failure. Narrower definitions of demand management involving the control of consumption using the price mechanism are unlikely to be effective. Leakage in distribution networks for England and Wales remains of concern, although it is behind that for Scotland and mercifully, the trend in all cases is downwards (Chapter 5).

Demand management is best implemented by regulation through managerial and operational means including repair of leaks, improved storage, conveyance, distribution systems and uses of water. Using less water by savings in consumption and increased efficiency in supply distribution is appealing. The operational, environmental and financial constraints of developing new supplies make demand management a fruitful area of research and policy development.

The development of effective demand management is favoured in several quarters including the EA, Ofwat, environmental pressure groups such as the Council for the Protection of Rural England and professional groups such as CIWEM (Rees and Williams, 1993; EA, 1997b). It is, however, with Ofwat that leakage targets are agreed. Efficiency is promoted by the Environment Act 1995, and this is incumbent upon water undertakings; under the Water Industry Act 1991, they have to produce water efficiency plans.

Table 6.1 is CIWEM's view of water efficiency measures, these are the well-known examples such as metering, legal sanctions, building regulations, re-cycling, advice and of course water education.

The table covers measures described as metering, switching to measured tariffs, education and information, bye-laws, building regulations, ecolabels, efficiency, advice, water re-use and re-cycling and planning and design. Absent from the table is the matter of licensing, discussed in Chapter 3. It is taken that Water Act 2003 deals adequately with abstraction licensing for present purposes.

Charging other than a flat rate tariff was once believed inappropriate because it could have adverse impacts on poor consumers. However, CIWEM believes that controlling consumption should be achieved through charging for water services and that this will deliver real reductions in household water use. Charging using metering (see section 6.4.5) should encourage society to place a greater value on water. Identified should be issues of water scarcity in particular regions, as well as access to water supply for poorer households. Metering should be implemented as soon as practical, alongside improved tariff structures and measures to protect households with low incomes (CIWEM, 2015).

The 'K-factor' is the annual rate by which each licensed water company can increase its charges annually on top of inflation determined by the Retail Price index (RPI). The K factor sets the maximum percentage by which total income can be increased for a basket of principal charges to all customers, whether domestic, business, metered or unmeasured. The K factors are set by Ofwat at a price review, generally over five years. Companies can defer K, and both Ofwat and companies can seek interim adjustments of price limits between reviews (OFWAT/Defra, 2006). The charging limits apply to a company's average charges. The formula is applied to a "tariff basket" of charges that are:

- charges for unmeasured water supply;
- charges for measured water supply;
- charges for unmeasured sewerage services;
- charges for measured sewerage services;
- charges for reception, treatment and disposal of trade effluent.

Within that basket, a company may increase some charges more than others, although the overall average must not increase more than RPI plus K (Ofwat, 2001).

Table 6.1 Strategic portfolio of water efficiency measures (CIWEM, 2015).

Measures	Who Initiates	Who Should Respond
Metering		
Universal metering of all users, where practical	Defra, Ofwat, WCs, Agency, CCW	All water users
Selective metering of high water users	Defra, Ofwat, WCs, Agency, CCW	All water users
Switching to measured tariffs	WCs	All water users
Introduction of innovative tariffs	WCs, Defra, Agency, Ofwat	All water users
Introduction of informative water bills	WCs	All water users
Education and information		
Targeted awareness campaign Public education programme Using water wisely roadshows	CCW, Waterwise, Agency, WCs, Local Agenda 21, Defra, Environmental Groups (RSPB), Going for Green, Global Action Plan, Ofwat Customer Service Committees, CIWEM	All water users
Water efficiency information targeted to specific water users	Agency, WCs, CCW, Waterwise, Professional bodies	Industry waste minimisation groups, Industry Associations, CBI, farmer groups, NFU, CLA, NHS Trusts, Local Authorities, Facility Managers
Water byelaws, building regulations and ecolabels		
Promotion of water efficient appliances	Consumer Association, Defra, Agency, CIBSE, BRE, BREEAM Manufacturers	Planners, Architects, Designers, Appliance Manufacturers, Retailers
Promotion & enforcement of compliant fixtures and fittings	Institute of Plumbing, Water Byelaws/Regulations scheme	Plumbers
Planning and design		
Infrastructure rebates for water efficient developments	WCs, Ofwat, Agency	Property developers
Regulation on appliances in new homes	CLG, Defra BRE, RIBA, CIRIA, BSRIA, CIBSE Envirowise	Planners, Architects, Developers
Promotion of best practice design standards	Defra, DTI, Envirowise, Agency, BRE, CIRIA, BSRIA, Academia	Planners, Architects, Developers
Promote research for new technologies	CLG, Defra	All water users
Product labelling for water appliances	Planners, Architects, Property developers	All water users

(Continued)

Table 6.1 (Continued)

Measures	Who Initiates	Who Should Respond
Re-use and recycling		
Promote greywater and reuse in all buildings and residential developments	Defra, Agency, CBI, Envirowise	All new buildings and users
Promote recycling in industrial usage	WCs, Ofwat, Agency, Defra	Individual businesses
Water efficiency advice and audits		
Promotion of water efficiency plans	WCs, Envirowise	All customers, including household, business, industry, farmers
Audit large volume users	WCs, Water UK, Private Sector funding	Industrial customers
Provide retrofit kits (e.g., cistern devices, variable flush devices, low volume showerheads or taps, water butts	WCs, Water UK, Private Sector funding	Targeted users in water stressed areas
Replace water wasting appliances and fittings	Ofwat, Agency, WCs	Targeted users in water stressed areas
Mains leakage reduction	WCs	WCs WCs, customers
Water mains		
Customer supply pipes		

Key to acronyms in Table 6.1:

Agency (EA)	The Environment Agency
BRE	Building Research Establishment
BREEAM	Building Research Establishment Environmental Assessment Methodology
BSRIA	Building Services Research and Information Association
CBI	Confederation of British Industry
CCW	Consumer Council for Water
CIBSE	Chartered Institution of Building Services Engineers
CIRIA	Construction Industry Research Information Association
CIWEM	The Chartered Institution of Water and Environmental Management
DEFRA	Department for Environment Food & Rural Affairs
CLG	(Department for) Communities and Local Government
OFWAT	Office of Water Services
RIBA	Royal Institution of British Architects
RSPB	Royal Society for the Protection of Birds
WCs	Water Companies

Econometric models have long been used in water charge forecasting (Ofwat, 1994), which are empirically developed for expenditure relating to operating, distribution, resource, treatment and 'business activities' (customer services, scientific services, regulation and doubtful debts). A range of variables are used in modelling, including mains length, pumping costs, number of billed properties, and treatment costs.

However, the long-term outcome has not kept water charges down. Research at the University of York has shown that since the privatisation of the water industry in England

and Wales in 1989, water prices have been increasing—faster than overall prices and faster than average earnings. If by 2011, all RPI items are calculated as around 230% those of 1987, average earnings are a little over 300%, then the RPI of water prices are close to 420% creating real concern over water poverty (Huby and Bradshaw, 2012). Simply increasing charges for water and sewage charges is a blunt and unfair instrument to limit demand.

6.4.1 Leakage Management

Many stakeholders, be they statutory, such as Ofwat and the EA, or environmental NGOs and pressure groups, expect that water service companies to reduce leakage, especially in stressed areas. Since 1989, water service companies are required to submit audited and certified annual calculations of distribution input and water delivered to customers; in this manner distribution losses are derived. It is perhaps a truism that waste puts pressure on supplies without benefiting either consumer, or any water service company. With treated water particularly, the cost of that treatment, pumping and maintain the pipes in the network the loss reflects not just a loss of revenue, but the cost of over-production.

Overall, the water industry has far to go. Companies failing to meet targets may face tougher regulatory action. Causes of leakage loss may be summarised (NRA, 1995d) as:

- age of the distribution system;
- pipe materials used;
- soil type (corrosive or not) and weather conditions (freeze-thaw, clay shrinkage in hot weather);
- operating pressures;
- water companies' commitments to leakage control;
- rate of renewal of the distribution system; and
- availability and cost of securing local water resources.

Losses of treated water occur through leakage and overflows from the pressurised pipes and fittings in the distribution system and in private supply pipes. *Distribution losses* include all losses of drinkable water between the treatment works and the highway boundary to a property, whereas *supply pipe losses* are leakage from the customer's pipes. In certain situations, leaked water will recharge groundwater, but this represents a loss of potable quality water which has been treated and distributed at cost.

In deciding the degree of leakage management through reduction of losses, decisions have to be taken largely on local conditions, by identifying priority leaks or bursts and considering the ability of the system to deliver adequate water in the face of demand and operating pressures. Eventually, decisions have to be made on the basis of operational and economic considerations in the face of appropriate local knowledge. Even if water undertakings can talk of an 'economic level of leakage', in extremely dry periods, the media, pressure groups and regulators to the matter of leakage reduction.

Water companies, in planning resource development and distribution, have to evaluate the combination of investment in additional supplies, leakage control and metering (Ofwat, 1993a). Pipework beyond the stopcock is not the responsibility of the water service company. Figures quoted for the distribution network are therefore a measure of the leakage control by the water undertaking. Leakage targets, set by Ofwat, are supposed to incentivise reduction in losses. In 2010-2011, 15 water companies met their leakage targets, reducing their leakage by a total of 15 Ml day^{-1} and six failed, namely Anglian Water, Dwr Cymru (Welsh Water), Northumbrian Water (North East operating area), Severn Trent Water, Southern Water and Yorkshire Water (Ofwat, 2015c). The regulator regards this seriously.

While the long-term trends for water lost in this way, like overall consumption has reduced for some time in England and Wales this represents around one quarter of losses put into supply (EA, 2008). Unfortunately since then the trend has increased with over half of companies reporting a reversal of the trend (CCWater, 2016b). For Scotland, leakage losses have been high, often due to ageing (and often Victorian) infrastructure, but in the decade to 2014 losses approximately halved which is encouraging (Scottish Water, n.d.).

In repairing existing water mains there is the issue of diminishing returns (it eventually becomes uneconomical to fix progressively smaller leaks). Increased operating pressures aggravate the situation, and finally there is the matter of inconvenience and bother of digging up the road and accruing public displeasure. Causes of leakage are many, but primarily there is the age and condition of the distribution network, often past its sell-by date with pipes made of such materials as cast, spun or ductile iron, asbestos, cement. Synthetic pipes made of plastic are now favoured by the EU. Pipe replacement is expensive, generally because leakage is unacceptably high or larger bore pipes are required. Pressure has to be maintained not only for supply purposes, but also for firefighting, although it is sometimes dropped at night, when demand falls, in order to conserve water otherwise lost to leaks. The London ring main, described above, is one effort to overcome such problems.

6.4.2 Domestic Metering

Metering for purposes other than domestic supply has been usual in England, Wales and Scotland, although domestic customers in England and Wales have long been able to opt to be metered, once at their own expense (Ofwat, 1991), while today installation is free to the customer. While industry has, under the influence of pricing, reduced consumption, the domestic consumer has continued to use unmetered water on a basis set by a presumed 'rate' for a property, rather than a charge per unit volume. At the time of writing nowhere in UK is metering compulsory because incentivisation is preferred. This makes the UK unusual in Europe because many households are *not* charged by consumption (NRA, 1995d).

There is economic logic to metering. Metering should prove to be economically efficient and, provided it does not cause impairment of domestic life in the process, should also prove acceptable to consumers, or at least to better off households. The logic being that other commodities (especially in the energy and fuel sectors) are charged by consumption. Furthermore, reduced industrial and domestic consumption reduces both the environmental stress due to abstraction and the overheads of treatment by the water undertaking. The former Campaign for Water Justice (BBC, 2009) had been concerned that poorer households are not compromised and they oppose universal water metering. However, the spirit lives on; Fairness On Tap is a coalition of organisations calling for a fair deal for both customers and the environment and, as a part of this, supports metering for at least 80% of England by 2020, specifically where there is the greatest pressure on both the environment and household economies, by 2020. The call is for fair water tariffs (Waterwise, 2016). In England and Wales, the provisions of the Water Industry Act of 1991 (as amended) specifically prohibits the disconnection of water supply to domestic premises for non-payment of charges (OFWAT, 2011a).

Against the practice of metering is the moral argument that water is a common good, a resource delivered according to domestic need for a flat rate tariff (the 'water rate'). Unit charging for water has the possibility of penalising poorer households, larger families and certain users with special needs for water to wash themselves and bed-linen; where the charge is too low, the exercise is self-defeating, where too high, under-privileged groups may suffer. The unit cost of resource for water has been low and hence may not warrant the cost of meter installation in all circumstances. However, the cost of water to the domestic consumer is rising. Over the long-term, the average household bill for water supply has risen since privatisation. The Environment Act of 1995 imposes a new duty on water companies to encourage the careful use of water. One way to achieve this is via tariffs, that is, low standing charges and slightly higher volumetric charges. Furthermore, on account of the EA's duty to secure the proper use and management of water resources (including meeting the legitimate demands), it supports in principle the use of domestic metering, particularly in the south and east of England, where there are potential resource shortages.

Does metering encourage the wise use of water? To some extent, it is showing itself to be a way of customer restraint on the basis of paying for volume consumed, hence it *does have* the potential of being instrumental in demand control. However, with water being a relatively inelastic commodity and unit charges relatively small, other considerations, notably leakage control come to the fore. Metering trials have been undertaken, one involved the entire Isle of Wight, with a study of around 50,000 properties over three years (NMTWG, 1993). The exercise was primarily to examine the reduction in resource demand (rather than customer satisfaction), but it must be borne in mind that the socio-economic makeup of the Isle is rather different from areas of the mainland. Some results of this trial highlight the positive and negative aspects of metering.

In this study, metering proved significantly more expensive than the existing rating system, averaging an extra £19.08 per dwelling per year in 1992/1993, over and above the rateable charge. The study considered the entire installation and operational process, including cost and complexity of plumbing and the cost (1992/1993) of installation averaged £165 per meter for around 95% of properties, mainly installed internally. Automated meter reading may make savings and rateable charging (the old system of charging for water services) is based on estimates, yet is billed in advance. Meter charges are in arrears, and this can lead to cash-flow problems for the service company. Yet, the majority of customers had lower bills compared with rating. The 'losers' were households with higher consumption, such as large families, or those with health problems requiring heavy use of water for the toilet, bath and laundering. Some of these consumers became hostile to metering.

Yet, in water resource terms (as distinct from the economics) the trial was judged a success with a reduction in demand, although it was found that some 20% of meters had a degree of inaccuracy, generally, at low flow rates so this was not a large overall problem. The figure for overall saving of 11.7% actually comprised 2.6% saving by lower customer demand, and the remainder was through repairs to service pipe leakage (Davis, 1994).

With the issues defined more than 20 years ago, the discussion over water metering continues. A recent study reports some 30% of UK households are metered, yet household consumption remains at around 150 litres day^{-1}, a figure slightly above Western European average but below certain other developed countries. One school of thought has it that metering is fair for it not only means the consumer pays for what is consumed, but it also provides incentive to reduce consumption. As the Isle of Wight study found, it should also incentivise a reduction in leakage, although this has its limitations because most leakage is hidden. The government generally promotes metering for the UK's 28 million households. However, irrespective of water justice issues, international comparisons suggest there is little saving where water metering is compulsory (Staddon, n.d.), yet another claim has it that on average, metering reduces water by 25 litres per person per day due to reduced leakage and consumption. With average consumption of water in the UK at around 150 litres per person per day, the saving would be significant (Fairness On Tap, 2011).

Arguments for and against therefore revolve on the one hand on 'water justice' and on the other around technical issues about then kind and operation of metering options. Water metering has long been on the agenda of environmental organisations however there is a need to balance this with water justice for the poor as there are issues for affordability and this links in to long-standing concerns about the commodification of water replacing that of public service. Water charges will inevitably be greater where more water is used, and this may be the case where there is a real need, including large families and medical needs. However, incentivising people to reduce water consumption has environmental benefits. With some companies, including Anglian Water, Southern Water, South West Water and Veolia Southeast have or are planning near-universal (or very high levels of metering) an element of the squaring of the circle is required. Well-targeted tariffs (reflecting household need) and help to waste less water may show the way forward (Fairness On Tap, 2011).

6.5   Irrigation Demand Management

Irrigation is the augmentation of soil water in order to benefit crop production. Worldwide, agriculture is still the greatest single user of water, and this includes water delivered directly as precipitation. Arable farming in Britain was effectively unirrigated until after the Second World War. Since the 1950s, technical developments such as efficient delivery systems (overwhelmingly the importation of U.S. technology for spray irrigation), equipment and physically based means to calculate requirement (based upon the Penman formula) has made irrigation a familiar scene in lowland agriculture. In economic terms, considerable investment has supported this growth and locally its draught on water resources has been felt.

6.5.1   Irrigation Delivery

We must separate issues of *irrigation application* or *delivery* from those of *scheduling*. The former involves considerations of agronomic practice, topography, engineering and indeed leakage, whereas scheduling is a matter of crop requirement, stage, soil type, climate and weather. Scheduling establishes the timing, frequency, rate and quantity of irrigation water application. Indeed, water consumption by crops is mostly transpired, only a small fraction being used in the photosynthetic reaction. Where rainfall is insufficient to meet the evapotranspirational demand on the crop, water enters the soil as irrigation water, and it may be applied to the soil in a number of well-established ways.

Gravity systems are where the water is delivered by gravity into a basin or via furrows between the rows of crops. Other systems include delivery via holes in pipes laid within the crop *(trickle irrigation)*, or delivery as *overhead* by means of a sprinkler system or raingun. An invisible means of water arriving in the root zone is via *capillary rise* from a shallow watertable in soils with suitable hydraulic properties; this is *subirrigation* (Barrow, 1987, p. 229; Doneen and Westcot, 1988).

Systems of gravity irrigation are the cheapest to operate but have a large labour requirement. They were once employed in Britain for meadow irrigation to warm the soil, in the winter and early growing season, control weeds and flush pastures with nutrients (particularly N and P) and oxygen dissolved from the atmosphere in the water applied during the irrigation or 'drowning' of watermeadows and catchmeadows (Cook, 1994; Cook *et al.*, 2004, 2007; Cutting *et al.*, 2003). Late season drowning eliminated the soil water deficit beneath pastures to enable hay crops to be produced. Gravity systems have the advantage that water will infiltrate into the soil if suitably led across the meadow or through the crop. Where surviving, or reinstated, watermeadow systems in England probably only produce a modest draught upon water supplies by diverting river flow or springs on to the adjacent floodplain meadows or valley-side meadows and then immediately returning the water to the local hydrological system during times of low evapotranspirational demand. By contrast, the environmental stress in hotter climates caused by flood and gravity irrigation systems may be considerable and also lead to soil problems, especially salinization.

Trickle irrigation relies on water trickling out of a perforated pipe lying along the surface of the soil. Such pipes are expensive to install, but have low labour costs and only require low water pressures to operate. Fixed and trickle irrigation delivery systems are comparatively rare in Britain apart from horticultural situations, especially in glasshouses, and are only justified for high value crops such as hops (MAFF/ADAS, 1982). Delivery within the canopy at the soil surface minimalises evaporative losses.

Overhead systems, common in developed countries, requires high water pressures to operate sprinklers or rain-guns mounted in semi-permanent or mobile lines, or centre-pivoted spray booms. These systems are generally much cheaper to install than trickle systems, although their labour costs are relatively high, but reduce with self-propelled systems (MAFF/ADAS, 1982). The main advantages of overhead systems are that they most closely imitate rainfall, are mobile and that they are applicable for most crops grown in Britain. A range of topographies may be irrigated, meaning that operation is not restricted to flat land. Spray equipment is capital rather than labour intensive (a situation which suits British agriculture) and portable, but there can be up to 25% evaporative losses through direct evaporation to the air and interception loss (Barrow, 1987, p. 243).

Subirrigation involves manipulation of a shallow watertable, so that it remains below the root zone, but feeds it through capillary rise in the subsoil. It requires nearly level topography and soil hydraulic properties which favour upward movement of water. In Britain, water may be applied through subsurface pipes. Capillary rise has the effect of increasing the effective root depth to below the physical root zone. The delivery system is generally an underdrainage system set to back up with water from its outfall channel during a dry spell. In flat, underdrained land such as reclaimed 'marshland' in grazing, horticultural or arable management, subirrigation an attractive option because there is no direct evaporative loss, once water has left the feeder watercourse. One example of this practice is in the silt Fens of East Anglia.

6.5.2 Consumption and Restriction

Agricultural uses accounted for just 0.7% of recorded water abstraction in England and Wales in 2011. As a percentage, of recorded water abstraction within each region, this varied between less than 0.1% in Wales and 5.1% in the Anglian region and can be higher on a daily basis during the summer (Observatory Monitoring Framework [OMF], 2014). The majority of water designated as agricultural in England and Wales goes to spray irrigation. The former National River's Authority (NRA) produced the 'Water Resources Strategy'. The Strategy required to ensure 'quality, reliability and continuity of produce at stable prices which are often essential marketing requirements', and where sufficient resources are not available, production in some sectors would be jeopardised (NRA, 1994 g). Furthermore, demand is at its highest precisely when river flow is lowest. For example, during the 1988-1992 dry period licences permitted some restrictions upon abstractions. Long-term demand will depend upon the progress of agricultural policy. The irrigated land area is only expected to rise in England and Wales. In Scotland, irrigation is largely in the east for potatoes and high values crops, but it is increasing and a charging system is being introduced (Scottish Environment Link, 2015).

Across the UK, demand for the irrigation of arable crops continues to grow. Historic contingency plans for increased water demand for agriculture (NRA, 1993c; NRA, 1994e) included:

- Restriction of licences, not issuing new ones in future or including restrictive clauses within them (this was realised under the Water Act of 2003 that allows for the amendment and retract licences to abstract).
- Scheduling systems designed to optimise the application and phasing of irrigation to match the specific crop need, specifically the introduction of data loggers (for recording farm consumption and assisting management decisions). There is an irony here; certain farmers apply less than a calculated soil water deficit, so better forecasting may lead to greater use.
- Conjunctive use of groundwater and surface water abstractions to minimise impact upon low flows and groundwater reserves (only where an individual farmer has both options).
- Cooperation between farmers: for example, between farms where there are compensatory discharges from groundwater systems and farms where low flow licence restrictions still apply. This would be settled commercially between the parties including the possibility of 'buying-out' of water licences.
- Better use of under-used sources where this will not have adverse environmental effects (given the stressed parts of the country this does not look to be a promising avenue for development).
- Development of winter storage reservoirs as a cooperative venture between farmers (to make it an economic proposal) where technically feasible.
- Better consumer (i.e., farmer) education and information on water conservation including cooperative action to enhance conjunctive use.
- More accurate and less leaky delivery systems (a technical solution).

Any means to support efficiency in irrigation delivery will assist in times when water scarcity is to the fore. In England and Wales, up to 20 cubic metres (4,400 imperial gallons) of water per day are exempt from licensing. To cap the permitted volume, economic instruments have been proposed, for example, the issuing of 'tradable' permits is similar to the way in which milk quotas are bought and sold. To date, only licences themselves are tradable and due to complexity this has been restricted (Utility Week, 2015). Other measures include more efficient delivery systems to the sprinkler that reduce leakage at any joint which both represents a loss of water and an unwelcome pressure drop (Barrow, 1987).

6.5.3 Irrigation Scheduling

Across the world, in flood irrigation, water allocation between users or fields is part of scheduling. Historically in England, meadow irrigation was generally scheduled non-quantitatively through legal agreements (Cook *et al.*, 2008). Spray irrigation, however, is more localised making calculations of the quantity applied to a specific field most important. This is especially important where sources are few and far between, or where there is overall pressure of water resources, for example, during a dry season. Hence, meteorological methods enable soil water depletion rates to be calculated, and then balanced against precipitation rates, calculate a shortfall used in irrigation planning. The resulting soil water balance enables a measure of irrigation requirement to be calculated.

Penman potential evapotranspiration from a short green crop (E_T) has been regularly available for many years, calculated from standard meteorological criteria (Wilson, 1983, Ch. 3). The method relies upon knowing air temperature, humidity, wind speed and sunshine or radiation. It can be employed to calculate *either* open water evaporation (Eo) *or* Et for a short green crop and has been shown to be accurate and hence useful for establishing reference crop evapotranspiration. The original Penman formula has been modified on a number of occasions, most notably by Monteith to take account of canopy resistance and aerodynamic roughness, and hence produce a calculation of actual evapotranspiration that is crop specific (Ward and Robinson, 2000, Ch. 4). Alternatively, actual evapotranspiration *(AET)* is calculated by the application of an empirically derived crop coefficient ('K_c'). This constant, which is continually reviewed in the light of continuing agronomic research, depends upon crop and growth stage. To obtain a figure for *AET* it is multiplied by a 'reference' figure E_T for potential evapotranspiration (e.g., Doorenbus and Pruitt, 1976, p. 39). K_c relates to a disease-free crop grown in an open field under conditions when the soil water supply is not limiting, and fertility is optimum. E_T is defined at the maximum level possible, under the prevailing weather conditions. The preferred definition is a green crop (such as grass) which is 12 cm high, with a fixed canopy resistance (70s m^{-1}) and an assumed albedo of 0.23, actively shading the ground (near 100% cover) when soil water supply is not limiting (Smith, 1992).

Where soil water content in the root zone falls appreciably, and soil water supply cannot meet the transpirational demand on the crop (i.e. the ratio between actual and potential evapotranspiration falls below unity), a further coefficient has been proposed, the *soil water coefficient, Ks* (Hanks and Ashcroft, 1980, p. 115). When a crop is irrigated according to an effective calculated schedule, then it has to be assumed that the crop is growing in close to optimal soil water conditions so the need for *Ks* may be minimal. Crop coefficients tend to be used in hotter climates. In Britain, the modelling of soil water balances is becoming more sophisticated, and by comparison there is less divergence between actual and potential evaporation.

Due to the complexity of establishing the empirical variables demanded by the equation, until recently the Penman-Monteith approach remained relatively unused, but represented a large step forward in conceptual understanding. However, the MORECS model developed by the UK Meteorological Office (Chapter 4) is based upon the Penman-Monteith (Prudhomme and Williamson, 2003) and is especially practically useful because it considers difference soil types; it is especially helpful to consider water shortage in terms of periods of 'low rainfall' leading to a 'hydrological drought'. Model construction is frequently updated. Scheduling which employs a 'one-step approach' incorporating data on canopy resistance, leaf area index and crop height, and crop albedo directly into the Penman-Monteith equation eliminates the need for a crop coefficient (Smith, 1992). Although improvements in accuracy are always welcome, inaccuracies in delivery systems frequently outweigh scheduling uncertainties.

Ideally, monthly estimates are used for irrigation *planning*, and 10-day figures for irrigation *design*, while daily scheduling requires daily estimates. An example of a scheme used in long-term irrigation planning is given by Dent and Scammell (1981) for Norfolk. For example, to achieve maximum yield for sugar beet on peat soils irrigation would be required for 8 years in 20; under sandy soils the figure is raised to 19, which is a dramatic difference. For potatoes, a crop vulnerable to water stress, the figures are raised to 17 and 20 years, respectively.

It has been common practice in Britain to draw up a balance sheet for irrigation scheduling which is in principle no different from a bank statement (Withers and Vipond, 1974, p. 103), with rainfall as income and evapotranspiration as expenditure. Evapotranspiration figures may often be available for 10-day periods, while daily rainfall figures are readily available on a farm. Daily figures for *AET* may be derived from mean monthly figures, then adjusted retrospectively as the figures for the preceding (real) period become available. Published irrigation guidelines, based upon trials, will advise the soil water deficit for optimal growth. This is seldom field capacity, and may be appreciably less, which is no bad thing where water supply is limited. For example, sugar beet, once established, grows well under soil water deficits of 25-50 mm, and supplementary watering may be beneficial during one year in three (MAFF/ADAS, 1982).

To avoid the problems of meteorologically derived calculation, direct, instrumental measurement of soil water is an alternative to using meteorological models (Muñoz-Carpena, n.d.). Direct measurement as a basis of scheduling has the advantage of not cumulating error over the season. However, it has a number of disadvantages, mainly because any point measurement is open to the criticism that it may not be representative of the entire field, given soil variability. Actually, the simplest (and standard) method is rather cumbersome. The investigator takes a known volume of soil, weighs it, dries it at 105 °C for 24 hours and re-weighing enables the volumetric water content to be readily calculated. It can be undertaken repeatedly in order to establish a time series, although not from the same place and there emerge problems of depth sampling. There is no absolutely problem-free and cheap method of measuring soil water contents and instrumentation can present significant costs to the farmer. Trials of crop water management, plant production and yield permit *threshold potentials* in the root zone to be identified below which crop performance, and hence production, is in some way inhibited (Cook and Dent, 1990), typically seen as reduced photosynthetic activity and reduced yields.

6.5.4 Unirrigated Soils and Water Conservation

Overall, the demands on water resources, delivery and methods of scheduling have been discussed. It remains to consider the possibility of managing the soil itself to encourage better use of soil water for crop growth. It is also important to appreciate the operation of unirrigated soils in order to assist land-use decision making. Cook (1987) used cluster analysis on selected crop and soil combinations for soils in East Anglia. Soils can then be grouped into 'hydrological families': groupings of soil series according to texture. This method anticipates the kind of approach used in the HOST soil-type classification (Chapter 2).

Identifying the ability of a soil to supply the transpirational need of a crop is there for central to understanding soil water uptake, water-stress induced and irrigation requirements of a crop. Another approach is to somehow reduce excess loss of soil water such that a crop can maximise the available water for its growth. Considerable research has gone into mulching and/or incorporation of surface-modifying substances in order to reduce bare-soil evaporation; effectively seeking to close off one hydrological pathway to the atmosphere.

Mulching (with a range of substances from plastic sheets to compost) and topsoil incorporation of organic matter may be a way of conserving soil water for crop use. Application of mulch materials to sandy soils may be very efficacious in tropical situations, as indeed in temperate environments (De Silva and Cook, 2003). Actually, the whole subject of mulching is complicated and tied up with other considerations including energy balances (and hence thermal properties) of a soil, soil fauna ecology, weed management, erosion protection and nutrient supply and balances (Movahedi Naeni and Cook, 2000; Baggs *et al.*, 2003).

Uson and Cook (1995) demonstrated that the available water capacity of a silt loam soil with composted organic waste incorporated at a rate of $60\,t\,ha^{-1}$ increased from 14.6% to 17.4% and that the mulch applied at the surface at the same rate increased soil water content in the surface 0.1 m until 80 days after sowing maize with a herbicide treatment to suppress weed growth. This effect was reduced in the presence of weeds. Mulching, however, has the capacity to reduce infiltration into the soil through interception of rainfall or irrigation water and its evaporation back to the atmosphere as the application becomes thicker (Cook *et al.*, 2006). Figure 6.6 shows the effects of compost mulch on soil surface layer water retention under a maize crop. Soil water contents in the top 0.1 m are plotted under maize with herbicide weed supressing treatments. Arrows show lines of significant difference ($p = 0.05$) between treatments. This finding shows a possible future for organic mulching material which not only potentially increases the organic matter content of the topsoil, but also provides a source of nutrients when it degrades. Furthermore, organic wastes (from farm and domestic sources) are readily available, although there is a cost involved in commercial composting.

Soil water conservation measures and sound understandings of land capability should be included in any lists of water conservation measures for farmers. Any means to physically reduce demand by better scheduling and more efficient delivery systems and crop water use are to be welcomed in regions where there is pressure upon water resources from agriculture, and may yet prove economically viable in that they reduce irrigation inputs. Efficient use of soil water and its conservation is therefore an integral part of any notion of 'sustainable agriculture', no matter how this may be defined.

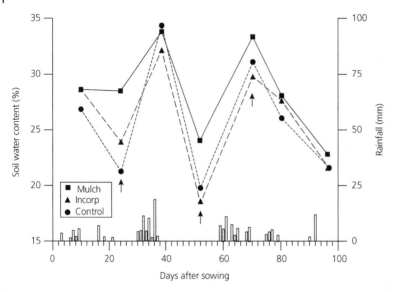

Figure 6.6 Effects of compost mulch on surface soil water retention under a Maize Crop (Redrawn from Uson and Cook, 1995).

6.6 Water and Effluent Re-Use

'The NRA will have regard to where water will return to the system when considering abstraction proposals.' Thus, more than 20 years ago, the then competent body for England and Wales was able to place water re-cycling centre stage (NRA, 1993d). There is encouragement for water to be returned after use and hence available for re-use and the statement also embraces licensing policy. There is thus 'negative feedback' in the abstraction-consumption-treatment-return cycle in order to reduce demand. Used water is to be returned to the river as close as possible to the point of abstraction.

Effluent re-use can be *direct,* whereby treated effluent is returned into the supply system pipe-to-pipe. It may also be connected to a water treatment system or discharge into a direct supply reservoir. Alternatively it may also be *indirect,* whereby effluent is discharged to a river system, aquifer or raw water reservoir and subsequently re-abstracted and treated (NRA, 1992g). There are no direct schemes in Britain, because the possibility of human error is large. It is anticipated that any new schemes would be by indirect use, the water environment acting as a self-cleansing buffer between returned and re-abstracted water.

It has been estimated that, in the Thames Basin as a whole, some 13% of river abstraction is used for public supply (NRA, 1992g); overall perhaps one-third of water is recycled in London, around 43% for the UK (Londoncouncils, n.d.). This proportion suggests a need for improvement through coherent policies on water re-use to reclaim water for consumers. At stake is securing supplied for the long-term in situations of climate change and rising demand. Actually, modern policy drivers arise from EU Directives, specifically the Urban Waste Water Treatment Directive, the Water Framework Directive, the Integrated Pollution Prevention and Control Directive as well as more local Catchment Abstraction Management Strategies. A House of Lords report has called for water re-use from 'sub-potable' treatment for industry (HOL, 2006, p. 69).

The long-term success of water supply efficiency measures relies upon 'evaluation', that is, periodic review of operations in the face of objectives in order to minimise inefficiency in water resource development planning. An assessment of plans should be made ahead of implementation (Barrow, 1987, Ch. 5), remembering that the success is only as good as the assessors. The entire process, which may be seen as *water resource management centred*, ideally comprises a *pre-project evaluation* of likely economic, environmental and social costs, *in-project evaluation* when a scheme is operating with the objective of improving operating practices, and a retrospective *post-project evaluation* so that mistakes are not repeated.

Water re-use can be either for potable or non-potable purposes and the practice can reduce both consumption as well as bills in metered households. Of the regulators, both Ofwat and the DWI continually review their operation following evaluation of procedures. The former reviews charging and includes public consultation, while the DWI is mostly concerned with monitoring procedures (DWI, 2015). The water service companies are part of the process through the periodic creation of five-year corporate plans including carbon use, economic efficiency, pricing, investment and customer relations, as well as practical considerations such as leakage, sewage services and demand (Thames Water, 2015). These plans are submitted to OFWAT. Evaluation is applicable to both supply and demand efficiency in resource management. This approach should be a normal part of policy review, and indeed of operation and policy review following, for example, metering trials.

6.7 Public Perceptions and Water Education

Public information was effective in restricting demand generally during the dry spell of 1975/1976 and during the period 1988 to 1992 in the Brighton area. However, sympathy for the water undertakers had run dry during 1995. Yet it is, entirely defensible that the EA should press for better public awareness, and at face value it is reasonable for water undertakings to produce educational material promoting efficient water use (NRA, 1995d), especially as it is a part of their duties under the Environment Act 1995 to promote efficient use.

There remains an historic problem. Pre-1989 most water services remained in public hands and the 'private' water only supply companies were heavily regulated and operated in concord with the public-owned services. Since then, the EA was created as a public body but no water undertakings in England and Wales remained in the public sector. In Scotland, by contrast, the industry has been re-organised effectively as a publicly owned industry.

Critics have often commented on the instinct of a private company to sell its goods and services, and find this at variance with the ethos of a public service industry. Indeed, the switch from a public service 'industry' to privatised utility has caused a 'commodification', for the resource itself, and the services provided (including treatment and sewage services) may be bought and sold (albeit regulated), while the undertakings themselves are available for takeover bids. Meanwhile, for consumers, the privatised water utilities have, with the regulation of Ofwat, passed on substantial charges to the consumer. Rightly or wrongly, pleas for public understanding of resource problems, especially losses from leakage, are met with criticism from pressure groups and media derision. This was especially true of Yorkshire Water during 1995. In such an economic climate, consumers may expect to purchase this commodity at will instead of controlling consumption. We may ask, in the public minds, is water in England and Wales effectively different from, say fuel for their cars? Furthermore, during the winter of 2014/2015, there was political concern, reflected in the media, around the cost of energy for domestic consumers; something analogous to arguments around public water supply that may yet come to bite the water industry.

Clearly, government is concerned about charges to the customer and, while effectively defending privatisation on account of a stated ability to capture investment in infrastructure, feels the need to set out principles for charging (Defra, 2013b). It can state there is a need to represent a fair deal for all customers, better reflect the costs of supply, increasing the long-term resilience and sustainability of our water resources, increase competition (including the facility to switch suppliers) and 'provide for new entrants to the market' (perhaps reflecting the practice in France of municipalities franchising out water services to private undertakings). Meantime, Ofwat will continue to use a price review process 'to protect households and ensure that the charges they pay are fair and transparent'. Neither will households pay for the 'implementation of the competitive market'. And a growing number of companies are 'implementing Social Tariffs to help vulnerable households at risk of affordability problems'.

One can, especially in the light of political concerns over domestic and industrial energy charges, detect a concern that water should not enter into a furore comparable with that of energy bills. Yet, the case is once more building towards a campaign to change the private status for the water utilities (Corporate Watch, 2013), while the idea of re-nationalisation does not go away. One reason may be a lack of transparency and dubious relationships (and reporting practices) between companies and regulators. While private investment has been good and little public investment would seem to support the status quo, the matter of ownership of this vital industry in England and Wales remains both far from the surface.

6.8 Summary

To meet the perceived increase in demand in England and Wales there are a number of approaches, summarised in, and developed from, the NRA's 'Water Resources Strategy' of the 1990s. These have continued into the present neo-liberal influenced political and economic situation south of the Border. Investment has enabled 'technical fix' kinds of solutions to remain a feature, taking the form of some bulk raw water transfers in preference to reservoir construction and groundwater recharge schemes. Actually, some 25 years from privatisation, the progress in this area has been steady but never dramatic. Water conservation measures have, on the other hand, continued apace. Irrespective of political stance, this has to be an agreed 'public good' to which further development is due.

Reservoirs remain central to existing arrangements, but with the importance now attached to environmental imperatives, care has to be taken not to upset the hydro-chemical balance of the receiving waters which may have adverse effects upon their ecology. The key to resource management is being able to balance abstraction with conservation, especially of river flow and groundwater levels. Here the saving grace is better use of information, especially modelling which permits conjunctive use.

Private investment has not led to any grand raw water transfer schemes between regions, from areas of excess water for supply towards the drier south and east of Britain. In policy terms the emphasis has shifted away from seeking further supply measures towards demand-based and use-efficiency measures. Most prominent is leakage management because it is treated water which is going to waste. Demand management places the onus clearly on the water supply and water service companies, that is, the private sector to which water charges go from the consumer. Both the EA and Ofwat continue to favour leakage management and, in appropriate areas, metering. Increased charges and shortages may now mean that a climate of sympathy towards water undertakings is difficult to achieve.

Following privatisation of the majority of the water industry in England and Wales in 1989, there have been a number of shifts of emphasis. For one, the sense of a loss of public service in favour of commodification remains strong and this is most manifest in the debate around fairness of tariff-setting and metering; although the latter is gaining currency in environmental terms, issues around water access and water justice has not gone away.

What was once 'our' problem, faced by predominantly public employees, is firmly 'their' problem, and we pay private companies to get it right, but is the right and

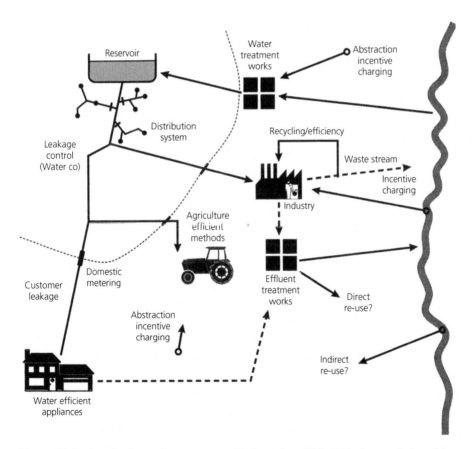

Figure 6.7 Options for demand management. (Redrawn from NRA, 1995c, by permission of the Environment Agency).

importantly will it remain thus? Education is always worthwhile, but long-term government policy is apt to set public cooperation against water restrictions, while in reality we are asking suppliers to encourage their customers to purchase less. By way of summary, Figure 6.7 shows diagrammatically the established options for demand management (NRA, 1995c). This may be laid alongside growing concern that the fundamental principles of the *status quo* for water supply in the political and economic arena may yet be challenged.

7

Water Quality Background Issues

Everyone knows that water from a tap is the stuff that falls as rain or snow, flows down rivers and comes up from wells. The difference is that we expect to be able to drink tap water neat.

Water in Great Britain, p. 13, *Celia Kirby (1984)*

7.1 Introduction: Point and Non-Point Source Pollution

This chapter examines the causes of contaminant loading in waters including not only chemical sources, but also 'physical' sources such as nanoparticles and heat pollution. First, the sources of chemical, physical and biological problems are explored then the hydro-chemical environments (where chemical transformations take place) are discussed. The following sections deal with agrochemical loading, acidification and urban water quality issues as introductions to later case studies. Pathways by which pollutants enter the hydrological cycle may be summarised as:

1) Spillages or discharges to surface waters (e.g., industrial effluents, silage leakage, untreated sewage).
2) Spillages on the ground which affect soil and groundwaters (e.g., spillages from toxic industrial substances such as solvents, pesticides).
3) Contaminants discharged directly into the ground (e.g., potentially toxic substances from landfill, septic tanks).
4) Sprays applied at the surface which may be washed into the soil, groundwaters and ultimately surface waters (e.g., pesticides).
5) Soil disturbance and/or agricultural practices involving disruption of a given management system and leaching pollutants (e.g., nitrate leaching following grassland ploughing, phosphate pollution associated with soil erosion).
6) Soil disturbance and/or agricultural practices involving applied agrochemicals (e.g., boosting nitrate levels in the plant-soil system through inorganic fertiliser application leading ultimately to leaching).
7) Pollutant loading of the atmosphere resulting in dry or wet deposition of pollutants (e.g., acidification of the soil and consequent mobility of metal ions in waters).
8) Changes (generally increases) in the 'natural' temperatures of waters, typically from cooling of power stations that can affect rivers, lakes and the shallow marine environment.

The Protection and Conservation of Water Resources, Second Edition. Hadrian F. Cook.
© 2017 John Wiley & Sons Ltd. Published 2017 by John Wiley & Sons Ltd.

Numbers 1, 2 and 3 constitute point source pollutants on account of the origin of the pollutant source being readily identifiable such as an outfall or discharge pipe or identifiable location of a spillage. Numbers 4 to 8 constitute diffuse pollution (both physical and chemical) on account of the source of pollution being 'diffused' over a wider area while number 7 might be extended to include the all-embracing matter of anthropogenic climate change. Both agricultural and urban land supply contamination including pesticides and nitrates, while acid deposition (with origins in industrial processes and from transport) is a problem of international dimensions.

Point pollution can readily be monitored and controlled via consents to discharge; diffuse pollution is less straightforward and its control potentially relies upon changing land management, reducing certain activities in urban areas and controlling emissions to the atmosphere. The Anti-Pollution Works Regulations 1999 are served under section 161A of the Water Resources Act of 1991, and were inserted by the Environment Act of 1995. These enable the EA to serve 'works notices' on polluters (or potential polluters) that require them to carry out works or operations to remedy or prevent pollution of controlled waters. There is an additional regulatory tool to help prevent pollution incidents. The works notice powers also allow cost recovery from whoever is responsible, or where no person are found on whom to serve a works notice (Defra, 2001).

7.2   The Chemical Basis of Water Pollution

The following is largely based upon White (1979), Moss (1988), Ward and Robinson (2000), Mannion (1991), NRA (1992c), appropriate chapters in Mitsch and Gosselink (1993), Battarbee and Allot (1994), Gray (1994) and Cook *et al.* (2009). Pure water has the composition H_2O, and at sea level it boils at 100°C and pure water freezes at O°C. The molecule is covalently bonded (meaning that electrons are shared between the oxygen and the two hydrogens) and it is structured with an angle between the bonds of the hydrogen ions of about 105°. Consequently, the water molecule is 'bipolar'. This polarity means that water molecules are able to break up the weaker, electrostatic bonds of many substances, which encourages solution in water.

7.2.1   Water Acidity

Water molecules also have the property of dissociation, whether other substances are present or not. This means that some will separate into hydroxyl (OH^-) and hydrogen (H^+) ions. This fundamental property, whether or not other substances are present, is measured by the *pH* scale. In formal terms, *pH* is defined as the negative logarithm of the concentration of hydrogen ions, or *-log [H^+]*. The pH of natural rainfall is about 5.64 on account of the solution of carbon dioxide (CO_2) in the atmosphere. Acid hydrolosis in the presence of CO_2 is as follows:

$$CO_2 + H_2O \leftrightarrow H_2CO_3$$
$$H_2CO_3 \leftrightarrow HCO^{3-} + H^+ \leftrightarrow CO_3{}^{2-} + 2H^+$$

The presence of the hydrogen ion affects the pH of the solution and a series of acidic reactions are possible, notably concerned with weathering through solution of calcium carbonate (see section 7.6.2). Solution of pollutants such as sulphur dioxide (SO_2)

and nitrogen oxides (N_2O from denitrification, NO_2 and NO from combustion) produces reactions which result in sulphurous and sulphuric acids and nitrous and nitric acids. For example:

$$NO + O_3 \rightarrow NO_2 + O_2$$
$$NO_2 + OH^- \rightarrow HNO_3^- \rightarrow NO_3^- + H^+$$
$$SO_2 + oxidant \rightarrow SO_3$$
$$SO_3 + H_2O \rightarrow H_2SO_4 \rightarrow SO_4^{2-} + 2H^+$$

Such reactions also reduce the pH of precipitation, and hence potentially acidify natural waters well below pH 5.5, though not usually below pH 3. 'Acidity' may be considered the ability of a substance to react with an aqueous solution's OH^- ions, and 'alkalinity' its ability to react with H^+ ions.

7.2.2 Hardness

Hardness is a property associated with the presence of divalent metal cations such as Ca^{2+}, Mg^{2+}, Sr^{2+}, Fe^{2+} and Mn^{2+}. Associated with these are anions such as HCO^{3-} (bicarbonate), SO_4^{2-} (sulphate), Cl^- (chloride), NO_3^- (nitrate) and SiO_3^{2-} (silicate). Permanent hardness is associated with non-carbonate hardness and is not removed by heating, as is carbonate hardness of the kind forming fur in kettles. A typical hardness reaction is:

$$CaCO_3 + H^+ \rightarrow Ca_2^+ + HCO_3^-$$

calcite   bicarbonate

Calcium carbonate goes into solution as the bicarbonate, and for the reaction requires carbon dioxide dissolved in natural rainwater, which is slightly acidic. A scale of hardness has been devised, where concentrations below 75 mg 1^{-1} are considered soft, while excessively hard waters have concentrations above 300 mg 1^{-1}.

7.2.3 Reduction-Oxidation

Heterotrophic bacteria will decompose organic matter and cause the oxygen concentration to fall. In doing so, a range of reducing reactions are made possible. The reduction-oxidation, or *'redox'* potential of water is an important variable in controlling chemical reactions in aquatic, soil and groundwater environments. The redox potential is a measure of electron availability in a solution and can quantify the degree of electrochemical reduction in wetland soils.

Oxidation can occur when:

1) Oxygen is taken up, for example, sulphide oxidation (see below)
2) Hydrogen is removed, for example, $H_2S \rightarrow S^{2-} + 2H^+$
3) When a chemical gives up an electron, for example, $Fe^{2+} \rightarrow Fe^{3+} + e^-$

Reduction is the opposite:

1) When oxygen is given up.
2) When hydrogen is gained (hydrogenation).
3) When electron gain occurs.

The redox potential (E_h) is measured in millivolts (mV), using an inert platinum electrode. Where free, dissolved oxygen is present ('aerobic' conditions), the range is generally in the range +400 to +700 mV; when there are reducing (e.g. 'anaerobic') conditions, the range tends to be between +400 and –400 mV at pH 7. In a waterlogged soil, where organic substrates are oxidised, and the redox potential falls (i.e., moves towards or into the negative range), electron gain occurs.

7.2.4 Nitrates and Ammonia

The nitrogen cycle starts with nitrogen fixation from the atmosphere by prokaryotic bacteria including some blue-green algae in soil, and the process is most effective in neutral to alkaline soils. Once fixed, nitrogen is available for plant growth in various forms including material recycled through organic material and artificial fertilisers. Ultimately, nitrogen is lost to the atmosphere through denitrification. A summary of the reactions are shown below:

$$NH_3 \rightarrow NO_2^- \rightarrow NO_3^- \rightarrow N_2O \rightarrow N_2$$

ammonia   nitrite   nitrate   nitrous oxide   dinitrogen

Nitrification (aerobic)   Denitrification (anaerobic)

The process of creating nitrate is termed nitrification, and that of reducing it denitrification. The reactions involved are microbial and governed by the redox potential. Ammonia can result in waters in catchments that have high organic matter inputs, especially agricultural catchments. It originates as a product of disassociation of organic compounds, which may be carried out by either anaerobic or aerobic bacteria, and will remain in waters (either free or within sediments and porous media) under anaerobic conditions. Ammonia volatilisation represents an important loss of nitrogen from agricultural land.

Nitrate is considered a problem in water supplies when it exceeds the EU maximum drinking water limit of 50 mg l^{-1}. It is the major factor causing eutrophication in coastal waters, and combined with phosphate it is a cause of eutrophication in lowland freshwaters including toxic blooms by cyanobacteria (blue-green algae) which may contaminate reservoirs. Its origins are agricultural runoff and sewage, although a small amount results from deposition and from natural sources including lightning fixation.

7.2.5 Phosphate

Most phosphorus arises from sewage effluent and fertilisers. Phosphorus (P) in soils has a most complicated chemistry. Orthophosphate ($H_2PO_4^-$ and HPO_4^{2-}) is released by mineralisation and adsorbed by soil particles, and held with differing degrees of availability. 'Labile' P is either available by being readily desorbed or held in solution. Availability is conditioned by soil chemistry, especially soil pH and exchangeable Al^{3+}. In situations with free aluminium and iron oxides predominant and in low pH soils which receive dressings of fertilisers, phosphorus may be released to solution and this will increase with rising pH. Where there is a naturally low phosphorus status in acid but clay-rich soils with low sesquioxides (hydrated iron and aluminium oxides) hydrolysis of Al^{3+} provides exchange sites for P adsorption and P compounds may

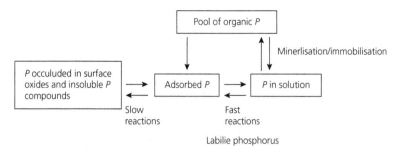

Figure 7.1 Phosphate Transformations in the Soil (after White, 1979).

become complexed within the clays and hence immobile, a situation at variance with that of acid but low clay content soils. In initially high pH soils, such as calcareous soils, P may be immobile as hydroxyapatite and octacalcium phosphate, essentially non-labile forms. Solubility will only increase as pH falls, so that availability to plants is greatest between pH 6 and 7.

In broad outline, phosphate transformations in the soil may be summarised in Figure 7.1.

Phosphorus enters water bodies as inorganic orthophosphate ions and as PO_4^{3-}: as organic phosphorus compounds and as inorganic polymers. Some forms are available for plants (including algae), while others are not immediately available, but microbial activity may make them so. Hence, a measure of total P in either soils or waters does not mean that it is immediately available, or potentially polluting. Infertile lakes may contain as little as $1 \mu g\ l^{-1}$, the most fertile $1000 \mu g\ l^{-1}$. One hazard associated with phosphate enrichment in waters is that, together with nitrate, it causes toxic algal blooms. Neither is phosphate mobility limited to solution; because it forms complexes with soil particles, it is mobile within the soil or may reside in the sediments of lakes to be later released. Persistence in soils is a major problem and raised levels may persist for decades, and indefinitely in lake sediments. Phosphate is generally quantified in terms of particulate phosphorus (PP) and soluble phosphorus (SP), which together comprise the total phosphorus (TP), or as soluble reactive phosphorus that is potentially mobile in soils and waters and hence available to plants including algae. Since about 1990 the European Environment Agency reports an overall decline in P in European waters, attributable to improving water treatment and drop in use of phosphorus in detergents (EEA, 2012b). There is furthermore a drive to ban phosphates in laundry detergents because this is a significant source of waterborne P (after sewage and agriculture [EC, 2014b]); however, there is so far no EU limit set for P in natural waters. The Environment Agency regards 'orthophosphate' concentrations above $0.1 mgP\ l^{-1}$ ($100\ \mu g\ l^{-1}$) as 'high' or more, those below 'moderate' or lower.

7.2.6 Pesticides

Pesticides (also called plant (or crop) protection products (or chemicals) consist of a complicated range of chemicals used as herbicides, insecticides and fungicides. There is a constant stream of these new products, making testing for them and predicting their behaviour in the environment and in organisms a problem, and they should be approved for use (European Commission, 2015).

These chemicals and their metabolites are increasingly causing concern with both waters and the wider environment. The first synthetic insecticides were used during the 1930s, and have become a hallmark of post-Second-World-War intensive agriculture, including assisting the 'Green Revolution' in the developing world. Certain pesticides are found in surface waters and increasingly in groundwaters. Although their concentration in aquifers is not yet thought to be sufficient to be of great concern, the true extent of pesticide contamination in all waters is only beginning to be realised, with this knowledge depending on good sampling strategies and improved methods of detection. Modern pesticides tend to degrade relatively quickly in soil, but older compounds can persist. In the Isle of Thanet chalk aquifer, pollution from diuron and its metabolites is a ubiquitous problem (Lapworth *et al.*, 2005).

The organochlorine group has been found to be persistent in soils for decades, especially DDT which is banned in many countries, and its metabolites DDE and DDD. As organochlorines (in particular) persist, being stored in the oils and fatty tissues of animals, they are prone to accumulate in the food chain, being transferred and *even concentrated* from one trophic level to the next.

Pesticide mobility in waters depends upon persistence (measured as 'half-life') and water solubility, which in turn affects their residence times in soils, and hence waters. Many pesticides show a strong affinity for soil organic matter, forming complexes with other soil constituents. Most degradation processes in soil are by chemical hydrolysis and bacterial oxidation. The behaviour of pesticides in subsoils is critical; short soil residence times will increase the chance of water contamination as the unsaturated zone of aquifers provides little chance of degradation. Neither is degradation straightforward, as some intermediate products can be extremely toxic. Others can accumulate in the fatty tissues of soil organisms (as indeed of other organisms), and are therefore only slowly leached from the soil profile.

In the UK, pesticides that have most frequently polluted water are no longer authorised. However among those that are currently in use, the most frequently occurring in surface waters are metaldehyde, propyzamide, carbetamide, chlorotoluron, 2,4-D, mecoprop-P and MCPA while in groundwater, the most frequently occurring products currently used are bentazone and mecoprop-P (Defra, 2012).

7.2.7 Other Elements

Chloride concentrations occur in natural waters as a result of saline springs from connate (original) waters in sediments, evaporite deposits, saline intrusion of groundwaters by sea waters, sea spray on soils and freshwater bodies in coastal areas, or from the effects of potassium chloride fertiliser. Concentrations ranging between 200 and 500 mg l^{-1} are common in the coastal chalk aquifers of eastern England; for comparison, the average chloride concentration in the world's oceans is 19 300 mg l^{-1}. The chloride ion moves in soils and groundwaters in a way that is analogous to nitrates; however, it is unaffected by bacterial processes such as denitrification.

Sulphates (SO_4^{2-}) occur in soils and rocks as a consequence of the oxidation of metallic sulphides, the breakdown of organic matter under aerobic conditions, or are the result of fertiliser and sulphurous precipitation. They are highly mobile and the common calcium sulphate minerals, gypsum and anhydrite, are readily leached.

Iron and manganese are both highly mobile in certain circumstances. Iron is insoluble under aerobic conditions, forming ferric (Fe_3^+) compounds, while reducing conditions cause electron gain, and the ferrous ion (Fe_2^+) is highly mobile in waters (during water treatment, oxygenation of waters precipitates the iron in the ferric state). The oxidation of sulphides in certain soils causes precipitation of ferric hydroxide where reactions with calcium buffered waters are involved. Manganese is similarly mobile under reducing conditions, but is unpalatable in waters long before any toxic thresholds occur.

7.2.8 Solvents

There are many *industrial solvents.* Although persistent, they are highly volatile in air, so much is lost before contamination of waters is possible. They are used for such purposes as paint stripping, degreasing, metal and plastic cleansing, aerosol propellants, dry cleaning and pharmaceuticals. They are organic compounds, potentially carcinogenic, and highly volatile, making them a great risk to groundwaters, either by accidental or illegal disposal. They are not very soluble and can sorb on soil particles. While soil bacteria can break some down, others will not yield. There are four main ones found in drinking waters: methyl chloroform or 1,1,1-trichloroethane (TCA), tetrachloroethene or perchloroethylene (PCE), trichloroethylene (TCE) and methylene chloride or dichloromethane (DCM). Chlorinated solvents furthermore are carcinogenic to mammals, and the WHO has set limits for their concentration in drinking waters.

7.2.9 Radioactive Substances

The occurrence in Britain of radioactive substances in waters is in part natural. Radon-222 occurs from the breakdown of uranium, and is associated with granitic areas such as Aberdeen and Devon. Radon is very water soluble, only to be released when water is agitated. Accumulation in the foundations of buildings is therefore more serious than the risk from drinking waters. Traces of radium-226 and 228 and the uranium isotopes 234 and 238 may also be found in waters in minute quantities.

Groundwater labelled with tritium following the atmospheric weapon testing of the early 1960s proved useful in elucidating the mechanisms of downward movement of solutes in the unsaturated zone of the chalk (Geake and Foster, 1989). Contamination from radioactive fallout from atmospheric nuclear bomb testing in the 1950s and 1960s and nuclear power disasters such as Chernobyl in 1986 had no known adverse impact upon British waters (unlike food products from animals in the uplands, and groundwater contamination in Ukraine).

Rowan *et al.* (1992) report caesium-137 and 134 accumulation in sediments at the Chelmarsh pump-storage reservoir which were transported by the River Severn, and attributed them to the Chernobyl nuclear reactor accident. Radiocaesium had been mobilised from the Upper Severn catchment, but as the element is strongly bound to sediment particles, the impact is not considered to present a danger to water supply. However, the mechanism of redistribution in catchments does need better understanding. In Scotland, raised levels of americium-241 and caesium-137 have been recorded in certain Scottish lochs.

Low-level wastes from the nuclear industry are incinerated, disposed of in landfill sites, and discharged to sewers where they are not considered to be dangerous.

There continues to be controversy about sites for disposal of higher level waste products from the nuclear industry, especially concerning the possible contamination of groundwaters, hence care must be taken to select geologically stable sites. For example, rock formations considered impermeable for water resource purposes may prove permeable over a timescale of thousands of years; certain isotopes remain active for a long time. There is also controversy regarding how such material may be stored; groundwater is capable of corroding canisters and borosilicate glass (a form of waste storage containment) in the long term.

There remains the possibility of drinking water contamination from reprocessing facilities and nuclear power plants. Dangerous substances are listed under the Dangerous Substances Directive (76/464/EEC) and the EC Groundwater Directive 80/68/EEC concerned uranium as a List II substance under the EC Groundwater Directive (80/68/EEC) that was repealed in 2013 (EU, 2006). A list of 'Hazardous Substances' replaced the old List I. There is an identifiable risk of bioaccumulation of radioactive material in the food chain and in sediments near to nuclear installations.

7.2.10 Thermal Pollution of Waters

The act of artificially altering (generally raising) the temperature of a natural water body (rivers, lakes or marine environments) arises from the waste heat generated by an industrial process. Thermal pollution of natural waterbodies may be associated with direct effluent discharges or from cooling processes in the power generation industry from both nuclear and fossil fuel burning power stations. In the UK it has been estimated that around one half of all river flow is used for cooling purposes leading to elevated discharge of higher temperature water (Hogan, 2012). Large cooling towers associated with power stations on the river Trent in the Midlands or shallow marine thermal pollution associated with the Dungeness power stations that affect fish populations (Spencer and Fleming, 1987) are well-known historic examples. Ecosystem outcomes typically involve a change or reduction in biodiversity and/or the introduction of previously alien species to water bodies.

7.3 Biological Aspects of Water Quality

Odour and taste are two properties which give an indication of water quality, and are more closely linked with biological aspects of water quality than with chemical. Organic materials may be soil particles (such as peat), decaying vegetation, microorganisms (such as moulds, actinomycetes, iron and sulphur bacteria, green and blue-green algae and diatoms), domestic sewage, food-processing effluents, carbohydrates, proteins and fats. Dead organic matter can be broken down by aerobic bacteria if there is enough dissolved oxygen in the water, and the end products can be relatively harmless and odour-free substances such as nitrate and sulphate, although phenolic compounds can create odours upon chlorination. Synthetic substances which produce odours include solvents, pesticides and benzene compounds.

Pathogens are agents capable of causing diseases in humans or animals. They are transmitted to natural waters from faecal material, human or animal, and are of three kinds: bacteria, viruses and protozoa. The term 'pathogen' does not apply to those organisms which are harmless, or indeed beneficial, to the environment or to higher animals, the term being entirely functional rather than specific to any particular group of microorganisms. Waters can transmit (viral) hepatitis A. The bacteria *Salmonella* found in humans and in farm animals, and present in human sewage sludge, causes typhoid and paratyphoid, while *Vibrio* causes cholera. *Escherichia coli* and *Pseudomonas aeruginosa* occur in the guts of warm-blooded animals and need not be harmful, although their presence in waters is an indicator of sewage pollution, can affect vulnerable individuals and some strains cause food poisoning. The hot springs at Bath in the west of England have carried protozoan carried amoebic meningitis in the past. The cysts of the flagellated protozoan parasite *Giardia* infects the small intestine and can have many hosts including man and wild animals. *Giardiasis* is a waterborne infection that can cause symptoms ranging from mild nausea to acute, severe intestinal distress and is found in water in north America and elsewhere. *Cryptosporidium*, similarly a protozoan which affects both humans and animals with digestive tract problems including stomach cramps and diarrhoea, arises from sewage contamination and its cysts should be removed or not introduced to water supplied. People who are HIV positive are especially vulnerable and there have to be plans to protect consumers from downstream contamination (United Utilities, n.d.) including challenges for the operation of water treatment works.

Algal toxicity can arise as a result of algal blooms in eutrophicated waters when phosphorus is not limiting, and algae can thrive. Green and blue-green algae produce discoloration of the water, adversely affect taste and odour, and increase the organic matter content of the water upon death of the organisms. This in turn increases the BOD of the water (see below), normally affecting water bodies such as lakes and reservoirs rather than rivers. Most concern arises from blue-green algae (*cyanobacteria*) pollution which in addition can sometimes produce toxic substances called *cyanotoxins*. Genera such as *Anabaena*, *Aphaoizomenon*, *Oscillatoria* and *Microcystis* regulate their buoyancy with gas vesicles, and their sinking enables uptake of nutrients from different layers in a water body, while rising to surface waters increases possibilities for photosynthesis. The problem is worst in summer, when thermal stratification of standing water bodies may lead to anaerobic conditions in the lower layers; blue-green algae are tolerant of such conditions enabling them to use H_2S as well as H_2O as a hydrogen donor during photosynthesis.

Anaerobicity may be followed by release of such substances as silica, phosphorus and ammonia, which act as nutrients for the algae and accelerate the problem. Restriction of light penetration to depth by a dense crop in the surface can furthermore cause blue-green algae cells from depth to rise rapidly to the surface and remain as a scum which, upon decay, will produce an odour problem. Treatment of affected waters is unreliable, and it is prevention of the development of blooms by reducing the nutrient loading of waters which is the best solution of the problem. One method is to introduce river water upwards by jets into a storage reservoir; mixing will reduce anaerobicity in the lower levels, and hence prevent the sulphide content making the water unusable.

Biochemical oxygen demand (BOD) is the amount of oxygen taken up by microorganisms in decomposing the organic material in a water sample stored in darkness at 20°C. Whereas 5 days is long enough a determination to consume the products of sewage, longer periods such as 20 days may be required for more persistent industrial wastes such as wood pulp. BOD relates to the quantity of natural organic material which can be discharged into a water body.

Chemical oxygen demand (COD) is a measure of the oxygen requirement to chemically degrade pollutants. It is quicker to establish than BOD. The result is then related empirically to produce a prediction of BOD using 'dissolved oxygen sag curves'.

7.4 Hydrochemical Environments

In the following sections, hydrochemical environments of groundwater, rivers, lakes and wetlands and their soils are considered.

7.4.1 Groundwater

Hopefully, as will be elucidated, groundwater will no longer be considered out of sight and out of mind, nor will it be regarded as somehow a magic and mysterious infinite source of pure water. We have spent a long time reaching the current position where it is a protected and valued asset for both human consumption and for maintenance of river flow and wetland ecology (Cook, 1999a).

The complexity of groundwater systems and their protection can never be understated; protection and rehabilitation of aquifers presents special problems. These may be summarised as:

- A matter of timescale. Polluted groundwaters may take decades to cleanse due to slow dispersal of contaminants, low horizontal permeability and slow horizontal transport rates.
- The quality of recharging waters. Problems can arise either from infiltration through contaminated land, or contaminated surface waters naturally or artificially recharging groundwaters.
- The ability of groundwater to 'self-cleanse'. Many biological contaminants degrade with time in, and hence travel through, underground strata. Such self-cleansing is less encouraging for solutes, for example nitrate in aerobic groundwaters can remain unaffected.
- Limits on abstraction set by a tendency for inundation of unsaturated strata by contaminated waters. There is a risk of saline intrusion where pumping lowers the watertable in coastal aquifers below the level to where horizontal movement from sea to groundwater occurs.
- Uncertainty regarding the origin of point pollution of aquifers. Old industrial areas may be prone to rising groundwaters intersecting dumps of toxic material.
- Uncertainty of hydrogeological modelling procedures leading to inaccuracies in the land areas identified for appropriate implementation of groundwater protection and conservation measures.
- Uncertainty around fracking operations

It is appropriate to describe the background hydrochemistry in groundwater systems at this point as found in temperate or humid zone groundwaters. Earlier geochemical research has established patterns of evolution of anions, cations and oxidation-reduction potential in temperate environments. In terms of cation constituents, concentration is characterised in terms of Ca^{2+}, Mg^{2+} and Na^+ plus K^+, or those with no dominant type. For anions, waters dominated by HCO_3^{3-}, SO_4^{2-} and Cl^-, or those with no dominant type, are identified (Ward and Robinson, 1990, Ch. 8). These lead to a sequence of natural groundwater chemistries found in temperate and humid conditions termed the *Chebotarev sequence* (Price, 1996, p. 181).

This sequence is primarily concerned with the evolution of meteoric (atmospherically derived) groundwaters. Meteoric waters are naturally slightly acid due to the solution of carbon dioxide, sulphur dioxide and NOx compounds from the atmosphere, and carbon dioxides from the soil atmosphere. Weak acidity encourages the solution of mineral substances such as calcium. Contact time in the soil and aquifer means that the more mobile ions are taken into solution at first, but will also rapidly be removed in drainage waters, springs and at boreholes. Minerals which dissolve to release chloride (e.g., halite) are more readily soluble than those which release sulphate (e.g., gypsum), which in turn are more soluble than those which release bicarbonate (e.g., as calcite and dolomite) (Ward and Robinson, 2000, Ch. 8).

The zone of active circulation (primarily the unconfined portion at outcrop) is dominated by soluble anions such as nitrates, sulphates and bicarbonates, while the total dissolved solids (TDSs) are relatively low. In the confined portion, first the bicarbonate portion increases and dominates the anions, followed by sulphates. Deeper in the 'dead zone' (where there is little circulation under confined conditions), chloride comes to dominate the anions. In these deeper, immobile areas, it is the more mobile and soluble ions which dominate. In general, the cations calcium and magnesium give way to sodium. Deep groundwaters are, therefore, sodium and chloride ion dominated in this ideal sequence, brine being the end-member of the sequence.

This model makes a number of assumptions. For example, water may not move deep enough, or remain in the aquifer long enough to reach a particular state. Additional loading of such solutes as chloride from saline intrusion and nitrates from certain agricultural practices may distort the 'ideal' sequence. Aquifer mineralogy will also affect the evolution; limestone and dolomite rocks will provide ample calcium and magnesium, but certain sandstones will not. Similarly, pH is affected by the buffering capacity of the strata.

In terms of oxidation-reduction potential, unconfined groundwaters at surface outcrop are oxidising, and in desert sandstones of the Permo-Triassic (dominant aquifers in the English Midlands) oxidising conditions are found well into the confined portion of the aquifer. By contrast, the Jurassic-age Lincolnshire Limestone of eastern England tends to become reducing about 12 km from the start of confining conditions. This is attributable to organic material in the limestone which acts as an electron acceptor and reduces groundwaters leading to the presence of hydrogen sulphide gas (Price, 1996, p. 181).

At the time of writing, fracking remains of potential concern for groundwater resources, especially if energy market conditions make its development viable for Britain. Detractors of fracking do therefore sight potential environmental problems as well as extending the usage of fossil fuels; yet the future would seem to rest somewhere between pressure from activists opposed to its development and perceived shortfall in energy supplies (Inman, 2014). From a purely water resources point of view one has to express concern for contamination of aquifer formations.

7.4.2 Rivers

Rivers present the most dynamic systems of all water bodies, and are no only complicated ecological and hydrological entities in the landscape, but have provided for transport, water supply and food. They are primarily linear features in the landscape, of variable

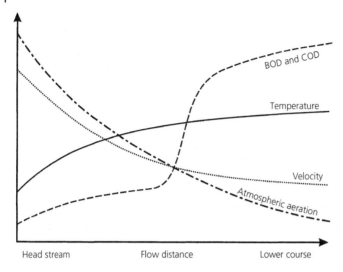

Figure 7.2 Zonation of aeration, temperature, BOD and COD along the length of a stream. (Redrawn from Ullman, 1979).

discharge with a flow-through time of minutes to weeks. Rivers may be shallow and fast flowing, multiple or single channel, meandering or straight (particularly for artificial channels) and display a range of sediment sizes and geochemical environments.

Alternatively, a river may be viewed as a means of transferring potential energy of water which lands on the surface of the land into kinetic energy which generates heat from frictional forces, moves sediment and solutes, and erodes a river bed (Schumm, 1977). It will transfer any pollutants from diffuse or point discharge sources along with the 'natural' suspended or bed-transported load and dissolved loads. A river will also have a (generally) turbulent upper surface which it presents to the atmosphere and this will permit gaseous exchanges to take place, most essentially the dissolving of atmospheric oxygen.

Figure 7.2 shows zonation downstream. From source to lower course, atmospheric aeration is reduced; meanwhile BOD and COD increase as do water temperatures. Actually, the velocity often increases with distance downstream, but this does not happen in all situations (Newson, 1994). The 'natural' chemistry of a river will have a lot to do with catchment geology and soils. Where acidic soils and rocks predominate, and there is little or no buffering capacity in a catchment, waters will be of lower pH. Calcareous soils and geology produce 'hard' waters buffered against acidity. An example is the chalk streams of southern England where pH values can be as high as 8 and water quality is naturally good. Introduced pollutants to rivers will affect the presence of a range of chemical substances, be they organic, chemical or physical.

Figure 7.3 shows the changes which may be expected downstream of a sewage outfall. Near to the outfall there is a fall in the dissolved oxygen, followed by a gradual rise downstream. This is the 'oxygen sag curve', the plot of dissolved oxygen (percent saturation) against the time of flow downstream. It is associated with a sudden rise followed by a gradual fall in BOD and total dissolved solids (TDS). Ammonia shows a similar

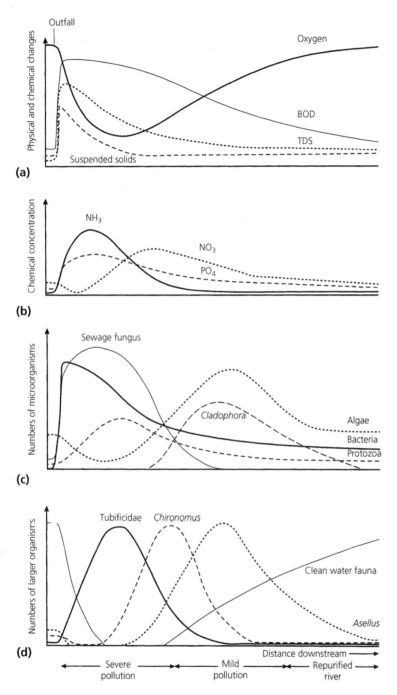

Figure 7.3 Changes downstream of a sewage outfall. (Brown *et al*. 1984, redrawn from Open University, 1984).

pattern, while there is an increase downstream in nitrate as a consequence of nitrification (see below) followed by a fall. Immediately after the outfall there is a rapid increase in bacteria, including pathogenic forms, followed by an increase in algae populations which flourish on the nitrate and phosphate produced by the breakdown of the sewage material. Finally, there is a sequence of changes in organisms, and eventually species of larger organisms return heralding cleaner water conditions. This property of a river ecosystem to recover after an input of organic pollutants on account of microbial activities is a self-purifying ability (Mason, 1991, Ch. 3). It is made possible by biodegradation, and is reliant upon the concentration of oxygen.

7.4.3 Lakes

Lakes are a feature of the British landscape often celebrated in art and literature, and for well over a century reservoirs constructed for water supply, less usually for industrial purposes, have become integral to the ecological and hydrological system. Respecting and enhancing their biodiversity or recreational benefits became a 'virtue from necessary' once the large reservoirs from the later nineteenth or early twentieth century became established. Because lakes generally represent large volumes of fresh water, these may be viewed as a more permanent store of water than are rivers. Damming may further improve and smooth out the supply of runoff water. Mean residence times ('renewal times') are calculated by dividing the mean volume of the water body by the replenishment volume per unit time. Values are typically of the order of a year or so for freshwater lakes found in Britain, although the Norfolk Broads are flushed out at times varying between a few weeks and annually (Moss, 1988).

The thermal structure of lakes is interesting; because water is at its densest at close to 4°C there exists a thermal stratification of a typical lake. Water warmed during the spring and summer exhibits a strong thermal stratification, with warmer, less dense water at the surface. This situation generally persists, despite turbulence from wind passing over the surface. The upper zone is termed the *epilimnion,* and is separated from the lower *hypolimnion* by a layer of transition, typified by a sharp drop in temperature with depth and termed the *metalimnion.* During autumn, the surface temperature of lakes in temperate environments falls towards 4°C (the optimum density of water) and the layers overturn on account of the surface being denser than the underlying layers, a process assisted by turbulence induced by the passage of the wind. In the winter, water at around 4°C may persist at the bottom, whereas the surface may cool towards 0°C and even freeze. As the surface waters warm in the spring towards 4°C, another 'overturn' may occur before strong thermal stratification occurs.

This unusual thermal behaviour of lakes has important repercussions for their ecology and their chemistry. Stratification may mean that the hypolimnion is anaerobic, with most productivity occurring in the epilimnion. Iron may be reduced from Fe^{3+} to Fe^{2+} at the sediment-water interface, while ammonium, nitrate and phosphate may be enriched depending on the redox potential of the water. These and other ions are returned to the surface by vertical mixing, should the water column in the lake overturn due to the changes in the density of the water (Moss, 1988). In general terms, lakes with a high nutrient loading (especially nitrate and phosphate) are termed eutrophic, whereas low nutrient systems are termed 'oligotrophic'.

7.4.4 Wetlands and Their Soils

Wetlands were once vilified for harbouring disease, especially malaria, or for impeding agricultural or transportation development in Britain as elsewhere. Although the rate of loss across Britain has varied in both space and time, since 1600 the loss has been phenomenal, and since 1940 once again a matter of great concern with the inexorable march of modern agriculture. This processes is thankfully being put into reverse. One negative consequence of this history has been the ecology of mythical creatures that inhabit wetlands. These had the potential for carrying off attractive women, causing scares in early Dr Who films or attacking hapless strangers; Conan Doyle has much to answer for. On the same theme, Black Shuck (the hound of, or the actual, god Woden) appears today on certain Suffolk marshes, generally soon after pub closing time.

Although there is no universally agreed definition of 'wetlands', primary (unaltered) wetlands represent valuable ecosystems of high animal and plant productivity in their own right. They are areas where the watertable is at or near the land surface which may be inundated at times. The significant *processes* largely taking place within soils and provide not only ecosystem *functions* (for example the removal of nitrate in water through denitrification) but also ecosysem *services* important to human society, for example water purification and biodiversity. The latter therefore becomes an economic proposition.

The Ramsar Convention concerning internationally important wetlands takes a broad approach in determining the wetlands which come under its aegis wetlands are defined as:

> "areas of marsh, fen, peatland or water, whether natural or artificial, permanent or temporary, with water that is static or flowing, fresh, brackish or salt, including areas of marine water the depth of which at low tide does not exceed six metres".

In addition, listed areas of internationally important wetlands:

> "may incorporate riparian and coastal zones adjacent to the wetlands, and islands or bodies of marine water deeper than six metres at low tide lying within the wetlands".

Ramsar recognises five major wetland types (Ramsar, n.d.), these are:

- marine (coastal wetlands including coastal lagoons, rocky shores, and coral reefs);
- estuarine (including deltas, tidal marshes, and mangrove swamps);
- lacustrine (wetlands associated with lakes);
- riverine (wetlands along rivers and streams); and
- palustrine (meaning "marshy"—marshes, swamps and bogs).

In short, wetland soils are said to be 'hydromorphic', that is their characters, be they primary or subsequently altered by drainage marshes are caused by the ambient hydrological regime. For example, marshes, have mineral alluvium dominated soil and may be high in silt content, where organic soils are dominated by peat and described as 'bog' where they are acidic, with a low pH, or 'fen' where there is a high calcareous content and hence a high pH.

Wetlands are valued for their ecosystem services. In flood alleviation they provide a temporary store of surface or soil or shallow groundwater. More modern considerations are keen to protect and enhance organic wetlands for carbon sequestration. They are naturally biodiverse and highly productive ecosystems. Wetlands are also sinks for pollutants such as excess nitrogen and phosphorus, pesticides and sediments, and the use of *Phragmites* reedswamps in sewage treatment, or reedmace in taking up 'heavy' metals by *Typha*, are common. They have many other, often non-market values for recreation, education, aesthetic value and human well-being.

The poor image of 'swamps' may not be helped in schemes to reinstate wetlands, for example, as pollutant sinks. Neither is the capacity of wetlands to cope with either sediment or chemical loading, infinite (Moss, 1988). Otherwise there is a potential for sediment build-up, and peat or other organic material accumulation. The capacity for the soils to retain metals in sediments is ecologically important because these may be otherwise mobilised into food webs via roots, or damage natural microbial processes in sediments.

Be they organic or mineral, when soils are saturated, oxygen cannot diffuse through the pores and anaerobic (or reducing) conditions rapidly result (Mitsch and Gosselink, 1993) because oxygen diffusion rates are some 10 000 times slower than in drained soils. A summary of wetland oxidation—reduction processes is given in Correll and Weller

---

Box 7.1 Below-Ground Reduction-Oxidation Processes in Wetlands

Wetland below-ground oxidation–reduction processes

(E_{h7} is redox potential at pH 7, these processes being affected by pH; oxidation–reduction is measured in millivolts (mV); 'CH_2O' stands for organic matter)

Reduction processes
E_{h7} approx. +400 mV: Manganate reduction
$$2MnO_2 + 'CH_2O' + 4H_3O^+ \rightarrow CO_2 + 2Mn^{2+} + 7H_2O$$
E_{h7} approx. +300 mV: Denitrification
$$NO_3^- + 'CH_2O' + H_3O^+ \rightarrow CO_2 + \tfrac{1}{2}(N_2O) + 2(\tfrac{1}{2}H_2O)$$
E_{h7} approx. −200 mV: Iron reduction
$$4Fe(OH)_3 + 'CH_2O' \rightarrow 4Fe(OH)_2 + CO_2 + 3H_2O$$
E_{h7} approx. −220 mV: Sulphate reduction
$$SO_4^{2-} + 2'CH_2O' + 2H_3O^+ \rightarrow H_2S + 2CO_2 + 4H_2O$$
$E_{h7} \leq -260$ mV: Methanogenesis
$$2'CH_2O' \rightarrow CO_2 + CH_4$$

Oxidation processes (where $E_{h7} > +400$ mV)
Respiration
$$'CH_2O' + O_2 \rightarrow CO_2 + H_2O$$
Sulphide oxidation
$$H_2S + 2O_2 + 2H_2O \rightarrow SO_4^{2-} + 2H_3O^+$$
Nitrification
$$NH_4^+ + H_2O + O_2 \rightarrow NO_3^- + 2H_3O^+$$

(1989) and Cook *et al.* (2009). This is summarised in Box 7.1. 'CH$_2$O' stands for organic matter. The kinds of chemical processes detailed in the box are of major significance in appreciating the fate of pollutants in surface and groundwater systems. An application is the emplacement of agrochemical-free buffer strips alongside watercourses, which are useful in attenuating and assimilating potential pollutants in shallow groundwater that issues to surface waters.

This chemistry forms the basis of many 'ecosystem services, provided by wetlands.

7.5 Agrochemical Contamination in Catchments

The last 30 years has seen ever-increasing concern over diffuse (non-point) pollution and the continuation of research into sources, and policies to combat, contamination that arises from farming systems particularly. This was already manifest by 1990, and remains of great concern, so a cynic would say that actually little has so far been achieved. The main culprits are considered in turn.

7.5.1 Nitrates in Surface and Groundwaters

Enhanced nitrogen leaching (and loading of aquatic systems) results from the ploughing of grassland, from artificial additions of nitrogen including the excessive application of both organic and mineral nitrogen fertilisers, and from leakage of certain materials such as slurries on farms. Non-agricultural sources are largely sewage breakdown, certain industrial processes and deposition from internal combustion and jet engines. In itself, nitrate is not a harmful substance. The health risk arises because reduction in the body leads to the production of nitrite from nitrate and there are two aspects: the risk of methaemoglobinaemia ('blue baby syndrome') in infants, a rare complaint in Britain, and a suggested link with stomach cancer in adults (Gray, 1994, p. 121). Atmospheric deposition provides concern for a range of human health issues, as well as contaminating both land and water (Defra, 2015a). Provided that piped supply is clean of certain bacteria, reported blue baby cases in Europe have been limited to contaminated supplies from wells. Otherwise there is concern over raised nitrate concentrations in surface waters which contribute to eutrophication.

It was once thought that the leaching of fertiliser was directly from solution of inorganic fertiliser. However, fertiliser application increases crop yields and add to the organic nitrogen pool in the soil which degrades in the late summer and autumn, following the cessation of uptake around harvest time. This organic nitrogen pool is available for microbial degradation, mineralisation and hence leaching (Croll and Hayes, 1988). Furthermore, nitrogen deposition from the atmosphere across Britain may be above 24 kg ha^{-1} yr^{-1} (Defra, 2015b), and significant amounts of nitrates have long been known to leach from permanent grassland under intensive management (Ryden *et al.*, 1984). Management of N is therefore complex and the management not only of agricultural systems is at issue, but so are other issues such as air pollution and sewage discharges to rivers.

The figures for N-application to land in Britain has been are staggering, due to the Second World War and subsequently due to pursuing policies of food security. By

1965, consumption in Britain was 545,000 t N yr^{-1}, peaking in 1984 at 1,499,000 t N yr^{-1} then steadily falling to 940,000 t N yr^{-1} in 2010 (AIC, 2015). The cause for the incredible rate of increase into the 1980s is not always clear, but there has been much 'trial and error' among farmers to optimise yields, and hence inefficient use of fertiliser (Sylvester-Bradley, 1993). The reason for subsequent decline is almost certainly economic as farm incomes decline and certain incentive schemes to reduce fertiliser application come into play. In certain situations the cessation of fertiliser application to land can cause a dramatic reduction in concentration of the soil leachate in under a decade (Moorby and Cook, 1993), but it is residence times in groundwaters which is a major cause of concern.

One report, DoE (1986), notes an increase in tillage between 1950 and 1985 which is dominated by cereals, while both temporary and permanent grassland decreased in area. Estimated leaching losses are as follows: winter cereals, 40% of nitrogen applied; other arable, 50%; peas, equivalent to 50 kg N ha^{-1}; cut grass, 10% of nitrogen applied; grazed grass, 13% per grazing head per hectare. Ploughing grassland releases 280 kg N ha^{-1}. Applied nitrogen rates in England average 107 to 182 kg N ha^{-1} for cereals, being higher for winter-sown oil seed rape and potatoes (274 and 203 kg N ha^{-1}, respectively) (Foster *et al.*, 1986). In addition, particulate organic nitrogen may form up to 20% of the total N load in streams, a source which is prone to nitrate release upon mineralisation of organic matter (Johnes and Burt, 1991). Absolute values in surface waters in England typically vary between 15 and 40 mg l^{-1} NO$_3$, the highest values being in East Anglia, however, the overall trends measured are that 77% of sites had a declining trend while 23% was increasing. Since 2000 nitrate levels have gradually fallen from 39% of river lengths exceeding 30 mg NO$_3$ l^{-1} to 29% in 2009 (Defra, 2010b). Arable catchments in England exhibit considerably higher concentrations in surface waters than for either Wales or Scotland. For groundwaters in England, although overall concentrations are lower than for surface waters, the picture is less clear, but provides little comfort (AIC, 2015).

Supplies exceeding the recommended EU limit of 50 mg l^{-1} nitrate for drinking waters have long been recorded in supplies for isolated wells, and this was generally attributed to local point sources of pollution such as raw sewage or farmyard drainage (Foster *et al.*, 1986). Long-term data from supply boreholes exist, but originate from many depths in the aquifer, under variable pumping regimes and from differing borehole constructions. The chalk and Jurassic limestones are affected, but it is the Permo-Triassic aquifers that are especially badly affected in this respect.

Extensive ploughing of grassland in southern and eastern England during 1939 to 1946 may well have caused the substantial increase observed in so many aquifers in the 1970s (NRA, 1992c). While this may have represented a kind of 'kick start' for groundwater contamination, there also exists a strong relationship between contamination and continual arable cropping in many unconfined aquifers since that time.

Figure 7.4 shows down-profile pore water concentrations for arable sites over chalk in East Yorkshire. Displayed are envelope curves for profiles determined for high and low porosity chalk, largely for the unsaturated zone. Chalk normally displays porosities of between 0.20 and 0.45; lower porosities cause deeper penetration of higher concentration waters, it is probably because solutes penetrating deeper on account of more rapid downward movement. Elsewhere, the chalk of Norfolk displays a highly

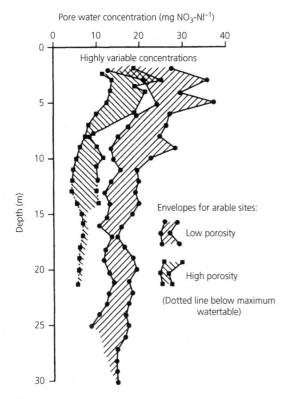

Figure 7.4 Down-profile nitrate concentrations in pore- water in the Chalk. (Redrawn from Foster *et al.*, 1986 by permission of the National Environment Research Council).

variable downward migration of groundwater, while in the Isle of Thanet in Kent, rates have been estimated at around 0.5 m per year (NRA, 1992c).

Many studies show pore water concentrations way above the EU limit (11.3 mg NO_3-Nl^{-1} in Figure 7.4, or 50 mg NO_3 l^{-1}) implying a potentially serious problem for the future. Notwithstanding problems with other aquifer systems such as Permo-Triassic sandstones, it is the chalk which gives the greatest cause for concern, not only because it is the largest single aquifer in Britain, but also because of its 'dual porosity' that is, the opportunity for nitrate to be retained in the finer matrix pores. The chalk immediately beneath the surface may be depleted of water by transpiration, and it has been shown that conditions are right for matrix pores to empty in summer (Wellings and Bell, 1980). Autumn re-charge with water containing nitrate as leachate can then replenish the subsoil. Dispersal of nitrate in chalk is evidenced by a tendency for a downward moving peak to become broader and flatter with depth, a process which would also be enhanced by changes in seasonal vertical movement of water in the matrix in the surface levels of a chalk aquifer. A range of solutes and isotopes (notably tritium and ^{14}C arising from thermonuclear bomb tests in the 1950s and 1960s) have been used to elucidate the mechanisms and speed of transport in chalk aquifers;

chloride transport is most likely to mimic nitrate except that little is taken up by vegetation and there is no nitrification or denitrification (NERC, n.d.)

Fissures in the chalk may provide a 'fast route' for pollutants to reach the water-table along with the rapidly recharging infiltrating water, and these will by-pass the intergranular pores in the unsaturated zone. The detailed mechanisms of solute movement in the unsaturated zone are complicated, and not well understood despite much research, and the detailed mechanisms seem to vary geographically (Geake and Foster, 1989). Groundwater systems take several years for a new equilibrium nitrate concentration in abstracted water to be reached following a change in nitrate loss from the agricultural soil; a figure in excess of 40 years is probable for deep chalk aquifers, but more typical figures for British aquifers would be 10 to 20 years (Archer and Thompson, 1993).

Denitrification is most important in shallow watertable systems where there is more likely to be sufficient organic matter to initiate reduction of dissolved oxygen and nitrate (Hiscock *et al.*, 1991). There is only limited evidence of denitrification in chalk. Although some studies have shown that denitrifying bacteria exist in the upper parts of the aquifer, there is a low level of dissolved oxygen and uncertainty as to the origin and nature of the necessary organic substrate. Other aquifers, notably the Lincolnshire Limestone, display an eastward decline in nitrate concentration which corresponds with a lowering of redox potential away from outcrop into the confined aquifer. The Hastings Beds aquifer of southeast England also displays anaerobic conditions, low pH values and even reduction of iron and manganese, and perhaps experiences high rates of denitrification (see also section 7.4.4).

7.5.2 Phosphate Loading in Waters

Phosphate is generally the controlling nutrient causing eutrophication; a property which essentially involves the re-distribution of the planet's nutrients (Bailey-Watts, 1994). Phosphorus is, in reality, a relatively scarce element in the Earth's crust, yet it is required for plant nutrition. Although locally abundant either through geological factors, or concentrated due to human activity, there is no single global store in the same way as there is for nitrogen, and unlike nitrogen, the global biogeochemical cycle of phosphorous has no biologically determined atmospheric component, although there is atmospheric loading from volcanic eruptions, fires and industrial processes that cause atmospheric deposition on terrestrial surfaces and in waterbodies. P has been accumulating in agricultural soils for decades that may erode, loading watercourses (Defra, 2004), however, it is furthermore thought that P-loading of rivers from sewage is greater than from agriculture (White and Hammond, 2006). In this study, estimates of the diffuse total P load to the waters of England, Wales and Scotland was estimated to be about 41.6 kt y^{-1}. The agricultural contribution to this TP load was 11.8 kt y^{-1} (28.3%), the household contribution was 25.3 kt y^{-1} (60.7%), the industrial contribution was 1.9 kt y^{-1} (4.6%) and the contribution from background sources was 2.7 kt y^{-1} (6.5%). While this does no exonerate agriculture, it means that it is appreciably less than that from urban wastewater, including human waste and detergents.

The chemistry of inorganic fertilisers which supply phosphates to the soil is complicated but can be divided into three categories: orthophosphates (including superphosphates), polyphosphates (higher molecular weight polymerised forms) and insoluble mineral and organic phosphates. The last-mentioned, being insoluble, are favoured by organic management systems. Ground rock phosphates in acid soils are appropriate for slower growing crops due to the slow rate of release (White, 1979).

The tendency for phosphate to become fixed in soils furthermore means that residual fertility is important, and a factor of significance in terms of habitat restoration where an oligotrophic state is desired in terrestrial ecosystems (Marrs and Gough, 1989). This is because the effects of phosphate fertilisers may reside in the soil for decades, a situation at variance with that for nitrates.

It has been noted that phosphorus enters water bodies in solution as inorganic orthophosphate ions and as PO_4^{3-}, as organic phosphorus compounds and as inorganic polymers. Orthophosphate may rapidly become fixed on soil particles, and remain almost indefinitely within soils and sediments, a feature discovered in lakes. Sources of phosphate include soluble forms, complexed forms on eroded particles, pig and other animal slurry, and sewage sludge. Frequently, the bulk of this comes from sewage treatment works; however, agricultural inputs and concentrations in fish-farm effluent are often considerable (NRA, 1992c). Phosphate stripping is possible at sewage treatment works (Moss, 1988, Ch. 6), and is desirable on account of controlling identifiable point sources, but agricultural sources, being diffuse, remain a major problem.

Figure 7.5 shows the processes involved in the transport of phosphorus from terrestrial to aquatic ecosystems. There are three routes for transfer from the terrestrial to the aquatic environment.

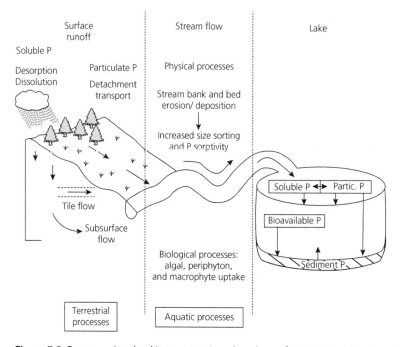

Figure 7.5 Processes involved in transporting phosphorus from terrestrial to aquatic ecosystems. (Redrawn from Withers and Sharpley, 1995, by permission of Wye College Press).

There are three routes involved in the transfer from terrestrial to the aquatic environment:

- *Surface flow and solutes.* Soluble phosphorus (SP) will travel with surface flows and originates from the desorption of phosphorus compounds from soil particles, dissolution and extraction of phosphorus from soil, crop and inorganic fertilisers, and from organic fertilisers and manures. Where there is little particulate matter, SP dominates, while from arable land it may only be between 5 and 25%.
- *Particulate transport.* Particulate phosphorus (PP) arises from transport of clays, colloidal and organic matter and greatly increases the total phosphorus transported. Upper limits for loss from agricultural fields are around $5\,kg\,P\,ha^{-1}$.
- *Subsurface flow.* Here we might expect SP concentrations to be relatively low; figures from drainage waters from mineral soils are typically $100\,\mu g\,l^{-1}$ or less in the UK. However, due to dispersed clay particles in drainage waters in fissures in dry (and shrunken) summer and late autumn soils, and in the presence of pipe underdrainage, concentrations may rise to as much as $10\,000\,\mu g\,l^{-1}$, including both the PP and SP fractions. Slurry applications can also greatly increase phosphorus in transit. The matter of management to restrict phosphorus losses is more complicated than that for nitrogen, essentially because it is more important to limit PP losses than SP losses, the latter being important where receiving waters are oligotrophic, but overall representing a relatively small proportion of total loading.

The overall application of phosphate-fertiliser in Britain between 1965 and 2010 ranged between 125, 000 and 480,000 t P_2O_5. The pattern is similar to that for nitrogen fertiliser but there was a peak in the mid-1980s and a dramatic and sustained decline thereafter (AIC, 2015). The environmental response in England and Wales is encouraging, with overall downward trends for dissolves ('orthophosphate') P reported since 1990 and since 2000, phosphate levels have gradually fallen from 62% of river lengths exceeded $0.1\,mg\,P\,l^{-1}$ ($100\,\mu g\,l^{-1}$) to 50% in 2009 (Defra, 2010b). In Scotland dissolved ('orthophosphate') concentrations are trending upwards, except in the south-east of the country (SEPA, n.d.a).

7.5.3 Pesticides in Waters

As with nutrient loading, there has been research into the environmental behaviour of these complicated chemicals for decades. In industrialised food production, pesticides are designed to control organisms which may interfere with crop biomass production. Pesticide contamination arises from diffuse pollution in both rural and urban land uses, as well as from spillages. The combination of persistence (i.e., rate of breakdown), interactions with soil constituents, and mobility (a function of solubility in water) determines their presence in the aquatic environment. Studies of pesticide occurrence in waters find that there is a readily identifiable 'basket' of substances, and their breakdown products, which are of concern in certain catchments, and increasingly in groundwaters. In a six-year study to the mid-2000s, EA data suggests that at least 95% of all surface water quality exceedances are caused by nine pesticides, three active ingredients vie for first place in the league table of pesticides found in rivers and reservoirs, at levels above the EC 0.1 parts per billion ($0.1\,\mu g L^{-1}$) drinking water standard. The worst offenders appear to be: isoproturon (IPU), mecoprop-p (CMPP-p) and diuron and six more cause concern are MCPA, 2,4-D, simazine dichlorprop-p, chloro-toluron and atrazine. These account for the vast majority of pesticides found in water (UK Parliament, 2005).

Degradation can produce some extremely toxic intermediate products. Furthermore, considered alongside pesticides *sensu stricto* are hormonal substances aimed at growth regulation in a crop, and other substances including wood preservatives. Although more-modern pesticides tend to degrade in soil, hence they are less persistent, certain older compounds persist by forming complexes with soil organic matter and in the lipids of living organisms. For example, DDT is banned in many countries on account of its persistence in the food chain. Pesticide use is controlled under statute and government can withdraw individual compounds or call for their reassessment at any time (HSE, n.d.). Most are organic compounds performing one or more of the functions of herbi-cide, fungicide, insecticide, molluscicide, acaricide, nematicide, rodenticide and plant growth regulator. It is herbicides, applied to arable, pasture, transport and urban land, and sheep dip (containing insecticides) which cause most concern for water quality.

Notable Acts of Parliament are the Food and Environment Protection Act of 1986, the Control of Pesticides Regulations of 1986, and the Control of Substances Hazardous to Health Regulations of 1988 made under the Health and Safety at Work Act of 1974. In handling pesticides there are regulations regarding their storage, loading and unload-ing, transport, disposal, equipment filling and application, including regard for wind-speed in order to prevent spray drift. Records must be kept regarding application, personnel trained, protective gear worn and equipment correctly maintained.

Soil properties defining pesticide behaviour are the key to pesticide behaviour in the wider environment. White (1979) identifies five main pathways of pesticide loss from soils:

- Volatisation, especially of surface-applied compounds.
- Chemical and biological decomposition.
- Absorption by plants and animals.
- Transport on eroded soil particles.
- In solution by leaching or in surface runoff.

It is the last two which are of greatest significance here, and together with risks arising from storage and accidental spillage, cause the most significant threat to the aquatic environment. Pesticides applied during the autumn or winter are more likely to leach than those applied during spring or summer because degradation is relatively slow during periods of lower soil temperatures.

Furthermore, because evapotranspiration is minimal during the winter, many soils become periodically waterlogged. Profiles in such a 'drainage condition' favour the transport of substances in solution, or complexed with soil particles. It is desirable to summarise the main groups of pesticides and their properties. The summary below is based on Hurst *et al.* (1991), Worthing and Ranee (1991), NRA (1992c) and Gray (1994).

Organochlorines (or chlorinated hydrocarbons) were developed early, and were applied during the 1950s and into the 1960s. Their persistence made them effective (insecticide substances such as DDT and its breakdown product DDE belong to this group), but also problematic because they are inclined to become concentrated in fatty tissues and animal milk (including human breast milk) as their solubility is low. DDT is banned in the UK. Another organochlorine, lindane (or gamma HCH, not banned), is a persistent pesticide used as a soil insecticide, and although like the others it is of low solubility and low mobility in waters, its persistence makes it widespread in surface waters. Aldrin and related substances are used in seed dressing, and can degrade by oxidation to dieldrin, which is the only 'drin' normally found in surface and groundwaters in Britain.

Organophosphates are used as insecticides and nematicides, and although dilute operate on the nervous system. Examples include malathion, diazinon, dichlorvos (used to control salmon lice in fish farming), triazophos, chlorpyrifos and parathion (no longer approved as a pesticide in the UK). Organophosphorus compounds have low solubility and low mobility and, having rates of degradation from weeks to months, they are less persistent than the otganochlorines. Carbamates are insecticides of low mobility and variable solubility and include tri-allate, dimethoate, carbofuran and aldicarb.

They are relatively expensive, but decay rapidly, and, being based upon natural substances, they are regarded as being relatively safe.

Chlorophenoxy acids are hormone-type herbicide compounds consisting of chlorine and various organic acids attached on a benzene ring. Examples affecting the water environment include mecoprop (CMPP), 2,4-D, 2,4,5-T and MCPA (a cereal herbicide distinguished by significant solubility and mobility).

Other hormone herbicides include dicamba (arenecarboxylic acid) and dichloroprop (2-aryloxyalkanoic acid).

Triazines are herbicides used in both agricultural and non-agricultural weed control. They include atrazine (now banned), simazine and terbutryn, and are overall the most widely used pesticide chemicals. Consequently they are most frequently found in waters.

Bipyridyl compounds are the ammonium compounds diquat and paraquat. Being total (as opposed to selective) herbicides they are popular with gardeners; however, they are tenaciously absorbed to soil constituents.

Urea-based herbicides include chlortoluron, isoproturon, linuron, metoxuron and methabenzthiazuron, and these are often detected in drainage waters.

Other pesticide substances detected in waters include propyzamide (an amide group herbicide), cypermethrin (a pyrethroid insecticide), triadimefon (a sterol biosynthesis inhibitor fungicide), and tecnazine (a chloronitrobenzene fungicide, used for dry rot control and for inhibiting sprouting in potatoes). It is found in washings from potato preparation. Pesticides entering waters and hence becoming available for human consumption and causing environmental toxicity are naturally a cause of grave concern. However, there is in many instances a lack of understanding of these most complicated substances and their breakdown products in both the human body and the environment.

Problems associated with pesticide use are also variable. More recently, synthesised substances apparently have few, if any, adverse effects when correctly applied because they readily break down (an example here would be the carbamates). Accidental spillages, exposure and ingestion of pesticides typically lead to nausea, giddiness, respiratory problems and may be fatal. Lethal doses, expected to kill 50% of a population of mammals, birds and fish, have been published (Worthing and Ranee, 1991).

However, it is chronic background levels of pesticide exposure which cause most concern in the wider population. This more diffuse form of pollution via food and water may lead to tumour formation, including cancer, allergies, psychological problems and even birth defects (Gray, 1994, p. 135). Lindane has been implicated in the last mentioned, and in causing cancer and has been banned in farming within the EU as well as the United States. The actual problem appears to lie with our ignorance of the real dangers.

Table 7.1 Pesticides in tile drain outlets at Boxworth.

Pesticides	Concentration range (μg l^{-1})
Herbicides	
Simazine	0.39–35.0
Tri-allate	0.00–0.25
Metoxuron	0.06–3.30
Isoproturon	0.32–2.20
Methabenzthiazuron	0.07–0.06
Insecticides	
Chlorpyrifos	0.00–0.06
Triazophos	0.00
Cypermethrin	0.00

(*Source:* Eagle, 1992).

For the loading of waters, three categories will be considered. These are based upon published studies, and deal with pesticides in soil drainage waters, in surface catchments and in groundwaters. Naturally the presence of a given substance in waters depends not only upon the history of pesticide use within the catchment, but also on the analysis employed. The behaviour of pesticide chemicals in subsoils remains particularly poorly understood.

Eagle (1992) describes the results of an analysis of pesticides from tile drains in clay soils at Boxworth (Cambridgeshire) during the winters of 1987 and 1988, comparing the effects of low level and high level pesticide use. His results are presented in Table 7.1.

Four out of eight of these substances (all herbicides) are capable of breaching the EC limit of 0.1 μg l^{-1} for individual pesticides, and the level of simazine is extraordinarily high (this compound, however, causes low toxicity). Although simazine is only slowly broken down in soils and toxic to laboratory animals, there is only limited evidence of carcinogenic properties through its breakdown properties (Gray, 1994, p. 140). Tri-allate, methabenzthiazuron, isoproturon and metoxuron may all have carcinogenic properties (Hurst *et al.*, 1991).

With the bulk of transport during periods of soil saturation, and with 'by-pass flow' (via macropores in the soil) invoked as a mechanism to carry pesticides from near the surface to tile drains in clay soils (Williams *et al.*, 1995), it is not surprising to discover a seasonality in pesticide contamination of surface waters. Concentrations in watercourses in winter and spring have been shown to be elevated (Gomme *et al.*, 1991). In general, surface catchments are all subject to risk from the storage and use of pesticides on farms and elsewhere. Surface runoff is prone to move pesticide chemicals in solution or associated with particulate material, in a manner similar to phosphate. Pesticides are commonly found in rivers with agricultural catchments, and in areas where pesticides are applied for non-agricultural purposes, especially for weed control on urban and transport land.

Furthermore, aerosol applications and rainsplash may cause them to be detected in rainwaters during times of applications. Gomme *et al.* (1991) analysed 20 pesticide substances in an agricultural chalk catchment in eastern England, and

Table 7.2 Pesticides detected in water from the river Granta at Linton in Cambridgeshire.

Pesticide detected	Number of detections	Range of values ($\mu g\ l^{-1}$)
Atrazine	6	tr–0.10
Chlortoluron	3	tr–0.13
Dichlorprop	1	tr
Dimethoate	6	tr
Isoproturon	16	0.07–0.76
MCPA	1	8.90
Mecoprop	11	0.03–2.70
Propyzamide	2	0.06–0.24
Simazine	15	tr–0.56
Triadimefon	1	tr
Tri-allate	4	tr

'tr' (trace) refers to the chemical being detected, but in too low concentrations to be confidently quantified.
(*Source:* Gomme *et al.*, 1991).

Table 7.2 shows some of their results of analysis of water from the River Granta at Linton in Cambridgeshire. Although both the frequency and concentrations vary, and the seasonal aspect of detection related to application has been noted, certain substances (especially isoproturon) were persistent in the river. Furthermore, those more frequently used were not necessarily those detected in the river water. Their presence here reflects (presumably) their environmental persistence and mobility. Most notable are the triazine herbicides (atrazine and simazine) whose concentrations were very low, while other intensively used chemicals were not detected at all.

Concentrations of pesticides vary through a flood event. In the study shown in Table 7.2, several concentrations, notably atrazine, mecoprop, simazine and propyzarnide, showed a good correlation with the heavy rainfall during the 24 hours prior to peak flow, while others correlate with peak flow. The interpretation given by the investigators was of (rapid) runoff from hard surfaces and application close to the river channel. Otherwise, second peaks and main peaks coincide with maximum river flow. Earlier studies had found, ominously, mecoprop, atrazine, simazine and isoproturon to be persistent in river waters while the latter three were detected in groundwaters. Variation in frequency and concentration of pesticide detection in the river is a function both of flood behaviour and seasonality of flow and groundwater contribution to baseflow. Within a small catchment (around $0.5\,km^2$) at ADAS Rosemund, Herefordshire, England, stream water was analysed for seven pesticides (isoproturon, MCPA, atrazine, simazine, carbofuran, aldicarb and dimethoate) by Williams *et al.* (1995). The history of pesticide use was well known, and there is fairly close coincidence of the catchment boundary with the farm. Rainfall events were found to give rise to short-lived pulses which generally preceded the hydrograph peak, a process inferred as being enhanced by 'by-pass' flow in clay soils to drains, and it is possible that as many as one-third of UK soils have a similar regime. However, the maximum accumulated load of pesticide measures at a sampling point (i.e., reaching the water) was small at only 0.6% of that applied. For comparison, Gomme *et al.* (1991) report that on average only 0.1% of pesticide applied reached the rivers, although the figure for atrazine was high at 1.52%. In the Herefordshire study, concentrations in the stream and drains were enormously variable, from traces at below 0.1 to 264 µg 1^{-1} for the insecticide carbofuran.

Most mean concentrations of individual substances during runoff events exceeded the EC limit of 0.1 µg 1^{-1} in drinking waters. The study also suggests that 'flushes' of pesticides in the headwaters of streams in arable catchments can be considerable and may have serious implications for aquatic habitats. There is considerable variation both between studies on individual formulations, and between different substances which are prone to leach to waters. For example, the half-life quoted for atrazine is long, at 60 days, but has a wide range between 18 and 119 days. Dimethoate, by contrast, has a half-life of 7 to 12 days. Rates of loss in soils and subsoils will depend upon such variables as stability of the compound, rate of decomposition (microbial and chemical) complexing with soil particles, temperature and moisture regimes in the soil, and rates of volatilisation and leaching to waters. Given the widespread occurrence of pesticides in surface waters, it is not surprising that they have been sought—and found—in groundwaters. This is a cause for concern because of the importance to public supply of waters in eastern England where aquifers are frequently overlain by intensive arable areas. Harmonised groundwater monitoring has shown from data collated for 2004 indicate that atrazine was the pesticide most frequently found in groundwater at levels above 0.1 µgl^{-1}, (approximately 3% of the samples) followed by bentazone (0.88%) and mecoprop (0.74%). Residence times in groundwaters remain uncertain for most pesticides (Defra, 2006a).

Atrazine, banned by the EU in 2004, has been a globally a significant herbicide, causes major concern because it may cause possible negative health effects, effects on aquatic organisms, levels in drinking water and the development of resistance in organisms where it is applied. Whilst it is becoming less widely used, the effects of its long-term persistence may still cause health and environmental problems in the future (Pesticide news, 2006). Since then, the EU is moving towards a ban on neonicotinoid insecticides on account of impacts on honey bee populations (Rabesandratana, 2013). This is proving controversial.

Long-term solutions to the pesticide problem include:

- Continuation of the harmonised monitoring in surface and groundwaters in order to assess the scale of the problem.
- More information on the effects of individual substances on the aquatic environment and on human and animal health (it does not follow that a particular substance in limited concentrations is necessarily harmful).
- Better knowledge of environmental persistence (organochlorines are especially bad in this respect, although not a problem in drainage waters or aquifers).
- Withdrawal of approval for use when some danger to health or habitat is detected (Atrazine being one such example).
- Improvement of water treatment to 'strip out' pesticides and their residues.
- Evaluation of differing land-use options including agrochemical-free buffer zones between land to which pesticide is applied and the receiving waters.
- Improved training of pesticide users (already required in the UK), including proficiency tests as required for the users of sheep dip.

- Better storage and retrieval of stored and unused material.
- Alternative means of pest management, based upon cultural and 'biological control' and genetically induced disease resistance in crops, however, fungicides are not really a problem in waters.
- Better testing at water treatment works due to the ever-increasing range and use of pesticide products.

Finally, in foodstuffs, the Food Standards Agency (FSA) accepts the use of pesticides in food production (FSA, n.d.) providing:

- regulatory bodies follow a precautionary approach when approving the use of pesticides.
- independent scientific advice says that the safety of pesticides is within acceptable limits
- acceptable levels can be set for residues in food.
- enough good-quality information is available to the regulatory bodies on which to base these decisions.

The FSA considers that current levels of pesticide residues in the UK food supply do not present a significant concern for human health. This would seem to chime with the view of most statutory agencies in respect of pesticide use.

7.6 Environmental Acidification

Acidification of soils and waters include both natural processes which, over time reduce pH, and anthropogenic activities such as the draining of certain soils, mining activities, industrial discharges, and sulphurous and other emissions to the atmosphere. Acid deposition is a serious *diffuse* pollution problem, discharges from minewaters and manufacturing processes being *point sources.* The subject is broad, complex, and once 'acid rain' may be seen as the political precursor to contemporary issues of global warming. The topic is amplified below.

7.6.1 Natural Acidity

This occurs apparently without human intervention, although human interference dates back into prehistory. Due to the dissolution of carbon dioxide in rainwater, and the complete or partial mineralisation of organic matter in soils, acidification of soils can occur. Natural precipitation is on the acid side, at around pH 5.5, and oxides of nitrogen are produced by lightning and sulphur dioxides from volcanic eruptions. Respiration by the soil biota can generate carbon dioxide and also boost acidity. In any situation where the soil becomes progressively depleted by the leaching of such elements as calcium and magnesium, certainly common in areas where the geology comprises rocks of all kinds deficient in 'base' elements, the profile will develop a lower than neutral pH.

Catt (1985) found ample evidence for acidity in Pleistocene soils, including podsolisation from interstadials, pre-dating not only the industrial age, but also the arrival of humans in Britain during the Holocene. Early Britons created further acidification due largely to land clearance for agriculture and a resulting loss of cations through leaching on vulnerable soils. Pollen records in tarn sediments including (among others) *Ericacae* and *Sphagnum* spores suggest that certain upland areas were acidified by the dawn of the industrial age. Although prehistoric peoples accelerated the acidification process through land management practices, acidification was probably triggered in the first instance as a soil-forming process. Acidification (as well as other forms of soil degradation including erosion, leaching and podsolisation) is associated with the creation of heathlands in acidic sandy soils in southern England and elsewhere since prehistory, and reaching well into historic times (Cook, 2008b; Tubbs, 2001, p. 51).

Under natural conditions, soil acidity is quite sensitive to the species composition of a stand, with the effects of conifers in producing acidity more marked than for hardwoods, although on already base-poor soils beech may enhance these effects, even producing podsols (Hornung, 1985). Coniferous vegetation is likely only native to Britain as Scots Pine in the north of Scotland (Rackham, 1986). It is regarded as a cause of podsolisation through the breakdown of organic compounds in litter, and the generation of abundant H^+ ions. This is primarily because the resulting litter is so base deficient, and favours the production of acid 'mor' humus over base-rich 'mull'. The concentration of calcium in the above-ground biomass and a low rate of magnesium cycling have been cited as a cause here. In commercial forestry, base cations are subsequently removed from the system by tree harvesting.

The timescale of the development of a fully podsolic profile is hundreds to thousands of years and soil acidification under coniferous forestry is well documented. It may be enhanced by deforestation. In true podsols, cations are mobile, particularly those of iron and aluminium which move down-profile under the influence of organic mineral complexes (e.g., Bloomfield, 1957). Within 2 m of the surface these are deposited as 'hardpan', typically organic matter, iron and aluminium compounds and complexes with clays in a dark 'spodic' horizon (FAO, 2001).

Buffering against soil acidity may be possible where calcareous waters drain to depressions in the land surface and lead to the formation of fen peats where vegetation develops. Major areas of organic soils are the Fenland in eastern England, now virtually all reclaimed, and the Somerset Levels in western England. A fen (calcareous) peat is said to become *ombrogenous* when it grows vertically through the accumulation of vegetation and becomes predominantly *rainwater* (rather than *groundwater*) fed. Leaching, loss of buffering of calcareous peat and acidifying rainwaters with vertical accretion led to a fall in pH. In Fenland, the vegetation succession was from reed-sedge fen through fen woods and carr to *Sphagnum* in the acidic upper layers (Godwin, 1978, p. 15). Low pH values in upper peats are a natural process with (in the Somerset Levels and Moors) a change to vegetation dominated by *Sphagnum, Eriophorum* and *Calluna* at around 4000 BP (Curtis *et al.*, 1976, p. 240), and acidification seems to have accelerated at the change to a wetter climate around 2500 BP in southern Britain. These bogs have (or had) much in common with naturally acid 'blanket peats' in the upland zones whose acidity arises from base-poor parent material, and grew by the same processes to form raised acid peatbogs.

7.6.2 Soil Acidity following Soil Drainage and Mining

This arises from the oxidation of pyrite (FeS_2) deposited under brackish conditions. Soils deposited under saline or brackish conditions contain pyrite derived from sulphate reduction, and these will oxidise following the lowering of the watertable after drainage (Dent, 1999). This will produce extremely high acidity on account of the production of sulphuric acid, unless there is calcium carbonate available to buffer the acidity, in which case gypsum is formed. In England, areas of great subsoil acidity occur in certain Fenland peats, reclaimed marshlands in Broadland, and the North Kent Marshes. In such areas as the Romney/Walland marsh areas of Kent and East Sussex, and the Somerset Levels, calcareous drainage waters from the surrounding uplands are effective in buffering against potential acidity.

Although the issue of soil reclamation from primary wetland soils is complicated, involving both natural and restoration-imposed changes in vegetation, dewatering and structural development of soils, essentially it is lowering the watertable by drainage operations which introduces aerobic conditions. Pyrite becomes oxidised to soluble iron and sulphuric acid along a complicated reduction pathway facilitated by the bacterium *Thiobacillus ferroxidans.* Accounts of the chemistry is given in Dent (1984, 1999) and George (1992, Ch 2):

$$FeS_2 + 14Fe^{3+} \rightarrow 2SO_4{}^{2-} + 15Fe^{2+} + 16H^+$$

The neutralisation reactions, which will only occur in the presence of calcium carbonate, are:

$$CaCO_3 + 2H^+ + SO_4{}^{2-} \rightarrow CaSO_4{}^+ + H_2O + CO_2$$

This will precipitate gypsum in the soil, and the iron is precipitated as ferric hydroxide ('ochre'), $Fe(OH)_3$ following a series of reactions with water.

Low pH in the root zone severely inhibits rooting in crops subsequently sown (Cook, 1986). Where there is no 'buffering' reaction, pH will typically fall into the range 2 to 4 and aluminium becomes mobile, which is highly toxic to plants. In order to artificially neutralise acid sulphate toxicity, enormous quantities of lime may be required, and where the oxidation of pyrite is progressive, the results may only be temporary. In extreme cases, entire under-drained systems can be rendered inoperative, following reclamation, due to the precipitation of ochre. Where the pH falls below 3, the mineral jarosite may form, causing there to be a striking yellow line marking the top of the acid horizon in the dyke edge. Drainage pipes and pumps installed may become blocked with iron hydroxides should acid waters become neutralised, a situation frequently found in the ditches.

There are a range of acids presenting a range of pollutants. The European Union sets limits for contamination on wastewater emission limits, including for acidity (Frost, 1999). Acids, such as chromic acid, hydrofluoric acid, phosphoric acid, nitric acid, hydrochloric acid, sulphuric acid, oleum (fuming sulfuric acid), sulphurous acids and carboxylic acids. Hence, in theory, these are controlled by consents to discharge provided that they are not accidental. One result is the increased mobility of many potentially toxic metals including aluminium that is enhanced by the acidity in waters.

Acidity from mine spoil essentially has the same chemical origin as soil acidity. It is the oxidation of pyrite (originating in 'solid' geological formations rather than the soil) which causes a drop in pH. The normal consequence is high concentrations of iron, aluminium and sulphate, and these may affect groundwaters as well as surface drainage (Todd, 1980, p. 329).

Pyrite may occur within shales from the Coal Measures, or it may have a hydrothermal origin in veins of metalliferous ores. Given the prevalence of heavy metals associated with mineral spoil, such elements as copper, zinc and lead may become mobile in acidified waters leaching from mine spoil. Such problems occur nationwide, from Cornwall (where metalliferous mining once was a key part of the economy) to the northeast of England (where former coal workings and spoil heaps may acidify water); in Wales, both problems have been manifest. In the rehabilitation of mine spoil, dressings of lime may be used to buffer against the acidity, while nutrient deficiency for re-vegetation may be approached by the addition of dressing of nitrate and phosphate fertilisers, although appreciable iron and aluminium oxides will bind the phosphorus making it unavailable to plants. Measures of remediation of acid mine drainage include draining acid waters through a limestone filters, use of settling lagoons or neutralise the low pH and a range of biological remediation strategies including wetland treatments aimed to generate alkalinity and immobilise metal that cause toxicity (Johnson and Hallberg, 2005).

7.6.3 Acid Deposition

Acidification of soils and waters from atmospheric acid deposition may enhance the effects of natural acidification or compound the problems created by other forms of industrial acidification. There has been a considerable literature generated since such acidification was first identified during the 1960s as a problem in Scandinavia, a problem caused by burning fossil fuels in Britain and Central Europe (Morrison, 1994). (The adoption of 'tall stack' policies disperses local acid precipitation only to 'export' the problem.) The worst offenders would appear to be coals with high sulphur content, especially brown coal, and mineral oils. There was furthermore evidence of power station ash from the English Midlands falling in the Cairngorm region, and several studies have shown acidification to be associated with recent soot particle deposition (Fowler *et al.*, 1985).

Factors other than sulphurous emissions, especially in Western and Central Europe, were the generation of nitrogen dioxide (NO_2) and nitric oxide (NO) from vehicle and jet aircraft emissions. Collectively these are known as NOx, they contribute to acid rain and furthermore present problems of eutrophication. Typical reactions were given in section 7.2.1. The worst-affected areas in Britain are parts of Wales, Cumbria, and western Scotland, especially Galloway (Fowler *et al.*, 1985). Wet deposition is the straightforward deposition of solutes in rainwater and snow, the most significant man-made acids in precipitation are sulphuric (H_2SO_4) and nitric (HNO_3) from the dissolution of NO_2 and SO_2 arising from industry, especially power stations. On the other hand, 'dry deposition' is deposition of sulphur dioxide and nitrogen dioxide at the surface. This may subsequently oxidise on the surface of plants or the soil or be taken up via stomata. Another kind, termed 'occult deposition', is the interception of mist by vegetation, has long been appreciated as an important route for coniferous vegetation where needles are efficient at intercepting water droplets. 'Critical loads' for both acidity and nutrient nitrogen deposition may be set for woods and forests, hopefully below which significant harmful effects on specified sensitive elements of the environment (vegetation, soils and waters) do not occur (Forest Research, n.d.a.).

Where the catchment geology lacks sufficient buffering base elements such as magnesium and calcium (and granites, 'acid' volcanic rocks, slates, most other metamorphic rocks' and clastic sediments are notorious for this), the receiving waters in streams and lakes will acidify, with serious implications for water quality and the associated ecosystems. The progressive acidification of lakes has been charted by the changes in diatom assemblies, which may be a sensitive indicator of pH conditions (Moss, 1988, p. 342) and have been used to establish aquatic pH records. There have also been cataclysmic events such as the 'fish kills' first observed in Scandinavia. Otherwise, the effects of enhanced acidification and a natural shortage of (especially) calcium may not be easy to disentangle. The effects are well known where the kinds of problems outlined above cause a fall in pH in natural waters, and are common to all affected areas. Changes in species composition and overall species depletion are due to factors such as the concentration of H^+ (affecting fish eggs) and the mobility of the aluminium ion. Fish mortalities, notably of trout, may be high, salmonoid fisheries are put at risk (through vulnerability of their eggs), and calcicole aquatic plants have been observed to be absent and the invertebrate fauna limited (Morrison, 1994).

The effect on the food chain may also adversely affect bird and mammal populations. Solutions to the problem of acid deposition ultimately rest upon limiting emissions of sulphur dioxide by reducing the amount of sulphurous fuels, 'scrubbing' the sulphurous emissions at source, that is 'flue gas desulphurisation' (DTI, 2000) and treating the symptoms. Howells *et al.* (1992) report efforts to control the problem by applications of limestone in the Loch Fleet catchment (southwest Scotland), where it is anticipated that applications greater than $20\,t\,ha^{-1}$ should contain a problem of acidity (pH < 5.1) over time. However, blanket liming may have adverse effects on otherwise naturally acidic soils and vegetation. To date, concern over acidification of waters is limited to the impact upon aquatic ecosystems, notably fisheries and recreation. There remains the strong possibility that supplies for human consumption may experience falling pH, especially in upland areas. Because of the need to avoid corrosion and dissolution of certain metals in the distribution system, the addition of alkalis to increase pH is common practice at water treatment works.

7.7 Particulate Matter and Nanoparticles

Wind and water erosion both play a part in soil erosion, and sediment reaching surface waters may be regarded as a physical pollutant. Serious sediment loading of surface waters occurs in urbanising areas and is a result of the erosion of soils in agricultural catchments. Arable farming of sloping land is a significant contributor to the problem, especially such practices as leaving soil bare in winter. Winter wheat cultivation is a problem here, as is ridge and furrow cultivation for potatoes and bulbs, the problem being most severe when cross-contour ploughing is employed. All sizes of particles from stones to colloidal material are potentially mobile. It is the silt and finer sands which, due to their lack of cohesion and small size, are liable to be picked up by air currents over the surface, while dry peat (because of its low bulk density) is also vulnerable to wind erosion.

Water erosion is usually first observed as rills on the surface of arable fields. These may cause a build-up of eroded material at the bottom of slopes (or other concavity); alternatively, the sediment may feed surface watercourses directly. As a gross generalisation, finer sediment, both silt (0.05-0.002 mm diameter) and clay and colloidal material (below 0.002 mm diameter) are capable of transport in suspension in quite low flow conditions. Coarser material (diameter greater than 2.0 mm) constitutes the 'bedload' and will only be carried under conditions of high flow, often by 'traction' along the bottom of the stream. The intermediate sand grade (0.05 to 2.0 mm diameter) is moved according to the carrying capacity of the stream (generally equated with velocity of the water, although this property varies with depth and distance from the bank) and may be suspended, or travel by saltation (jumping) in response to eddies in the flow.

Soil erosion has been a consequence of British agriculture for some 6000 years since the wildwood clearance (Cook, 2008b). Population pressures or shortage of imported food coupled with mis-management of ploughland lead to down-slope erosion, flood-plain illuviation and loading of watercourses. Exact causes may not always be clear, but it has been chalk and limestone slopes which are most vulnerable (Boardman, 1992). Since the eighteenth century, and especially since 1940, the conversion of heath, forest and permanent pasture to arable land has been blamed for erosion of sediment (Evans, 1990). In the late 1970s and early 1980s there was a sharp rise in the number of recorded cases of erosion on agricultural land in Britain. In water resource terms, adverse effects of sediments on waters are many.

In excess it may cause turbidity in waters, affecting supply and aquatic ecosystems including fisheries. Choked watercourses increase the risk of flooding in overloaded channels, and there may be serious problems with siltation of reservoirs and loss of storage capacity. In chemical terms, sediment may carry with it such contaminants as pesticides, phosphates and heavy metals. Suspended sediment is more mobile than bedload, and it can even penetrate gravels with adverse effects for aquatic life such as salmonoid fish, as well as causing eutrophication of open water bodies from phosphate pollution.

Concern has been expressed about the release of engineered nanoparticles (ENPs) from industrial processed to air and waters, these range between 1 and 100 nanometers in size and may be present in contaminated drinking water. A range of metal, metal oxide and organic-based ENPs have been identified and worse case predicted concentrations in drinking waters were in the low to sub- μgl^{-1} range and more realistic estimates were tens of ngl^{-1} or less. For the majority of products, human exposure via drinking water is predicted to be less important than exposure via other routes. The particles contained included the elements Ag and Al as well as TiO_2 and Fe_2O_3 and carbon-based materials. Although predicted concentrations of these materials in UK drinking water are low, there is concern for the future as the UK develops markets for these materials with significant uncertainties concerning the risks of nanoparticles to human health from environmental exposure pathways (Tiede *et al.*, 2011). Ironically, while ENPs present problems for wastewater treatment, nanoparticles may come to play a significant role in treatment technology, nanomaterials can also remove toxic metals and dangerous organic molecules from water to be put into supply (NERC, n.d.).

7.8 Urbanisation and Water Quality

While the effects of urbanisation on volume and timing of runoff are dramatic, impacts on water quality are also profound, with sewage the single most important problem. This is the primary concern of the EU Urban Waste Water Treatment Directive (91/271/EEC)

Chapter 6. Both existing urban and urbanising areas pose threats to water quality, for surface and groundwaters, with both frequently in demand for public supply. Recycling effluent waters frequently sustain dry weather river flow, for example in the Lower Thames catchment (Chapter 5). In a small (0.26 km^2) catchment near Exeter, Walling (1979) found that sediment yield increased five- to ten-fold during building activities in a hitherto undisturbed portion of the catchment (about 25% of the total area), while storm runoff volumes and peak flows increased between two and four times. Once building has ceased, sediment loads tend to decline below the levels expected from a rural catchment.

Subsequently runoff will contain substances characteristic of urban, industrial and transport surfaces. In a review by Andoh (1994), urban storm water runoff was found to contain raised BOD and COD levels and faecal coliform and a range of contaminants including suspended solids, total N and nitrate N, phosphate and metals (including copper, lead and zinc). Concentrations were variable, but tended to be greatest in more heavily trafficked areas. He concluded that quality issues are more complicated than considerations of water yield; there is a need to appreciate the stochastic (random) nature of the problem, and a need to employ low-level technology which mimics natural processes, paralleling ideas such as reedbed emplacement to control nitrogen pollution in rural areas. Low energy consumption, low cost and maintenance are extremely desirable in schemes designed to control pollution, but, of course, where possible, prevention is better than cure.

Runoff concentrated on impermeable urban surfaces not only leads to urban water channels, but also to soakaways which drain into the unsaturated zone of an aquifer. This provides both a convenient source of groundwater recharge, and a potential source of pollution from abrasion, de-icing (e.g., glycols) and defoliation (herbicides) chemicals, oils and various accidental spillages, all unacceptable in drinking water. The sources of land and groundwater pollution may be summarised (Todd, 1980; Lloyd *et al.*, 1991):

1) Industrial chimney fallout; contaminants, usually metals, or through acid rain.
2) A wide range of contaminants in shallow unsaturated-zone disposal sites including domestic landfill, foundry wastes and phenolic compounds from coking. These may be reactivated by redevelopment.
3) Disposal pits for fluids, industrial discharge contaminants (typically hydrocarbons), solvents and sludge wastes, above or below the watertable.
4) Disposal in major excavations such as mines or quarries including backfills of solid, domestic and many other wastes. Mines can produce a range of groundwater contamination, including hydrocarbons and heavy metals. The greatest single problem is acidification of drainage waters.
5) Disposal of liquid wastes to disused wells or mine-shafts, often to the saturated zone.
6) Small-scale pollution from leaking sewers. This occurs where sanitary sewers are not watertight (a situation common in older sewers which were poorly constructed), decay or are disrupted by tree roots, construction activities and settlement. Concentrations of nitrate, organic chemicals, viruses and bacteria are released to groundwaters.
7) Where industrial concerns release effluent to sewers, heavy metals such as arsenic, cadmium, chromium, cobalt, copper, iron, manganese and mercury may enter waste water.

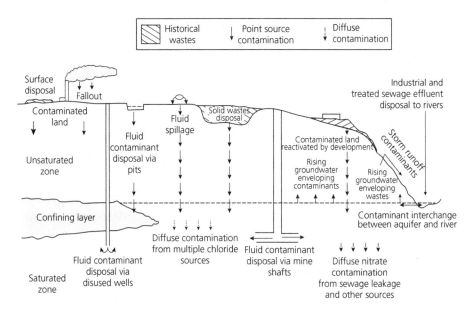

Figure 7.6 Sources of possible urban groundwater pollution. (Redrawn from Lloyd *et al.*, 1991 by permission of Oxford University Press).

8) Manifold chloride compounds capable of producing diffuse chloride distributions.
9) Accidental spillages, mostly of fluids, tank and pipeline leakages, can provide a range of hazards to surface and groundwaters.

Figure 7.6 shows the context of possible urban groundwater pollution; certain kinds date from the dawn of the industrial age and include coking and foundry wastes. Solvents are widely used in industrial processes, and a low level of chlorinated solvent and oil contamination has been detected in the unsaturated chalk aquifer of the Luton and Dunstable area in both liquid and vapour phases. Here, for example, a groundwater 'clean-up' was not considered practical and treatment of water at abstraction source is the preferred option (Longstaff *et al.*, 1992).

Cooling waters may be softened prior to being used in order to prevent scaling, and if discharged as such salinise groundwater, as does the salt accompanying the gritting of roads in winter. From pipeline and tank leaks, oils and petroleum products are the usual contaminants; they are immiscible and less dense than water. Hydrocarbons will percolate vertically through the unsaturated zone, and spread at the top of the watertable, migrating laterally with groundwater flow. Where contaminants are lost from landfill sites, the chemistry can be complicated.

Raised BOD, COD, iron, manganese, chloride, nitrate, water hardness, and trace elements occur. Mostly landfill is domestic rubbish, although some 15% of sewage sludge is disposed of in this way in the UK (Matthews, 1992). Originally disposal was by the backfilling of holes remaining after mineral extraction (albeit requiring detailed hydrogeological surveys to ascertain a site's ability to 'dilute and disperse'). There is a tendency for a vertical plume of contaminants to develop in the unsaturated zone, with a horizontal plume developing in the dominant transport direction of the saturated zone

(Todd, 1980). Contamination from any dump is aggravated by increased rainfall (where it is permitted to leach through the infill) and shallow watertables. Gases such as ammonia, carbon dioxide, hydrogen sulphide and methane may also be by-products. The last-mentioned is an explosion hazard. Trends in the UK are towards lining both the cavity and covering the top of landfill dumps to reduce leachate. Occasionally 'permeable liners' of fine-grained materials have been used to increase the opportunity for attenuation reactions beneath the landfill, and limit the release of leachates. There has been research into anaerobic and aerobic methods of biological control (Williams *et al.*, 1991).

However, cover with an impermeable clay cap has led to the undesirable effect of methane migration beyond the landfill boundary. The ability of the unsaturated zone to attenuate pollutants through physical, chemical and biological processes became the focus of much research in the 1970s. However, subsequent studies call for a more cautious approach. On the Sherwood Sandstones, Nottinghamshire, sequential profiles of chloride and total organic carbon beneath the Burntstump landfill site (for controlled wastes) have been taken. They show a downward migration with time in the unsaturated zone, as did a zone of reduced pH (Williams *et al.*, 1991). This aquifer exhibited a low buffering capacity (due to its limited carbonate content) and consequently the groundwaters received a high loading of organic acids and dissolved carbon dioxide in their interstitial waters. The chalk, by contrast, has a high buffering capacity which encourages microbial metabolisms, causing a reduction in readily degradable organic compounds with time.

In Britain there are large reserves of water beneath urban areas such as London, Birmingham, Coventry, Manchester and Liverpool, and these are valuable sources of water for local supply. Pressures include loss of infiltration through urban cover, but net importation from surrounding areas may more than compensate for this. This factor, considered with the decline in abstraction for certain industries, means that it is quality rather than shortage which provides problems. Rising watertables have already been noted as a problem, and examples of groundwater contamination in the West Midlands are given in Chapter 7.

Because of uncertainty, and concern for resource sustainability enshrined in EU directives and national legislation, heavy reliance upon physical/chemical/biological attenuation of pollutants or dilution and dispersion in groundwater systems may prove unacceptable (Price, 1996, p. 229–230). Groundwater is a precious resource and, finally, the legislation concerning it is catching up with that for surface water protection. The complexity of the contamination in urban areas puts these issues into sharp focus. Surface waters, on the other hand, are served by 'end of pipe' type controls and consents to discharge. Problems of surface waters *per se* in urban situations are under control provided that resources are made available in consenting, monitoring and policing.

It is heartening that the EA and NRW find that river water quality has generally improved over the past couple of decades in terms of chemistry and biology (see also chapter 4). There has been a fall in the amount of nutrients in rivers over this time. More specifically, using the WFD classification system, results for assessed rivers in England and Wales show that for overall ecological classification 26% of rivers are good or better, 60% are moderate, 12% are poor and 2% are bad. The situation for groundwater is perhaps less encouraging. For assessed groundwaters, 65% meet good quantitative status (in relation to groundwater abstraction pressures) and 59% meet good status for chemicals. In surface-water–dominated Scotland it is likewise pleasing that statistical trends in variables including biochemical oxygen demand, ammoniacal nitrogen (NH_3-N), lead and sulfate, have all declined reflecting improved treatment, cleaner technologies and tighter regulation and changed in agricultural practice. However, impacts of climate change may be evident are increases in water temperature and river flow at many sites and also increases in total organic carbon in many northern rivers (Anderson *et al.*, 2010). Overall, it is diffuse pollution from agriculture that remains the single most intractable problem.

7.9 Summary

From the policy angle, and despite decades of trying, diffuse pollution of agrochemicals remains a particularly difficult. Implementation of WFD by 2015 has not been achieved in Britain (among other states) and this will be explored in Chapter 10. The trends across Britain for water quality (and also for 'ecological status') are encouraging, but no more. Overall, impacts of environmental acidity arising from acid rain show signs of being under control.

There remain considerable threats to water quality from diffuse pollutants, varying from hydrocarbons and solvents in urban areas to agrochemicals, acid precipitation and eroded soil elsewhere and there nanoparticles present an new threat although ironically, nanotechnology also provides potential wastewater treatment opportunities. The growing diversity of chemical substances including pesticide products presents problems for regulation and monitoring as it threatens water for human consumption, aquatic, soil and other ecosystems as well as human and animal health.

To make matters worse, residence times in lakes and groundwater systems may be sufficient to cause much concern and are difficult and costly, if not impossible, to rectify. In recent decades, some serious issues of contamination in urban areas have come to the fore. However, a sound appreciation of hydro-chemical environments permits the special qualities of soil and aquifer systems and self-cleansing properties of surface waters (e.g., constructed reed beds) to be used in environmental protection schemes. Technological optimism reigns in places, especially water treatment technology, but it will be argued there generally no substitution for prevention of contaminant loading at source.

8

Environmental Issues of Water Quality and Quantity

At Folkestone, you don't so much swim, as go through the motions.

Former Kent sea bather

8.1   Complex Problems That will Not Go Away

This chapter discusses examples of water quality and quantity problems, especially those that have their origins in land-use practices. Solutions are made complicated because the causes may be multiple and sources of contamination diffuse. Most issues go beyond concern for public supply and encompass the wider water environment, in terms of both pollution loading and changes in the hydrological regime. Environmental commentators talk of 'Wicked Problems'.

This is never easy, for a wicked problem, where we can frame it, may describe social and environmental planning issues that seem difficult or impossible to solve because of incomplete, contradictory, changing requirements and difficulty in recognition. The term 'wicked' is used due to resistance to resolution. If that is not bad enough, complex interdependencies may reveal or create other problems (Rittel and Webber, 1973; Ludwig, 2001). In summary, they present water resource planners with both societal and technical uncertainties galore and has to operate in a changing governance environment. In a modern context, the matter of 'civil society' is where the citizen is free to act, away from coercion and is 'enabled' to operate within the context of rights and responsibilities (Chapter 10).

For example, a simple example of reducing nitrate concentration discussed in Chapter 9 relates to nitrate loading of the chalk aquifer around two sources operated by the former Eastbourne Waterworks company. Drastic reduction of agricultural activity reduced concentrations of nitrate in re-charging water showing a technical fix would work, yet serious questions remain as to whether such drastic land management changes are either economically desirable or politically acceptable when food production is at stake. Nitrates have become a wicked problem.

Problems are complicated for they revolve around impacts on waters from agriculture, industry (including mining), urbanisation (particularly sewage management), and afforestation, and influence both water quality as well as volume abstraction. Some of the most serious challenges to integrated catchment management affect lowland hydrological systems, and there are legal requirements to improve matters. Historically, drainage

The Protection and Conservation of Water Resources, Second Edition. Hadrian F. Cook.
© 2017 John Wiley & Sons Ltd. Published 2017 by John Wiley & Sons Ltd.

authorities were obliged to have regard to the preservation of natural beauty and the conservation of flora, fauna, geology, archaeology, access and amenity during the course of their work under the Water Act 1973 (Part II, Section 22). The Water Resources Act 1991 Section 16(1) confers a duty to 'further the conservation of natural beauty, flora and fauna', while the Land Drainage Act 1991 Section 12(1) confers the same duty on Internal Drainage Boards (IDBs) (NRA, 1995f). Abstraction regulation is dealt with under the Water Act of 2003 (Chapter 3) and has been noted represents a significant revision of licencing arrangements for England and Wales, although this remains under review by the regulators as potentially not being fit for purpose (Ofwat/EA, n.d.) and there are environmental abstraction regulations in Scotland (SEPA, n.d.b).

The WFD requires that 'good ecological status' be achieved for all European water bodies, originally by 2015 and to achieve this there must be agreement regarding what actually constitutes that status. The complicated modelling process of standardising this important policy question is referred to as 'intercalibration'. While this sounds an ugly piece of euro-speak, it is reflects the imperative that at least there is a common assessment system for seeking 'good ecological status' wherever a particular water body is located in the EU, although naturally that is relevant to geographic location, incorporating aspects of the natural environment. For example:

> 'The species of fish, invertebrates and plants in Baltic rivers differ from those in Alpine rivers, which in turn differ from those found in Mediterranean rivers' (WISE, n.d.).

It is a truism to say there should be sufficient water for the well-being of a given aquatic habitat. Somehow, the EU (and EA) seems to say less specifically about how this may be achieved, than do other agencies. For example, UNESCO (Speed, 2013) are greatly concerned about allocation within basins and do pay heed to environmental requirements, including fisheries-in this case, the guidelines intended for international use, modern management quite reasonably requires there is (among other things) a review of human activity and its impact including abstraction and particular needs, and means to protect that body including balancing abstraction needs while recognising the importance of groundwater (Europa, 2010). Actually, it is probably because 'water allocation issues' have their basis in legal rights to water, specifically in balancing the needs of upstream and downstream users that allocation for in-stream purposes have not hitherto come to such prominence, at least in European policy. Furthermore, the historical problem of setting 'Environmentally Acceptable Flow Regimes (Chapter 5) may indeed make British policy makers wary.

For Oklahoma in the United States there is a Stream Water Allocation Modelling Program that supports the appropriation, allocation, distribution, and management of stream water. The program is self-described as a 'pioneer state-wide initiative in the United States for the assessment of both water availability and reliability as well as a supporting tool for the adjudication and effective management of water rights in Oklahoma' (Oklahoma Water Resources Board, 2015). Computer modelling is being in order to establish water availability at any location on a seasonal basis, establish the unappropriated portion, determine areas particularly prone to water shortage, and review policy. Aquarius software used by the United States Department of Agriculture (Diaz, 2015) supports system components that allocate for reservoir storage, hydropower plants, agricultural water use, municipal and industrial water use, in-stream recreation water use, reservoir recreation water use, fish habitat protection and flood control areas.

As far as Britain is concerned, there are issues for 'in-stream allocation of water', simply on the basis that there are major issues for balancing abstraction against (not only) up-stream and down-stream users, or where appropriate maintain sufficient depth for navigation, but also maintaining the general ecological status and fish populations in particular. An example of problems of water allocation relates to flow in the chalk rivers of south and eastern England.

An important report 'UK Biodiversity Action Plan Steering Group for Chalk Rivers' was produced jointly by NE and EA in 2004. This followed a period of consultation, monitoring and modelling (EA, 2004). The stated vision is that "Chalk rivers should be protected or restored to a quality which sustains the high conservation value of their wildlife, healthy water supplies, recreation opportunities and their place in the character and cultural history of the landscape." To make this vision a reality, it is required to:

- maintain and enhance the characteristic habitats, plants and animals of chalk rivers, including winterbourne stretches;
- restore water quality, flows and habitat diversity;
- identify cost-effective means of restoring damaged river reaches.'

In total, 161 streams were identified, and while by no means all are all are in serious danger, long-term objectives include restoration of habitats and control run-off, co-ordinating diffuse pollution control (among other factors to control water quality), flood risk and aquifer recharge, advice to farmers, maintain river flows (including setting target flows should be set for all chalk rivers that take account of the needs to protect their ecology, protect wildlife, review the outcomes of management actions and manage non-native invasive plants and animals.

8.2 The Problematic River Wandle

The poor old river Wandle that flows northwards to the Thames through south London not only personifies a wicked problem, but its massive and long-standing issues reflect all that is problematic about urban catchment and river channel management; so much that the concept of 'Wicked Problems' might have been framed here! Actually, the Wandle has been a victim of a total historical failure of environmental governance over perhaps 200 years of downright abuse. This chalk stream lost its status as a river, let alone be developed as an urban amenity, let alone remain a trout stream. It became, by the mid-twentieth century a metaphor for environmental neglect. In the longer-term, lessons have been learned and there are real efforts at the time of writing to improve the Wandle and its environs.

If good can come from bad, then it is helpful to summarise this river's contribution in environmental management. Specifically, 'integrated river basin management', or a crying need for it is now being met via partnerships including the Wandle Trust in the River Wandle Catchment Plan (Wandle Trust, 2014).

Historic themes that have been considered are (Cook, 2008a, 2015):

1) Transition from trout stream to foul urban sewer
2) An exemplar of serious industrial pollution

3) An early case study in environmental auditing
4) Innovations in water mill technology
5) Problems of sewage treatment and experimentation
6) Pressures on surface and groundwater
7) Abuse of the channel morphology through economic development

Space does not permit all of these themes to be explored but the story of the Wandle and sewage is of general historical and environmental interest. First it had to undergo a transition borne of numerous negative externalities. Praise for the Wandle's water quality and fish population seems to have lasted until the middle nineteenth century. In the sixteenth and seventeenth centuries, it was a clear trout (*Salmo trutta*) stream, yet trout disappeared from the Wandle in the mid-nineteenth century as 'London turned its back on the Wandle' (Pike, 2004).

By 1801, Croydon's population was 5 743 and growing for it was in transition between Surrey market town and a suburb. The population had grown to 16 712 in 1841, reaching almost 80 000 by 1881 (Goddard, 2005). There was inevitably a consequence of the Victorian 'suburbanisation' in the Croydon area that would put pressure on open space, on groundwater resources and sewage treatment (Goddard, 2007), yet, in 1833, the Wandle had been found to a 'pure and unadulterated stream' (Rafferty, n.d.).

In the 1830-1840s, Croydon, Mitcham and Carshalton experienced sewage problems, the river Wandle and its ponds were found to be 'open sewers' and there were outbreaks of typhus and cholera. Then in 1860, the Croydon Health Board culverted the river, Beddington Sewage Works was created, but effluent discharges still made the river Wandle probably most polluted river in London. Braithwaite (1861) found discharges from industry and sewage in both the Carshalton and Croydon branches, although he was especially damming about the latter "the stench is only kept in check, by low temperature"… And "the chief cause of this, is the filth from Croydon, which finds its way "through the wheel at Waddon mill".

The Croydon Local Board of Health had a problem with sewage disposal and the Wandle carried all the sewage from that town. After 1860, a new Sewage Farm at Beddington struggled to improve the situation. The Beddington irrigation meadows, instigated by the Croydon Local Board of Health, had been celebrated by two tufts of ryegrass on the Arms of the Former County Borough of Croydon (Goddard, 2005, pp. 132-148).

Clearly, sewage was emerging as a problem and any reduction in channel flow could never help the situation. To address this, the local authority had purchased agricultural land at Beddington that could be irrigated with sewage effluent; this was then 'cleansed' by processes of filtration and bacterial oxidation. Braithwaite reported that sewage was being carried over about 60 acres (24 ha) of land at Beddington, and found this sufficient in most weathers assisted by a de-odourising process using 'McDougal's Fluid', a distillation product of gas tar rendered alkaline by quicklime. However, certain owners of the mills and fisheries thought the tar was injurious to the fish.

The Croydon Board of Health was evidently amenable to improving the treatment of sewage. There were, however, complaints on account of the irrigated land being held 'in a sodden and offensive condition', a 'sewage marsh'. The irrigated area expanded to 560 acres (228 ha) by the 1880s with another area developed at Norwood, and there was a campaign against the smell (Goddard, 2007, p. 123; Clapp, 1984; Braithwaite, 1861, p. 236). Eventually, new sewage technology was adopted with the building of the

Wimbledon works (1877–1971) and Wandle Valley Sewage Works - again sited on riparian meadows (Nicholson, 2001). In the 1960s, the Beddington Sewage works in the upper catchment were re-designed and this went some way to improving issues of water quality in subsequent years (Thames Water, 2014).

8.3 The River Kennet

The river Kennet rises on the Marlborough downs as a typical of an English 'Chalk stream', flowing 70 km through some iconic chalk landscapes on southern England including the heart of the archaeological province of 'Wessex' to become a tributary of the Thames at Reading. It is an SSSI and it is linked with the Somerset Avon via the Kennet and Avon canal, completed in 1810 and the river supports a brown trout population, making it a good starting point to discuss rural southern rivers.

The former NRA produced a catchment management plan in March 1993 (NRA, 1993e) outlining the issues facing the river including pesticide pollution and over-abstraction affecting river levels in the upper catchment, while urbanisation and sand and gravel extraction affect the lower catchment. Sadly, problems of pesticide pollution persisted in the river (EA, 2013f). Then, in December 2012 the EA produced the Kennet and Vale of White Horse Catchment Abstraction Licensing Strategy (EA, 2012b), incorporating a Catchment Abstraction Management Strategy for the Kennet and its tributaries. Water abstraction in the Catchment Abstraction Management Strategies (CAMS) area is from both surface water and ground water. In case of extreme conditions, a licence may still be granted with conditions which protect the low flows. This is the Hands Off Flow (HOF) condition on a licence which requires abstraction to stop when the river flow falls below a certain amount (EA, 2013b, chapter 5). The majority of abstractions in the catchment are from groundwater, abstraction for public water supply predominates and a water resource availability plan was created to provide a strategy. In this process, river lengths of segments are designated and may be summarised in a manner that reflects control of licenses as listed below:

- High hydrological regime (more water is available than is required to meet environmental needs but licencing may be restricted to protect waterbodies).
- Water available for licensing (new licences can be considered depending on local and downstream impacts).
- Restricted water available for licensing according to a resource allocation for the environment that is defined as a proportion of natural flow, known as the Environmental Flow Indicator (EFI). Full licensed flows fall below the EFIs. If all licensed water is abstracted there will not be enough water left for the needs of the environment. No new consumptive licences would be granted.
- Water not available for licensing. Recent actual flows are below the EFI. This scenario highlights water bodies where flows are below the indicative flow requirement to help support Good Ecological Status (as required by the Water Framework Directive).
- Heavily Modified Water Bodies (HMWBs and/or discharge rich water bodies). These are water bodies that have a modified flow that is influenced in turn by reservoir compensation releases or they have flows that are augmented. These are often known as 'regulated rivers'. They may be managed through an operating agreement, often held by a water company.

Statistically speaking sources may be defined in term of whether their consumptive abstraction available less than 30%, at least 30% of the time, at least 50% of the time, at least 70% of the time, at least 95% of the time or otherwise not assessed. This information may be expressed in terms of maps of the river system. The development of CAMS (Chapter 2) not only enables control of abstraction licencing, but also develops a clear picture for strategic planning of water abstraction from a catchment.

8.4 The River Itchen

An extremely detailed three-year study was on the River Itchen in Hampshire that suffered from a range of problems arising from its proximity to urban areas, it enjoyed considerable support by local authorities. This was the detailed River Itchen Sustainability Study (Eastleigh Borough Council, 2012) that took place over three years and was co-ordinated by the Halcrow Group Ltd. It was funded by the Environment Agency and the water companies (Southern Water and Portsmouth Water) which abstract from the River Itchen. It was overseen by a steering group, consisting of these organisations, plus English Nature, Hampshire County Council, Eastleigh Borough Council, and Defra.

The purpose of the study was to assess the impact of water company abstractions on the River Itchen. This was achieved by establishing the current condition of a range of environmental parameters (such as water quality and flow) and a key range of habitats and species in the river. In 1998, the River Itchen was notified as a candidate Special Area of Conservation (cSAC) under the EC Habitats Regulations (English Nature, now Natural England), while the EA set about reviewing abstraction licencing, and consents to discharge. Another driver behind the study was the water industry Asset Management Plan (AMP) process, by which water companies prepare capital plans for necessary improvements to their infrastructure facilities and thereby are invited to make bids to Ofwat for support for these capital improvements. One example would be sewage treatment.

The main problems of the Itchen relate to a high level of abstraction from this chalk-dominated catchment (EA, 2011b) that requires flow of sufficient quality and quantity to support fish populations, maintain quality and support abstraction for watercress beds and fish farms as well as the ubiquitous problems arising from sewage management (see also Chapter 5). Unfortunately, despite some good efforts, the SSSI status (designated in 1981) and international recognition, habitats in the Test and Itchen catchments are an unfavourable condition. Key reasons cited for the unfavourable condition include historical modifications to the physical structure of the channels, banks and riparian zone (Skinner, 2012). In cases of both the Itchen and the nearby river Test (another valued chalk stream) there is an emergency programme to restore the rivers' banks, bed and associated in stream ecology. Water quality overall it good, but in times of low flow is improved by groundwater augmentation pumping.

The Test and Itchen Abstraction Licensing Strategy includes measures to combat problems of both rivers. These include, not issuing any uncontrolled licences, time limits on new licences, concern of both marine and freshwater bodies and their ecology, controlling abstraction from the Chalk, the possibility of HOF being acted upon, abstraction and storage is encouraged at times of high flow, modification of existing licences to abstract and to encourage the traditional operation of watermeadows to divert high flows and promote early growth of grass, with appropriate licencing arrangements. The catchments are suitably mapped for consumptive abstraction (EA, 2013g).

The famous chalk streams of England are but one example of a valued kind of habitat that is under threat from a variety of directions, and solutions are not forthcoming. This kind of complexity sets the scene for two regional examples of areas that are under threat. The next category of catchment problems can be characterised as hydro-regional problems because they occur within regions that are actually defined by wet habitats.

8.5   The East Anglian Broad Land and Waveney Valley

8.5.1   Origins, History and Nature of Broadland

The region of East Anglia known as Broadland is not only one of Europe's most important wetland areas, but hydrological intervention has made it extremely complicated. Before the Flandrian rise in sea level -which followed the melting of the North Sea ice - Broadland rivers in the immediate post-glacial era would have been tributaries of the Rhine. As sea level rose, the lower river valleys became flooded, causing a progressive landward migration of the estuarine environment. Essentially, a sequence dominated by silty and clayey calcareous marine alluvium with peat deposits was deposited in these valleys. Coarser material in the form of marine and windblown sands was deposited largely on the seaward edges of Broadland. These deposits, dating from around 8500 BP, came to overly a range of Pleistocene and older formations such as the Pleistocene 'Crag' and the chalk. Today there are two dominant soil types: poorly drained marine silts and clays which form soils at the seaward end of the valleys, and fen peats which predominate near the upland margins and in the upper valleys. The ages of soil parent materials are between 2250 and 1610 BP.

Geographically, Broadland comprises the lower valleys of the Rivers Yare, Bure and Waveney (Figure 8.1). The region is defined by valley-bottom land, together with the silted-up estuaries of the rivers Bure, Yare and Waveney inland from Great Yarmouth. Most of the area is below 2 m OD, with extensive areas of drained marshland below mean high tide level.

Water from the west (known as 'upland'!) enters the region via the main rivers Bure, Wensum, Tud, Yare, Tas and Waveney, predominantly groundwater-fed streams. Both diffuse and point pollution from industrial and agricultural sources or sewage effluent, affect their water quality. The modern climate is continental with warm, dry summers and cool winters. Annual effective rainfall (the difference between actual evapotranspiration and rainfall) can fall below 100 mm. Nonetheless, the low-lying nature of the land makes drainage difficult, and the surface catchment area of the rivers is considerably greater than the Broadland region itself (Figure 8.1).

Broadland supports a variety of wetland environments. All are essentially semi-natural and merit conservation on account of their ecological and landscape value. The Broads are open-water, shallow lakes; there are also fens, alder carr and reed-swamp following hydrological interference. 'Marshlands' are secondary wetlands that are embanked and criss-crossed with drainage dykes and developed on both peat and mineral alluvium soils.

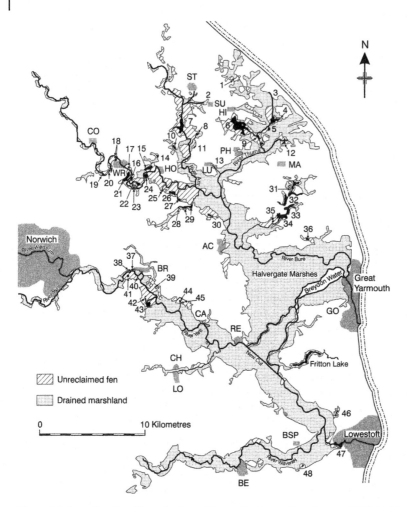

Figure 8.1 Broadland and its main rivers. (*Source:* Redrawn from George, 1992, inside cover).

KEY to Figure 8.1

BROAD	TOWN
1 Calthorpe Broad	AC Acle
2 Sutton Broad	BE Beccles
3 Waxham Cut	BR Brundall
4 Horsey Mere	BSP Burgh St Peter
5 Blackfleet Broad	CA Cantley
6 Hickling Broad	CH Chedgrave
7 Barton Broad	CO Coltishall
8 Catfield Broad	GO Gorleston
9 Heigham Sound	HI Hickling
10 Alderfen Broad	HO Horning
11 Cromes Broad	LO Loddon
12 Martham Broad (North and South)	LU Ludham
13 Womack Water	MA Martham
14 Burntfen Broad	PH Potter Heigham
15 Hoveton Little Broad	RE Reedham

16 Hudso's Bay
17 Snapes Water
18 Bridge Broad (East and West)
19 Nortons Broad
20 Belaugh Broad
21 Wroxham Broad
22 Hoveton Great Broad
23 Salhouse Broad
24 Decoy Broad
25 Cockshoot Broad
26 Ranworth Inner Broad
27 Malthouse Broad
28 Sotshole Broad
29 S. Walsham Broad
30 Upton Broad
31 Ormesby Broad
32 Rollsby Broad
33 Ormesby Little Broad
34 Filby Broad
35 Little Broad
36 Mautby Decoy
37 Brundall Broad (Inner)
38 Brundall Broad (Outer)
39 Strumpshaw Broad (East and West)
40 Bargate Broad
41 Surlingham Broad
42 Wheatfen Broad
43 Rockland Broad
44 Buckenham Broad
45 Hassingham Broad
46 Flixton Decoy
47 Oulton Broad
48 Barnby Broad

ST Stalham
SU Sutton
WR Wroxham

The landscape is described as either 'enclosed' (broads or rivers enclosed by carr, fen or trees and grazing marsh with trees such as poplar or willows) or 'open', more or less treeless grazing marsh, extensive open water and reed fen. The Broads result from medieval peat cutting between the twelfth and fourteenth centuries. An estimated 25.5 million cubic metres has been dug from over 30 excavations to an average depth of 2.4 m (George, 1992, Ch 4), although today the Broads are appreciably shallower due to siltation. Relative increases of sea level with respect to the land would have left the excavations liable to serious flooding from around 1300 onwards, so that by the fifteenth century they were flooded, much as they are today (Dymond, 1990). Around these artificial lakes are a range of fringing habitats, reedswamp, alder carr, fen and grazing marsh. Some present-day fens were grazing marshes in the last century but reverted when drainage systems broke down.

On Halvergate Marshes, grazing has occurred since Saxon times (Dymond, 1990), although it is probable that most reclamation took place between the thirteenth and seventeenth centuries. Considerable improvements to drainage were effected through windmill technology driving scoopwheels to lift the water in the eighteenth century. There were efforts to extensively reclaim the area for arable agriculture during the 1980s. Risks of flooding occur from breaching of the North Sea defences, from saline waters overtopping banks during tidal surges, and from freshwater river floods. The function of water management systems is to remove water to a degree which permits a specified land use, for example reedbed cultivation or grazing (Cook and Moorby, 1993).

Major ion chemistry shows that rivers entering Broadland are alkaline (around 300 mg 1^{-1} HCO^{3-}) and chloride concentrations low. Figures for Hickling Broad and Horsey Mere in the northeast are both about 150 mg 1^{-1} for HCO^{3-}, but 1740 and 2840 mg 1^{-1}, respectively, for chloride. Concentrations reflect the influence of the chalk aquifer to the west, and sea spray and other saline intrusion from the east (Moss, 1988, Ch 3) The fens are irrigated with base-rich water (pH between 5.5 and 8.0) while chloride concentrations in chalk groundwater can exceed 100 mg 1^{-1} in eastern Broadland (IGS, 1976), limiting borehole abstraction in the area.

8.5.2 Environmental Problems of the Region

The landscape has been much altered by human activities, including peat abstraction, drainage works, agricultural development and (more recently) tourism and water-based recreation. Upstream there are problems arising from sewage works discharging into the main rivers. Water quality problems affect not only the open water bodies, but also environments in hydraulic continuity with open waters and rivers for example. Certain fen waters have been shown to contain raised concentrations of nitrogen and phosphorus. The problems of the alluvial marshland areas are soil based, and primarily affect the dykes adjacent to the fields. Difficulties only arise on conversion to arable which involves improvements to regional drainage, deep underdrainage, ploughing and the application of agrochemicals. Soil acid sulphate toxicity following conversion to arable is locally serious (Chapter 7).

Acid sulphate affected soils are found fringing many of the valleys, but not in their upper parts where deposition of the fen peats was under freshwater, calcareous conditions. In deep-drained marshland reclaimed for arable, acidity may inhibit rooting, and where acid drainage waters become neutralised when they meet buffered waters in the ditches, drainage pumps and pipes become blocked with 'ochre'. Residual salinity in calcium-deficient clayey soils also affects drainage. Such soils experience a structural collapse, and hence severe drainage difficulties following reclamation due to loss of macroporosity and the blocking of tile drains. Originally recognised in the North Kent Marshes, the remedy is to add calcium in copious quantities, generally in the form of gypsum, in order to replace the sodium in the exchange complex of the clays. The quantities required are considerable, hence the cost is great (Hazelden, *et al.*, 1986). Salinity of itself will affect the ecology of water bodies (Hazelden, 1990), and is undesirable for many other uses, such as drinking or irrigation purposes.

In Broadland, phosphate and nitrate may be washed into water bodies and enter sediments where denitrification can occur, and phosphate forms complexes with sediment particles. Phosphates can arise from phosphorus-containing fertilisers, and arrive in waters forming complexes on eroded soil particles. They occur in soils as insoluble iron, calcium or aluminium phosphate, and are released only slowly. However, in Broadland, the major source is from sewerage works, and a major contributor here is phosphate detergent, in use since the 1940s.

Nitrate is leached from arable and intensively managed grassland (Chapter 7); in Broadland, as elsewhere, agricultural runoff is the main source. The nitrate ion is negatively charged, as are the majority of soil particles, making it mobile and prone to leaching. Nitrate is not usually a limiting nutrient in natural aquatic systems, however its strong fertilising effects play a role in increasing the biomass of algae (see below). It has also been implicated in the decline of reedswamp in Broadland rivers and lakes by causing the plants to become over-fertilised and produce top-heavy growth, making them susceptible to damage by wind and wave action while their rhizomes become weakened (Crook *et al.*, 1983).

It is estimated that about 60% of nitrate in waters in England originate from agricultural land. Under the EU Nitrates Directive Nitrate Vulnerable Zones (NVZs) have been designated in the area, these are areas where regulations include a limit on the amount of nitrogen fertiliser that can be applied to farmland. Nitrate pollution also originates from non-agricultural sources including sewage treatment works, industry and various diffuse sources such as transport and urban drainage systems. Investment by water companies has led to reductions in pollution from sewage treatment works particularly through the Urban Waste Water Treatment Directive.

In the Broadland waterways, there is a strong seasonal pattern in nitrate concentrations with peaks in the winter months. Only the River Waveney displays nitrate concentrations that are above the Nitrate Directive limit of 50 mg/l NO_3 for the majority of the time, however since the early 1990s, nitrate concentrations have remained at similar at most monitoring sites (Broads Authority, n.d.a). In practice, most constraints on productivity arise from phosphorus limitation, and much research effort in aquatic ecosystems has been into phosphorus behaviour and options for its limitation. In most Norfolk Broads, the phosphorus concentrations often exceed 60 μg 1^{-1} total P, while water supply problems might be expected above 30 μg 1^{-1} (Moss, 1988, Chapter 6). Under natural situations, phosphorus is the major limiting nutrient because the amount of biologically active phosphorus is small. In Broadland, research has reported mean annual total concentrations of phosphorus between 100 and 1000 μg 1^{-1} in the Broads and in the River Yare, and here soluble reactive phosphorus levels as high as 2000 μg 1^{-1} have been recorded during the summer (George, 1992, Chapter 5), a condition described as 'hypereutrophic'. In a study published in 2002, phosphorous stripping at sewage works at Fakenham and East Dereham on the river Wensum upstream of Broadland (that has its confluence with the Yare at Norwich) were shown to be highly effective in removing soluble phosphate from river waters (Demars and Harper, 2002). At the time of writing, there is active land use and river nutrient loading for the River Wensum under the River Wensum Demonstration Catchment (Wensum Alliance, 2010).

In Broadland proper, Barton Broad, which in the 1950s was rapidly losing its underwater ecosystem and becoming badly turbid due to sewage discharges and agricultural nutrients, reaching a low point in the 1970s, has been recovered by phosphate stripping on the river Ant, and dredging 1996-2001 with the dredgings being returned to agricultural land. Overall, between 1979 and 2004 the total phosphorous in the water fell by around 80% and there was a corresponding dramatic fall in chlorophyll-a, an indicatior of algal presence in the water column also fell dramatically (Broads Authority, n.d.b) similarly Wroxham Broad on the river Bure is improving. Overall, the condition of the Broads and their associated rivers is improving in this respect. Both nitrate and phosphate remain a problem in the hydrological system of the Broads.

Organochlorine pesticides have been detected in sediment cores from Hoveton Great Broad which is in contact with the River Bure, and hence, its arable catchment.

Also affected are Upton and Martham South Broads which are, respectively, hydro-logically isolated from the river system and surrounded by grazing marsh, at least from the late 1970s. In sediment cores from the Hoveton Great Broad, residues of dieldrin, DDD, DDT derivatives and TDE were found (Stansfield, *et al.*, 1989). Dieldrin was present but its concentration was less, and other pesticides were rare or absent in cores from the other lakes. Eutrophication is the enrichment of waters by inorganic plant nutrients, and where it is the consequence of human activity the term 'artificial eutrophication' is preferred.

Flooding associated with Broadland's low gradient rivers has always been a problem. The Broadland Flood Alleviation Project (BFAP) commenced in 2001 as a public-private partnership with stakeholder engagement important in planning. It aims over 20 years to improve and maintain 240 km of flood defences. The project is to protect and enhance the biodiversity and also provide an improved service level of flood defence. The mitigation strategy adopted to manage the biodiversity within defined engineered 'flood compartment' considers notable species including mammals such as water vole, otter and bats, reptiles (adder, common lizard, grass snake and slow worm), invertebrates (including rare molluscs and dragonflies, both over-wintering and spring-nesting birds, as well as bittern and marsh harrier, and rare plants, be they aquatic or terrestrial. An important process has been the use of 'setback', that is re-aligned flood defences and create new scrapes, open water, reedbeds and other wetland habitats. The network of drainage ditches acts as 'wet' fence lines to manage the livestock and also to provide aquatic habitat. In some BFAP flood compartments, defences are setback up to 30 m behind an existing bank, creating new bank in the adjacent grazing marshes, there will be post-project monitoring in re-engineered areas (Dodgson, *et al.*, 2011).

In summary, eutrophicated aquatic ecosystems experience a decline in species richness, change in dominant biota, increase in turbidity (cloudiness of the water) and increase in sedimentation. Anoxic conditions may develop in sediments, and in earlier stages increased vegetative growth may choke waterways while in later stages water plants may be lost in favour of phytoplankton (algal) growth. Four 'phases' of eutrophication may be recognised (Mason, 1991, Chapter 4; Broads Authority, 1993):

Phase 1. This situation existed in the Broads before 1900. Open water was clear, covering low-growing plant species, with a transition of vegetation from alder fen and open swamp fen through to reedbeds around the margins. In 1993, only three of the Norfolk Broads were in this category.

Phase 2. This situation, dominant up to the mid-twentieth century, still featured clear water, although with presumed increases in external loading of nutrients. Plant species tended to be taller than Phase 1 with low-growing macrophytes becoming disadvantaged by low light intensity. Invertebrate populations increased in density because of the added cover for their egg-laying, and this led to large populations of fish and waterfowl.

Phase 3. By the 1950s, much higher nutrient availability led to abundant phytoplankton. The water became turbid, with a poorly developed benthic fauna dominated by tubi-ficid worms and chironomids. A switch from Phase 2 to Phase 3 corresponds to annual mean concentrations of total phosphorus in water of around 100 μg l^{-1}.

Phase 4. Nutrient enrichment conditions may lead to a dominance of blue-green algae, and these can produce large blooms during hot summers such as 1989. Bluegreen algae can produce toxins harmful to mammals.

Macrophyte communities stabilise via their uptake of nutrients which deprives the algae. Furthermore, they secrete chemicals preventing growth of plankton, shed leaves with heavy epiphyte burdens and shelter populations of grazers on phytoplankton from their fish predators. On the other hand, the phytoplankton community may be stable by virtue of factors including an early growing season, vulnerability of herbivores in the absence of aquatic plants, and the production of large, inedible algae (Irvine, *et al.*, 1989).

The role of organochlorine pesticides in eutrophication and the switch from submerged plant dominance to phytoplankton dominance have been suggested by Stansfield *et al.* (1989). Examination of pesticide and microfossil remains in datable cores showed a link between pesticide residues and zooplankter decline. These chemicals were used liberally in the 1960s and 1970s, and a decline in the *Cladocera* community (such as *Daphnia* sp.), which grazed the phytoplankton, may have led to a decrease in macrophytes to the advantage of the (no-longer grazed) phytoplankton.

These came to dominate the affected Broads, bringing about Phase 3 conditions. An effect of the switch from Phase 2 to Phase 3 is an increase in sedimentation rates on account of the dramatic increase in the rain of dead phytoplankton to the lake bottom and, in some instances, the breaking up of fringing vegetation, such as the natural reed-swamp. In Barton Broad (a Phase 4 Broad in 1993), between one and four millimetres of sediment was laid down each year between 1400 and 1950, whereas a figure of 12.8 mm has been reported for the 1960s.

8.5.3 Management Authorities, Structures and Strategies

In the post-war period it became apparent that the Broads, of international importance, were under pressure from visitors and habitat degradation due to agricultural development. In 1978, the Broads Authority (BA) was established, and its function was to tackle a range of environmental problems. Since April 1989 it has been invested with the executive and statutory authority (appropriate to its own problems) to effectively function as a National Park, the first in lowland England and the area is also designated under the Ramsar Convention. The major issues may be summarised as:

1) The over-abstraction of water for public supply and irrigation from both rivers and groundwater, with associated threats to both river flow (affecting navigation) and water quality. Groundwater abstraction affects the baseflow of rivers, threatens wetland habitats, and increases the risk of saline intrusion.
2) Issues of flood prevention from breaching North Sea defences, from saline waters overtopping banks during tidal surges, and from freshwater river flooding. This continues to produce controversy.
3) Problems of agricultural intensification such as conversion to arable cultivation and land drainage. This may cause a contamination of waters, especially from the leaching of nitrates and pesticides, soil and water acidification, and increased sediment loading from erosion.
4) Contamination of rivers and Broads from sewage.
5) Loss of habitats associated with fens or grazing marshes and loss of intrinsic landscape value.

6) Habitat degradation in open water bodies.
7) Direct problems arising from visitor pressure, needs of holiday accommodation, angling, water sports, access on foot, toilets, showers, launderettes, litter disposal and cars.
8) Mechanical destruction of river banks and ronds (shelf-like river banks caused by the setting back of flood embankments) which support the fringing vegetation of Broads. The causes are largely the wash from motor boats, but also results from the consequences of agricultural operations and animals. The BA and EA are producing joint policies to tackle the water resource, quality and flood protection issues (Broads Authority, 1993), while Nitrate Vulnerable Zones (Chapter 9) cover the main areas of the Broads.

In March 1985, the Broads Grazing Marsh Conservation Scheme was launched following protests about the threatened ploughing and deep drainage of Halvergate Marshes. Funded by the Countryside Commission and MAFF it provided for farmers to be compensated where they retained low-input livestock management (Purseglove, 1988, Chapter 8). What had been a one-off plan aimed to conserve a landscape of historic and ecological importance, proved to be the precursor of all Environmental Sensitive Areas (ESA), now replaced by Environmental Stewardship schemes (Natural England 2015). In spring 1987, Halvergate and eight other areas (including the Somerset Levels) were designated ESA with agrochemical application and grazing density limited and financial incentives offered to revert to grass. Under 'Tier Two' of the ESA, incentives were offered for raising waterlevels, while shallow pools in late winter and early spring were created to benefit over-wintering and breeding birds. An even 'wetter' option is Tier Three (introduced during 1991), which aims to hold watertables high well after the grazing period (Broads Authority, 1993).

Since then, the advent of Water Level Management Planning (Chapter 4) (Water Management Alliance, n.d.) has sought to balance the needs of agriculture and conservation and of necessity engages with the Internal Drainage Boards, as well as the statutory agencies and Defra. In Broadland, WLMPs are formal agreements over water management and are prepared on a whole IDB district basis, satisfying statutory obligations under the Norfolk and Suffolk Broads Act 1988 and Defra targets. The majority of plans will be produced by Internal Drainage Boards through consultation and aim to integrate the water management needs of different interests, including agriculture, flood defence and conservation. With an ironic echo of grant aid for land drainage in the mid-twentieth century (Cook, 2010a), grants towards capital costs for implementing actions identified in WLMPs are available. It is hoped that, given time, management systems will meet the objectives of all parties and optimise the benefits for wildlife (Broads Authority, 2001).

8.5.4 Environmental Protection and Restoration

Early efforts to improve water quality involved isolation of individual Broads from the main river system with the prime objective to 'turn off the phosphorus tap'. In 1978, following similar work in the United States, Alderfen Broad was isolated from the main system thereby cutting off the supply of sewage effluent. However, total phosphorus concentrations in the summer rose from around 200 µg l^{-1} to almost 900 µg l^{-1} by August 1979 and blue-green algae predominated (George, 1992, Ch 13). The reason was release of phosphate from the upper layers of sediment during the summer, and the loss of the ability to control this pollutant by water circulation. This pattern was repeated in subsequent years; clearly this was not by itself the solution.

Around 1982, Cockshoot Broad was isolated from the River Bure, with the objective of reverting the Broad to a Phase 2 condition from Phase 3. The Broad was then suction dredged in order to remove the upper layers of sediment which contained phosphorus which was able to exchange with the water above, and the dredgings deposited in a bunded lagoon. So far, the prospects for ecological recovery have looked encouraging (Broads Authority, 1993). Since that time, Belaugh Broad (already benefiting from phosphate stripping) and Hoveton Little Broad have been suction dredged, and dredging continues with ecological monitoring in progress. Other methods have to be investigated. One such idea is 'iron dosing'. At Ranworth Inner Broad, experimental injections of ferric chloride into sediment (which have the effect of inhibiting the release of phosphorus) have been tried. Proven as a method, phosphate stripping of contaminated waters is a very attractive idea, because most phosphorus is derived from sewage works.

Figure 8.2 shows the effect of adding ferric sulphate to sewage effluent discharged to the River Ant at Stalham and North Walsham. Loading is reduced by around 90% with around 80% of phosphate reaching Barton Broad estimated to be from sewage, the remainder from agriculture. Concentrations in the Broad rose from around 100 μg 1^{-1} in winter 1980 to 290 μg 1^{-1} by September 1981, attributed to release of phosphorus from sediment stored in the Broad and not sewage. Phosphate stripping is now a major feature of water quality policy in the Bure catchment (Broads Authority, 1993). Ecosystem responses are less clear; reduction of external phosphorus loads may not, in itself, be sufficient to restore aquatic communities to their prior state.

A method of nutrient stripping which takes advantage of the natural ecology of the region is the use of specially constructed reedbeds in sewage treatment works in water purification. This approach uses a reedbed's ability to uptake phosphate and nitrates, and to encourage denitrification in the root masses. Managing relationships in an ecosystem is called biomanipulation. Methods include the temporary removal of fish populations, thereby encouraging populations of zooplankton which, in turn, consume phytoplankton. Experiments at Cockshoot and Belaugh Broads aim to monitor the return to clearer waters, and hence Phase 2 conditions. Water plants have to be protected from grazing wildfowl.

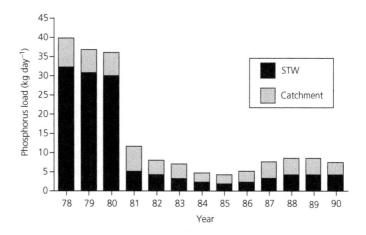

Figure 8.2 The effect of ferric sulphate stripping of phosphorus in the river Ant. (Redrawn from Broads Authority, 1993).

Dredging is needed in many watercourses because accelerated siltation over some 60 years has reduced navigability and reduced volumes discharged, hence increasing flood risk. Dredging is now carried out in order to ensure a minimum of four feet (1.2 m) in mid-stream for main rivers, and both the BA and EA have the objectives of improving the habitat, including creating dredging profiles that are appropriate to the re-establishment of reedbeds on the ronds, affording extra protection from backwash.

Weed cutting, a direct by-product of eutrophication, is also frequently required in order to maintain navigability. Protection measures from visitor pressures have to be addressed. These include limiting river craft speed (the maximum speed is now 6 mph [9.6 kph]). Modified hulls designed to minimise the wash from river craft are also encouraged. Electric craft have the advantage of being quieter and produce no pollution. Other measures include artificial protection and strengthening of affected banks.

Ronds provide nesting for birds and spawning for fish, as well as sites for reedbeds; they are also corridors for wildlife. Over-fishing should be tightly controlled through the issuing of rod licences, and damage to banks and their vegetation minimised by the construction of fishing platforms. Holiday developments, camping and caravan facilities, boats and day visitors all generate requirements for more toilets, and hence further sewage treatment. Vessels should be fitted with holding-tank toilets to avoid further water contamination.

Flood defence and drainage issues in Broadland have a long history on account of the region's low-lying nature; therefore land drainage and coastal defences are considered together. Three management issues which continue to require attention are (i) land drainage (associated with agriculture), (ii) freshwater flooding and its control, and (iii) sea flooding. The need to maintain local surface or groundwater levels for wetlands, grazing marsh or arable land uses is intimately tied to flood alleviation strategies.

While the EA deals with overall strategies for water management, it is the Internal Drainage Boards (IDBs) which deal with such issues at the local level. IDBs have ancient origins and retain much power (Purseglove, 1988, Ch 8), and hence have much influence in wetland environments. Major watercourses have been, since 1989, under the EA, main drains and pumps are often the responsibility of the IDB, while the smaller ditches are the responsibility of local farmers and landowners. The IDBs effectively represent the interests of agriculture, being primarily composed of farmers; hence they played a key role in the threatened deep drainage and ploughing of Halvergate in 1984. Many IDB members, on account of their land-holding interests, or ability to purchase undrained land, stood to benefit from deep drainage and conversion to arable.

Since the late 1980s, with the advent of the ESA, the future of the area is looking more secure. Arablisation has largely ceased over the grazed areas, and marshes managed by the Royal Society for the Protection of Birds (RSPB) now benefit through watertable management and earth dams. These enhance winter flooding for the benefit of birdlife. George (1992, Ch 9) describes the reorganisation of pumping arrangements for the Eastfield level near Horsey in the Thurne catchment (Figure 8.1), which, unlike the Halvergate area, does not contain sufficient calcium carbonate to buffer acid sulphate problems that occur where the watertable has been lowered. Conversion to arable in the 1960s caused so much ochre to be discharged by the pumps operated by local IDBs, such that the Gibb's outlet and the Waxham Cut (into which it discharged) silted and made the passage of larger craft difficult. In 1987, to alleviate the problem, the Anglian Water Authority created a new outfall channel from the Eastfield pump to protect the Brayden marshes, and built an overflow sluice near to its outfall to reduce the particulate material discharging into the Waxham Cut. The ESA provides an opportunity in such situations to raise watertable levels and provide some compensation for farmers.

Flooding has long been a problem. A disastrous event in January 1953 caused low-lying areas to be engulfed, with loss of life. There was damage to riverside properties, damage within the town of Great Yarmouth, and salinisation of agricultural land (most serious for arable areas). The development of a flood alleviation strategy was announced in July 1991 (NRA/Binnie and Partners, 1991) to investigate options for riverine and marine flood alleviation for some 21 300 ha of land and 1700 properties (NRA, 1994d).

Originally, flood defences were built to withstand flood events with a 1:25 year return period, but locally this had deteriorated to 1:5 years, especially around Haddiscoe Island between Reedham and Great Yarmouth (Figure 8.1). Existing measures were reviewed, hydraulic models developed, cost benefits of options for flood defences and alleviation were considered, and a full environmental assessment carried out. Freshwater flooding of rivers, although not so serious as saline incursion, has none the less been a problem for centuries and was a major obstacle to land drainage. In response, there is to be major bank strengthening, with the bulk of the cost coming in grant aid (NRA, 1994d), of some 240 km of banks, involving most of the three main rivers and their tributaries.

There have been controversial plans for the construction of a barrier either across the River Bure or Yare, within the town of Great Yarmouth. A Bure barrier would prevent saline intrusion of the northern Broads, while that at the mouth of the Yare would protect the rivers Yare and Waveney. An additional feature would be the creation of a washland between the two rivers which, on account of its projected flooding three or four times a year, would be lost as grazing marshes.

Maintenance of groundwater levels protects springs, river flows and wetland sites. Furthermore, because of the risk of saline incursion (IGS, 1976), particularly of Horsey Mere (POST, 1993), protection of potable supplies is reduced. Rivers also intimately connect with the Broads, and flush them between every two weeks and a year (Moss, 1988, Chapter 7). This situation has long-term ramifications for water quality and ecology and is not helped by rising sea level and saline incursion.

Much licensed abstraction is close to the fens within the Broads (Broads Authority, 1993). There is a need for extreme caution when abstraction is proposed close to spring-fed fens, and existing licences may need to be reviewed. Similarly, land-spring drains dependent on spring flows at the upland margins are significant in maintaining water levels in grazing marsh. The modem practice of setting SWQOs for waters, bearing in mind the special problems of the area, is under way.

An interesting example occurs outside the Broadland area west of Diss in the Waveney Valley. Redgrave and Lopham Fen is an extensive area of valley-bottom fen, with 123 ha designated as an SSSI (POST, 1993). It was never enclosed, although fuel-cutting allotments were made at this time for the benefit of the poor ('Poor's Land'), and turbary rights have been shown to have existed in medieval times (Wheeler and Shaw, 1992). This is consistent with what is known of the origin of the Broads proper, and peat extraction occurred on Lopham Fen. Otherwise common rights existed over grazing and mowing.

Although the observation of fen drying and consequent reduction in the number of individuals of certain rarer species was observed at Redgrave and Lopham Fens as early as 1901, the onset of serious damage seems to date from the 1960s. Six species of vascular plant were lost between 1958 and 1990, 34 other fen species experienced reductions in abundance and 77% of invertebrate species have been lost. The adjoining Waveney was deepened, and the effects of a public supply borehole in the adjoining chalk pumping 19 hours per day since 1957 were realised (Wheeler and Shaw, 1992; POST, 1993) This would have been treated as a 'Licence of Right' following the Water Resources Act 1963. Pumping was stopped for an experimental period in 1989, and the recovery of groundwater levels was dramatic, the SSSI having been within the cone of depression. The Suffolk Water Company relocated a borehole away from the site, and augmented supply to ponds where the great raft spider lives.

A precedent was set for wetland support at Chippenham Fen in Cambridgeshire where, in consultation with other agencies including English Nature, the NRA funded wetland support through abstraction from a nearby borehole over several years (NRA, 1994e). To protect wetlands, the EA will presume against issuing abstraction licences on or close to SSSIs, National Nature Reserves or in the Broads Authority Area unless the impact upon ecology is demonstrated to be negligible.

The Broadland Abstraction Licensing Strategy (EA, 2013h) recognises the complexity of an area that, by British standards, experiences low rainfall, high visitor and societal abstraction pressures. Broadland has many areas of high conservation value and comprise some 3 188 km^2 of relatively flat land predominantly located within the county of Norfolk. Indeed, many parts of the Broadland Rivers CAMS area are designated for their variety of habitats and species. These include Broadland Special Protection Area (SPA), The Broads Special Area of Conservation (SAC), Norfolk Valley fens SAC, River Wensum SAC, and The Broads Environmentally Sensitive Area. High baseflow of many rivers (fed by the Chalk and gravels) makes the area especially vulnerable to over-abstraction, and demands on water supply for domestic purposes and for agriculture (especially spray irrigation) are high, while wetland habitats including grazing marshes remain vulnerable to abstraction and other pressures. Other issues in the Broadland CAMS area are eutrophication from nitrate and phosphorous (especially in the broads), saline intrusion, invasive species (Chinese Mitten Crabs, Killer Shrimps, Signal Crayfish, Asiatic Clam and *Corbicula* and *Zebra Mussel Dreissina polymorpha*) are ubiquitous. Diffuse pollution continues to cause concern, particularly nitrate and phosphorus which result in eutrophication.

8.5.5 The Future for the Region

It is clear that there is now a whole range of options for conservation and restoration of the water environment, habitats and landscapes in Broadland. These include both agri-environmental policies targeted in catchments, and operational measures operated by water undertakings. Upstream of Broadland, notably in the Waveney Valley, measures are being taken to control pollution from nitrogen, while improvements in the treatment of sewage in the Norwich area (including phosphate stripping) should reduce up-catchment loading. Remedial measures to clean up polluted water bodies, and biomanipulation are under way for the Broads. Care is increasingly being taken to manage water levels in appropriate ways, notably for wetland and grazing marsh conservation, and for birdlife. The establishment of the Broads National Park is recognition of the recreation and heritage value of the area, however problems remain.

Eutrophication has not yet been solved, and the perennial issue of 'sustainable tourism' arises from problems of tourism and boating remains.

Nonetheless, active riparian zone policies such as the emplacement of agrochemical-free zones parallel to watercourses are yet to permeate into the region. This is perhaps surprising, given the range of land uses and occurrence of fringing vegetation which is itself under threat. The effort continues, and the Broads Authority aims to assist this through sediment removal, with four main objectives (Kelly, 2008):

1) Reduce internal phosphorus loading from relatively recently deposited nutrient rich sediment;
2) Maintain or create sufficient water depth for submerged water plants;
3) Increase water depth for navigation purposes; and
4) Remove toxic substances associated with the sediment.

In recent years, communication with stakeholders has been considered especially important. An Ecosystem Health Report Cards are available for the upper Thurne catchment. It provides a snapshot of the physical, chemical biological and nutrient status of waterbodies within the catchment. This assists in communication such problems with the public, official bodies and the farming community (RELU, 2009).

Flooding remains a persistent problem in Broadland (Broads Authority, 2011). The single greatest threat comes from changes in sea level that might hold back water trying to drain from the rivers. Furthermore, if a combination of wet weather and high tides causes a surge in the North Sea, flood defences would be under particular threat. Higher sea levels and climate change would make the north-east Norfolk coast particularly vulnerable sea defences being overtopped or breached, with inundation and increasing incursion of saline water into a predominantly freshwater ecosystem. Impacts might be seen in terms of habitat change, increase in non-native species and changes in visual landscape character. The Broadland Flood Alleviation Project remains very much a work in progress, including both bank strengthening and bank setback to increase channel capacity.

8.6 Somerset Levels and Moors

8.6.1 Physical Background

Known popularly as the 'Somerset Levels', this area provides a comparison with the East Anglian Broadland because it represents another example of a major reclaimed British wetland. The mosaic of grassland and ditches are of international importance, being recognised under the 'Ramsar Convention'. Located in the southwest of England, the Levels experience higher rainfall than Broadland; the mean annual figure between 1981 and 2010 was 755 mm at Cannington near Bridgewater, Somerset compared with 636.4 mm for Great Yarmouth, Norfolk for the same period (Met Office, n.d.b).

The Levels are surrounded by upland areas and also contain 'islands' and ridges of solid geological formations which cross the region. These include Glastonbury Tor and the Polden Hills (main trend WNW-ESE). Geologically these are Triassic marls and Jurassic 'Lias' (alternating bands of limestone and clay). To the north, the Mendip massif is predominantly Carboniferous Limestone, while the Quantocks to the west comprise

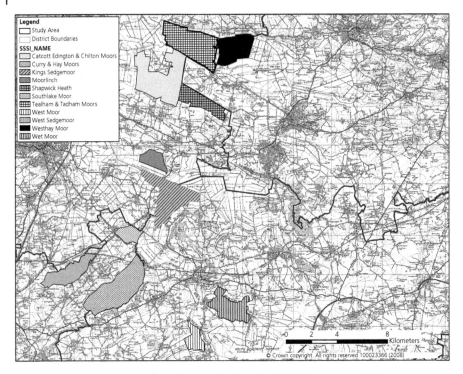

Figure 8.3 The Somerset Levels and Moors SPA showing SSSIs (Covington, 2009).

a range of Palaeozoic grits and shales. Between these upland areas are the claylands and peat moors (Cook *et al.*, 2009). The disposition and age of the Quaternary deposits is complicated and their geological detail uncertain.

Surrounded on three landward sides with uplands and the Bristol Channel on the fourth, Figure 8.3 shows the areas with conservation designations and within, the reclaimed area, compiled under the EC Habitats Directive (92/43/EEC). Established in 1987, there is a Somerset Levels and Moors ESA covering around 27 000 ha (Royal Haskoning, 2009). The area comprises the following SSSIs:

- Catcott Edington and Chilton MoorsCurry and Hay Moors;
- King's Sedgemoor;
- Moorlinch;
- Shapwick Heath;
- Southlake Moor;
- Tealham and Tadham Moors;
- West Moor;
- West Sedge Moor;
- Westhay Heath;
- Westhay Moor; and
- Wet Moor.

The peat moors vary between 3 and 6 m OD and include the King's Sedgemoor, Shapwick Heath and Tadham Moor which are surrounded by areas of mineral alluvial soils. 'Clayland' soils dominate the coastal regions (the Somerset Levels *sensu stricto*), which tend to be of higher elevation. The result is that the lower reaches of the rivers, which have imperceptibly low gradients (Hawkins, 1982), are often below high tide level and have to be controlled by tidal sluices in order to prevent tidal inundation; pumping up and out of the coastal clay belt is required when the lowest land can be 4 m below high tide (Purseglove, 1988, Ch 8). Regional drainage is via the main rivers and gravity drainage is assisted by major pumping stations (Parrett Internal Drainage Board, 2010).

Embankment generally constrains the river on both sides of its main channel and water has to be transported across the low-lying areas, there being no 'catchwater' drains around the edge of the region. Figure 8.3 shows the major rivers, Tone, Parrett, Axe, Cary, Isle, Yeo and Brue, whose floodplains coalesce to form this extensive lowland. Water is carried across the lowlands by engineered rivers acting as high-level carriers, for example, the River Cary is canalised for much of its length in the King's Sedgemoor Drain, while the Huntspill River is wholly artificial. Water resource management has for long involved a balancing act between flood management, drainage for agriculture and wetland conservation and reinstatement (Cook *et al.*, 2009).

8.6.2 Landscape Origins

The Somerset Levels and Moors landscape is dominated by 'rhynes' (drainage ditches and canals), willows and meadows punctuated by 'islands' and ridges of upland has, like Broadland, an intrinsic landscape value. The earlier Holocene deposits date from the post-glacial rise in sea level which flooded the area between about 10 000 and 6500 BP, depositing silts and clays (with peat forming where there was a temporary lowering of sea level) around upland areas and sand islands (Coles and Coles, 1989). After this time, the sea retreated and reedswamps led to the formation of peat over the marine clays. The vegetation was calcareous in character, on account of the predominance of limestone in the upper catchments of the rivers. By 5500 BP there was carr (comprising birch, willow and alder) in addition to reedswamp and pools. This led to the development of raised bogs which became acid due to being rainfed rather than fed by (calcareous) surface or groundwater.

The Levels are well known for Neolithic and Bronze Age prehistoric wooden trackways which cross the region, linking the uplands, a practice largely abandoned by the Iron Age (after about 2500 BP) due to flooding in response to climatic change. Peat growth then ceased after 2400 BP and flooding increased a situation which continued into historic times. Important archaeological remains would be threatened not only from the repeated operations of arable management such as ploughing, but also by deep drainage through the drying out of important organic remains such as the preserved wooden trackways.

There was also a peat cutting industry from early times to the present. However, despite the peat digging no features comparable with the Norfolk Broads were created, possibly because of the greater availability of wood as a fuel source. Regional drainage improvements date from the Middle Ages. There was a particularly active time between about 1770 and 1840 (Williams, 1970). The landscape today is dominated by 'secondary' wetlands reclaimed by ditches and embankments; otherwise there is limited deep drainage for arable cropping (Cook *et al.*, 2009). The relatively recent period of drainage activity is associated with enclosure and the creation of approximately rectangular fields.

The low-lying Moors provide floodwater storage and normally protect the towns of Bridgewater and Taunton. Flooding, particularly in winter, is also valuable from the standpoint of wildfowl. The Wessex Water Authority had major plans to improve drainage on the rivers Brue and Parrett between the mid-1970s and early 1980s

(WWA, 1979; Purseglove, 1988, Ch 8). All this was *despite the requirement on the RWAs and IDBs to have regard for the natural beauty and conservation under the Water Act 1973* (NRA, 1991b). Pumping is locally employed to lift water from the artificial field ditches, up to the embanked arterial watercourses. Otherwise, much of the drainage from these to the sea is by gravity, while tidal inundation during the large tidal range experienced in the Bristol Channel is controlled by sluices. Unlike the 'Black' (peat) Fens, or similar areas in the Netherlands, peat wastage through oxidation and shrinkage has not led to a significant regional lowering of the land surface and has not been a serious problem. There has only been concern on West Sedgemoor. This is presumably due to the wetter climate, lower arable acreage and maintenance of higher watertables than in Fenland (Williams, 1970, p. 246).

On account of the higher rainfall, the drainage density of ditches is around twice as high as in the drier east of England, typically $15\,km\,km^{-2}$ (Cook and Moorby, 1993). Certainly the result has been a pattern of small fields, small holdings and fragmented ownership (Purseglove, 1988, Chapter 8). Relaxation of flood measures to promote conservation during the twentieth century entails lower engineering costs and gains in ecological 'goods', but a loss in agricultural production. This is all reminiscent of Broadland, but with less arable land the emphasis is different. Flood events across Europe, notably those of the winter of 1994/95 and again in 2013/2014 provided a reconsideration of the natural function of floodplains and a political battle over flood defence priorities.

During the 1980s, controversy raged over the Somerset Levels as well as Halvergate Marshes in Norfolk and drainage plans focused attention on the landscape and conservation potential of the area. Although horticulture and forage maize growing are practical, widespread arable conversion never occurred due, most likely, to a perception of climatic unsuitability and there was also an ochre problem on West Sedgemoor. Otherwise, eutrophication of watercourses, pesticide contamination and peat wastage might have become serious issues, as in Broadland and the Fens.

8.6.3 Problems

Rahtz (1993, p. 14) suggests that the extent of flooding in the early medieval period was considerable, affecting the entire coastal strip and the valleys of the Parrett, Brue and Axe to many kilometres inland. The risk of freshwater flooding from upstream probably lead to a perception of the place as unfavourable marsh and swamp before the thirteenth century when monastic houses commenced drainage improvements (Williams, 1970).

The persistence of spring and autumn flooding, fear of loss of commoners' rights to cut peat and low agricultural prices may have caused more later major reclamations in the Levels than in eastern England, and the flood problem persisted into the nineteenth century. Improvements between 1930 and 1970 (including the cutting of the Huntspill River in 1940), and improved pumping efficiency, extended the drainage season by four months, from between May and September to between March and November (WWA, 1979). In summer, the drainage system is reversed by causing the rhynes to be augmented through gravity-fed pumping from the main rivers to keep them bank-full and operating as wet fences for livestock and sub-irrigating the grassland.

Key issues for the Somerset Levels and Moors which bring with them attendant problems for the long-term are not unlike those for the Broads (EA, 2014b; English Nature, 1997):

- Increased risk of flooding (55-60 significant events in 2012-2014) and associated balancing needs of management to protect people, property, agriculture and conservation.
- Securing investment in infrastructure in order to maintain complicated infrastructure including the pump-drained systems. This is shifting away from protecting purely agricultural land towards people and property. Maintaining main rivers, especially through dredging, especially the Rivers Parrett and Tone, including establishing 'local ownership' of problems.
- Establishing working partnerships with local authorities, the Internal Drainage Board and others to raise funding for dredging.
- Planning for climate change, predictions over the next 100 years indicate a significant increase in the risk of serious flooding in Somerset from both increased rainfall and rising sea levels.
- Overall, the future of agriculture on the Levels and Moors remains a serious issue (including supporting a local community of predominantly livestock farmers).
- supporting local industry ('withies' or coppiced willow growing) and managing a reduced peat digging industry.
- managing and encouraging tourism and recreation, especially fishing.
- nature and landscape conservation including restoration of species-rich fen meadow or flood pasture and bird habitats through extensive farming practices.

While perhaps the pressure from tourism is less intense than in Broadland when viewed as a conservation matter, the extent to which flooding is a 'problem' becomes a relative and a matter of user expectation until a serious event occurs. What is appropriate for low-input grazing is not appropriate for intensive grass or arable; these require lower watertables and stringent flood controls. Debating land-use issues involves consideration not only of the desired outcome, but of economic inputs. We should ask whether the area is to be managed for semi-natural habitats, floodplain or intensively managed agriculture, and to what extent these land uses are compatible. The county council is opposed to the peat digging on account of wildlife and habitat conservation.

Unlike Broadland, the rivers do not experience serious pressure from visitors, and sewage disposal is not a threat to the aquatic ecology. Public water supply is from aquifers to the north while high rainfall does not normally create serious problems of low flows. Tourism is an important industry in Somerset. The Levels also represent extremely valuable habitats for migratory birds, wildfowl, otters, fish (especially eels), insect life, and aquatic and meadow flora. Much of the wildlife interest is restricted to SSSIs and parts of the Moors are Special Protection Areas for Birds under EU legislation as well as a Ramsar site (Natural England, 2013). Coarse fishing is the most important water-based recreation within the area, and locally elver netting is important. Protection of fisheries from agrochemicals is not a problem, although silage effluent, slurries and yard washings are potential pollutants. Care has to be taken to avoid inappropriate ditch clearance (particularly disturbing the bottom of ditches); the slow-flowing rivers present problems for aeration and disposal of sediment.

Live peatlands have a great capacity to sequester carbon. However, climate change recently caught up with the Somerset Levels and Moors in a direct and cruel way. For while a modicum of inundation is to be expected on what is basically one coalesced floodplain ('Somerset Levels') with drained raised bogs ('Somerset Moors'), the flooding of February 2014 was unprecedented in modern times. Several interesting points emerged from the angry discourse that ensued, generally associated with damage to livelihoods and to homes. The undignified slanging match reported in the preface to this volume has its roots in the following considerations, for better or worse, be they true or false:

1) It should never have happened
2) Somebody should take the blame
3) Economic losses have been enormous
4) Individuals, families and communities were traumatised
5) Lowland (largely livestock) farmers blamed their upland counterparts for arable development that would be expected to reduce infiltration into cultivated fields
6) 'The Government' (actually in the form of the EA) should have dredged the rivers and then it would have not occurred
7) Climate change may actually be real

Dredging was undertaken on the Parrett and Tone and a twenty-year 'Flood Action Plan' prepared, under-written by central government, for:

> 'We see the Somerset Levels and Moors in 2030 as a thriving, nature-rich wetland landscape, with grassland farming taking place on the majority of the land. The impact of extreme weather events is being reduced by land and water management in both upper catchments and the flood plain and by greater community resilience.' (The Somerset Levels and Moors Flood Action Plan, 2014)

Risk reduction actions proposed dredging and river management, land management that recognised what happens in the upper and mid catchment has an impact on the lowlands and management of urban runoff. Resilience will be built into both infrastructure, and communities, but the most cost effective mixes of management, infrastructure and resilience actions, is unclear. Resilience is to cover a range of factors, from road access (several communities were cut off) to WLMPs and sluice installation and operation and construction materials, especially for roads and protection of power infrastructure, create detention areas and wetland restorations.

But are flood defence investments worthwhile? Actually flood defences deliver fantastic value for money, preventing £8 in future damages per £1 invested (Committee for Climate Change, 2014). Flood victims would hang on to this statistic, for after all 'The Government' must invest wisely, particularly where the flooded taxpayer is concerned. Sadly, there will always be a law of diminishing returns, and even if EA subsequently embarked on a programme of dredging, as with land-use issues up-catchment the real likelihood is that the impact would be marginal, given the intensity of the flooding. The flooding resulted from 170% of the average winter rainfall (1981-2010) (Huntingford, 2014) and the economic cost to the country probably in excess of £1.5bn (admittedly this includes other areas, particularly the Thames valley and large areas remained under water for weeks.

Ultimately, the point is not so much about statistics as about expectations. The points listed above attempt to summarise the outrage and public debate that ensued after winter 2013/2014 in Somerset and the Thames valley particularly. It did happen, and ultimately 'who is to blame' lies more in the realm of theology than geology. The area is reclaimed from natural wetlands, yet the infrastructural arrangements, for a range of reasons, could not cope with the imposed hydrological surcharge. The question is how much to invest next time and how to make that effective. And who might be excluded and why. Here we have less of a perfect storm, more a perfect wicked problem.

8.6.4 Conclusions

The problems of the Somerset Levels and Moors in general are less severe than those faced by the East Anglian Broadland, specifically those related to water quality, bulk water supply, soil erosion and bank erosion from boats. Due to factors including climate, small land holdings, concerns over costs of drainage, and innate conservatism (or scepticism) of some of the farmers, the area did not go so far down the road of arable conversion as other reclaimed wetland areas. Many landowners and farmers stand to derive income from environmental management, rather than entirely from agriculture. The positive aspects of flooding (in terms of hydrological management), lowering economic inputs to flood defence and habitat maintenance and creation are now being actively pursued, efforts re-doubled by actual and modelled climate change impacts. There is a need for multi-functional stakeholder action in balancing the needs of conservation, agriculture and flood defence as indeed will the communities affected by flooding and those, notably farmers, whose activities impact upon water quality. Meanwhile, farming has to remain viable if semi-natural habitats are to remain and be managed in an appropriate fashion.

8.7 Afforestation and Woodland Planting

Forestry Commission statistics had shown a decrease in area of scrub and a slight increase in broadleaved woodland between the 1950s and 1980s (Locke, 1987). The implication is that some scrub areas, once probably marginal agricultural land, is growing to mature secondary woodland. On a more systematic basis, Farm Woodland schemes encourage the planting of broadleaved woods as an alternative crop. Between 1980 and 1998 the area of broadleaved woodland increased by 34% and the relative proportion of broadleaves to conifers increasing from 35% to 39% (Forestry Commission, 2003; Forestry Commission, n.d.a).

In 2010, the forest and woodland area of England was estimated to be 9.9%, that of Scotland 17.8% and Wales 14.3%. The trend since the early 1990s is therefore upwards for England and Scotland, with a small estimated decrease for Wales. In Britain overall, some 2 982.5 thousand hectares were under broadleaved or coniferous wood in 2010, or about 13% of the land area. These estimates are indicated to have a gently upward trend. By the turn of the century conifer woodland was dominant, representing 49.0% of all woodland. Broadleaved woodland represents 32.1%. Mixed woodland 7.9% and open space within woodlands 8.1%. (Forestry Commission, 2011a,b,c).

Under the defunct Farm Woodland Scheme, farmers were encouraged to plant broadleaved woodland with the objectives of reducing agricultural surplus, diverting land from agriculture, enhancing the landscape, supporting farm incomes and encouraging timber production. Since then it has been replaced by the Woodland Grant Scheme (WGS) that offered grants towards the costs of establishing and maintaining woodlands. The Farm Woodland Premium Scheme (FWPS), offers annual payments to compensate for agricultural income foregone. Administered by the Forestry Commission both are closed to new applicants The Community Forest Scheme ran for 15 years to 2005. It funded extensive tree planting adjacent to cities, aimed to engage local communities, improve the landscape (including planting on derelict land), improve sport and recreational activities, develop educational initiatives and 'establish a supply of timber and other woodland' (Countryside Agency/Land Use Consultants, 2005). Such schemes are drivers to shift tree cover in Britain towards a greater proportion of broadleaved woodland over conifers, as well as increasing the overall proportion of woodland and forestry cover.

8.7.1 Afforestation and Water Quality

Forest organisations are closely associated with legislation (Forestry Commission, n.d.b). Management of all kinds of forestry will affect water quality, water yield and sediment yield. The planting, rate of growth and felling time and regime also impact hydrology and there are parallels here with hydrological changes associated with catchment urbanisation (Chapter 4). Afforestation affects both water yield and quality which has long been a concern of the Forestry Commission. Water quality will be affected by eroded catchment sediment, and any pollutants in sedimentary complexes, or leached from forestry land or otherwise carried in solution. Forestry is not exempt from pollution laws and EU legislation that affects specifically water may be summarised (Forestry Commission, 2011b):

- Water Framework Directive 2000/60/EC
- Drinking Water Directive 98/83/EC
- Groundwater Directives 80/68/EEC and 2006/118/EC
- Nitrates Directive 91/676/EEC
- Floods Directive 2007/60/EC
- Habitats Directive (Council Directive 92/43/EEC) and Directive 2009/147/EC on the conservation of wild birds 'The Birds Directive'

In turn, these are represented in a range of relevant regulations and Acts of Parliament for England and Wales and Scotland. In pursuance of Good Practice, the Forestry Commission publishes best practice guidelines for a range of activities associated with forest operations and planting. Of importance here is 'Forests and Water' concerned with water and conservation (including buffer strip emplacements, pesticide management, fish populations, waste materials including sewage sludge and acidification (Forestry Commission, 2011d). Other publications concern biodiversity, climate change, historic environment, landscape, people and soil.

Where deficient growth is observed, nutrients are added to forestry typically six to eight years from planting. Generally this is rock phosphate, but nitrogen and potassium may also be added (Best, 1994). Potassium is most mobile, then nitrogen, with phosphorus the least, and elevated levels of potassium may be detectable in drainage waters over the first year of application. Where forested areas drain into oligotrophic lochs, levels may rise and lead to eutrophication (Bailey-Watts, 1994). There is also concern for drinking water quality.

Broadleaved planting is subject to the same kinds of concerns where sewage sludge, fertiliser and pesticide are used, and possible contamination of waters remains an issue. However, it is probable that much broadleaved woodland will not be so intensively managed, particularly if broadleaved woods are incorporated in protection zones. Short-rotation energy coppice may be an exception here as productivity of biomass is the main objective. Furthermore, lowland soils are far more likely to be calcareous and hence buffered against acidity. Deciduous litter fall represents an effective recycling of elements, especially calcium and magnesium, while a switch to deciduous afforestation from intensive grass or arable implies a loss of land otherwise prone to agrochemical application.

Work in the United States (Jordan *et al.*, 1993) shows the role of riparian forest, a feature sadly missing from the British landscape, as an effective sink for certain pollutants of agrochemical origin, while the planting of broadleaved woodland has been proposed as a water protection option for Nitrate Protection Zones (DoE, 1986). Abandoning intensive agricultural land to 'low-' or 'zero-' input woodland over several decades may result in a reduction in agrochemically derived diffuse pollution, and as a by-product improve the amenity and conservation benefit of the countryside.

8.7.2 Afforestation and Water Yields

Tree cover and forest management have been shown to affect both the yield of water to watercourses and reservoirs and seriously affect runoff, creating changes in both peak flows and sediment transport. The literature regarding comparative studies between forestry and more open areas is well-established. At Emmental in Switzerland in 1890, a comparison of paired catchments, one largely forest, the other largely pasture, showed that the forested catchment recorded a loss of 861 mm, and the pasture 696 mm (Rodda, 1976). During the first years of the twentieth century, it was hypothesised that deforestation was linked to increased flood frequency in the River Ohio basin. Eventually, by comparative catchment studies, it was demonstrated that overall water loss was reduced following felling (Schiff, 1962).

In Britain, catchment water yield from afforested catchments compared with other land covers (such as short grass) has been a cause of concern since the 1950s, although more recent research points towards water yield reduction from upland catchments being by far the greatest problem. In Britain, transpiration per se has been found to be typically at 310 to 350 mm yr^{-1} for most stands of trees (Hall and Roberts, 1989); it is interception loss that produces most problems.

Over a forest, for a set time period:

Total water use = Transpiration + Interception loss,

and for a given catchment:

$$P = I + \left(S + T\right) \pm dC$$

Where: P is gross precipitation (captured in raingauges above the canopy); I is interception loss (calculated from the other terms); S is stemflow (captured from the trunk of the trees); T is throughfall (captured by raingauge network on forest floor); and dC is the change in storage within the catchment. $(S + T)$ is sometimes referred to as 'net precipitation', while 'interception loss' is expressed as intercepted and evaporated water as a percentage of the gross figure (e.g., 'percentage of annual rainfall', or for a shorter period), also termed the 'interception ratio'. Modelling procedures can then extrapolate these variables at the catchment scale.

Law (1956) found that water yield is appreciably reduced by coniferous afforestation during a lysimeter experiment in a spruce plantation in the Pennines. Long-term monitoring at Plynlimon established that coniferisation reduces water yield. Typical interception loss may be of the order of 35% of gross rainfall in the uplands (Calder, 1990). Coniferous afforestation over a chalk aquifer beneath Thetford Forest (where annual precipitation is low, averaging around 550 mm) in eastern England has also been shown to reduce aquifer recharge; drainage was 44% less than beneath grass, this was also attributed to increased interception loss (Cooper, 1980).

Newson (1979) summarised the 10-year studies on Plynlimon. Here comparison was made between the Upper Wye catchment (area 10.55 km^2) which was predominantly grazed (forestry 1.2%), and that of the Upper Severn (area 8.7 km^2), of which some 62% is coniferous. Otherwise accounts of reductions in water yield are summarised from Law (1956), Gash and Stewart (1975), Insitute of Hydrology (1976), Cooper (1980) and Wilson (1983, p. 54):

- At Plynlimon it is probable that the total annual loss (i.e., transpiration loss and interception loss together) from forest (about 850 mm yr^{-1}) was about twice that from grass (average 405 mm yr^{-1}). Annual rainfall was around 2300 mm.
- In drier parts of the country, where annual rainfall was about 600 mm, calculated interception losses comparing grass and coniferous forest vary. Some studies claim that there is only a marginally greater yield from grass (most significant during the winter period of aquifer recharge), while others claim appreciably greater drainage under grass in the longer term, meaning that interception loss is considerable.
- The reason for the loss over the forest is largely attributed to the interception and subsequent evaporation of rainfall, transpiration figures showing little variation. Results need to be interpreted in the context of the climatic, and specifically the rainfall, regime.
- Potential evapotranspiration (E_T), as derived from the Penman formula, gives good agreement for actual losses from grassland catchments. It underestimates the figure for forested catchments in wetter parts of the country and during high rainfall periods in drier parts of the country.
- From lysimeter experiments, it has been found that slightly over half of rainfall falls through to the forest floor, while a further 12 to 23% reaches the ground by stemflow; 22 to 38% of gross rainfall evaporates from the canopy.

Increasing interception loss inevitably means less water for aquifer recharge and for river flow. The effects on river flow in Britain have been modelled and Figure 8.4 shows a relationship between runoff and rainfall for differing percentages of afforestation in a catchment. Best (1994) reports that at Plynlimon 85% of gross rainfall on grassland reached streams, while the figure was only 65% in the afforested part of the River Severn catchment. Afforestation in Scotland may produce a decrease of only 20% when compared with grassland.

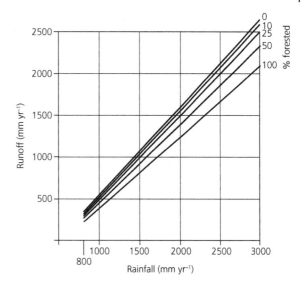

Figure 8.4 The relationship between runoff and rainfall for differing percentages of tree cover. (*Source:* Redrawn from Best, 1994, p. 420).

8.7.3 The Balquhidder Catchment Studies

Research by the Institute of Hydrology (Scotland) evaluated the effects of upland afforestation in situations of more mixed land use/cover and marginally higher rainfall than are experienced over much of England and Wales. Two catchments were selected in the Balquhidder Glen, Highland Scotland, situated some 60 km to the north of Glasgow (Johnson and Law, 1992; Johnson, 1995).

The two catchments are:

- The Monachyle catchment (area 7.7 km²), which is predominantly moorland, although 14% was recently planted with trees. Moorland may be further subdivided into heather and grass (both at 36%), bilbury, moss and bracken cover.
- The Kirkton catchment (area 6.85 km²) has 36% mature forest with 50% of this recently clear-felled. Of the 'moorland', grass dominated at 30%, and there was an additional 5% for forestry roads.

The objective was to provide a comparison with the Plynlimon studies for Scottish conditions of climate, topography and soils. Although land-use contrasts between the two catchments are less striking, the pattern of clear-felling and re-establishment of coniferous forest and differing moorland covers provides a more realistic dynamic of land-use change for long-term studies.

At between 200 and 900 m OD, the altitude range is great. Geology comprises Dalradian schists, with soils (rankers, peats, peaty gleys, surface water gleys, podsols and brown forest soils) reflecting the nature of the drift material and blanket peats. Short-term catchment average annual precipitation 1983-1989 was 2799 mm for the Monachyle and 2372 mm for the Kirkton; there was additionally a marked upward trend over the period concerned, and the 1985 summer was relatively wet, with the 1987/1988 winter wet.

Figure 8.5 shows the stream network and instrumentation of the catchments. Instrumentation permits water balances, including interception loss, to be calculated

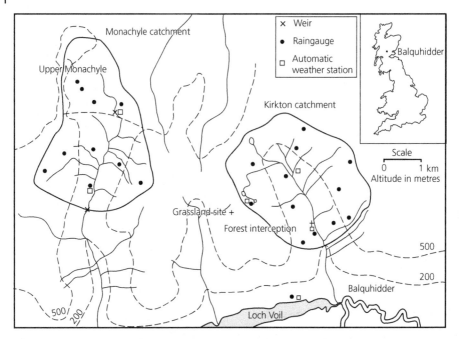

Figure 8.5 Stream network and instrumentation, Monachyle and Kirkton catchments. (*Source:* Redrawn from Johnson, 1995, p. 4, by permission of the Centre for Ecology and Hydrology).

for grass and mature forest (50-year stand). Micrometeorological equipment and two lysimeters were installed at the grassland site, and the forest site recorded precipitation, throughfall and stemflow. Stream discharge was recorded from weirs at the lower edge of the catchments defined in the experiments.

Forest interception loss (percent of gross precipitation) at Balquhidder averaged 28% in 33 months (summer and winter averages were 34 and 26%, respectively). There was otherwise no clear seasonal trend, and the average figure is towards the lower end of the range quoted for coniferised upland catchments in the UK (range 25 to 49%). Data from the grassland site suggested that summer use of water was close to the Penman E_T figure, falling to half of the calculated figure as the grass became dormant. Consequently, annual water use may be estimated at 0.75 Et. Precipitation and streamflow data enable the annual catchment water balance to be calculated using a water balance approach:

$$P = (Q + E) \pm dS$$

where P is catchment mean precipitation (from the raingauge network) and including a significant percentage of snowfall; Q is streamflow leaving catchment (from weirs); E is total evaporation from the catchment (the unknown term to be solved); and dS is the change in the catchment water storage term over a given time period. The term dS comprises snow, soil moisture and groundwater terms.

Groundwater was not measured, and in any case the solid geology and flow duration curves (Figure 8.6) suggest an impermeable and 'flashy' catchment with negligible groundwater flow (a small amount of storage in the drift and peats is possible).

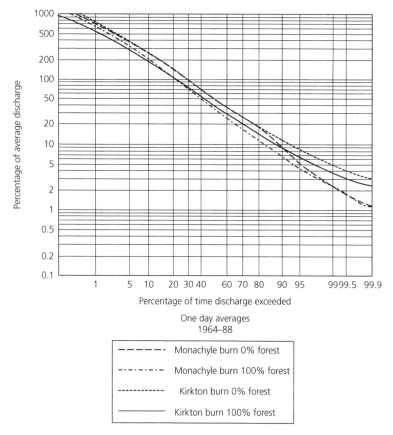

One day averages
1964–88

- - - - - - Monachyle burn 0% forest

- - · - · - · - Monachyle burn 100% forest

- - - - - - - - Kirkton burn 0% forest

———————— Kirkton burn 100% forest

Figure 8.6 Flow duration curves for the Balquidder catchments. (*Source:* Redrawn from Johnson, 1995, by permission of the Centre for Ecology and Hydrology).

In the long-term dS becomes negligible as the terms cancel out throughout the year. Solving for E gives a total evaporation figure of 22% mean precipitation in Monachyle and 17% for Kirkton.

The land-use implication here is (apparently) that:

- The moorland (heather and grass) dominated Monachyle catchment experienced a mean annual water use of 621 mm (22% of precipitation), which is higher than the average annual Penman E_T estimate for potential evaporation of 551 mm.
- The (predominantly forest and grass) Kirkton catchment only used an annual average of 423 mm (17% of precipitation), which is lower than the Penman E_T figure of 484 mm.
- By contrast, the Plynlimon studies showed grassland to use around half of the average annual water compared with the forest, with losses from grassland showing good agreement with the Penman E_T figure.

The Balquhidder study, which also employed hydrological modelling, concluded that forested areas use more water than 'moorland' land covers, all other things being equal. The apparent contradiction, that heather used more water than forest was shown to be an artefact of location and climate. In short, heather was dominant at middle and higher altitudes in the Monachyle catchment which experienced both higher precipitation and higher potential evapotranspiration, whereas Kirkton had grass dominant at higher elevations which have a relatively low level of evaporation (annually 0.75 E_T). Lower elevations experienced less precipitation and evaporation under forest. The combination likely explains why heather land cover 'used' more water than the grass and forested catchment.

Hydrological modelling permitted simulated flow duration curves for the two catchments in the Balquhidder catchment to be plotted for zero and 100% forest cover (Johnson, 1995); these are shown in Figure 8.6. They predict that afforestation displaces the curves in the direction of reducing discharges so that the 95 percentile flow reduces from 5 to 4% of average discharge at Monachyle, and 8 to 6% at Kirkton. The low figures for the 99.9 percentile flows suggest low groundwater contributions. Summer baseflows can be increased by the cultivation and drainage of otherwise wet soils, a condition which may persist for 15 to 20 years when canopy closure is approaching, when the reverse obtains.

8.7.4 Comparisons with Broadleaved Afforestation

Experiments near Micheldever in Hampshire and at a site in Northamptonshire between 1988 and 1990 were established in order to look at the water relations of small, mature, broadleaved plantations (Rosier *et al.*, 1990). Such plantations might result from take-up of farm woodland plantation and, should it occur, abandonment of agricultural land. The species chosen were ash and beech. Essentially it was found that the interception loss averaged only 12% during the leafed period, and 11% during the un-leafed period. Transpiration was found to be below the Penman potential rates, especially marked during times of high evaporative demand. These findings are consistent with those for conifers; both kinds of tree are capable of operating physiological control over water loss.

However, maximum water use by broadleaves is more concentrated in the summer months when consumer demand upon water resources is highest, and interception loss marginally higher. There remain uncertainties with small broadleaved plantations; differences in regional rainfall regimes, species composition, evapotranspiration losses from understorey and 'edge effects' on small plantations add further complications, and therefore greater uncertainty, as in management. The effects of coppicing on soil water status depend upon stool densities and age since cutting, and become important in drier years (Cummings and Cook, 1992), further indicating how differences in the management of broadleaved woodlands affect their water relations.

Reasons proposed as to why interception loss is greater for conifers rests upon the interaction between meteorological conditions and canopy architecture. It is probable that coniferous trees, with their needles able to hold individual droplets, are able to store more water than other canopies, but, more importantly, their open canopy and greater roughness presented to air flows encourages interception. Darker colours also absorb more incident energy which may then be used to evaporate water on the leaf surfaces that for lighter canopies. Values of high interception loss (above 35%) from British coniferous forests have been variously interpreted as a consequence of monitoring high trees and small stands with an open-structured canopy (Johnson, 1995).

Table 8.1 Range of annual evaporation losses (mm) for different land covers receiving 1000 mm annual rainfall. No irrigation is assumed.

Land cover	Transpiration	Interception	Total water use
Conifers	300–350	250–450	550–800
Broadleaves	300–390	100–250	400–640
Grass	400–600	No data	400–600
Heather	200–420	160–190	360–610
Bracken	400–600	200	600–800
Arable	370–430	No data	370–430

An illustrative, albeit generalised, summary has been published by the Forestry Commission (2005). In Table 8.1, broad ranges of different land covers for Britain are summarised from a number of studies of tree cover compared with other land uses.

The summary suggests that the water use by coniferous land cover would be expected to be well in excess of the other land uses, and while overall broadleaved afforestation may be in excess of agricultural land it may replace, the outcome in terms of water use is closer to losses from arable or grass. While transpiration rates vary little between coniferous and broadleaved forest, with annual losses generally 300–350 mm, although Harding *et al.* (1992) report that for southern England, higher annual transpiration losses may occur for broadleaves between c 360–390 mm. Such information is useful in land-use planning in regard of water resources.

8.7.5 Runoff, Sediment and Afforestation

Water yield is an important part of the picture, but it tells us little about timing of flood peaks. The ploughing and drainage of entire catchments has shown an increase in peak flows by as much as 30%, but reducing to 10% after 10 years, as the forest drains become clogged. During this time the decrease in time to peak flow is up to one-third (Forestry Authority, 1993). The effects of ditching on the unit hydrograph are shown in Figure 8.7.

This shows an increase in peak hydrograph flow and reduction in time to peak immediately after planting. Insensitive planting involving soil compaction and the ploughing of ridges/installation of surface drains up and down a hill can lead to a considerable increase in rapid runoff and erosion, especially in the first few years following plantation (Binns, 1979). Such events are prone to increase sediment loading which may destabilise channels, increase flood risk and may cause siltation in reservoirs.

The hydrological impacts of non-intensively managed, broadleaved plantations are limited when compared with coniferous stands, and can only support further development of woodland on marginal and abandoned land. Coniferisation has real implications for water supply, conservation (particularly of fish populations), reservoir recharge and timing of runoff (depending on stage in the forestry cycle). These impacts are important, especially because much of the water which falls in Wales is impounded for transfer eastwards. In Scotland, water yield for hydroelectric schemes is a major consideration. A balance must be sought between commercial forestry interests and the needs of bulk water supply.

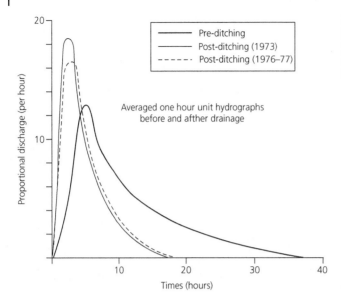

Figure 8.7 Unit hydrograph before and after forest ground preparation, Coalburn catchment. (*Source:* Redrawn from Best, 1994).

8.8   Contamination of Groundwaters in the West Midlands

Studies of the groundwater quality in the English Midlands show there is extensive contamination from both organic and inorganic sources, notably beneath the industrial cities of Coventry and Birmingham. Where the groundwater is not likely to be used for drinking, it may be that cleansing the aquifer may not so stringent, nonetheless, aquifer cleaning is underway (UK Groundwater Forum, n.d.). A review of urban groundwater issues in Coventry and Birmingham has been published (Lerner and Tellam, 1993), and the following is based upon their work.

Coventry has developed over a complicated aquifer system of Permian and Carboniferous strata overlain locally with glacial drift. The main aquifers are sandstones with some conglomerates, often in discontinuous units. The city of Coventry and the town of Kenilworth abstracts about one quarter of its supply from groundwater into public supply. Although inorganic concentrations of determinands (NO_3, Cl, SO_4, Ca, Mg and metals) are raised beneath these urban areas, levels do not cause concern. Chlorinated solvents, however, present a problem, as almost all urban boreholes contain over 10 $\mu g\, l^{-1}$.

Five chlorinated solvents are present, and they are listed:

- TCM, trichloromethane (chloroform)
- TeCM, tetrachloromethane (carbon tetrachloride)
- TeCE, tetrachloroethene (perchloroethylene)
- TCA, 1,1,1-trichloroethane, no limit set
- TCE, trichloroethene

TCM has been subject to past industrial leakage, but mains leakage raises levels in groundwater because it is a by-product of chlorination. TeCM (rarely used today) is extremely common, while TeCE, TCA and TCE are used at the rate of thousands of tonnes per year as solvents in cleaning (drycleaning of clothes, metal cleaning and degreasing, and component cleaning in the electronic industry). The solution to the problem is treatment at the point of abstraction. One public supply and several industrial supply boreholes exceed the three limits. All industrial sites sampled had chlorinated solvents present in groundwaters beneath them.

Birmingham experiences contaminated groundwater that influences not only drinking water quality, but fluxes of contaminants towards the river Teme. It is calculated to contain mostly copper, nickel, sulphate, nitrate, chlorinated solvents, e.g. trichloroethene, and their biodegradation products (Ellis, 2002). The larger metropolitan area of Birmingham presents problems for inorganic contamination. Situated over a largely unconfined Triassic sandstone aquifer, the city represents one of the oldest industrial centres in the world. Figure 8.8 (Lerner and Tellam, 1993) shows the basic geology, major, minor and trace ion concentrations and concentrations of total metals according to industrial use of site. Water consumption for industry fell from 55 to 11 Ml day^{-1} between 1940 and 1989, and groundwater levels are rising as a consequence. Furthermore, water quality problems have led to the abandonment of certain sources for public supply. Most sampling is from production boreholes, usually greater than 100 m deep. Although major ion chemistry is surprisingly good, and generally below EU limits (nitrate being an exception with median values close to the limit), there are some problems for minor and trace ions.

Figure 8.8 shows a close correlation between metal-working establishments (past and present) and total metal concentration. The problem, as with the chlorinated solvents in Coventry, is localised, but there are other considerations. Mixing of deep (low concentration) waters with contaminated higher-level shallow groundwaters in boreholes may be reducing the apparent severity of the problem. On the other hand, the Triassic sandstone aquifers have high sorption capacity (the ability to retain ions due to electrostatic bonding with clays), hence aquifer overloading (especially above the watertable) may only result from high loadings of metals. Acidification arising from the rise of groundwaters into the more polluted saturated zone may mobilise metal ions, and falling pH has been observed. Finally, the low transmissivity and high storage coefficients of the sandstones may help to exaggerate the localised nature of the pollution.

Although the high sorption capacity of these aquifers may assist in the protection of groundwaters from certain pollutants, the localised concentration of the problem linked with rising watertables and falling pH values may yet produce serious problems. Reducing further pollutant loading is the key to groundwater protection, and the NRA 'Groundwater Protection Strategy' (NRA, 1992d). Birmingham's contaminated groundwater suffers chemical contamination and there is an added a risk from viruses from treated wastewater injected into an aquifer, in addition to water quality issues from sustainable urban drainage systems (Darteh *et al.*, n.d.).

8.9 Soil Erosion

8.9.1 The National Picture

Soil erosion is but one serious problem affecting an overall soil degradation that is manifest in Britain. The others include compaction (adversely affecting infiltration, drainage, aeration, root penetration etc.) organic matter decline (affecting soils stability and productivity), podsolization, nutrient leaching and soil acidification. Soil erosion,

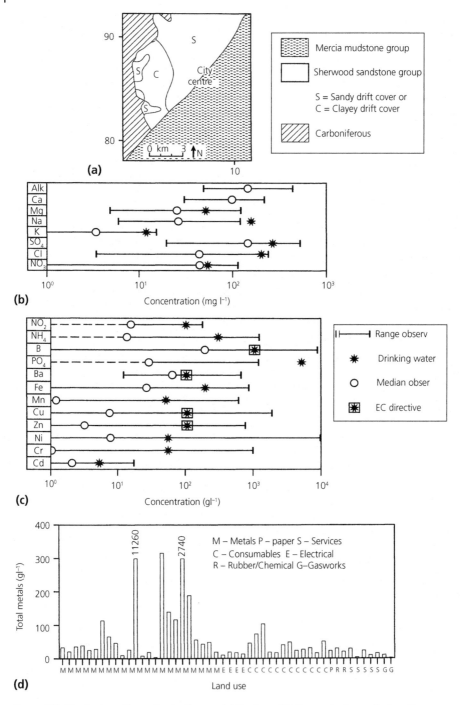

Figure 8.8 Birmingham Groundwater Survey: (a) Geology; (b) Concentrations of minor/trace determinands; (d) Total concentrations of all metals listed in (c). (*Source:* Lerner and Tellam, 1993).

mostly from arable land, not only degrades agricultural productivity but it also mobilises unwelcome sediment that frequently enters water bodies. It is likely that, so serious was this problem since the introduction of agriculture in the Neolithic, that it caused accelerated alluviation on floodplains. The result is that river channels in 'Wessex' are liable to adopt a resulting multi-channel geometry (Cook, 2008b).

By no means has the problem ceased into modern times and Boardman (1990) found a real increase in cases of erosion on arable land since 1970. Erosion occurs under a range of crops; for example sugar beet and potatoes render land vulnerable through spring and summer rainfall events, while winter-sown cereals make land vulnerable to autumn and winter erosion and due to the large acreages involved. Urban soils should also be sustainably managed (Hollis, 1974).

Wind erosion is a particular problem in eastern England where it affects valuable farmland (Boardman, 1986). The readily disaggregated and transported silty soils are especially vulnerable. Sandy and peaty soils in areas like the Vale of York, the Fens and the 'Suffolk sandlands' are known to be especially vulnerable (Boardman, 1988). Water erosion may occur as a consequence of either sheetwash or rill formation (not uncommon in England and Wales). Disaggregated soil particles are thus transported downslope to be either deposited at a lower portion, or may enter water courses. During the mid-1980s, it was estimated that some 73 800 km^2 (or 9% of the total agricultural area of England and Wales) had been actually affected by erosion, with long sloping land and steep valley sides under arable crops being particularly at risk (Evans and Cook, 1986). Or more recently about 37% of specifically arable lands shows effects beyond those considered 'tolerable' (EC, n.d.) For upland areas alone, around 2.5% is affected (McHugh *et al.*, 2002).

Degraded soils inevitably cause losses in profitability on farms. Costs borne by society (as opposed to commodity pricing) as economic externalities of soil erosion may distort the market by encouraging activities that are costly to society even if individual benefits are large. Such costs are often neglected; often occur with a time lag; often damage groups whose interests are not represented; the identity of the producer of the externality is not always known and they result in sub-optimal economic and policy solutions (Inman, 2006). Impacts affect drinking water supply, fisheries, in-stream ecology, banks (through erosion) and navigation among other factors. In 2009, it was calculated that soil erosion due to wind and rainfall results in the annual loss of around 2.2 million tonnes of topsoil across the UK at a cost to British farmers of £9 m a year in lost production. Furthermore, climate change has the potential to increase erosion rates (should drier conditions cause soils to become susceptible to wind erosion) and, if coupled a greater intensity of rainfall, water erosion would increase (Defra, 2009b).

Commenting on the situation, Boardman (2013) finds that Britain probably has more information about soil erosion than any other country and it is a process that continues to be monitored, especially where land use changes and there is a trend in recent decades to shift from on-farm to off-farm impacts of erosion, notably loading of watercourses, important in the context of the Water Framework Directive and the emergence of catchments as a whole.

Appraisal of land specifically at risk from water erosion under winter cereals has been undertaken using a modification of the soil erosion model designed by Morgan *et al.* (1986). This classifies land in England and Wales as at 'very high', 'high', 'moderate', 'slight' or 'negligible' risk of soil erosion according to soil texture, annual average rainfall and slope (Palmer, 1993). In broad terms, 12.7% was considered to be at slight or greater risk, with a further 4.3% likely only to have localised problems. If land which is climatically limited for cereals or urbanised (and hence not capable of sustaining regular winter wheat production) is excluded, the percentage of vulnerable land rises (Palmer, pers. comm., 1995). A modelling approach has been used to predict the sites most at risk over the South Downs, and this might be extended countrywide (Boardman, 1988).

It is the link between field and river which is so important in determining on the one hand how much soil is 'yielded' into a drainage basin, and on the other how much is lost from productive land. Boardman and Favis-Mortlock (1992) note the significance of this link - soil eroded may affect water quality while water in the stream channel may transport eroded soil, agrochemicals and urban or industrial wastes. Although the figures calculated from suspended sediment load for lowland arable land rarely exceed $0.5\,t\,ha^{-1}\,yr^{-1}$ for soil eroded, the figures could be considerably greater where there is storage in the system, for example, at the bottom of slopes or within river channels (Boardman and Favis-Mortlock, 1992). Appreciable volumes may be stored on the river floodplains themselves (Evans and Cook, 1986). This suggests a poor delivery ratio of eroded sediment to river channels which may be exaggerated by a shortage of coarser material actually reaching the channels due to storage *en route*. Small plot studies suggest that field-scale values for a range of soil types under arable management vary between 0.3 and $55.3\,t\,ha^{-1}\,yr^{-1}$. Highest values are for sandy, silty and loamy soils (Walling, 1990).

More specifically, it is possible to highlight a number of individual policy instruments that could be used to address the issue of soil erosion, including (Inman, 2009):

- Regulatory Instruments
- Whole Farm Planning
- Farmer Self-Help (Voluntary Initiatives)
- Co-operative Agreements
- Grant Aid

One recent and innovative approach to the control of soil erosion comes from the National Trust (NT) (Jarman, 1999). Since 1996 NT started to investigate the importance of soils to its work and in 1999 published a Soil Protection Policy. Essentially, this recognises not only the importance in soil to support a range of habitats, but also seek to protect the archaeological resource on its land. For soils under productive forestry or agriculture, minimum cultivation and reduced inputs of agrochemicals are promoted. There is also an aim to reduce the impact of over-stocking and over-grazing in compaction ('poaching') and erosion.

Soil degradation presents as yet another classic wicked problem affecting *inter alia* water quality physical hydrological problems as well as and geomorphological problems. Research has been active for decades but, despite best efforts, no easy solution is presented.

8.9.2 The South Downs

The South Downs extend through southern parts of the counties of Hampshire, East and West Sussex and incorporate some $1,600\,km^2$ of a National Park designated in 2010 (South Downs NPA, 2011). Although there is an appreciable area of 'Wealden' geology within the Park boundaries to the north of Petersfield and Midhurst, the area is dominated by a chalk upland that rises to over 200 m and has a relief range of around 150 m (Boardman, 1990) including dissecting dry valley systems. The soils of the South Downs include stony rendzinas with high proportions of silt (60 to 80%) arising from loess of Pleistocene age with typical textures silty clay loams or silt loam. These are rarely thicker than 250 mm. Cultivation dates from prehistoric times with soil erosion commencing during the Neolithic, but it is likely that during the middle Bronze Age (1500 BC to 1000 BC) the rate of wildwood clearance and resulting accelerated erosion of topsoil were especially high. Subsequent erosion rates are surmised most serious following arable conversion; examples being the Roman period and the second half of the twentieth century (South Downs Joint Committee, 2008).

Since the Second World War appreciable ploughing of former pasture has occurred, and since the late 1970s the spread of winter cereals eventually accounted for over half of the cultivated area; even slopes typically 5 to 25° went under cultivation. Conscious management decisions have lengthened slopes by removal of hedges, banks, lynchets and walls. Suspended sediment yields in associated catchments can be great, from 2 to 148 t km^{-2} yr^{-1} (Walling, 1990). However, field-scale data (expressed as volume or weight loss per unit area) are useful in pinpointing the geographical origins and magnitude of the problem. On the South Downs, most erosion occurs within the first three months of the growing season (during the winter for winter cereals) when there is maximum bare ground cover exposing wheelings and rolled drilled surfaces prone to rilling (Boardman, 1990).

Rates of loss measured in the range 0.01 to 5.0 m^3 ha^{-1} yr^{-1} are not considered great, however there are occasions associated with heavy storms that may exceed 200 m^3 ha-1 yr^{-1}, on individual fields, and these can result in 'muddy floods' that have affected urban areas (Boardman, 2003). Projected increase in flood frequency resulting from climate change could mean these serious events become more frequent and mitigation measures are required (Butler, n.d.)

Relatively wet winters in 1982/1983, 1987/1988 and 1990/1991 and 2000 caused flooding and deposition of eroded soils on roads and in residential properties, the cost of which was met largely by individual householders and local authorities rather than by landowners or farmers (Boardman, 1995). Most erosion occurs in wetter winters, but it is intense rainfall events (typically 30 mm over two days) which initiate rainfall. While flooding is a problem, and affects urban areas, most of the eroded material remains within the dry valley systems. For a study area of the ESA in the vicinity of Brighton, some 70% of arable land deemed as being 'highly vulnerable' on account of steeper slopes and heavier rainfall could potentially be targeted for grass restoration or other less damaging land uses affording protection from both soil erosion and aquifer contamination (Cook and Norman, 1996). Other, non-land use options, include groundwater pumping, increasing infiltration, construction of dams and increased storage for floodwater (Boardman, 2003). In water resource terms, some 1.2 m people receive water from the South Downs National Park, both from ground and river water and these currently fails EU standards due to pollution including sediment derived from farmland that becomes deposited in the rivers that also interrupts river flow and contributes to flood risk. One land-use response to soil erosion event is to plant native tree species in vulnerable areas (South Downs National Park Authority, 2014).

8.10   Siltation of Reservoirs

Soil erosion is recognised as a problem in lowland Britain. Mapping procedures and spatial models predict where (if not when!) soil may move downslope, and the characterisation (by dynamic models) of transport mechanisms of sediment towards channels is well established. There was by contrast, insufficient data on the siltation of lowland river channels and receiving waters. This could be because the problem of siltation in British reservoirs was never considered great by global standards. However, siltation reduces capacity with uncertain engineering consequences and reduction in volume below 25,000 m^3 for which there are legislative implications in reducing capacity and also there are safety implications in the event of dam failure of loaded dams (DETR, 2001).

Predictably, the impact of eroded soil from agriculture tends to be seen as an issue for phosphate and pesticide transport and for its ecosystem effects such as eutrophication and impact on salmonoid fisheries (NRA, 1992b) rather than as an 'engineering' problem causing the siltation of water bodies. In Chapter 5.4 it was noted that reservoirs regulate not only the flow regime, but also sediment transport. The reservoir acts as an effective trap where faster flowing streams meet a largely still water body favouring the precipitation of a wide range of sediment particle sizes and organic material.

There is a long legacy of reservoir construction in Britain, yet the popularity of any further construction since the 1980s has been open to question; for example the construction of a reservoir at Broad Oak, near Canterbury in Kent and in SW Oxfordshire remains a long way off, that is if they will ever be completed (Chapter 1). Nonetheless, not only are reservoirs a major part of the landscape of upland Britain but the importance of protecting upland catchments as Water Protection Zones (WPZs) will be noted (Chapters 9 and 11). Certainly protection of reservoirs from 'siltation' is a major component of that policy (JNCC, 2010).

Butcher *et al.* (1992) report on the problems of reservoir siltation and discoloration in the southern Pennines. The cause is not ploughing of upland soils for afforestation of slopes, but rather the erosion of upland peatbogs and mineral subsoils which feed the headwaters of the recharging streams of the reservoirs. Blanket bogs were formed during the prehistoric era and subsequently dissected with serious erosion after 1770, the start of the modern industrial age. It is probable that peat formation stopped and peat erosion accelerated due to air pollution. Other sources of sediment here include footpath erosion and fires (Mannion, 1991, p. 282) and overgrazing. Otherwise Butcher *et al.* (1992) note the incised gulleys in many headwater streams. The study catchments are shown in Figure 8.9.

Both catchments are in the Carboniferous 'Millstone Grit' series of shales and sandstones overlain by impermeable sandy clay head. Here, blanket bog peats dominate the soil, yielding acid drainage waters.

This study is an example of an overall problem of environmental degradation specifically:

- The loss of grazing land in the uplands.
- The loss of upland peatbog.
- The loss of water storage capacity by siltation.
- Deterioration of water quality. It is probable that water colour is irreversibly raised when eroded peat is dissected during periods of drought.

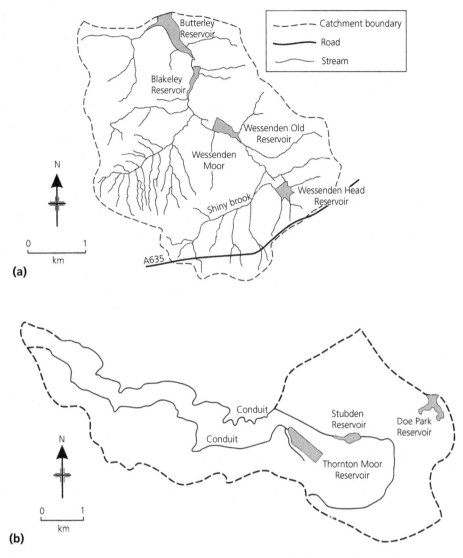

Figure 8.9 Study catchments in the south Pennine reservoir siltation study. (*Source:* Redrawn from Butcher *et al.*, 1992).

Direct measurement of reservoir sedimentation provides a measure of catchment sediment yield, and is the best measure because erosion is episodic and concentrated in storm runoff which is difficult to measure. Discoloration is most likely due to microbial activity releasing long-chain humic substances from the peat to the water.

Topographic studies of reservoirs were undertaken either when they were drained for maintenance, the reservoir dry, or else using sonar techniques while full. Comparing the bathymetric maps produced enables there to be a measure of capacity loss. All six reservoirs were constructed between 1835 (Wessenden Old) and 1907 (Butterly). Annual loss of capacity varied considerably, being between 137 m^3 (or 3.7% per century) for Wessenden Head and 2736 m^3 (75.2% per century) for Blakeley. Figures for soil erosion rates in the South Pennines are variable; measured in terms of weight of soil, loss can be as much as 252 t km^{-2} yr^{-1} (Walling, 1990), which is among the highest recorded for England and Wales. Data for reservoir siltation rates from Butcher *et al.* (1992) were re-calculated to annual volume of eroded soil loss, giving the following figures: Wessenden Moor catchment, 362.66 m^3 km^{-2} yr^{-1}; Thornton Moor, 174.87 m^3 km^{-2} yr^{-1}; and Stubden catchment, 191.58 m^3 km^{-2} yr^{-1}. This is consistent with other studies; however, the high Wessenden Moor figure confirms the observational evidence for serious peat erosion in the catchment. This study noted that treatment works adjacent to upland reservoirs were often inadequate to deal with significant changes in water quality, something exaggerated during drought years due to increased usage of reservoirs.

In general reservoir siltation is only of local importance in Britain, however, in the Peak District is one area where a serious effort to restore the peat and not only moorland ecology (potentially in the face of climate change) but also reduce peat and sediment loss from gulleys and hence reduce sedimentation. Since 2010, the Moors for the Future Partnership won EU funding for the MoorLIFE project. The aim is to restore the blanket bog in the South Pennines Moors area over five years (National Parks, n.d.) by:

- stabilising areas of bare peat
- restoring moorland vegetation on these and other previously stabilised sites
- turning these areas back into 'active' blanket bog forming landscapes
- reducing peat and water flow through gully blocking

8.11 Summary

By reference to specific examples, this chapter demonstrates the enormous shift of emphasis towards concern for the wider environment and away from only considering only water supply issues in isolation. Examples quoted reflect the triple developments over the past 250 years of major urbanisation, industrialisation and agricultural intensification. Into this mix comes a mature suite of habitat and biodiversity conservation measures.

Virtually none of the case examples given studies has an easy solution. Preventing certain environmentally damaging activities requires resources for monitoring and policing. The knock-on effects have economic implications for industry or agriculture or potentially require increased charges to the consumer. An example of this is where improved treatment of sewage effluent is required.

Urban rivers or rivers with a significant amount of urban area within their catchment remain a focus of concern; this does not, however, mean that rivers on rural areas are exempt from problems. Far from it, for not only does agriculture bring problems of pollution (largely diffuse) but agriculture is linked with loss of habitat and biodiversity simply because it represents a change in land cover towards monoculture and associated environmental modification including changes to drainage in historic times. Serious soil erosion dates back into prehistory.

Table 8.2 A summary of some of the complex (or wicked) problems faced by water resource managers in Britain and a list of key land use and policy solutions.

Issue for water resource management	Generic scope of problem	Land use or technical solution	Policy solution
Over abstraction of surface and ground water	Low and zero flows, concentration of pollutants, damage to habitats	Careful siting or re-siting of abstraction points	Good abstraction controls through monitoring and licencing alternative sources
Drainage of wetlands	Loss of habitat, flow regulatory measures and other ecosystem services of wetlands	Wetland conservation, creation and re-creation	Specific policy instruments such as SSSI, HLS and EU designations
Canalisation of rivers	'Improved' drainage dissociates river from floodplain, its associated wetlands and loss of in-stream habitats	Restoration of channel to 'natural' condition, removal of artificial channel lining materials - as appropriate	Multiple agency action, engagement with landowners, good technical information and measures
Pollution of rivers, lakes and wetlands	Chemical, physical and biological degradation of water damage to organisms, habitats, human and animal health	Strategic emplacement of buffer zones, water protection zones, adoption of 'sustainable' farming practices, good sewage and drinking water treatment	Wide range of monitoring, compulsory, and incentivising operational measures that control land use, industrial processes underpinned in law that interprets WFD. Codes of Good Practice for a range of industries, engagement with farmers, etc
Soil degradation	Erosion, compaction, acidification and leaching of potential pollutants. Loss of upland peatlands	Range of practices that reduce impacts from agriculture, forestry and urbanisation, also acid drainage waters. Reversal of certain drainage practices	Soil conservation policies including Catchment Sensitive Farming and engagement with farmers, etc
Groundwater contamination	Largely chemical contamination that remains in groundwater bodies, often for years	Surface protection zones, cessation of damaging practices, remediation and water treatment (both expensive)	Adoption of EU and UK policies that specifically protect groundwater
Flooding	Inundation of farmland and urban areas, danger to infrastructure and health	Adoption of both 'hard' and 'soft' engineering measures the prevent or mitigate effects of flooding, land use measures to prevent runoff	Adoption of EU and UK policies that specifically protect against flood
Afforestation	Loss of yield of water to reservoirs. Impact on water quality. Changes in flood behaviour, especially linked with clear-felling	Careful planting and rotation of trees to minimise impact. Broadleaved afforestation is best option if appropriate	Adoption policies that reflect Good Practice

Multiple causes of water and environmental problems require multiple and flexible solutions and, finally, the integration of water issues into the formal planning process (Table 8.2). The solution to such problems is the subject of the following chapters, in the absence of a national water strategy. Examples include controlling over-abstraction, soil protection measures and habitat protection and enhancement and control of diffuse pollution. Yet there is no cause for complacency, for diffuse pollution along with 'greenhouse gas' emissions remain serious problems.

9

Towards Solutions

Land Use and Technical Fixes

From how much back-breaking travail for horses and arm-aching labour for men does this obliging torrent free us, to the extent that without it we should be neither clothed nor fed. It is most truly shared with us, and expects no other reward wheresoever it toils under the sun than that, its work being done, it be allowed to run freely away. So it is that, after driving so many noisy and swiftly spinning wheels, it flows out foaming, as though it too had been ground and softened in the process.

> A Description of Clairvaux. *From* The Cistercian World: Monastic Writings of the Twelfth Century, *translated by Pauline Matarasso, Penguin Books (1993)*

9.1   Introduction: Approaches

Solutions offered in the area of water resource conservation have broadly fallen into three categories: *administrative measures, standard setting* and *land-use controls. Technical responses* then include optimisation of available information, modelling and water treatment (Cook, 1993). These may result in policy development for land-use change and/or targeting and zoning. Developments such as zonation of land areas to protect waters are pre-emptive and pro-active. The 'technical fix' of water treatment, while often essential, is seen as a reactive response to damage already done.

In getting to solutions, it is often easy to state a problem then proffer a technical method. Reaching a practical solution may be far more complicated. In SW England, 'Upstream Thinking' implements land use and land management measures in protecting sources of water requires involvement of the private sector, voluntary sector - as well as statutory regulators. Other water company initiatives include SCaMP in upland England and the catchment management programme funded by Wessex Water (Chapter 10). This chapter, however, deals with land-use measures and technical responses to problems while administrative measures were covered in Chapter 3. It is proposed to demonstrate the role of modelling of hydrological processes in order to achieve sustainable development of water resources.

The Protection and Conservation of Water Resources, Second Edition. Hadrian F. Cook.
© 2017 John Wiley & Sons Ltd. Published 2017 by John Wiley & Sons Ltd.

9.2   Land-Use Change and Water Management

In urban areas, much of the required legislation and infrastructure to control point pollution and water quality is in place. Additionally, land use is effectively controlled through the statutory planning process. It is therefore appropriate to turn our attention towards controlling diffuse pollution, notably from diffuse agricultural sources. Unfortunately, this is a problem that just will not go away and is arguably the single most important driver for the socio-political measures to be discussed in Chapter 10. The problems, as classically framed, are outlined in the next two sections.

9.2.1   To Divert Land or Extensify Agriculture?

To polarise the issue, there are two solutions to the diffuse pollution from agriculture:

- We may take land out of production and put it to some other less damaging use, including the employment of buffer zones to protect waters. This is *'land diversion'.*
- We may modify existing land management practices so as to reduce the environmental impact of agriculture. This incorporates notions of *'extensification'.*

In the UK, 'Land Budgets', which match population requirements for food with the available farmed areas and incorporate estimates of future productivity, have been calculated since the 1960s. Research at the former Wye College, University of London, the University of Reading (CAS, 1976) and elsewhere predicted a substantial 'land surplus' in British Agriculture by 2000. This finding was 'counter Malthusian' in that management of the agricultural resource-base caused an outstripping of demand for certain agricultural products. Estimates vary, but a 'ball-park' figure of 1 million ha of agricultural land released from agriculture was estimated if production continued to rise (Potter *et al.,* 1991, p. 70). Cereal production formed the largest component.

In the face of such blatant market failure, over-production can be addressed in a number of ways, such as reducing the support price paid to farmers, reducing acreage (e.g., by setting land aside from agriculture) or reducing the intensity of agricultural operations. 'Managed change' had to consider three basic objectives:

- Production reduction (aimed to match food supply with demand).
- Income support (supporting the economy of rural areas in some way by preventing farmers going out of business).
- Providing environmental benefits (especially reducing adverse impacts upon soil and waters).

This situation, applicable to both the United States and EU, is the consequence of a marriage between agricultural technology and economic strategies directed at food security-hence the process is counter Malthusian.

Such a situation is unlikely to arise in the developing world where there is frequently a perennial land shortage. The situation fluctuates in Europe, from 15% in 1993, compulsory set-aside, the area reduced throughout the 1990s becoming zero by 2008 (UK Agriculture, 2007). Although farmers can rotate their set-aside land, fixing it gave opportunity for 'agri-environmental' schemes, for example, where land was liable to soil erosion, flooding or it can be deemed to possess intrinsic conservation value. Needless to say, farmers, given the choice would generally select areas that are problematic or otherwise give poor economic return. The option remains, however, for the selection of land areas for inclusion in a policy of one kind or another. Targeting specific geographic areas emerged as a key factor in environmental conservation, yet it requires appropriate criteria (Cook and Norman, 1996).

Targeting need not only be geographical, there is a case for targeting on other criteria such as farmer group or farming systems whereby economically vulnerable farmers are targeted for support. The case for precise land targeting for agri-environmental policies was made for the United States, and it is persuasive in that it saves public money and increases policy effectiveness (Batie, 1983). A similar case was also made for the UK (Potter *et al.*, 1993).

Since the 1930s, the United States has operated a land diversion programme aimed not only at reducing over-production, but also at easing serious problems such as land degradation and water pollution (Potter *et al.*, 1991). The amount of land enrolled in simple set-aside schemes has varied over time as this kind of policy came in and out of fashion. Farmers agreed to reduce acreage in surplus crops (such as wheat, corn, tobacco, peanuts and rice) in return for payments to fallow certain land areas.

In 1956, in the United States, there was established the 'Soil Bank', which had a component called the Conservation Reserve. This involved larger areas enrolled over a longer period of years. Significantly, enrolled farmers maintained an approved conservation cover on reserved land, thereby preventing further soil degradation. Subsequently, efforts were made to introduce specific land targeting, that is, identification of land to take out of production, and to improve the efficiency of the scheme. In 1985, a 'Conservation Reserve Program' (CRP) was established which had the objectives not only of reducing production, but also retiring land vulnerable to erosion and hence any knock-on effects on waters. This matter will be returned to later in this chapter.

Alternatively, the adoption of 'extensification' has been favoured by some. This refers to the possibility of the present acreage of land remaining in production, but with reduced inputs. Livestock extensification would include reduction of stocking density and changing grazing regimes. Under arable extensification, farmers may be subsidised to reduce their fertiliser use, pesticide application and mechanisation (Bowers, 1988). In England and Wales, all the main forms of artificial fertiliser application ('NPK') to crops and grass have been in decline since around 1990 (AIC, 2015) reflecting lowering economic inputs to farmland; perhaps more 'low-input' systems, or a swing towards 'organic' farming may yet result from changing market demands in favour of restrictions on agrochemical use? In 2013, on percentage terms of farmed area, organic farming represented only around 3.5% England, 6.5% for Wales and 2.6% for Scotland. In England, specifically, the trend for fully organic farmland (after conversion) was steadily upward from about 2000, peaking in 2010, with similar trends for Scotland and Wales (Defra, 2014c).

It has been argued that organic farming systems reduce soil erosion, pesticide and phosphate pollution where these would otherwise arise from arable farming. Nitrogen management under organic farming systems can produce benefits in terms of reducing overall pollutant loading to waters (Stopes and Phillips, 1992), a situation possibly attributable to higher bioactivity and organic matter levels which cause greater denitrification and immobilisation in undisturbed soils.

However, care must be exercised before uncritically proposing philosophically satisfying catch-all solutions. Specific causal factors should be identified which enable manipulation of a farming system in the interest of environmental quality. For example, Armstrong Brown *et al.* (2000) investigated soil organic matter in paired organic and conventional farms. Although 'organic' production tended to improve the organic matter status (and by implication the stability of the soil and its resistance to erosion) on both arable and horticultural farms, organic farms tend to apply more manure as a matter of course. Under organic management, nitrogen losses have been found to be locally high, and especially at times in the rotation when leys are ploughed (Davies and Barraclough, 1989). Achieving 'sustainability' in farming systems with respect to water quality ultimately relies upon a detailed understanding of management practices. For example, nitrogen management associated with organic converted agriculture within groundwater protection zones in Lower Saxony, Germany remains problematic (Aue, 2014).

Extensification has long had its supporters among the 'sustainable agriculture' faction (e.g., Reganold, 1995). It is probably true that such an approach would favour agricultural employment, overall environmental, food and water quality, and answers the criticism that land diversion may cause intensification *away* from land not targeted. Currently, however, the balance of evidence is in favour of targeting 'environmental hot spots'. The debate between 'targeters' and 'extensifiers' continues and should be central to the development of policy for agricultural development and land use. Arguably, the 'extensifiers' have the more difficult task, because implicit in their philosophy is a radical shift in attitudes towards land management incorporating issues of food policy, economic support and even the social fabric of the countryside. Meanwhile, many would take exception to incorporating 'organic' production in this category; inputs in terms of labour and natural alternatives to agrochemicals can be high in such systems. The present dominance of land targeting reflects the administrative convenience of identifying specific areas and it also means the number of farms thus identified is restricted, with obvious implications for the public purse when compensation is an issue. It is probable that targeting will remain because it is both defensible in the planning process and (ideally) scientifically justified.

Yet the population of Britain is growing, and it comes as no surprise that issues of food security are once more coming to the fore. Maximising domestic food production with the objective of feeding the population of the country and minimising food imports has historic precedents (notable during wartime) and will be perceived by some to be at odds with sustainable food production. At the time of writing, the UK may be exiting the EU. Along with the concerns around the future of continuing to honour EU Directives, such issues are thrown into relief. This is primarily because of the continuation of direct payments to farmers to farm in an environmentally sustainable way as well as funding for the wider rural economy that has been channelled via the EU. If UK leaves the EU, we may ask who, if anybody, pays farmers to farm in an environmentally sustainable way? Political battle lines are being drawn, but it is likely this burden will return to the Government.

Actually, the drive towards more sustainable food production is wider, and incorporates emotive issues of food quality, food safety and animal welfare which are outside the scope of this book.

9.2.2 Source or Resource Protection?

It will be shown that policy development in the arena of diffuse pollution control has elements of both the targeting and 'sustainable agriculture' (or sustainable land use) approaches. No single approach seems to have produced a 'magic bullet', yet it is important that we are able to draw lines on maps in order to determine what may be permissible on either side.

Lines on maps enable areas more or less prone to a specific problem to be defined, ranked and mapped for vulnerability assessment. Division of land areas has to be achieved, which provides the basis for a consenting system, for planning or land-use decision making. Where farming systems are to be targeted, there is a parallel with manufacturing industry where specific industrial processes are regulated for the purpose of integrated pollution control.

Vulnerability is the susceptibility of the resource to pollution (a function of the physical and biological environment), whereas 'risk' *sensu stricto* arises from an assessment of an activity proposed at a specific location which takes account of natural vulnerability and the scale of preventative measures proposed (NRA, 1992d). This might be the outcome of *risk assessment*, a process that considers all of hazard identification, hazard assessment, risk estimation and risk evaluation and in general is the study of decisions subject to uncertain consequences (Hiscock, 2005, p. 265-266).

An important distinction has long been made between *resource protection* and *source protection* (Adams and Foster, 1992). This distinction is especially helpful in the protection of groundwaters. There are two elements:

- Division of the *entire* land surface on the basis of (aquifer) pollution vulnerability in a way that relates to (ground) water protection.
- Creation of a series of special protection areas for individual sources in which various potentially polluting activities are either prohibited or strictly controlled.

The starting point in developing an aquifer protection policy is to identify the extent of the unconfined aquifer outcrop and assess the importance of the strata in resource terms, especially water yield. From there, information on semi-confining beds (such as boulder clay) and soil type give an indication of where protection measures may be required (Norman *et al.*, 1994) and finally proximity to the point of abstraction suggests how severe the required protection measures may be (NRA, 1992d). More integrated methodologies have to be developed based upon mapping soils, superficial and solid geology and moving on to more detailed hydrogeological considerations such as depth to saturated zone and proximity to abstraction points (Cook, 1991).

Once there is more detailed information available that incorporates more detailed information, for example in relation to aquifer characteristics in terms of solid geology such as a chalk aquifer, or spatial and vertical differences in superficial differences in superficial deposits such as boulder clay, models can be suitably refined. For example, the transmissivity of chalk is overall lowest on the interfluves and highest along valley sides, while the boulder clays will contain many lenses of sand that provide for infiltration (Rushton, 2003, p. 352). Following site-specific investigations involving aquifer characteristics and abstraction rates, zonation becomes possible based upon an understanding of the catchment of the spring or borehole (Skinner, 1994). Ranking land areas according to vulnerability of waters beneath (or adjacent to) enables a flexible approach to land targeting (Cook, 1991).

Resource protection of surface waters from pollutants, by analogy, has to encompass the whole catchment. In the United States, the Tennessee Valley Authority (TVA) represents an early and influential example (Chapter 11). More recently, in the UK, Catchment Management Planning has arisen from integrated catchment management into proactive management schemes. Source protection would be on a smaller scale, for example the emplacement of a reedbed to reduce nitrate or heavy metal pollution (Chapter 8). Changes in agricultural policy has given opportunity to protect the rural environment, and are increasingly considering water protection and conservation in the implementation of policies so that water protection and management is a highly important factor in shaping future land use in the British countryside.

9.3 Mapping and Modelling

Modern process modelling is based on the operation of computer software written with a specific purpose in mind. In common with all branches of science or engineering, modelling has a considerable modelling requirement. Modelling is differentiated from empirical science (which draws conclusions from the real world through observational and experimental means) because it is in the business of *prediction and scenario construction*. One example would be a hydrological model for the Thames basin that links the behaviour of both groundwater systems and surface flow (BGS, 2015) that is used to explore scenarios of climate change, abstraction changes and flooding.

These are based upon *best available data inputs* and *model construction* designed to mimic natural processes; credible scientific assumptions must be made, and the models validated against independent data. Any abstraction from the real world which expresses distribution of a resource and/or mechanisms of the operation of processes relevant to that resource may be viewed as a model.

A map, no matter how complicated, is only as good in terms of the description of the area portrayed as is the information supplied, and represents the first form of spatial model. Maps are cast in terms of *theme* (e.g., topography, soils, vegetation, watercourses, geology, etc.) and in terms of *detail* (e.g., scale, data resolution and data quality). Data thus assembled as *attributes* are referred to a system of coordinates (such as the National Grid, or longitude and latitude), in order to give them spatial reference. Not only are the conventional datasources of ground survey, thematic soil and geological mapping, air and satellite imagery used, but in recent years drones have been employed in flood risk mapping flood mapping (WWT, 2015).

For water protection, a software-based and systematic approach to land mapping is required. This improves the implementation of schemes through the planning process. There are two major kinds of model, non-time-dependent *spatial models* and time dependent dynamic, or *process models*. These are physically based and are by far the most varied group. Spatial models are considered first.

9.3.1 Spatial Modelling

Geographical Information Systems (GISs) are software systems which enable the electronic management of spatial data, hitherto referenced and drawn in conventional (i.e., paper) map form.

A GIS is normally able to:

1) Capture and verify mapped data.
2) Store and manage spatial and other data.
3) Transform and manipulate the data.

4) Output the data.
5) Interact with the user.

Data capture (point 1) may be by digitising or scanning, database importation or derived from remotely sensed images. Many image processing packages also have impressive GIS capabilities. A GIS is innovative not because it deals with 'georeferenced' data (Computer Aided Design and Automatic Cartography have long had these kinds of capability), but because it has the ability to bring datasets together and create new spatial information (point 3).

Where a GIS is employed, differences in scale and projection are soon resolved by the software, and overlay algorithms permit the inclusion or exclusion of certain land areas. This apparent panacea for data processing in environmental research and planning was at first slow to spread in the general planning process, despite extensive research. Probable reasons have been:

- Differences in *data structure,* for example between 'vector' (based on lines) and 'raster' (based upon squares) in both the original mapped form and in electronic datasets.
- Differences in *quality and resolution* of data, be it the scale at which original survey is collected or compatibility of mapping in adjacent areas. All spatial data are inherently inaccurate, and may thus introduce inaccuracies that will adversely influence GIS operations and these may be unpredictable. The data may typically have been captured from a scan of manual digitisation of a dataset on paper maps or plans, inducing sources of error from not only the operation, but also the underlying map that my change as its fabric ages.
- *Cost of datasets,* which is often linked to draconian UK copyright laws. These permit intellectual property to be tightly controlled by certain agencies, a situation at variance with the approach in the United States and Canada which have a concept of the public domain. Apart from paper maps and conventional air photographs, other sources include LiDAR and satellite imagery. Modern forms of data collection are more easily georeferenced and imported into software. There is generally a cost.
- *Time requirement* (and hence cost) in capturing, editing and combining datasets into usable formats within a single software system. A project manager faced with limitations on specifications, cost and time may opt for conventional manual overlay techniques requiring only the skills of drafting and cartography where the number of overlays is limited. However, GIS methodologies are now widespread in water planning in the EA regions, while the completion of the digital storage of the Ordnance Survey dataset for the entire country has been achieved, creating an electronic archive which is regularly updated. Basic geological mapped data is also available online.

Modern developments involved accurate satellite determined Global Positioning System (GPS) enabling field-referencing as well as representation of topographical mapping to give three-dimensional representation. Furthermore, the great advantage of GIS in water resource planning has been not only to enable spatial analysis of mapped data (climate, soils, aquifers, watertables, etc.) but also provide accessible output formats for process models of surface and subsurface flow.

It remains all too easy to run away with a sense of technological optimism borne of satellites, GIS and geostatistics, to say nothing of advances in process models.

The reality is that much environmental data was collected over a century or more of field survey and depended on earlier methods including conventional analogue mapping, or air photographs without ready georeferencing.

For example, the main limitations to vulnerability mapping for groundwater protection were described by Gosk *et al.*, (1994). These are:

- a lack of representative data and their relation to the scale of the map;
- inadequate description of the system (specifically the inability to accurately map all hydrogeological variables);
- a lack of a generally accepted methodology;
- verification and control of vulnerability assessment.

Similar criticisms could be levelled at mapping for surface water protection, although hydrological variables would inevitably be better appreciated in most situations. Scientific criteria in vulnerability mapping ideally include both the level of uncertainty in the model or the data, and its impact on the management actions (NRC, 1993). Between the two hydrological domains is the soil and efforts to characterise it hydrologically for resource planning purposes (notably characterising flood behaviour) include HOST (Boorman et al., 1995, Chapter 2).

Vulnerability for land-use planning purposes may be assessed to target policies, and a GIS has an obvious relevance in screening land areas either to identify areas where further modelling effort is required for the protection of waters or screening for eligibility for enrolment in a particular scheme. This is because a GIS has the potential to examine an area in detail. Designations of Nitrate Vulnerable Zones (NVZs) screening for enrolment schemes (such as Higher Level Stewardship (section 9.4.4) have obvious applications here; however to date it has usually been the conventional printed map which provides the basis for definition in virtually all agri-environmental schemes.

By far the most important problems in defining the boundary arise from the spatial data employed as input. Available input data dictated by, and selected for, eligibility criteria are derived from a range of agencies, and these data vary in terms of availability and spatial resolution. Consequently, each map layer in the model has its own resolution or accuracy dictated by the specifications of the original survey. It is problematic to decide how these may compound into the final model and problematic when map scales are 'blown up' to larger scales. Error propagation cumulates, and although statistical techniques of uncertainty analysis exist, few published vulnerability assessments account for uncertainties from either model or data errors (NRC, 1993).

However, 'hard boundaries' remain politically most desirable and planners prefer field boundaries or roads for practical purposes. Thematic interpretation of input data for a specified purpose is often required. In any case, all maps have a limit to their predictive accuracy. Matching map scale with the requirements of a targeting project is extremely important. Presenting data at a scale of 1:100 000 may be appropriate at the regional level for establishing environmental vulnerability, but it is inadequate for working at the level of farm or local conservation area. Larger scales are required with more stringent details of data; this scale of operation is indicated by the term 'microtargeting'.

The main objection to overlay maps is the unknown weightings to be applied to such spatial variables as soil leaching vulnerability, aquifer fissuring and pollutant attenuation in the unsaturated zone. However, it is possible and desirable that, given time, the requirements of microtargeting for source protection zones or nitrate pollution. Ideally, this depends of a detailed re-evaluation of local conditions, but the cost of investigations would be considerable. Maps showing pollution vulnerability and environmental tolerance to contaminants are increasing in number and complexity. The practice of mapping land areas for combined soil erosion liability and nitrate contamination of groundwater has been discussed by Cook and Norman (1996).

The USGS model DRASTIC (Shirazi *et al.*, 2012) has been around for over 20 years. DRASTIC method generally used seven hydrogeological factors to assess groundwater vulnerability:

1) Depth to groundwater table (D),
2) Net recharge (R),
3) Aquifer media (A),
4) Soil media (S),
5) Topography (T)
6) Impact of vadose zone (I) and
7) Hydraulic conductivity (C).

Each layer is weighted according to its relative significance to contamination, so that:

$$\text{Pollution potential} = D_R D_W + R_R R_W + A_R A_W + S_R S_W + T_R T_W + I_R I_W + C_R C_W$$

The subscripts refer to rating ($_R$) and weighting ($_W$) from 1 to 5 respectively. Each factor is ascribed first to a range of values depending on the impact on pollution, then a single rating (1 to 10) is ascribed. Aquifer media is a ranking based on aquifer permeability and soil media is a ranking based upon transport of contaminant from soil to aquifer. The vadose zone is that which is intermittently saturated and unsaturated as the watertable fluctuates. The higher the DRASTIC score the greater is the potential to pollute groundwater.

It has been concluded that (Stokes, 2001):

1) DRASTIC can be used to model groundwater vulnerability.
2) Results of applying DRASTIC model must be used carefully. This applies a framework but does not account for all the particulars of the chemicals released. A detailed study of a particular spill must incorporate the chemical properties of the contaminant.
3) GIS can help make the results of a complicated model more clear through visual representation, thus providing an applicable tool for decision makers.

In the UK a similar model soil or geological maps have to be interpreted for hydrological or hydrochemical purposes such as potential for nitrate leaching (Cook, 1991). Models of groundwater vulnerability for England and Wales are produced by combining soil and aquifer information and are essentially aimed at predicting leaching potential over an aquifer, and hence protect SPZs (see below). These take account of both solid and drift geological conditions and are assigned to major, minor and non-aquifers (EA, 2009d). The outcome, for planning and public information purposes, is available online (EA, 2014c) with areas defined as inner zone, outer zone and total catchment.

Key factors that define the vulnerability of groundwater are:

- presence and nature of overlying soil;
- presence and nature of drift;
- nature of the geological strata;
- depth of the unsaturated zone.

The Aquifer extents are not displayed at scales greater than 1:75,000 (derived from the Ordnance Survey 1:250,000 scale) as the data was only modelled to this level and is not accurate pass this. Revision is underway in the light of WFD (EA, 2013e).

Methodologies based in overlay application are absolutely important in resource protection, but pragmatism requires that account is taken in terms of proximity to a source. This takes protection beyond a consideration of soil and geological vulnerability (as originally mapped nationally, see below) but no available on line, generally accessed by postcode) towards consideration of proximity to an active source and direction of groundwater travel. Much as upstream pollution of a surface water source be avoided at all costs, the 'out of sight, out of mind' mentality must never be allowed to prevail for groundwater boreholes. Furthermore, mistakes in management mean that groundwater contamination takes a long time, frequently decades, to clear (Cook, 1999a).

In UK water planning, there are two kinds of vulnerability:

- Intrinsic vulnerability of a location depends on a number of factors including the soil type, presence of drift and the characteristics of the rock ('resource protection'). Mapped data required includes soil and geological maps, borehole information and information of subsurface flow.
- Specific vulnerability at a location includes: the nature of the activity under scrutiny and the characteristics of the potential threat of a contaminant. In this case we may also consider if a proposed activity results in the removal or bypass of soil or drift and (importantly) the unsaturated zone.

9.3.2 Water Protection Zones (WPZs)

As far as groundwater is concerned, Water Protection Zones (WPZs) are used around sources identified as being at high risk as a 'last resort' when other mechanisms have failed or are unlikely to prevent failure of WFD objectives. The concept arises from that of SPZs. Here are applied specific statutory measures over and above existing ones to manage or prohibit activities that 'cause or could cause damage or pollution of water', that relates to particular pollutants - or to polluting activities. Strong measures are required.

Drinking Water Safeguard Zones (SgZs) These are surface and groundwater catchment areas that influence the water quality at drinking water abstractions which are at risk of failing the drinking water protection objectives. Affected are locations where there is known deteriorating water quality or where existing measures should be strictly enforced for particular pollutants and activities and where there can be a focus on additional new voluntary measures. Partnerships are sought by EA, with water companies when designating SgZs and implementing actions. There are around 200 groundwater SgZs and there exist plans to review/designate more during the second river basin planning cycle from 2015 to 2021. They are non-statutory (EA, 2014c).

Source Protection Zones (SPZs) A general level of protection for all *drinking water* sources is administered via computed Source Protection Zones (SPZs). These are the basis for other controls within defined safeguard zones (see above). SPZs are identified close to drinking water sources 'where the risk associated with groundwater contamination is greatest' and are an important tool for the identification of sensitive groundwater areas, and for focusing control - or advice - beyond the protection applied to aquifers as a whole. Under the Water Resources Act of 1991 which implemented the Groundwater Directive, the EA has powers to prevent discharges, prosecute polluters and designate water protection zones (NRA, 1992d). That Directive was subsumed in WFD 2000 (Chapter 3). This approach permits not only the division of entire land areas on the basis of the vulnerability of the underlying aquifers to pollution, but also the definition of special protection areas for individual sources (Adams and Foster, 1992). Unlike the schemes designed to protect waters from nitrate pollution, the SPZ approach is non-pollutant specific (Cook, 1993).

The aim is to prevent deterioration in groundwater quality that *could harm abstractions intended for human consumption* but it does not mean drinking water standards are applied to untreated groundwater. Standards are applied regardless of the present water quality for they are relevant to the quality of water supplied to the consumer. There are approximately 2000 specific SPZs around major abstraction sources – both boreholes and springs - that are used for human consumption or required by the food industry. It is assumed that all sources intended for human consumption that do not have a bespoke SPZ have at least a default SPZ1 of a 50-metre radius (see Figure 9.1). For small sources (abstraction less than $20\,m^3$ per day) it becomes the responsibility of the abstractor to locate the source and define a 50-metre protection zone.

SPZs have three subdivisions (Figure 9.1; EA, 2009e):

- SPZ1 inner protection zone is defined as the 50-day travel time from any point below the water table to the abstraction source. This zone has a minimum radius of 50 metres. SPZ1 represents the immediate area around a borehole where remediation of pollution is unlikely to be achievable within available timescales, such as in less than 50 days.

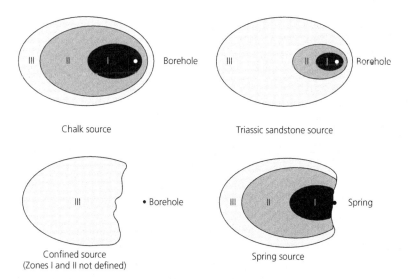

Chalk source

Triassic sandstone source

Confined source
(Zones I and II not defined)

Spring source

Figure 9.1 Example groundwater source protection zones (SPZ) surface subdivisions (NRA, 1992d). Zone I is black, Zone II is dark grey and Zone III is light grey.

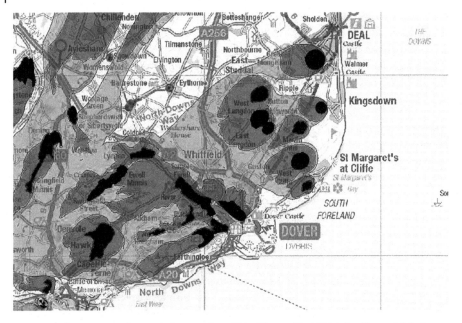

Figure 9.2 Source protection zones in part of east Kent (*Source:* EA, 2014c). See Figure 9.1 for key to zones.

- SPZ2 defined the outer protection zone – defined by a 400-day travel time from a point below the water table. This zone has a minimum radius of 250 or 500 metres around the abstraction source, depending on the size of the abstraction.
- SPZ3 is the wider source catchment protection zone – defined as the area around an abstraction source within which all groundwater recharge is presumed to be discharged at the abstraction source. Or to put it another way, this is the surface-defined area required to support abstraction in the long-term.

In areas of karstic groundwater flow and recharge, such as the carboniferous limestone of the Pennines or Mendips, there has been provision for zones of special interest (SPZ4). Historically defined, SPZ4 usually represents a surface water catchment that drains into the aquifer feeding the groundwater supply. In future, this will either be incorporated into one of the other zones (SPZ1, SPZ2 or SPZ3) or become a SPZ in its own right. They are non-statutory although the ideas is that their definition enables the EA 'to respond to various statutory and non-statutory consultations in a consistent and uniform manner (EA, n.d.c).

Figure 9.2 shows real groundwater Source Protection Zonation (SPZ) on the Dover-Deal area of east Kent. See key for details. SPZ 1 is the most critical, as any contamination is expected to take 50 days to enter the water supply. The influence of groundwater flow distorts the concentric pattern, and the elongation of Zone 1 in valleys reflects the increased fissuring in the chalk at these locations.

Figure 9.3 shows source protection measures for the Mendip Hills and surrounding areas. The fissured nature of the Carboniferous Limestone around Cheddar means that the inner Zone I dominates, for it is calculated than much of this massif produces a travel time of less than 50 days to a given source (EA, 2014c). The shaded areas are where only subsurface activities are considered potentially damaging.

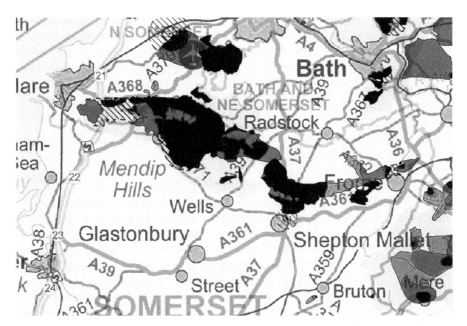

Figure 9.3 Groundwater source protection zones for the Mendip hills in Somerset (EA, 2014c). See Figure 9.1 for key to zones.

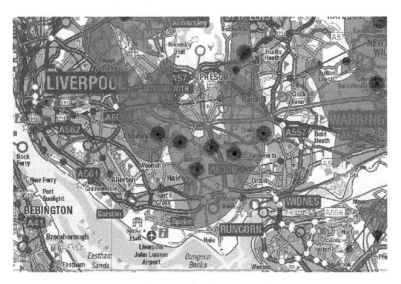

Figure 9.4 Groundwater source protection zones, permo-triassic aquifer in the liverpool area (*Source:* EA, 2014c). See Figure 9.1 for key to zones.

Figure 9.4 shows Groundwater Protection Zones in the Liverpool area (EA, 2014C). The major aquifer here is the Permo-Triassic aquifer and the relatively small Zones I and II are due to low aqufer transmissivities (Chapter 2). In urban dominated areas there are a lot of potential groundwater contaminants and mapping for specific substances would not be viable, let along enforceable. For example, pesticides are complex chemicals and their environmental behaviour is uncertain. Options within vulnerable areas would include selective bans, change in land management and more selective application.

Another unknown in groundwater SPZs is fracking. While it is difficult to be precise as to the degree, depth and impact of fracking in general (Chapter 7), the practice has the potential to violate any surface-based/land-use protection measure, not only might it create physical damage to aquifer that may alter its water storage and transmission characteristics (Chapter 2), but the introduction of chemical substances during the process may affect water quality.

Confined aquifers. The default zone for confined SPZ1 is a 50-metre radius. This provides a buffer protection for the infrastructure around an abstraction borehole. SPZ2 is *not* generally defined for confined aquifers however SPZ3 is used for the supporting catchment area.

9.3.3 Hydrological Process Modelling

In contrast to their purely spatial counterparts (for example those defining groundwater vulnerability alone), *process models* (also termed *'dynamic models'*) have applications where time-dependent variables are required. Since the 1970s, these have generally taken the form of digital computer models which are essentially a mathematical description of the processes involved. These normally include some sophisticated applied mathematics and statistics associated with surface and/or groundwater flow (the two are commonly linked in a model) and require highly specialist knowledge in their writing and a sound practical knowledge in their application to practical situations.

Examples of variables modelled dynamically include aquifer yield, river discharge, pollutant concentration, and pollutant path or travel time (Bland and Evans, 1984). Where relationships between variables are derived from experimental observation, the term *'empirical model'* may be employed (otherwise the terms 'statistical', 'probabilistic', 'stochastic' or even 'black box' have been used, depending upon model construction).

Models need to be calibrated against the real world. This is easiest in the case of process models where there are already historical datasets including measurements of the variable to be modelled. Validation should be undertaken in order to assess performance against an independent dataset (where available) so as to test calibration independently of the data used in model building. Once calibration and validation are undertaken, scenarios can be set so that the model is used to predict what may happen under differing conditions, such as modelling responses to the loading of a particular pollutant.

There are two kinds of approach in hydrological process models: the *'lumped parameter model'* and the *'distributed model'* (Bland and Evans, 1984). The former tend to treat interception, the soil, slope processes, the aquifer and surface flows as a series of tanks with overflows and transfers between compartments taking place according to parametric rules. These tend to be more probabilistic (Ward and Robinson, 2000, p. 348).

An example of a lumped parameter would be a value for recharge, such as effective rainfall, or discharge at springs. Lumped parameter models adjust the parameters until best agreement with historical data is met, a process called 'optimisation'. Output can only represent average groundwater levels within each compartment; none the less,

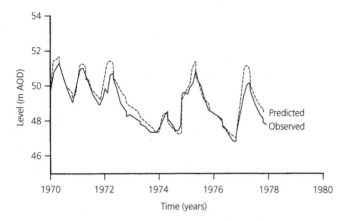

Figure 9.5 Observed and modelled groundwater levels in the Rhee chalk (Cambridgeshire). (*Source:* Redrawn from Bland and Evans, 1984).

good agreement with observed data can be achieved, as is shown in Figure 9.5. Distributed models tend to treat their subject as a system governed by physically based process equations, and consider a catchment in a geometric sense.

Two examples used in surface flow simulation of catchment models are the multi-nationally developed Système Hydrologique Européen (SHE) (Refsgaard *et al.*, 2010) and the Institute of Hydrology Distributed Model (IHDM); both are physically based (Calver, 1992). SHE requires that a catchment is divided into grid squares for which data are individually specified. The typical data requirements are surface elevation, vegetation, soil properties, meteorological data, channel and surface flow resistance data, impermeable bed levels and phreatic (watertable) surface levels. Subsurface and surface flow conditions are then modelled. IHDM, rather than using grid squares, requires that the catchment be divided into hillslope and channel elements and is typically used to model precipitation, snowmelt and evaporation by reference to topography and vegetation.

Distributed models give better spatial resolution in terms of output than lumped parameter models, but require a huge amount of inputted information. This is only available when there is adequate river monitoring or when an aquifer is fully developed. In practice, sparse information often makes distributed modelling difficult, making elucidation of processes in the top 50 m of the saturated zone (the most active part of a groundwater system) somewhat difficult (Grey *et al.*, 1995, p. 28). Distributed models have, however, proven useful studies of resource balances including the design of the Great Ouse Groundwater Scheme in the Cambridge area.

Component equations describing groundwater systems are based ultimately upon Darcy's Law and continuity equations. Although mathematically imprecise (compared with analytically solved flow-equations), in real terms the results are very useful and the flexibility provided to the modeller has proven invaluable modelling two-dimensional flow in aquifers. Two forms of distributed models are widely used in groundwater studies. *Finite difference* models set out a regular grid defining zones which may be triangular, polygonal or square (preferred) comprising nodes at which calculations may be undertaken (Todd, 1980, p. 400). Zonal boundary widths, zonal area and flow path length are defined by the grid, while data for storativity, transmissivity and net external flow (the balance between measured outflows at wells, boreholes and springs and recharge of the aquifer) are estimated using hydrogeological information on the aquifer. The variations in hydraulic head and its distribution over time and in space are then modelled. This is achieved by replacing a partial differential equation for water balances and the continuity equation with one based upon finite differences (i.e., real changes over time) and solving it numerically (i.e., by repeated computations). The approach requires that a series of simultaneous equations be generated, and solved for the unknown terms. The model is 'optimised' (calibrated) by means of adjusting storativity, transmissivity and net external flows for historic data on well water levels.

Once calibrated, scenarios of groundwater management may be explored. Models capable of using irregular grids use nodes at the corners of quadrilaterals or triangles which are constructed at the discretion of the modeller to fit the geometry of the flow problem. These are termed *finite element* models (Hiscock, 2005, Chapter 5). Finite element models reflect the reallocation of parts of the system such as observation and pumped wells in the case of groundwater studies. They, too, are numerical in their solution of groundwater flow equations.

The development of predictive modelling and mapping procedures in evaluating the likely outcomes of land-use change (rural and urban) on groundwater quality and quantity has been identified as a research objective in the UK (Grey *et al.,* 1995). In broad terms, modelling already plays a crucial role in the strive for sustainability in land use that move towards resolving water resource issues, including both water quality and flood prediction. In defining groundwater source protection zones (SPZs), data availability will vary with location, however where there is sufficient information, conceptual hydrogeological models have been created (EA, n.d.c).

One dynamic groundwater model is FLOWPATH, which can be used to calculate steady-state hydraulic heads and draw-down distributions, velocity distributions, water balances, time-related pathlines (of particle travel for water), and injection and capture zones. It has been used for two-dimensional analysis of aquifer properties to calculate a range of hydrogeological variables, most notably capture zones and time-related pathlines through the saturated zone for protection zone definition (Skinner, 1994). Boundaries, however defined, will have to incorporate a margin of error; this is consistent with the 'precautionary principle' of water resource planners enshrined in the NRA 'Water Resources Strategy' (NRA, 1994g). In practice, land-use planners guided by modelling follow 'hard boundaries' in the landscape (e.g., field boundaries) for administrative purposes, but these must be defendable in the planning process.

Typical applications of FLOWPATH II (Scientific Software Group, 2015):

- Determining remediation well capture zones.
- Delineating wellhead protection areas.
- Designing and optimizing pumping well locations for dewatering projects.
- Determining contaminant fate and exposure pathways for risk assessment.

MODPATH is a particle-tracking three-dimensional flow paths using output from groundwater-flow simulations based on MODFLOW, used by the USGS finite-difference groundwater flow model. It allows an analytical expression of a particle's flow path for each finite-difference grid cell; the particle path is computed by tracking from one cell to the next until it reaches a boundary, an internal sink/source, or satisfies another termination criterion (USGS, 2014).

Resulting zones are stored in a digitised format which can be displayed as paper maps or in output in a GIS (EA, n.d.c).

9.3.4   Data Problems with Modelling

In practice, there are elements of both dynamic and spatial models used in land targeting. The future will inevitably involve bringing these together in microtargeting for source protection. Zonation established from unsaturated and saturated zone characteristics, soils and even climatic variables (Cook, 1991) will be brought together with process-based approaches.

Boundary definition will inevitably have to incorporate error derived from input data and spatial and process modelling. Soil maps can have poor predictive accuracy at the soil series level (Beckett and Webster, 1971). While geological boundaries are easier to pin down, there are inevitably problems in characterising details of great hydrogeological importance, especially fissuring and porosity of aquifers. These too will vary over comparatively small distances, both horizontally and vertically. Uncertainty for surface waters is also an issue; not only are unconstrained rivers prone to meander, but specified flood events will vary in effect according to interference on floodplain hydrology and upstream river regulation will affect flood frequency.

Mapped data coverage is uneven, and it is estimated (Potter *et al.*, 1993) that, whereas the whole of Britain is covered by Ordnance Survey data at a scale of 1:10 000 or less, and major aquifers are generally covered by hydrogeological maps at 1:100 000, uniform soil survey coverage for all of England and Wales only occurs at a scale of 1:250 000. Only some 22% is covered at larger scales giving better resolution and mapped detail. In many cases, drift geological maps may be used as a surrogate to provide boundaries. In practice, policy makers may find it necessary to commission new surveys where detailed information on habitat, soils, geology, land use and hydrology is found to be lacking.

The precautionary principle, now enshrined in water resource planning, has to incorporate uncertainty of model and data inputs. It may be summarised as 'if in doubt play safe'. It furthermore applies to both temporal water budgets and to geographical areas. In the Anglian EA region, for example, there is a planning margin of 10 percent built into assessments of need (NRA, 1994e). In Chapter 6, it was shown how a considerable margin of error is built into aquifer management. Otherwise, the NRA response to problems in groundwater protection zone definition was to define zones within an uncertainty framework.

9.4   Agri-Environmental Policy and Water

Realisation that land use can adversely affect water quality is far from new. Furthermore, British concerns over water quality have been biased towards diffuse pollution from contaminants than from pathogens; the latter being the focus of water quality concerns in the United States. The presumption being that pathogen control during water treatment is adequate. Land-use control to protect water supply dates to the late nineteenth century when Birmingham Corporation acquired commons in the catchment of the city's water supply in upland Wales. Protection of groundwater dates from the period following the First World War when Brighton Corporation purchased land to be maintained under grazing around sources of public supply (Sheail, 1992). By the 1970s it was realised that a policy for groundwater protection was required by the Severn Trent RWA (Selby and Skinner, 1979) and likewise for the Southern RWA (Headworth,

1994). These formed the basis of the modem approaches to groundwater protection from both specific and non-specific pollutants.

Within the subset of diffuse contaminants, it is those from agriculture that have long outweighed those from (for example) wastewater discharges, or point source discharges from industry. The following is a summary of UK agri- environmental policies is viewed from the angle of benefits to water quality, of habitat and of soil conservation.

9.4.1 Best Management Practices, Codes of Practice and 'Sensitive' Farming

In north America (chapter 11), Best Management Practices (BMPs) implies relativity - 'best' but not 'good' or even 'ideal'- and hence while the measures taken would be appropriate (in delivering environmental goods that protect such as soil, water, habitats or communities), there remains a probability they will be effective. Hence monitoring measures is implicit. Installation or adoption of BMPs self-evidently implies co-operation of catchment stakeholders, something that may involve voluntary, legislative, political, economic and behavioural instruments within a framework where BMPs present, in any case, an element of uncertain outcome.

BMP implementation is very important, but it is but one example of tackling wicked problems. While it provides 'technical uncertainty', unquantifiable factors such as stakeholder confidence launch the managers of catchments into considerations of political analysis – questions such as 'where might real power and influence lie?' – are commonplace. It is therefore potentially easy to take catchment communities into blind alleys, because a measure found to work in one catchment may not be technically, politically or culturally transferrable to another. Analogous guidelines are the 'Codes of Good Practice' issued by government departments of agriculture in the UK that are issued for landowners, growers and farmers with the objective of protecting the environment such as soil and water (Defra, 2009c).

Means of implementing such Codes requires schemes that encourage individual farmers to take appropriate action. The clumsily named England Sensitive Catchment Farming Delivery Initiative (ECSFDI) aims to 'safely manage soils, fertilisers, manures and pesticides to reduce diffuse water pollution from agriculture'. The aim of Catchment Sensitive Farming (CSF) is to reduce pollution from farming in surface, ground waters and other aquatic habitats in a catchment and further downstream through moderate capital grant schemes. Funding may be for fencing and farm infrastructure that is aimed at the reduction of diffuse water pollution from agriculture, delivered in partnership by NE, EA and Defra (Defra, 2014d). There is targeted, free advice, training and information in 80 priority catchments in England. It also provides access to the CSF Capital Grants Scheme. There are networks of Catchment Sensitive Farming Officers and these foster both local and national partnerships.

9.4.2 Nitrate Protection Zones

The specific issue of protection from nitrates in the chalk was brought to the fore by two studies, one on the chalk of Thanet, the other around the 'Cornish Source' operated by the (then) Eastbourne Waterworks Company in the Eastbourne Block of the South Downs (DoE, 1986; Headworth, 1994, Jones and Robins, 1999).

For Eastbourne, the Friston source provides a large part of the supply for Eastbourne and around 30% of the catchment was under commercial forestry. The majority of nitrogen introduced to the catchment is presumed to derive from agricultural land and nitrate concentrations remained moderate between the 1950s and 1990s, typically between 18 and 27 mg l^{-1} NO$_3$. The Cornish source is different. Around it, the soil is mainly light, flinty loam and land use either permanent grass or a long rotation of grass and cereals. Hydrogeological investigations had identified fast fissure flow as the main transport mechanism. During the period 1969 to 1972 the nitrate concentration reached a maximum of 71 mg l^{-1} NO$_3$, falling to 35 mg l^{-1} NO$_3$ by the end of the decade.

In 1952, the Waterworks Company and the local council drew up a fertiliser agreement, restricting restricting N-application rates to 63 kg ha^{-1} in an inner zone and 75 kg ha^{-1} in an outer zone. Fertiliser applications were revised in 1968 and maintained. The quantity of N-fertiliser applied in a restricted area was limited to 37.5 kg ha^{-1} and to 63 kg ha^{-1} over the remainder of the catchment. Slow release formulations where possible replaced soluble fertiliser and applications of gas liquor were prohibited. By the late 1990s, data from both sources shows stablilisation, although Cornish remains over 30 mg l^{-1} NO$_3$ (Jones and Robins, 1999).

Over some 100 to 120 ha, restrictions were possible through management agreement because the local authority either owned or managed the land. Restrictions over 70% of the borehole catchment commenced in 1952, and, in a particularly far-sighted way, there was an inner and outer zone identified, with more stringent measures in the inner zone. Although the pattern of fertiliser use in the area could not readily be linked to either the hydrogeology or the observed levels, the outcome was beneficial and a pioneering example of cooperation between water undertaker and landowner.

The Isle of Thanet is an area of intensive agricultural and horticultural land use locally exceeding 80% of land areas (Cook, 1991). Here, concern over nitrate concentrations at source had been expressed since at least the middle of the nineteenth century. Then it was sewage that was likely the major source of contamination. More recent investigations, including modelling, have linked agriculture to raised levels of nitrate in the unsaturated zone. Fertilised grassland displays higher levels than unfertilised, fertilised arable still higher, and abrupt changes of land use (such as ploughing) produced highest peaks of all. The rate of downward movement of nitrate through the unsaturated zone is low, around 0.5 m per year, while predictions for levels in the saturated zone show levels from sources set to increase well into the next century, and already they exceed the EU limit in many instances (Headworth, 1994).

Responses to this problem include closing sources, source blending (see below), and recently a Nitrate Vulnerable Zone has been defined. The response to land-use changes in this situation would be expected to be measured in decades rather than years. Spatial models have been proposed which aim to map the vulnerability of groundwaters to nitrate pollution. They include factors such as soil type, aquifer drift cover, effective rainfall, unsaturated zone geometry, depth to watertable and proximity to abstraction points (Cook, 1992). To control nitrate pollution, the NRA favoured a two-tier approach with fertiliser application at recommended rates over large areas but with sources being initially protected as Nitrate Sensitive Areas (NRA, 1992c).

Much essential information relating to the management of chemical and organic substances on farms is to be found in the *Code of Good Agricultural Practice ['GAP']* *for Protecting our water, soil and air*, and produced by Defra. Such are issued regularly by government and this one specifically applies to England. The code is concerned with practical advice to farmers, contains advice related to the management of fertilisers, pesticides, manure and slurries. While nitrogen has been of concern for many years following the 'EC Nitrates Directive (91/676/EEC) actually requires Member States to introduce a Code of Good Agricultural Practice to control nitrate loss and to protect against nitrate pollution, which all farmers should follow on a voluntary basis.' Modern versions relate to the implementation of WFD and replaces separate Water, Air and Soil Codes published by statutory agencies (Defra, 2009c). UK supporting legislation is, understandably complicated, but The Water Resources Act of 1991 contains provisions which are designed to prevent water pollution. 'Controlled waters' include groundwater and all coastal and inland waters, including lakes, ponds, rivers, streams, canals and field ditches. Where relevant sections are not adhered to, EA will not prosecute or take civil proceedings, but the Agency 'will take this into account when issuing discharge prohibition notices under Section 86 of the Act and when exercising powers conferred on them by regulations under Section 92 of the Act'. Neither is following the Code a defence to the offence of polluting controlled waters, but it could be taken into account in deciding on enforcement action, penalties and mitigation. In Nitrate Vulnerable Zones (NVZs) that are designated under UK legislation that enforces the Nitrates Directive, farmers must comply with mandatory measures or rules that may be similar to, or sometimes cases stricter, than basic Good Practice guidance.

In NVZs, specific regulations include nutrient budgeting so that the crop requirement is not exceeded by application of manufactured fertiliser, that should not applied to grass between 15 September and 15 January and to other crops between 1 September and 15 January (unless there is a specific crop requirement). Regulations that nitrogen application applied are (Defra, 2013c):

- An upper limit on the amount of nitrogen from organic manure and manufactured fertiliser that can be applied to the major crop types. These include horticultural crops as from 1 January 2014.
- A limit of 250 kg ha^{-1} of total N from all organic manures that are applied to land in any 12 month period, The organic manure N field limit includes a limit of 170 kg ha^{-1} in any calendar year of the amount of nitrogen in livestock manure that is applied and
- Restrictions on the timing and rates of nitrogen applications.

For example, organic manures should not be applied when:

- The soil is waterlogged, flooded, frozen for more than 12 hours in the previous 24 hours,
- Within 50 m of a spring, well or borehole, or,
- Within 10 m of surface water, except,
- On land managed for breeding wader birds or as species-rich semi-natural grassland and under certain circumstances (stated) or,
- When using precision manure spreading equipment to apply slurry, sewage sludge or anaerobic digestate, in which case manure may be spread 6 m from surface water. Otherwise:
- Applications should be as even as possible on the ground.
- Should not be made on rock or uneven surfaces.
- Appropriate care is furthermore required on sandy or shallow soils.

Comparable regulations apply to manufactured nitrogen fertilisers and the Code also requires not only management to ensure appropriate balances of nitrogen, but also phosphorus, potassium, magnesium, calcium and sulphur and sodium as well as adequate availability of trace elements such as iron, manganese, boron, copper, zinc, molybdenum and chlorine. In farming systems, nitrogen escapes to the air as ammonia (especially from livestock manures) and as nitrogen gas and nitrous oxide through natural soil processes. Care should therefore be taken to manage to minimise loss of ammonia and nitrous oxides. Similar Codes are available for Wales and for Scotland.

However, early responses in terms of reduction in nitrate in at least one Nitrate Sensitive Area (NSA- Chapter 1) over the Jurassic (Great Oolite) Limestone aquifer was not encouraging (Evans *et al.*, 1993). In such a fissured aquifer, it is supposed that responses to land use changes would be faster than, for example, over the micro-porous Chalk aquifer. Monitoring continued. Results for the winter of 1995/1996 were encouraging for land in unfertilised, ungrazed grassland and woodland, but were variable for arable crops (MAFF, 1996). The NSA scheme was expanded after 1994, then subsumed into Nitrate Vulnerable Zones (NVZs). By 2006, N-concentrations in surface waters in England had stabilised, apparently in line with stabilisation and decline in fertiliser inputs. This had been attributed to the implementation of environmental measures including NVZs; and in some areas, concentrations appear to be reducing slightly. For groundwaters around one quarter of sources displayed statistically significant rising trends, so there is no cause for complacency (ADAS, 2007).

The *Nitrate Vulnerable Zone* (NVZ) scheme is designed to meet the EU 'Nitrate Directive' (91/676/EEC), which aimed to reduce water pollution by nitrate from a range of agricultural sources (MAFF/WOAD, 1994). The Directive requires Member States to:

- Designate as NVZs all known areas of land which drain to waters where nitrate concentrations exceed, or are expected to exceed, the EU limit of $50\,mg\,1^{-1}\,NO_3$.
- Establish action programmes which will become compulsory in these zones at a date to be agreed between 1995 and 1999.
- Review the designation of NVZs at least every four years.

Designation concerns itself with both vulnerable surface and groundwaters, and with eutrophic freshwater, estuarine, coastal and marine waters. It takes account of physical and environmental characteristics (such as a reasonable proportion of the theoretical catchment required for recharge), and considers especially the behaviour of nitrogen compounds, and current understanding of the impact of remedial action. Regular monitoring of waters is required. For convenience, the proposed boundaries are 'hard features' such as field boundaries or roads. At the time of writing almost 70% of England and Wales was under NVZ designation (Meat Info, 2008).

In Scotland, four NVZs defined by the Scottish Government in significant agricultural and horticultural areas (Lower Nithsdale, Lothian and the Borders, Strathmore and Fife and Moray, Aberdeenshire, Banff and Buchan). Originally, there was one only a few hundred hectares around a groundwater source at Balmalcolm in Fife, where intensive horticultural land use has raised groundwater nitrate levels above the EU limit. A second, by came to comprise the entire Ythan catchment in Aberdeenshire, just under 70 000 ha. Unusually for Scotland, the Ythan derives on average up to 75% of its flow from groundwater, largely from glacial deposits and the Old Red Sandstone (NERPB, 1994) on account of up-catchment loading from a variety of nutrient sources. There is about 12% of the land area in Scotland is in NVZ designation (Farming Scotland Magazine, 2012), but the total area is continually open to review.

Designation of groundwater NVZ (and former NSA) involved hydrogeological modelling, including accounting for uncertainty by designating a Zone of Confidence and a Zone of Uncertainty and 'Best Estimate Zone' using different modelling scenarios (Burgess and Fletcher, 1988). Designation of NVZs for surface waters is relatively simple when compared with the modelling procedures required for groundwater source protection. It involves monitoring of the nitrate concentration at least monthly over a 12-month period (MAFF/WOAD, 1994). In this way the lower extremity of a polluted water body is determined as the point where an abstraction exceeds the EU limit (in accordance with the procedure set out in the Surface Water Abstraction Directive (75/440/EEC).

The upper point is either the first upstream sampling point which complies with the limit on a 95 percentile basis using five years' data, or the source of the headwaters, or both where all sampling points fail. The vulnerable zone is the land draining into the polluted water and is defined by the boundary of the natural catchment but upstream of the abstraction point. Special arrangements are made for interference with the drainage pattern, and NVZs are to include the earlier, compensated, 22 Nitrate Sensitive Areas designated in 1994 in order to incorporate groundwater protection. This development shows a convergence of policies with the NSA measures being more stringent but voluntary while NVZ measures provided compulsory back-up for those who do not enrol. In designating areas for nitrate vulnerability we may see a clear direction in policy formulation: earlier Nitrate Advisory Area compliance had been entirely voluntary and advisory, and it soon disappeared. NSAs were heralded as a major leap forward because not only was there to be monitoring of soil leaching on nitrate, but this voluntary enrolment scheme was compensated for presumed losses in agricultural production. Finally, with European Commission legislative backing, NVZ definition and compliance is compulsory. NVZ designation is mandatory and uncompensated.

Certain EU states (the Netherlands, Denmark, Belgium and Germany) have applied comparable measures across all of their territories. In Britain, there is an alleged risk of land prices falling in designated zones; a kind of 'environmental planning blight'.

The designation of the Ythan catchment in Scotland to prevent estuarine eutrophication falls within a broader remit than surface water protection. The absence of blanket restrictions may, in any case, continue the pollution of certain potential water sources, threatening the long-term protection of public supply (Watson *et al.*, 1996). Otherwise, NVZ designation in England and Wales has been criticised for focusing too much on water for human consumption (Osborn and Cook, 1997), while the suitability of GAP in preventing pollution has been called into question. Yet more criticism may be levelled at the NVZ scheme on a land management basis. Unlike the former NSA scheme, it lacks options for the conversion of arable land to zero or low inputs of nitrogen fertilisers (no radical changes of land use were initially proposed), and there is an omission of measures to establish 'green covers' over winter or directly prevent the ploughing of pasture. It is also highly desirable that bans on sludge application to arable and grassland at vulnerable times of the year be implemented.

9.4.3 Non-Specific Protection Zones for Groundwaters

The NVZ approach is pollutant specific. While this enables land area targeting on the basis of a scientific assessment of the contaminant in question, groundwater may be contaminated by a wide range of substances – and pathogens. Therefore resource protection is essential to protect groundwaters from wide-ranging damage, for without resource protection, the long-term sustainability of individual sources is not possible, nor is the wider aquifer likely to remain unaffected.

A dual approach to groundwater protection requires both spatial modelling (required for resource protection) and dynamic models (for source protection). For general protection of the aquifer resource and its waters, originally a series of maps are being produced at a scale of 1:100 000, numbering 53 in all and covering all of England and Wales. These maps recognise three kinds of aquifer (major, such as chalk; minor, such as the Hastings Beds sandstones; and non-aquifers of negligible permeability). Soils overlying the aquifer formations are then classified as high, intermediate and low according to their supposed potential for leaching a wide range of pollutants, and these are further ascribed to three subclasses based upon soil depth, soil texture and ability to adsorb certain pollutants (Skinner and Foster, 1995). The effect is a matrix combining soils and groundwaters recognising 'extreme', 'high', 'moderate', 'low' and 'negligible' classes of vulnerability (Table 9.1).

An early vulnerability sheet was produced for East Kent (NRA, 1994m). While such maps are most welcome, limitations due to the complexity of contaminant behaviour, compromises made in the resolution and precision of the mapped data presented and geological uncertainty mean that they can only be of generalised use in planning purposes.

Low-permeability, non-water-bearing drift deposits occurring at the surface and overlying major and minor aquifers, including clay with flints over chalk, are indicated by stippling but omitted from the classification. Provided that they remain undisturbed, certain superficial deposits play a role in limiting pollutant loading by the partial confining of parts of aquifers. The inclusion of such deposits has been a central part in vulnerability classification in other schemes (e.g., Cook, 1991; Zaporozec and Vrba, 1994).

In practice, pollutant-specific land targeting would be very complicated because the designation of multiple pollutant-specific zones would create extreme administrative difficulties, and there is often a lack of detailed technical knowledge. SPZs began to be defined following public consultation in 1992, and are designed to control both potential point and diffuse sources of pollution (NRA, 1992d). Zones are identified with boundaries based upon supposed groundwater 'travel times' to an abstraction point from any point below the watertable in the manner NVZ designation. The main impact of SPZ designation on agri-environmental matters seems to relate to the storage and disposal of farm wastes and agrochemicals other than nitrates that are controlled under NVZ.

Zoning could provide a concrete basis for planning even though, in common with NVZ designation it does not currently include soil type or unsaturated zone variables such as fissuring or depth to watertable. For example, in Zone One (close to water sources) land uses as well as consents to discharge might be severely restricted. Only broadleaved woodland, low intensity grazing, or set-aside might be appropriate. Zone two may have low input farming targeted so that pesticide input is reduced. Only in Zone three might more conventional farming remain, but with agrochemical and organic waste application according to GAP.

9.4.4 The Emergence of 'Environmentally Sensitive Areas'

Water-specific schemes are accompanied by a growing array of policy options aimed at conserving the countryside, protecting the wider environment or reducing agricultural surplus. Many of these will also benefit water quality, even where this is not a stated aim.

Table 9.1 (a) The classification of aquifers and soils used in the UK vulnerability maps. (b) Practical significance of relative classes of aquifer pollution vulnerability.

(a)

Geological classes (unsaturated zone and confining beds)

Major aquifer	Minor aquifer		Non-aquifer
Highly permeable formations usually with known or probable presence of significant fracturing. Highly productive strata of regional water resource importance, often used for large potable abstractions	Fractured or potentially fractured strata but with high intergranular permeability. Generally support only locally important abstractions	Variably porous/permeable but without significant fracturing. Generally support only locally important abstractions	Formations with negligible permeability – only support very minor abstractions, if any

Soil leaching-potential classes (excluded in urban areas)

High H		Soils with little ability to attenuate diffuse source pollutants and in which non-adsorbed diffuse source pollutants and liquid discharges will percolate rapidly
Subdivided into	H1	Soils which readily transmit liquid discharges because they are either shallow or susceptible to rapid by-pass flow directly to rock, gravel or groundwater
	H2	Deep permeable course-textured soils which readily transmit a wide range of pollutants because of their deep drainage and low attenuation potential
	H3	Coarse-textured or moderately shallow soils which readily transmit non-adsorbed pollutants and liquid discharges but which have some ability to attenuate adsorbed pollutants because of their large organic matter or clay content
Intermediate I		Soils which have a moderate ability to attenuate diffuse source pollutants or in which it is possible that some non-adsorbed diffuse source pollutants and liquid discharges would penetrate to the soil layer
Subdivided into	I1	Soils which can possibly transmit a wide range of pollutants
	I2	Soils which can possibly transmit non- or weakly adsorbed pollutants and liquid discharges, but which are unlikely to transmit adsorbed pollutants
Low L		Soils in which pollutants are unlikely to penetrate the soil layer because water movement is largely horizontal or they have the ability to attenuate diffuse pollutants

(b)

Vulnerability class	Definition
Extreme	Vulnerable to most water pollutants with relatively rapid impact in many pollution scenarios
High	Vulnerable to many pollutants, except those highly absorbed or readily transformed, in many pollution scenarios
Moderate	Vulnerable to some pollutants but only when continuously discharged/leached
Low	Only vulnerable to conservative pollutants in long term when continuously and widely discharged/leached
Negligible	Confining beds present with no significant groundwater flow

(*Source:* Skinner and Foster, 1995).

The former Environmentally Sensitive Area (ESA) scheme sought to conserve and re-create habitats by compensating farmers, for example, for not ploughing pasture or for re-establishing grass on former ploughland (MAFF, 1993). The restoration of pasture at low stocking densities and zero fertiliser regimes will reduce the risks of pollution of waters and soil erosion. ESA target areas include chalk and limestone landscapes (such as the South Downs), and waterside landscapes (such as the Somerset Levels and Moors or Norfolk Broadland) including low intensity grazing marshes and riparian areas. Protection of landscape features such as archaeological and historic monuments from agriculture was also an objective.

ESA targeted on specific land areas. Farmers who enrolled were paid to undertake such measures as restoring pasture at low stocking densities or employing zero fertiliser regimes over a period of 10 years. A development of the ESA scheme, the Countryside Stewardship (CS) scheme, offered farmers contracts to manage, restore and re-create natural features in the wider countryside, it is not targeted and enrolment was also for 10 years. An initiative of the Countryside Commission in collaboration with English Nature, MAFF and English Heritage (Countryside Commission, 1994), CS specifically offered contracts to manage, restore and re-create natural features in the wider countryside including landscape features on a 'menu' system designed to match closely prescription to conservation requirements (Potter *et al.*, 1993).

Management of funded agri-environmental schemes passed to MAFF then Defra and more recently Natural England, Scotland Rural Development Programme and in Wales they have been incorporated into a single whole farm scheme called Glastir. Eligible areas include waterside lands and historic landscapes and there is an option for 'creating new wildlife habitats and landscapes'. These permit conservation of wetlands, marshes, riparian land and land around lakes and ponds (Countryside Commission, 1994; Cook, 2010b). An especially interesting development is the possibility of restoring 'historic irrigated watermeadows' through a special annual payout. At the time of writing, Environmental Stewardship (ES) is the current scheme for protection of both habitats and historic landscapes in England (Defra, 2014b). There are three tiers:

- Tier 1 Entry Level Stewardship (ELS) goes beyond the basic effective land management required for Single Farm Payment. About 60% of England's farmland is enrolled and the requirements are to enhance hedgerow management, support low input permanent grassland, appropriate ditch management, management of field corners, the sowing of wild bird seed mixture, buffer strips for water courses on cultivated land and the maintenance of skylark plots.
- Tier 2 Organic Entry Level Stewardship (OELS) provides for both upland and lowland farmers to manage their land organically, including buffer strips on intensive (organic) grasslands.
- Tier 3 Higher Level Stewardship (HLS):

    HLS aims to deliver significant environmental benefits in priority areas. It involves more complex environmental management requiring support and advice from our local advisers, to develop a comprehensive agreement that achieves a wide range of environmental benefits over a longer period of time. HLS agreements are for 10 years.

    The Single Farm Payment (see Tier 1, that requires enrolment in ES seeks to exceed) is the main agricultural subsidy scheme in the EU (Gov.uk website). It was replaced by the Basic Payment Scheme in 2015. The aim here is to achieve Good Agricultural and Environmental Conditions (GAEC) through 'cross compliance' that simply requires meeting legal requirements for public, animal and plant health, the environment and animal welfare, under the Common Agricultural Policy ('CAP'). Breaking older sub-sidy payments between production and economic support in favour of environmental and welfare considerations, deleterious impacts on soils such as erosion, loss of organic matter, compaction, salinization, landslides, contamination and soil covering are to be avoided. In support of this there is 'Catchment Sensitive Farming'). Other measures include the emplacement of buffer strips and maintenance of biodiversity and landscape features and special mention is made of pig and poultry operations in respect of soil degradation.

Tier 2 presumes that organic agriculture is less degrading than other forms, and this is largely the case although care must be taking in nitrogen management (Chapter 8). Tier 3 has far more wide implications for landscape, water and biodiversity.

An example of HLS enrolment is the Harnham Water Meadows on the river Nadder in Salisbury, England. The Trust manages some 34 ha of historic watermeadow landscape located close to Salisbury Cathedral between the rivers Nadder and Avon (Cook *et al.*, 2008; Cook and Inman, 2012). Dependent in part on volunteer labour and in part on paid professional services, it raises funds through private donations, a membership scheme and is in receipt of UK Government agri-environment financial support. As from August 2008, HLS enrolment (funded by Defra) for 10 years helps with improvement to infrastructure (hatches and channels, etc.), pollarding trees, boundary restoration and ditching as well as modifying grazing management in order to improve the sward diversity on an SSSI within the area of watermeadow (HWMT website). Here, benefits here are multiple, and historic watermeadow operation has been demonstrated to improve water quality (Cook *et al.*, 2015).

9.4.5 Others

Sites of Special Scientific Interest were established under the National Parks and Access to the Countryside Act (1949) that established the statutory Nature Conservancy (later 'Council' was added) and was the forerunner of Natural England. The definition of public National Nature Reserves and SSSIs on private land was possible. There are around 4100 SSSIs in England designated on either geological or biological criteria, most were re-notified under the Wildlife and Countryside Act 1981 Natural England and Defra, 2013; Natural Resources Wales, 2014). Special Areas of Conservation (SAC) are 'strictly protected sites designated under the EC Habitats Directive'. Article 3 requires the establishment of a European network of important high-quality conservation sites. The listed habitat types and species are those considered to be most in need of conservation at a European level, excluding birds (JNCC, 2013).

Whole farm planning is promoted by the Farming and Wildlife Advisory Group –'FWAG' (n.d.) who can assist in planning for habitat, soil, water, manure, farm waste and fertiliser management. Linking environment and Farming –LEAF (n.d.) :

'A whole farm policy aiming to provide efficient and profitable production which is economically viable and environmentally responsible. It integrates beneficial natural processes into modern farming practices using the most appropriate technology, and aims to minimise the environmental risks while conserving, enhancing and recreating that which is of environmental importance.' produce Whole Farm Conservation Plans'

The Farm Woodland Scheme (FWS) ran between 1988 and 1992, it encouraged the planting of broadleaved woods as an alternative crops and grants for planting were available from the Forestry Commission. It was replaced by the current Farm Woodland Premium Scheme in 1992. Enrolment is for 40 years (and grants for planting have been available). It applies in England, Scotland and Wales.

England's Community Forests aim at extensive tree planting adjacent to urban areas, especially on derelict land, with the primary objective being amenity. Dating from 1990 the programme was commenced by the Countryside Commission, now subsumed in Natural England and equivalent organisations in Scotland and Wales (Community Forests, 2005). Benefits to soil and water conservation are likely to be real, but as a by-product of diversion of land from arable. At the time of writing, achievements include over 10 000 ha of new woodland have been planted with more than 27 000 ha of exiting woodland brought under management around 12 000 ha of other habitats have been created or enhanced and 1200 km of hedgerows have been planted or restored. Community Forests have moved towards greater financial independence from national funding bodies.

In the voluntary sector, the Community Forest Trust supports community forestry in England. It works with Community Forests and other community forestry initiatives 'to create healthier and more attractive places in which to live and work'. It is believed that the new forests are transforming local areas to involve people, improving the environment, creating business opportunity and provide spaces for people and wildlife (Community Forest Trust, n.d.).

Sites of Special Scientific Interest (SSSIs) are specific measures aimed at conservation of existing habitats, and notification has been under the Wildlife and Countryside Act 1981. They have been described as efforts at 'Fortress conservation'. SSSI designation may contain elements of soil and water conservation, depending upon the site. Their integration into wider designations such as Catchment Management Plans and WFCPs is now of paramount importance, as is their relationship with wider conservation schemes such as CS and ESA.

Set-aside schemes seek to reduce production and, as such, are not geographically targeted. Farmers have been paid an annual amount per unit area of land and the CAP price reforms in May 1992 led to a certain amount of all arable land being enrolled in compulsory set-aside. What emerged was 15% compulsory rotational set-aside, or 18% non-rotational (i.e., permanent within the period of enrolment), subsequently reduced. From 2008, compulsory set-aside has been set at zero. Any environmental benefits of set-aside are accidental. As a policy, set-aside has always been controversial and the debate continues over such issues as whether former (un-ploughed) set-aside upon which there has been habitat creation might be incorporated in ES (UK Agriculture, n.d.).

9.5 Vegetated Buffer Strips

Reducing agrochemical use, changing tillage practices and timing of cultivation to reduce the impact on waters are clearly the most desirable means of environmental protection from contamination and loading from detached soil particulate material. Targeting measures involving the imposition of a buffer zone (e.g., at arable margins) or in wetland areas, in principle as with zoning for groundwater protection, is an alternative measure employed to separate a watercourse (or vulnerable habitat) from agricultural or forestry activities that may generate water quality problems. Where agriculture is intensifying involving the increase of application of livestock waste as well as application of N and P compounds and where land drainage itself has been intensified, buffer zones may have the effect of buffering or reducing the flux of nutrients to the wider environment (Martin *et al.*, 2009).

Vegetated buffer zones and wetlands thus provide a promising partial solution to the protection of vulnerable watercourses from intensive land use or periodic damaging operations such as forestry establishment. The idea originates in the United States where the Conservation Reserve Program targets land alongside rivers for diversion out of crops (Potter *et al.*, 1991), including the maintenance of riparian forest, a relatively rare landscape feature in Britain, yet is something being promoted by statutory agencies, most notably in Scotland (Scottish Natural Heritage, 2006). The National Trust is promoting the use of buffer strips along lake margins in the English Lake District (National Trust, n.d.).

The Forestry Commission has published general guidelines:

The riparian buffer should reflect stream size and the natural dimensions of the riparian zone. Minimum widths for either side of the stream channel are:

- 5 m for streams <1 m wide,
- 10 m for streams 1 – 2 m wide,
- 20 m for streams >2 m wide.

Where the natural riparian zone exceeds these widths, the dimensions of the buffer area should be increased, up to twice the minimum recommended width.

Greater widths should be considered where there is scope to restore native floodplain woodland. However, buffer widths greater than 20 m on either side of a watercourse are unlikely to result in further significant benefit to the aquatic zone (Forest Research, n.d.b).

Such guidelines recognise the restoration or conservations of (semi-natural) vegetation. However, soils and hydrogeology are also all important. The establishment of buffer zones in shallow watertable areas is one possible solution to control non-point pollution of surface waters. Strips of varying width, designed to operate as filters to reduce pollution of watercourses from forestry and agriculture, are today commonplace. The accumulating body of information is most encouraging and the literature abundant (e.g., Muscutt *et al.*, 1993; White *et al.*, 1998; Martin *et al.*, 2009, Chapter 19), but the full complexity of the subject needs to be explored.

Variables related to the operation, and hence, efficacy, of buffer zones may be summarised as follows:

1) There is a partition of precipitation at the soil surface between surface runoff and subsurface drainage arising from infiltration into the soil. This partition, in accordance with classic hydrological theory, is a function of variables such as soil porosity, antecedent soil-water conditions, ground cover and intensity and duration of rainfall, all of which will affect the infiltration rate and capacity of the soil. The route by which runoff, and hence pollutants, reach a surface water body is critical. Surface water runoff may carry with it particulate material which may become deposited (coarse fraction first) because it is trapped by the vegetation.

2) Buffer zones may be relatively ineffectual at attenuating dissolved pollutants. On the other hand, subsurface flow will carry little or no particulate material. Shallow

groundwaters provide an environment where chemical reactions that inhibit the passage of pollutants may occur. One such is the microbial denitrification under low redox potentials.

3) Soil type will have an important influence, especially in terms of drainage (largely a function of structure and texture), which will affect leaching rates and hence loading of potential pollutants in groundwaters. Composition, especially clay mineral content and active organic matter fractions, will have an impact upon complexes between soil particles and contaminants and breakdown of agrochemicals. Soil pH is another factor of importance, especially because it influences the oxidation-reduction status of soil water. The relationship between immobilisation and mineralisation in soil is the key here to considerations of pollutant attenuation and release to waters.

4) Otherwise hydrogeological conditions will affect the transfer of nutrients via the saturated zone to the watercourse in question. The presence of 'flow horizons' such as coarser horizons in layered soils and sediments, or fissures in rocks or clays, height of watertable, precipitation and seasonality factors will all affect the loading and transmission rates of potential pollutants. Deep percolation of waters may 'bypass' a buffer strip. This could make near-surface measures of limited value. Artificial underdrainage is another factor which can render a buffer strip ineffective. A knowledge of site hydrology is therefore essential prior to the establishment of strips.

5) Rates and timing of pollutant assimilation by vegetation or microorganisms, complexing in soils and rocks, and breakdown of potential pollutants by chemical or biological means (such as denitrification) will affect the tendency of waters entering a buffer zone to affect an adjacent watercourse. Vegetation will uptake nitrate and phosphate; hence vegetation cropping will have the effect of a net loss of biomass which may have assimilated nutrients. Conversely, an unmanaged strip may tend to recycle nutrients and organic matter, becoming less effective as a sink over time. It also follows that timing and rate of fertiliser or pesticide application in a catchment is critical and in general should avoid periods when the probability of leaching is high.

6) If buffers are operational in attenuating (or even mobilising) potential pollutants, then clearly their width will have an effect. The reason is that the 'contact time' spent in the soil/ground water is critical in terms of the time available for the chemical and biological processes to operate. This appears to be fairly critical if a specified result is desired, and should take account of ability to assimilate pollutants at critical times such as periods of enhanced loading following (for example) ploughing or heavy rainfall.

7) While it is generally assumed that emplacement immediately adjacent to watercourses is the best location for a buffer, this may not be appropriate universally. For example, where it is deemed important to protect the entire riparian zone from the adjacent arable land of the valley sides, it may be better to locate the buffer on or adjacent to the junction with the alluvium at the foot of a footslope. In detail, the positioning within a catchment (specifically in headwater areas, middle catchment footslope or floodplain, or lower catchment floodplain) will affect the operation of a wetland buffer zone or buffer strip (Martin *et al.*, 2009).

8) The efficacy of the buffer will also depend on the vegetation. Riparian forest is employed in parts of the United States, while agrochemical-free grass strips have been favoured in Britain. Chemical-free crop buffers also remain an option. Where possible, vegetation management should involve cutting or mowing. Nutrients immobilised in plant tissue are thereby removed from the site.

There are other aspects of buffer strip or riparian vegetation emplacement to consider, that are all arguably beneficial. Acreage reduction, provided that there is no corresponding intensification on the adjacent land, will lead to reduction in land area under the plough or in intensive grazing. This will have the twin benefits of reduction in production (where this is applicable) and reducing gross agrochemical use, including reduction of pesticide spray drifting into waters by keeping machinery away from the water's edge. The option of creating habitats and corridors for wildlife and colonisation of wetland plant species is also attractive. In an intensively farmed landscape, the creation of improved access for agricultural operations, watercourse maintenance and recreational purposes may prove practically useful.

Specific mechanisms of contaminant stripping are not always well understood.

However, many studies now show their benefits, even if the tendency of researchers in the early stages of enquiry has been to treat a buffer as a 'black box' in which substances mysteriously vanish as water flows across or through a buffer zone. As may be seen from the above, it is largely wetland soil functionality that not only potentially removes excess nutrients and agrochemical contamination, but may also sequester carbon, provide aquatic habitats and assist in general water retention during periods of high discharge (Cook *et al.*, 2009). Increasingly fenced buffer strips are being employed to protect banks from erosion by grazing animals and prevent animal waste entering a watercourse (Cook *et al.*, 2014).

Some classic work will first be considered, because it begins to elucidate the mechanisms by which pollutant attenuation occurs.

Correll and Weller (1989) present a simple conceptual model of the below-ground processes which dominate the nitrogen dynamics of a riparian forest wetland beside Chesapeake Bay, Maryland, United States, shown in Figure 9.6. Readers may also like to compare this account with the description of processes in Chapter 7.

In this agricultural catchment, continuous disturbance of uplands adjacent to the forest by cultivation fosters high rates of nitrification throughout the year with an average concentration of around $27\,mg\,l^{-1}$ NO_3 in shallow groundwaters which move towards the forest. High concentrations of hydronium ions are also released by the denitrification, lowering the pH of waters entering the wetland to between 3.8 and 4.5, but this rises again to an average of 5.5 upon leaving the forest. The normally wet soils

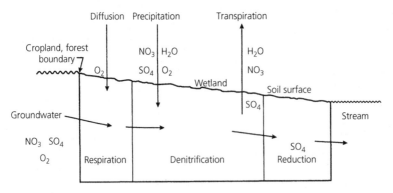

Figure 9.6 A conceptual model of below ground processes beneath a riparian forest. (*Source*: Re-drawn from Correll and Weller, 1989, Figure 1).

reduce nitrate concentrations to less than $4\,mg\,l^{-1}$ NO_3 by processes of denitrification and assimilation of nitrate by trees. Peaks in groundwater baseflow were correlated with peaks in nitrate over several years.

Mean dissolved oxygen measurements also declined within the first 20 m from the forest. Averages of 86 percent of NO_3^- and 25 percent of SO_4^+ inputs were removed in the wetland. There is a 'respiration zone' at the point where groundwater enters the forest from the cropland. Respiration dominates, due to the dissolved molecular oxygen in waters entering this part. The second zone is dominated by denitrification and transpiration. Here the hydronium ion is the electron donor (from organic matter) and the electron acceptor is nitrate. High transpiration by the trees may also cause high uptake of nitrate. In the third zone, feedback mechanisms such as pH rise occur, caused by the reduction of sulphate, and low NO_3 concentrations slow the rate of denitrification. The widths of these zones vary seasonally, and the nitrate concentration in the stream increases with high baseflow rates because groundwater flow rates and soil 'contact time' are inversely related.

At another site, Jordan *et al.* (1993) again found strong attenuation of nitrate beneath a riparian forest. They note that oxidation of sulphides as particulate iron sulphide may have occurred, and suggest that the process can drive denitrification. This study noted an increase in dissolved phosphate near to the stream bank, which may have been a result of the reduction of iron oxyhydroxides that bind phosphate to soil particles, not an encouraging discovery given the role of phosphorus in eutrophication of waters. They also report the discovery by several workers that riparian forests trap sediment and particulate nutrients in overland flow.

Muscutt *et al.* (1993) reviewed the potential of buffer zones in UK agriculture, examining some previous work on the three pollutants nitrate, phosphorus and pesticides. It is apparent that the attenuation of phosphorus (as sediment P, sediment PO_4-P, soluble P, soluble PO_4-P and total P) measured in studies of buffer strips between 1.5 and 36 m wide and on slopes between 2 and 16° varies considerably. This echoes the findings of Jordan *et al.* (1993) for dissolved phosphate. Although in some instances attenuation was above 90%, with the most encouraging results for total phosphate (strips narrower than 5 m can be effective in trapping coarser sediment and phosphorus moves in both organic and inorganic sediments), the results for soluble and extractable forms were variable. Processes such as reduced infiltration, satisfaction of vegetation requirements, loss of sorption sites, re-erosion/transport of sediment and mobilising of soluble forms are blamed for low attenuation results. As a general principle, plot studies have shown that the wider the strip, the better the attenuation.

The fate of P and sediment in soils and channel sediment across historic floated watermeadows of Wessex has been investigated by Cook *et al.* (2004, 2015) and Cutting *et al.* (2003). Whereas soil texture changes point to a tendency to trap all but the finest silt and clay size particles, plant-available P in topsoil was observed to reduce with passage across these ancient meadow irrigation systems. Indeed, the potential for the attenuation of nutrients, sediment and carbon sequestration for a range of freshwater riparian environments, be they 'natural' (primary marsh, peatland, bog or riparian woodland), semi-natural (reedbeds, secondary grazed peatland or marsh and grazed flooplain) or constructed (tertiary drained marshland, peat moor and floated watermeadow) has been reviewed by Cook (2010b) with a view to their wider inclusion and restoration in agri-environmental schemes.

Pesticides have been reported to be reduced by grass channels, because many are carried with the sediment load, not unlike that reported for phosphorus. The complexity of these chemicals, their variable behaviour in respect of degradation (chemical and microbial), complexing on soil particles and environmental mobility mean that little detail is appreciated. Griffiths *et al.* (1995) investigated the efficacy of 10 m wide, grass strips set at the bottom of a slope under permanent pasture in south Devon. The choice of a 10 m untreated strip derives from the Codes of Good Agricultural Practice (GAP) Defra (2009c), which relates to the application of slurry and manures on land adjacent to watercourses. The experimental plots were treated with inorganic fertiliser, cattle slurry and manure upslope of the buffer zone, the surface runoff and subsurface flow being monitored for species of nitrogen and phosphorus fractions. In summary, surface runoff was found to transport considerable quantities of nitrogen and phosphorus. The former was predominantly in organic forms and ammonium from slurry and manure, or nitrate from organic fertiliser; the latter predominates as phosphate, or organic phosphate from the manure treatment. Residues of slurry and manure remained on the surface longer than inorganic fertilisers, which rapidly dissolve.

Although total surface transport depends upon amount and intensity of rainfall, the role of the 10 m strip in reducing surface transport of nitrogen and phosphorus was demonstrated. It must not be assumed that merely transferring measures of nitrate mitigation to phosphates will be successful. Withers and Sharpley (1995) found that sustainable phosphorus management not only restricts overall use of inorganic fertiliser and manure, but importantly maximises its availability in plant-available forms and minimises losses from the soil system, especially through erosion. The key is in preventing accumulation in topsoil, because phosphorus in soils will build up in the long term (Marrs and Gough, 1989), while considering the chemical and physical properties of the soil and site hydrology.

Because over-application is common in Britain, manure and slurry applications should be spread as wide as is possible. It may be that strategies to limit phosphorus application are more important than limiting nitrogen. These might incorporate the application of fertiliser in reduced amounts, identification of erosion-vulnerable areas, and also emplacement of buffer strips. Employment of zero-tillage systems, winter cover crops and grass emplacement in valley floors are all options, and are all the kinds of solutions offered for nitrate management (Morgan, 1992; Muscutt *et al.*, 1993).

Surface crop residue accumulation should be avoided because the interaction of this and inorganic phosphorus fertilisers provides a source of phosphorus to runoff which is problematic. Liquid slurry applications in cracked soils should also be restricted. There is no evidence for a process analogous with denitrification for phosphorus, so unless soluble phosphorus forms new complexes with soil and subsoil materials it is potentially mobile, especially in low pH situations. Results of field trials are variable and inconsistent, and long-term phosphorus accumulation in buffers may cause increases in soluble phosphorus issuing from a buffer zone (Dillaha *et al.*, 1987). P in dissolved 'orthophosphate' form has been found mobile in watermeadow systems during irrigation, whereas the grass traps a proportion of particulate bound P (Cook *et al.*, 2015).

Of all the diffuse pollutants, nitrates are probably the best understood, and arise from the application of inorganic nitrogen fertilisers, organic fertilisers including manures and slurries, and the breakdown of soil organic matter and crop residues from ploughing.

Muscutt *et al.* (1993) report a number of studies from the United States, southern France, northern Germany, New Zealand and southern England which all show dramatic reductions of between 86 and 100% in groundwater concentrations entering and leaving the buffer zones. Vegetation varied across the studies, but woodland dominated and widths varied from 19 to 150 m while sampling times were throughout the year. Most notable is that, even at maximum pollutant loadings (giving concentrations around 90 mg l^{-1} NO$_3$), the strips were effective. There are probably three mechanisms of nitrate attenuation. When vegetation is actively growing, there is take-up, otherwise denitrification and assimilation in the microbial biomass are important.

The carbon substrate for denitrification comes from soil organic matter, plant decomposition and root exudates. In a study on a small grassed floodplain in the Cotswold Hills, where Great Oolite limestone is overlain by calcareous clay loam rendzinas with rapid subsurface flow, Haycock and Burt (1993) showed that there was rapid reduction in winter nitrate concentration over about 8 m when groundwater was moving towards the stream. They argue on circumstantial evidence that there was unlikely to be an extensive mixing of high and low nitrate concentration groundwaters to produce the observed reduction in concentration. The most probable explanation is loss due to denitrification and microbial assimilation of nitrate in the floodplain soil. However, deep percolation into aquifers carries with it the risk of nitrate transport over considerable distances where there is no attenuation.

A refinement which caters with outfalls from tile drains is to construct a 'horseshoe wetland' (Petersen *et al.*, 1992). These are vegetated, semi-circular excavations within a buffer strip with dimensions around 10 m x 8 m, which may prove effective in reducing nitrate loading of waters which pass under the strip.

Conversion to arable on the Romney and Walland Marshes in Kent and East Sussex (Cook, 2010a), and on the East Anglian marshes has been considerable. While ploughing destroys any floristic diversity or animal habitats associated with low intensity grassland systems, there is frequently a relict ecosystem associated with the dykes which is prone to eutrophication following conversion to arable. Buffer strip emplacement parallel to watercourses following conversion would seem appropriate. To this end, 5 m wide unfertilised grass strips were established on the Walland Marsh SSSI in the mid-1980s. English Nature (now Natural England) established management agreements with local farmers which compensate for the loss of land and reduction of yield through reduced fertiliser inputs on both grazed and arable land. From the Walland Marsh, Moorby and Cook (1992, 1993) report substantial reductions in nitrate concentrations from dipwells sunk to a depth of 2.5 m in reclaimed land where watertables were typically maintained at between 1 and 1.2 m throughout the winter. Subsoils were predominantly silt loams and silty clay loams with occasional coarser layers acting as preferential 'flow horizons'. White *et al.* (1996) report similar findings for coarser, stratified predominantly sandy loam subsoils which would display more rapid transport times.

Figure 9.7 shows dipwell nitrate-N concentrations (between October 1993 and July 1994) for arable sites with and without a permanent, 5 m wide grass strip.

During the winter groundwater elevations in the dipwells indicated dominant movement from the fields to the dyke enabling the efficacy of the strips in attenuating nitrate to be evaluated. Concentrations in the winter wheat crop in dipwells 10 m from the dyke edge are of the same order, suggesting comparable loading of the saturated zone with nitrate. Mean concentrations in dipwells at the edge of the field with the grass strip are

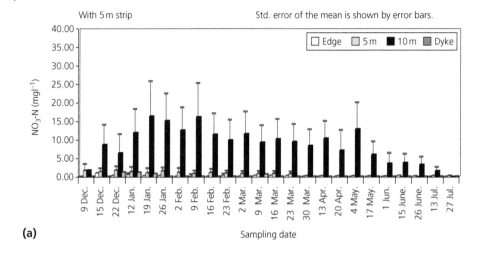

(a)

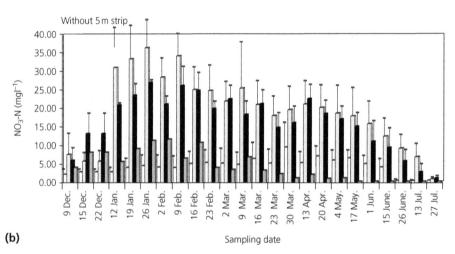

(b)

Figure 9.7 Reduction in groundwater nitrate concentrations beneath an agrochemical grass strip on Walland Marsh, East Sussex. Error bars show the standard error of the mean. (*Source:* Re-drawn from White *et al.*, 1996).

negligible, suggesting that the strips were effective in removing this pollutant as it moved in the shallow groundwater to the dyke. Those located where cultivation was close to the edge showed considerably higher concentrations than where the strip was present (the white bars). These results indicate that 5 m strips may be extremely effective and the likely mechanisms are anaerobic denitrification, immobilisation in the soil and vegetation uptake. With appropriate political will, and sufficient resources, buffer strip emplacement should become commonplace to protect watercourses from nitrate pollution, and indeed bring other benefits to the field margin including conservation and access.

The utilisation of wetland process alongside the other impacts of creating buffer zones for wildlife or for constraining the application of agrochemicals is now central to recommendation by statutory agencies. For example, in 'Making Space for Water', the call was for the restoration of riparian environments (Defra, 2005). Across ordinary arable and intensively grazed land, buffer strips are a mainstay of land use policy (HGCA, 2005). The preferred minimum width is 6 m for arable margins and riparian zones (Defra, 2011b).

9.6 Water Treatment and Blending

It is not the purpose here to describe the process of drinking water treatment at source. In Britain filtration is now universal (the last unfiltered supply came from Loch Lomond in Scotland) and all water into public supply will undergo an anti-pathogen treatment. Raw water from surface and ground is pumped, screened, flocculated, sedimented, filtered (through a sand bed), chlorinated to kill bacteria, de-chlorinated then put into supply (Anglian Water, n.d.). Certain sources, such as chalk groundwater may only require a moderate chlorination in order to meet EU drinking water standards. Other processes include pH balancing (generally increasing pH from acidic sources) (United Utilities, 2015), removal of pesticides and their residues using Granular Activated Carbon technology and also ion exchange systems employing resins to remove nitrates in treatment (Lenntech, n.d.). One major problem with advanced systems of water treatment is cost. The philosopy remains both in UK and also North America that source protection is not only cost-effective, but wider environmental benefits will accrue.

For groundwater, a few options are open to water undertakings in order to prevent, mitigate or remove contamination problems. The classic approach that of 'pump and treat' involves extraction of groundwater, treat it and if possible remove the source of the problem. Clean water may, subject to consent, be returned to the aquifer or discharged to surface water bodies (Hiscock, 2005, p. 250). Other possible options are to only selectively abstract groundwater, scavange contaminated ground water, treat the water in some way or blend high and low contaminant sources.

Closing down boreholes which have high or rising concentrations in favour of low nitrate sources is only an option where water can be diverted from elsewhere. In eastern England, where water budgets can be locally tight, this option can only be of limited use. Alternatively, where the nitrate concentrations arc higher in the surface layers of certain unconfined aquifers, especially in the chalk, the exploitation of deeper groundwater may be an option (Foster *et al.*, 1986).

Figure 9.8 shows the groundwater flow and quality conditions for part of the chalk aquifer in Norfolk. Here the distribution of nitrate (together with oxygen and tritium) in confined and unconfined portions is shown. Screening of the upper parts of boreholes in the unconfined aquifer may be appropriate, as may be increased pumping of the confined (and hence low nitrate) sources where resources permit. Alternatively, scavenge pumping removes high concentration water. Purpose drilled boreholes would be pumped in order to remove shallower, higher nitrate concentration groundwater. Water pumped in this way could be discharged to waste, or recycled through unfertilised grassland plots which would take up the nitrate.

There have been experiments with the artificial denitrification of groundwaters (Hiscock *et al.*, 1991) that utilise 'natural' processes of denitrification. There are two main kinds of treatments that involve stimulating artificial denitrification by the

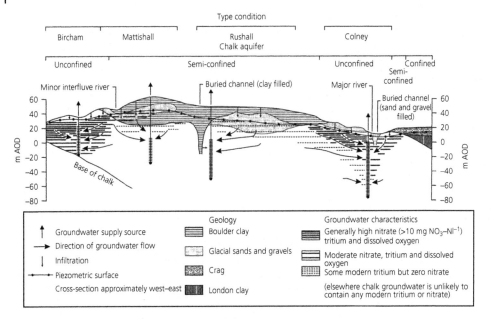

Figure 9.8 Groundwater flow and quality conditions for part of the Chalk aquifer in Norfolk. (*Source:* Redrawn from Foster *et al.*, 1986, p. 84).

injection of required nutrients. The first group are above-ground reactor units which perform denitrification followed by conventional secondary water treatment to remove excess biomass carry-over and for re-aeration. The second group are underground in situ techniques that use the aquifer to support denitrification and also filter and re-aeration of water prior to abstraction. A third option combines the first two, whereby waters are denitrified above the ground and recirculated underground. Interest in groundwater *in situ* remediation continues (EA, 2005).

Above-ground methods include passing waters through media such as gravel, floc blankets and fluidised sand beds. Bacteria coat the grains while such systems employ substances such as methanol, ethanol or acetic acid as carbon sources. Ion exchange systems require water to be passed through an anion exchanging fixed resin bed, where the nitrate is replaced with bicarbonate or chloride. Regeneration of the resin produces high nitrate, sulphate and chloride or bicarbonate concentrations and underground denitrification provides a simultaneous secondary treatment. Carbon sources (such as methanol or sucrose) are injected via a recharge well, and abstraction is via a second well.

Alternatively there are several injection wells located concentrically around the abstraction well. Using the aquifer for further treatment leaves the possibility for further denitrification. Underground systems can involve the complete interception of contaminated water in a contaminated groundwater plume, or the process may be facilitated by barriers that constrain flow enabling water as 'permeable reactive barriers' that contain the reactor for de-contamination to be treated before re-entering the groundwater store (Hiscock, 2005, p. 255).

Where waters of high and low contaminants from different sources are available, blending is an option although it may not be a permanent solution in situations where concentrations are rising, because of the requirement for increasing volumes of low nitrate concentration waters. Such an approach has been used in supplies from the Isle of Thanet in Kent, where concentrations from certain boreholes exceed the EU limit due to both leaky sewers and agriculture, water from outside the area is blended (Lapworth *et al.*, 2005). Concern has also been noted about continuing pesticide levels in the Thanet chalk and, wider afield in Britain, around other contaminants in ground-water include (Lapworth *et al.*, 2011). Those listed are nanomaterials from sunscreen, pharmaceuticals human and veterinary, illicit use, insect repellents, industrial additives and by-products—dioxane, phthalates, bisphenols, Methyl tert-butyl ether (MTBE), dioxins, musks, food additives, wastewater treatment by-products, flame retardants, surfactants, alkyl phenols, hormones and sterols—estradiol, cholesterol, Ionic liquids, nicotine and caffeine.

Phosphate treatment is an option for phosphate pollution control other than buffer strip emplacement. There are more uncertainties in this technique than is the case for nitrate attenuation. Effective sewage treatment, low phosphate detergents and stripping all play a role; alternatives to buffers to control eutrophication are even more important than for nitrate pollution. In Chapter 8, the phosphorus stripping in rivers feeding the Broadland ecosystems was described, and the results found were favourable. Precipitants other than ferric sulphate are available; dosing with certain aluminium salts and ferrous ammonium sulphate have been proved workable elsewhere (Moss, 1988, p. 191-192). Iron salts are generally preferred on account of the toxicity of aluminium compounds which cannot be subsequently used as farm fertiliser.

Thames Water plc has built a phosphate removal plant at Slough sewage works in Berkshire. It recovers phosphorus suitable for agricultural use from a £2 m nutrient-recovery reactor at a time of rising world costs of producing fertiliser P from other sources. Claimed as the first in Europe, it employs technology from the Canadian firm Ostara 'the UK's first eco-friendly, ultra-pure fertiliser phosphorous-based fertiliser' (Water Briefing, 2013).

Although generally a soft-water problem due to low pH values, plumbosolvency or lead solution in waters is similarly treated. In hard waters (such as occur in the Anglian EA region), in the presence of certain organic compounds, lead can be stripped by the addition of orthophosphates to produce a lead phosphate precipitate (Calling *et al.*, 1992). The WHO places a limit of $10 \mu g l^{-1}$ on lead in drinking waters, while the higher EC limit has been reduced to the same is under review (DWI, 2013). Young children and expectant mothers are considered most vulnerable.

9.7 Summary

Land-use measures and technical responses have been reviewed, mostly in the context of controlling diffuse contamination of waters. Point sources of pollution are, at least in theory, covered by the consenting and monitoring systems. Agri-environmental measures may be legally driven by EU Directives, but there is little denying that implementation is assisted economically by the need to balance essential agricultural production with conservation, anti-pollution and flooding imperatives. Modern concerns for water quality embrace both human and environmental health issues.

Over time a range of land-use measures and controls on discharges to waters (implemented through schemes such as NVZ, drinking water SgZs and ground water SPZs) have been adopted and the period of policy evaluation and review is upon us. There is a considerable technical background to diffuse pollution, scientific research continues and yet the challenge is to make things happen both 'on' (and quite literally 'in' the ground). Measures including not only land use, but non-agricultural sources and artificial and natural denitrification within aquifer systems are involved. At the time of writing, the picture for groundwater nitrate, pesticides and other contaminants remain uncertain even if surface water contamination overall is showing signs of improvement.

Parallel research into pesticides, their degradation and transport is also identified as a research need, although we may now be clear which licenced substances are presenting problems. Phosphorus, while not really being a driver of bespoke surface water targeting, has very much entered into the mindset of policymakers and due to its role in eutrophication shared with nitrate, features increasingly in both agri-environmental research and also in policy prescriptions such as Codes of Good Agricultural Practice. Numerical modelling of surface and groundwater behaviour has reached maturity and provides a methodological basis for pollutant specific designations (such as NVZ) and also the more generic groundwater protection sought by SPZs. Indeed the number of available models in hydrology is truly daunting.

Increasing the broadleaved forest and woodland area has been achieved at the expense of some arable and also derelict land. This provides both market and non-market opportunities for improving livelihoods, and while planting has been opportunistic and largely at the whim of farmers and planners, there are bound to be positive knock-on effects for water quality due to the cessation of previous industrial or agricultural activity. Unlike coniferous forestry, impacts on water yield to rivers, reservoirs and groundwater causes no concerns while there would be positive benefits in flood mitigation.

The last 30 years has also seen the widespread adoption of buffer zones including wetland restorations in agri-environmental policy. This is proving successful in not only reducing diffuse pollution but also pathogen and sediment loading of watercourses. While such emplacement may be very successful in solving local water quality problems, it should not be forgotten that Good Practice should be abroad in the wider countryside. To that end, ES, that contemporary doyen of agri-environmental schemes, demands not only a basic 'cross compliance', but also organic farming conversion and the more complicated Higher Level Stewardship for ecologically or historically valued landscapes. Older designations, notably SSSIs are often nested within later enrolled schemes.

10

Framing Water Policies

Emerging Governance Arrangements

> *Let not even a drop of rain water go to the sea without benefiting man.*
> Parakrama Bahu the Great, King of Ceylon, 1153–1186 AD

10.1   Accentuate the Normative

Arguably, we have enough technical knowledge ultimately to sort out many water resource problems. For one thing, legal and administrative arrangements should be sufficient to prevent point source pollution so that industrial processes are capable of being regulated, especially if 'end of pipe' measures are called for. We also know a fair amount about how water, sediment, carbon, nutrients, pesticides (and more) behave in the environment. We have a potential to adjust land use and write management prescriptions for pollution prevention, in parallel maintain food production and seek to enhance biodiversity. With appropriate investment in water purification plants, concentrations of problematic substances can be reduced - or effectively removed.

The modern geological concept of the 'Anthropocene' (whereby humans have actually changed the physical and chemical nature of the planet) reminds us of the truism that there would be few 'environmental problems' without humans. For catchments, a balance is sought that does not compromise the very 'provisioning' ecosystem services society relies upon and there are issues around fair distribution of resources – that is environmental justice. We may talk of 'win-win situations', whereby there is an outcome that leads to certain rewards for all, so we enter the political (with a small 'p') arena. The objective is generally 'to achieve behavioural change' so that technical measures may be enacted and the fruits of inter-disciplinary research recognised. This is a normative argument in which we *think* we know what the Common Good should be.

If a post-'Green Revolution' in agriculture has led to increasing food production, and more and more people have been fed globally, that is a good thing. This is despite the warnings of 'dismal scientists' (economists) anticipated by the Revd. Thomas Malthus FRS around two centuries ago. He 'theorised' that production of useable resources (i.e., food) would not be able to keep pace with population growth, without technological advances. Malthus famously advocated restraint in producing children, yet critics find that he overlooked the impact of technology, despite writing at a time of industrialisation (Abramitzky and Braggion, n.d.).

The Protection and Conservation of Water Resources, Second Edition. Hadrian F. Cook.
© 2017 John Wiley & Sons Ltd. Published 2017 by John Wiley & Sons Ltd.

Malthus's position remains a point of debate. For example, the American economist Julian Simon argued that increasing wealth and technological advances improve resource availability. In support of this view would be the rise of re-cycling to counter the decline in 'finite resource' availability and with it comes new markets (Regis, n.d.). Although Simon was not a natural scientist and his detractors would place him in much the same camp as climate change deniers, the argument that economic development could progress even if humanity took a dramatic turn towards renewable and sustainable technologies contains some truth. Yet common sense that tells us there is only one Earth, and that an ever-populous humanity continues to consume, pollute and reduce the delivery of environmental services. Neither is it true that our knowledge-base, combined with a collective will, enables instant 'technical fixes' that ameliorate resource depletion, pollution or habitat destruction. Finding economic and political ways and means are difficult. This conundrum is what makes such problems 'wicked (Chapter 8.1).

10.2   Pure Wickedness?

Before we can address questions about objectives in catchment management we must first address the matter of *how things might be*. This context is invoked because of the difficulties around water resource issues, and in catchment management, the societal uncertainty including institutional uncertainty is as manifest as are technical problems that seemingly require technical solutions.

Wicked problems (Chapter 8) may have the following characteristics (Rittel and Webber, 1973; Ludwig, 2001):

- complex and location specific.
- dynamic, uncertain.
- diverse legitimate values and interests.
- no definitive problem formulation.
- many externalities.
- multiple trade-offs.
- intractable for a single organisation.

Britain has codes to be followed for the application of Good Practice (including Good Agricultural Practice). The United States (Bureau of Land Management, 2015) has similar 'Best Management Practices (BMPs)'. These greatly assist the catchment management process and such codes, to be proved effective, require a degree of proofing and applicability in given situations so that environmental risk may be assessed.

This information is founded in 'positivist' science that expresses an expected outcome for a specific action, for example an assessment of reducing agro-chemical loading from land use change, or the handling of a specific potentially toxic substance in agriculture or industry. However, technical knowledge is only part of the story, because stakeholder involvement (e.g., a farmer, local authority, conservation group, industrial concern) is required. Wickedness by definition, involves societal and economic factors, and effecting behavioural change involves both ethical considerations and adoption of normative statements that express conceptions of the desirable. By definition these are value-laden and shed some light on the values that inform the object and the process of inquiry, whether social or natural. And within catchments there is no clear demarcation between natural and social science or engineering measures. Hence, problem solving actually involves 'normative ethics' in an attempt to derive its standards of right and wrong from subjectively interpreted social behaviour (Ethics, 2013). From this proposition will follow questions like 'What is the purpose of river basin management?', 'In whose interests might it operate?' 'Who is responsible?' What are their motives?' And, economically speaking, 'Who should pay for improvement?

10.3   A Very Short Excursion in Environmental Economics

From the purely human point of view, water may be viewed as a 'common pool resource' (Hardin, 1968). Past failures in environmental regulation are not only physically manifest, such as the stubborn persistence of diffuse pollution or habitat degradation, but can be identified as economically problematic. One example would be the full costing of monitoring and maintenance of existing schemes by a regulator enacting catchment management planning (Cook, 1998, p. 280). Such an omission would never solve the problem of 'externalisation' of damaging environmental factors within economic transactions.

River basins provide a focus for targeting many services, some being referred to as 'Ecosystem Services' (ES). ES represent benefits that people obtain from ecosystems, as opposed for example, to a service obtained through bulk water abstraction or river transport, although the water environment of a catchment should be seen in holistic terms. ES include *provisioning services*, such as food, clean air and water; *regulation services* such as managing floods, drought, land degradation and disease or carbon sequestration; *supporting services* such as soil conservation and nutrient cycling; and *cultural services* such as recreational, tourism, spiritual and religious benefits (Natural England, 2012).

Subsequent re-focussing on Ecosystem Services may be differentiated from ecosystem processes, or even ecosystem functions, because they provide a direct service, benefit or social good as an outcome of natural processes (Cook *et al.*, 2009). The 'environment' is thereby conceptualised as instrumental in serving the needs of human society (the 'anthropocentric' view), yet we may ask how to address environmental justice. For example, is it through a political agenda that is democratic, inclusive and looks to manage effectively with a sense of progress and, by implication, review? We require a normative framework.

Elinor Ostrom identified eight 'principles' for managing common resources. She was concerned with how communities succeed or fail at managing common pool (finite) resources including grazing land, forests or water resources (On The Commons, n.d.):

Ostrom's eight principles ('rules') for managing common pool resources:

1) Define clear group boundaries.
2) Match rules governing use of common goods to local needs and conditions.
3) Ensure that those affected by the rules can participate in modifying the rules.
4) Make sure the rule-making rights of community members are respected by outside authorities.
5) Develop a system, carried out by community members, for monitoring members' behaviour.

6) Use graduated sanctions for rule violators.
7) Provide accessible, low-cost means for dispute resolution.
8) Build responsibility for governing the common resource in nested tiers from the lowest level up to the entire interconnected system.

Ostrom's view is based in pragmatism and tells us how to do it, but environmental improvement comes at a cost. For example, one answer to 'who pays' is to invoke the Polluter Pays Principle (PPP) requiring the costs of pollution be borne by those who cause it. It aims at determining how the costs of pollution prevention and control must be allocated where the polluter must pay. It is thus a means of 'internalising' the (environmental) externality of pollution. In principle, PPP promotes efficiency, it promotes legal justice, has the potential for the harmonization of international environmental policies and it may define how to allocate costs *within* a State. It is also a principle of international environmental law.

An alternative approach is that of 'Payment for Ecosystem Services' (PES). In this approach, the perceived beneficiary pays. In economic terms, this might be the water services provider. Hence, for long, the water environment has been seen as an economic 'externality' by most users that discharged urban wastewater as well as industrial or agricultural wastes in both diffuse and point-source contamination. Concretely, there has not been a market for these waste products. For, if the phosphorous recovered in water treatment, or the nitrogen leaching from agricultural land could be somehow re-packaged and sold on, in theory these forms of pollution would tend towards elimination.

However, the burden of cost-sharing between polluter and user of water is an arena that requires, in a fully political sense, careful handling. Ostrom's rules see the water environment as an 'open access common property resource', intrinsically difficult for establishing any clear sense of ownership, yet still having to identify boundaries, be as inclusive as possible, and yet establish arrangements to deal with transgressors. One way of conceptualising such relationships is termed a 'Political Contracting Framework (PCF)'. This enables political analysis of institutions and their economic relations in a fashion comparable with commercial contracting, although an actual exchange of finance need not occur (Sabatier *et al.*, 2005; Benson *et al.*, 2013a). An implication is that formal and legally binding agreements might be developed, or where they exist, be analysed. In watershed management, it may be that avoiding water resources being subject to the 'tragedy of the commons' and that effective allocation of property rights is missing.

It has been stated the main problem (but by no means the only one) in environmental economics is how externalities are handled. By illustration, a conservation scheme may, as a by-product, improve water quality or a species habitat (positive impact), intense agriculture or waste water treatment may still cause environmental damage in the form of water pollution, a 'negative externality'. Internalising an externality may be the act of making a change in a company's private costs or benefits in order to make them equal to the company's social costs (Business Dictionary Online, 2015). In essence the *total* cost of an activity such as agriculture, manufacturing or exploitation of the water environment should be internalised - as real costs – in economic transactions. Environmental cost may be seen as a form of social cost (particularly where negative externalities are concerned) and hence loaded economic and ethical questions are asked.

The economist Ronald Coase aimed to address this issue. 'Coase's Theorem' states that, where trade in an externality is possible, and there are no transaction costs, bargaining will lead to an efficient outcome regardless of the initial allocation of property rights. This is unlikely in the real world for in practice, obstacles to bargaining or poorly defined property rights can prevent that efficient bargaining that may reduce transaction costs, and minimise negative externalities (Coase, 1960). To clarify, the concept of 'property rights' does not literally refer to private ownership (of land, water, infrastructure etc.), rather it implies incentive by an individual or organisation (including government) to use the resource wisely and thereby conserving it in the long-term, although a perceived conflict between property rights and human rights are open to political interpretation (Alchian, 2008).

Policy drivers are generally linked to legislation including EU Directives, typified by the WFD, that become translated into national laws and policies. To stereotype the argument, regulation to achieve compliance is expensive on account of requirements for monitoring, 'policing' pollution and for legal action should the system break down. Often, a combination of mistrust of officialdom by certain stakeholders, of economic recession, and a neo-liberal distaste for direct spending through state agencies leads potentially to involvement by the private sector in order to protect its own interests (section 10.12). Unfortunately there is no easy answer from economic theory, for apart from the desire to internalise externalities is the problem of market failure.

10.4   Market Failure and Water

One problem for conventional economic treatment of water services as a commodity is the low unit cost of water, making it a relatively inelastic commodity. For example, we may imagine that a hike in fuel price will reduce aggregate road distance undertaken, or an increase in the cost of a luxury good (alcohol, chocolate etc.) should reduce consumption. Water, however, is a necessity, regarded as being a 'public good' and generally abundant in England and Wales where a litre of tap water - supplied and taken away - costs less than one half of one pence (Ofwat, 2011b). This includes the cost of treatment, supply and disposal. A view commonly held among water professionals is that water is a commodity, unlike any other and most people have little idea of the cost of the resource.

Figure 10.1 illustrates inelasticity in demand for water that behaves close to an 'inelastic commodity' (Martinez-Espineira, 2007). In Figure 10.1, the steeper the curve between P (price) and Q (quantity), the more inelastic is the demand for a quantity of water, because for a stated change in price the corresponding change in quantity consumed is small (Demand Curve 1). Demand Curve 2 is flatter, suggesting that consumers maybe sensitive to changes in price; in that unlikely case, a hike in water charges would produce a noticeable drop in consumption. This is very elastic demand so that a small increase in price from A to B results in a large decrease in quantity of Q1-Q2. Alternatively, a (far larger) hike in price from A to C (Curve 1) achieves the same decline in demand as Q1-Q2.

To return to diffuse pollution as an externality, it can be considered that it may not be costed into a production process. In this instance, there is 'market failure' and it

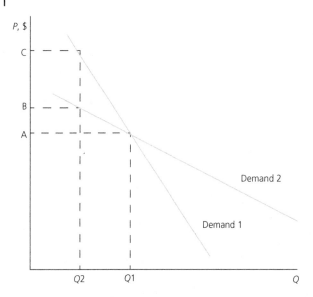

Figure 10.1 Inelastic and elastic demand in economics.

takes account of neither social costs, nor social benefits, nor any environmental impact of production and consumption:

$$SOCIAL\ COST = PRIVATE\ COST + EXTERNALITY$$

There is divergence between the private and social costs of production and the externality may be signed, wither positive or negative, depending upon its perceived value. 'Social cost' then includes *all the costs of production* of the output of a particular good or service. A negative externality imposes a higher social cost on society in general, because of the clean-up costs, health costs, loss of amenity, ecosystem services, flooding, and so on.

A marginal curve shows the cost, or change in cost, per unit increase in output. Marginal cost curves will cross the 'Demand Curves' (these are shown in Figure 10.1) and this point determines optimal efficient output. Figure 10.2 shows two hypothetical Marginal Cost Curves. The private marginal cost output curve shows lower production cost values because it is presumed externalities are not taken into account. The higher values belong to the marginal social cost curve, basically because all costs of production are taken into account. Or to put it another way, in calculating the private costs to the manufacturer, output may cost less than incorporating the social optimum cost level of production. If only private costs are taken into account, the producer creates externalities that *do not take account of the social and environmental impacts into the calculation.* These are represented by the difference between the two curves.

Figure 10.2 considers a fertiliser manufacturer supplying the agricultural industry. Market failure is said to occurs when marginal social cost > private marginal cost. This good produces pollution through manufacture and use in agriculture, for example diffuse pollution may be a by-product of food production – a negative externality. Profit-maximising level of output is at Q1. However the socially efficient level of production

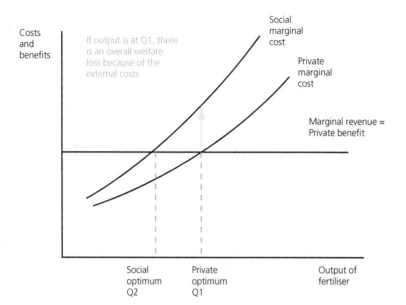

Figure 10.2 Market failure arising from negative externalities (Tutor2u, n.d.).

would consider the external costs too. The private optimum output is when where private marginal benefit = private marginal cost, giving an output (Q1), effectively the costs and benefits are increased by the upwards arrow, the cost of externality. As a whole, the social optimum is where social marginal benefit equals the social marginal cost at output (Q2). Alternatively, the social optimum output level is lower, at Q2 (Tutor2u, n.d.).

Dealing with the problem of market failure (thus termed because market economics have failed to take account of all real costs), therefore, has to consider costing in pollution (by that we generally understand diffuse pollution) in some way that effectively mimics the missing market forces (for example capping total pollution that is considered tolerable and employing tradable permits), a consenting system that can ultimately punish a polluter (regulation), charging for the act of polluting (charging for consents to discharge) or raising pollution taxes with the aim of reducing consumption (one example often proposed would be fertiliser taxes).

10.5 Integrating water Resource Management in WFD

Many of the issues outlined above are addressed in the WFD. 'Integrated Water Resource Management' (IWRM) as essentially 'integrated catchment management', is not novel. Efforts to manage catchments on a holistic basis date from the nineteenth century, although conceptualised in a full sense in the 1970s as outlined in Chapters 3 and 9 (GWP, n.d.).

Promoted by the Global Water Partnership (GWP) whose vision is for a water secure world importantly that 'is to advance governance and management of water resources for sustainable and equitable development'. Echoing The Dublin and Rio statements of

1992, the Millenium Development Goals (2000) and World Summit on Sustainable Development (2002), there is a Plan of Action, embracing IWRM and Water Efficiency plans. GWP adapts and elaborates these principles to reflect international understanding of the 'equitable and efficient management and sustainable use of water'.

The guiding principles are:

- Freshwater is a finite and vulnerable resource, essential to sustain life, development and the environment.
- Water development and management should be based on a participatory approach involving users, planners and policy makers at all levels.
- Women play a central part in the provision, management and safeguarding of water.
- Water is a public good and has a social and economic value in all its competing uses.
- Integrated water resources management is based on the equitable and efficient management and sustainable use of water and recognises that water is an integral part of the ecosystem, a natural resource, and a social and economic good, whose quantity and quality determine the nature of its utilisation.

The WFD 2000 (Chapter 1) may be summarised as follows:

- expanding the scope of water protection to all waters, surface waters and groundwater.
- achieving "good status" for all waters by a set deadline.
- water management based on river basins.
- "combined approach" of emission limit values and quality standards.
- getting the prices right.
- getting the citizen involved more closely.
- streamlining legislation.

For detail, see Box 10.1.

Box 10.1 The Water Framework Directive (2000/60/EC)

Under the Water Framework Directive, the basic management units for river basin management planning are the River Basin Districts (RBDs) that may comprise one or more river basins (and include as appropriate lakes, streams, rivers, groundwater and estuaries, together with the coastal waters into which they flow).

The WFD requires that river basin management plans (RBMPs) must be developed, and reviewed on a six-yearly basis, specifying the actions required within each RBD to achieve set environmental quality objectives. RBMPs must identify discrepancies between the existing status of rivers and other water bodies, and that required by the WFD, so that a programme of measures can be put in place to achieve the desired goals. The planning process should include an economic analysis of all water uses in each RBD, as well as determining the pressures and impacts on the water environment. A key element of this process is public information and consultation.

All water bodies are assigned to one of the Directive's five status classes: high, good, moderate, poor or bad. The WFD requires that all inland, estuarial and coastal waters within RBDs must reach at least good status by 2015. This is based on an assessment of ecological, chemical and quantitative criteria. There are more limited criteria for assessing the status of heavily modified and artificial water bodies; groundwater status is assessed on quantitative and chemical criteria alone.

There are over 30 differing criteria for assessing the status of rivers, lakes, transitional waters and coastal waters, but they all include consideration of biological quality, including presence or absence of various algae, plants, fish and invertebrates; physical and chemical quality, including oxygenation and nutrient conditions; environmental quality standards for levels of specific pollutants, such as pesticides; and physical aspects that support the biological quality of the water body, such as the quantity and dynamics of water flow (hydro-morphological quality).

If part of a water body fails on any one of the criteria monitored, it will fail to achieve or lose good status. This is described as the "one out all out" approach.

There are a number of exemptions to the general objectives, including disproportionate cost, that allow for less stringent objectives, extension of the deadline beyond 2015 or the implementation of new projects. For all these exemptions, strict conditions must be met and a justification must be included in the river basin management plan.

Member States were required by the WFD to prepare RBMPs by December 2009; key subsequent deadlines are 2015, 2021 and 2027. In each case, environmental objectives set out in RBMPs should have been achieved six years after the RBMPs were prepared, and improved objectives should be specified for achievement over the next cycle of river basin planning of the following six years. Catchment Flood Management Plans, produced to meet the requirements of the Floods Directive (2007/60/EC), must be co-ordinated and synchronised with RBMPs.

Member States' governments have to designate organisations to act as "competent authorities" within their territories for taking forward implementation of the WFD. The UK Government have designated the Environment Agency (EA) to carry out this role in England and Natural Resources Wales (NRW); the Scottish Environment Protection Agency in Scotland; and the Environment and Heritage Service in Northern Ireland. A map showing the UK River Basin Districts is included at Appendix 5.

(http://www.publications.parliament.uk/pa/ld201012/ldselect/ldeucom/296/29605.htm)

From its inception there has been great enthusiasm expressed from water professionals, but it has equally been linked with scepticism regarding the implementation period for such a wide-ranging, supra-national Directive. It does state: 'The best model for a single system of water management is management by river basin - the natural geographical and hydrological unit - instead of according to administrative or political boundaries'. This creates a potential problem for governance; however it makes complete sense in practice providing care is taken to accommodate groundwater behaviour that (as can be the case with the chalk aquifers in southern England) often does not coincide with surface-defined topographic boundaries. In short, it is the best we can do.

Implementation remains at issue. For one, a clear understanding of the outcome 'Good Ecological Status' needs to be understood across member states by a process of comparison called 'intercalibration'. WFD is driven by outcomes, requiring prescriptions for such goals as 'good ecological status' and 'good chemical status'. Targets for groundwater quality and quantity (prior to WFD this important matter was not regulated at EU level) are strict and are set in the context of the creation of 'river basin management plans'. The cycle of implementation to date ends with a third cycle of management planning in 2027. Ominously for the time of writing, the year 2015

Box 10.2 Excerpt from the House of Lords Select Committee on the implementation of the EU Water Legislation published in 2012 (House of Lords Select Committees, 2012)

24. Overall, while our witnesses recognised that there had been difficulties with aspects of implementing the Directive, they shared a consensus that the WFD had been a force for good in EU water resource management. They tended to agree that it did not need to be significantly changed at the current time, not least because of the long time period needed to secure acceptance and implementation of the Directive, filtering down from national levels to the local levels where it has to be applied.

25. The German Government said that it was a successful Directive, providing an EU-wide coherent and systemic approach to water management, and that its "added value" was demonstrated by the much-improved co-operation between Member States on water resources, particularly where rivers cross international boundaries. Its representative agreed that the WFD's targets were very ambitious and that the target that all water bodies should have good status by 2027 was "impossible … nevertheless, the Directive gives us the push to get better. I think there will be real improvements." Her own view was "no more directives, but do not change the existing system".

26. Evidence provided to us by the Environment Agency (EA) shed light on the extent to which some Member States had achieved "good status" for their water bodies in 2009, and the ambition which they had shown in planning for improvements by 2015. For France, the 2009 figure was 40%, planned to rise to 67% in 2015; for Germany, the 2009 figure was 22%, rising to 29% in 2015; for the Netherlands, the figure in 2009 was 4%, and an improvement to 20% was planned; for the UK, the 2009 figure was 24%, and a rise to 37% was planned for 2015. We asked the EA to comment on the reasons why France planned an increase in "good status" from 40% to 67% from 2009 to 2015, while the UK would only go from 24% to 37%. In written evidence, the EA said that some of the French data might be "skewed by uncertainty. Some 30% of their water bodies had uncertain chemical status. By 2015, the French expect the majority of these waters to turn out to have good chemical status, and where this coincides with existing good ecological status then overall status will then be classed as good.

http://www.publications.parliament.uk/pa/ld201012/ldselect/ldeucom/296/29605.htm

---

required the meeting of environmental objectives (http://ec.europa.eu/environment/water/water-framework/info/timetable_en.htm). In 2012, the House of Lords could report (following a consultation exercise) less than encouraging and variable results across Europe (Box 10.2).

10.6   Water and Politial Science

The aim is to describe water 'governance'. The word has Latin origins that imparts a notion of 'steering'. We seek to address wicked environmental problems that affect water and identify a number of possible normative pathways through the resulting complexity. To map these, we have to appreciate who does what, as well as asking what everybody wants. WFD sets some bold objectives, all be they of a general nature; the outcomes are easier *said* than *done*. Certainly Ostrom's 'rules' are applied, for it is a common pool resource. Not only does this allow us to ask questions about success or failure, but it is a vehicle for asking questions such as 'What makes a particular catchment partnership successful?'

Where partnerships are successful, resulting transaction costs (costs of participating) expressed in economic terms should be relatively low (Sabatier *et al.*, 2005; Benson *et al.*, 2012). We may then analyse conditions where a particular set of measures may travel, investigate the flexibility of institutions, stakeholder participation and trust. Influences of state policy cannot be ignored and the environmental case requires elaboration. Hence 'governance' is used in contrast to the established 'top-down' command and control style of getting things done (Benson *et al.*, 2010; Cook and Inman, 2012). Historically in Britain, legislation and central governmental institutional arrangements have driven water management issues. Governance paradigms imply 'power to' (including governmental powers to make and implement policy) rather than 'power over', yet the concept includes all of the included politics, rules, decision making and institutions.

In a modern democratic context of 'civil society', the citizen is free to act, away from coercion and is 'enabled' to operate within the context of rights and responsibilities. This action is in accordance with democratic imperatives of rights and responsibilities, bringing into question notions of 'top down' regulatory activity, perhaps redolent of the past democratic centralism of communist countries. Neither should the new paradigm be seen as a tenet of extreme 'neo-liberal' thinking. Even if the state is retreating from direct involvement its role remains as a backstop in a position of last resort. Empowering the citizen against large interests, be they public or private, and addressing market failure in such matters as diffuse pollution belong more to a co-operative and *uber-demokratie* pathway than historic top-down action. Actually, the citizen becomes embroiled in a policy environment rooted in collective and voluntary action that flourishes on the back of a (very English) tradition of philanthropy and mutual aid (Deakin, 2001, Ch 1). The third sector seems to arise from the same motivations as state welfareism in the nineteenth and early twentieth centuries (Sheail, 2002, p. 16). In the inter-war period, it was actively engaged in campaigning (typified by the Campaign for the Protection of Rural England), its role was officially acknowledged during the second half of the twentieth century and there is a plan to converge voluntary action and public policy. Modern commentators describe modern governance arrangements achieved in a shift away from state direction towards a conceptually more complicated regulatory framework based in hierarchies, networks and markets but may also be delivered through communities (Pierre and Peters, 2000, Ch 1).

Analysis also provides for investigation of multi-level environmental governance with policy transfer issues between institutions that extend beyond peer-to- peer networks of national governments or their agencies (Benson and Jordan, 2011). In context, 'hierarchy' may not actually sound desirable but the proportionate involvement bodies such as EA, NE in cases of environmental degradation of some kind, or of regulators such as Ofwat for privatised water companies (who also seek to act for the consumer) and the Consumer Council for Water the Consumer Council for Water (CCWater, 2015) operates on behalf of the consumer (Chapter 3). For voluntary sector bodies in water governance, the Charity Commission may be a positive appeal to 'top down regulation'.

Figure 10.3 shows the relations between the policy approaches that might be adapted from the above, specifically with farmers in mind. In essence, while the outcome is

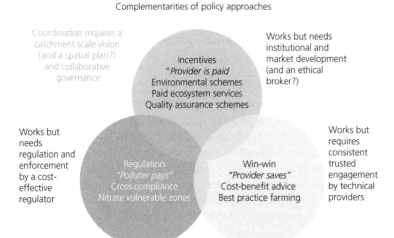

Complementarities of policy approaches

Coordination requires a catchment scale vision (and a spatial plan?) and collaborative governance

Incentives
"Provider is paid
Environmental schemes
Paid ecosystem services
Quality assurance schemes

Works but needs institutional and market development (and an ethical broker?)

Works but needs regulation and enforcement by a cost-effective regulator

Regulation
"Polluter pays"
Cross compliance
Nitrate vulnerable zones

Win-win
"Provider saves"
Cost-benefit advice
Best practice farming

Works but requires consistent trusted engagement by technical providers

Figure 10.3 Example of relations between policy approaches (After L. Couldrick, WRT and L. Smith, reproduced by permission).

'win-win' whereby the provider (farmer) gains from payments for the beneficiary, historic instruments remain very much in place, that is specifically incentives and regulation, but the latter is in the background. In this, the long-established 'regulation' approach is joined by new categories of 'win-win,' so that all stakeholders stand to gain something be it financial or in terms of fulfilling some societal or environmental aspiration, and incentivisation of parties, pitched such that water management goals are reached.

Major causes of complex environmental problems therefore arise from market failure applied to common property, open access resources affected by diffuse pollution and related environmental degradation problems. A second problem is the high transaction costs usually manifest in regulation of such problems that generally cost the public purse, hence the need for full costing required by the WFD. Experience demonstrates that cost-effective environmental management solutions seldom flow directly from managerial regulation (including protocols backed by legislation) or from economic measures such as pollution licencing. Furthermore, alongside rising costs, stakeholder relations are often aggravated when legal sanctions are taken. It is into a new democratising policy environment that wicked problems are pitched - we may ask whether this is better for proffering solutions. Caught up are the stakeholders of catchment management. Grimble *et al.* (1995) defined stakeholders as:

> 'those who affect, and/or are affected by, the policies, decisions, and actions of the system; they can be individuals, communities, social groups or institutions of any size, aggregation or level in society. The term thus includes policy-makers, planners and administrators in government and other organisations, as well as commercial and subsistence user groups.'

Stakeholder analysis falls into one of three broad categories: **descriptive, instrumental, or normative** (Mitchell *et al.*, 1997, Donaldson, 1999). **Descriptive** stakeholder analysis provides a snapshot of a situation at a particular point in time and could help managers or other stakeholders understand the context in which they are working. **Instrumental analyses** seeks to understand relationships between stakeholders and define the relationships that are needed to reach particular objectives, includes negotiation towards economically, ecologically, and socially acceptable outcomes for the parties involved. **Normative** stakeholder analyses are used as a tool to work towards the way a situation ***ought to be***: what might be called a 'visioning tool' or a 'change management' tool (i.e., they seek to effect behavioural change).

The above tells us that we may simply describe who is out there and has an interest, gather information (presumably with a view to change behaviour) and finally equip ourselves with a vision that includes most, if not all stakeholder interest groups. Accentuating the normative had been realised as acceptable in the social sciences, enabling at least goals perceived desirable by technical experts to be sought, and in doing so ethical considerations must be addressed. These ethical considerations range from issues of 'water justice' to simple questions requesting wicked answers, such as 'who pays' or 'who benefits' or should we be entirely anthropocentric?

Incorporating upfront stakeholder engagement must be approached through network building and elucidating novel mechanisms (including economic mechanisms) that link research, best practice, community opinions and stakeholder needs with anticipated positive outcomes for a catchment. Transitions 'on the ground' may already exist and will have to be evaluated alongside current governance arrangements, yet be evaluated for improvement in future, inclusive, governance practice. Behavioural change may be required in modes of living and land-use, adaptation and mitigation responses that impact upon the capacities of infrastructures (settlements, transport systems, energy, water, etc.), in all sectors, businesses, households and local communities.

Catchment management has made leaps and bounds, although there remain important areas whose full inclusion remained either part of a wish-list, or else remained murky. Since the 1970s, the UK had moved towards managing the interactions between land-use and water on a catchment basis. And that the catchment is a rational unit for environmental management because it contains within it the possibility of protecting water quality, flooding to a degree and also abstraction. As outlined in Chapter 9, it was the NRA (1989-1996) that saw the real birth of ideas around integrated and 'holistic' catchment management planning.

An inside commentator (Gardiner, 1994) recognised the following stages in a planning environment, specifically where holism is evolved to cover all aspects of planning:

1) Model policy formulation.
2) Catchment management plans.
3) Strategic land-use planning.
4) Development control.
5) Conservation and enhancement of the total river environment.
6) Achievement of sustainable restoration and conservation.

While this provided great insight back in the day, there are with hindsight, omissions. The first four are achieved through planning liaison rather than by statutory means as part of the Environmental Assessment process. Pursuing sustainable development through technocratic agencies would bring a return to criticisms of the former RWAs that they were ultimately unaccountable. Consultation with all potential stakeholders (including other agencies), it was hoped, would remove the undemocratic face of 'quangoism'. Compared with the policy-making process model proffered in the 1980s (Chapter 1), 'evaluation and review' is lacking. One major professional body, the Chartered Institution of Water and Environmental Management (CIWEM) is supportive of the notion of IWM, yet nervous about its application, for:

> 'Climate change, demographic change, economics and environmental legislation such as the Water Framework Directive all necessitate a more integrated approach to the management of water. CIWEM believes integrated water management (IWM) should be a process whereby human interventions can work within the natural water system, rather than disrupting it in ways that are ultimately unsustainable.' And yet:

> 'In general CIWEM believes our institutional arrangements tend not to support cross-sectoral decision-making, with organisations and regulations having specific and potentially diverging remits which limit the ability to manage the water environment at a wider level. Significant changes are required in terms of how problems are framed and how different institutions and sectors work together to achieve common goals' (CIWEM, 2011).

The status quo is found lacking, and the key issues mentioned appear to be 'adaptive management' (as evangelised by the USEPA) and increasing the role of the Third Sector (also termed NGO sector) that addresses a perceived shortfall in resources, financial or otherwise. Desirable in this context is a template approach to assist in plan formulation (Rickard, 2012, pers. comm.), which is the kind of technical format advocated in New York State (Smith and Porter, 2009).

Intuitively, on-the-ground neutral mediation may be seen to become ever more difficult as the number of stakeholders increases. The problems with catchment management are complicated, for example they involve multiple statutory regulators (Cook, 1998) and now a move towards a somewhat 'spontaneous' voluntary (or third) sector creation of catchment bodies such as the Rivers Trusts and there is a developing literature analysing success of failure of catchment groups (Cook *et al.*, 2012; Benson *et al.*, 2013).

The counsel of perfection (Figure 10.3) results in win-win situations. Required is a system that is iterative, that is it is a dynamic process and capable of renewal, to achieve that end. Based on the U.S. Environmental Protection Agency handbook, the scheme enables partnership building ahead of any technical specification process that starts with gaining knowledge of a watershed, only then commencing instrumental work with goal identification (Figure 10.4).

This is it the capacity to adapt that will influence the adoption of new ideas in river basin management, and this potentially affects all economic sectors and institutions. Such a process will have to consider political considerations and ecosystem services such as:

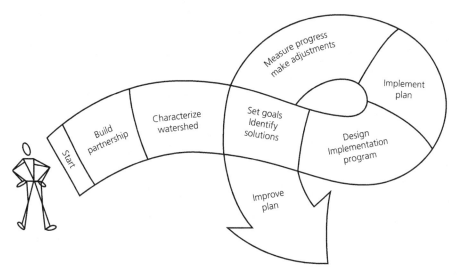

Figure 10.4 An adaptive management cycle for catchment planning and process implementation. (*Source:* USEPA, 2008).

- Climatic mitigation (especially carbon sequestration).
- Water quality (the fate of nutrients, sediments, pathogens etc).
- Flood control and mitigation (including flood control relaxation measures).
- Ecosystem health.
- Economic efficiency.
- Water resource planning.
- Land use change.
- Political enablement.
- Institutional flexibility.
- Sustainable funding.

The capacity to adapt affects all sectors of UK society and economy, societies, businesses, regulatory authorities, governments and communities. As the Rural Economy and Land Use Programme saw it (RELU, n.d.), they will have to adapt to environmental change in 'integrated perspectives in pursuit of more holistic solutions.' There are a range of natural processes that regulate the climate, protect from floods, purify water, and provide aesthetic and recreational value. These deliver 'ecosystem' (sometimes 'environmental') services, while at the same time conserving or improving 'ecosystem health.' Implicit is not only multi-functional agriculture and land use but also ecological and economic efficiency and this includes reduction of emissions from agriculture and other rural land uses alongside adaptation. Issues of food security and water sustainability (quality, abstraction and flooding) must also be addressed as land use change progresses. Two factors have been omitted, or under-played, these are public participation and its twin 'adaptive management.' These serve to underplay what is one-way, technocratic decision making and implementation.

Figure 10.4 presents what may also prove to be idealistic. On the other hand, Cook *et al.* 2012 (Figure 10.5) identify various 'disjunctures' in water governance for England and Wales, especially that relate to the matching of water institutions, including the

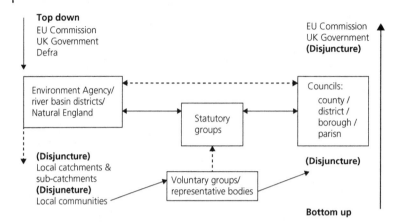

Figure 10.5 A view of water governance for England and Wales: the *status quo*? (*Source:* Cook *et al.*, 2012).

NGOs horizontally in terms of thematic coverage and vertically in terms of the democratic process. In short, local government, including county councils and unitary authorities have no real remit in water resource management including treatment at both 'clean' or 'dirty' ends of the industry, or in delivery. While this has not always been the case (Chapter 3), a long-term devolving of responsibilities separates water management and societal supply from much of the democratic process. Compared with France, where municipalities have the responsibility of hiring and firing private water undertakings, or the Netherlands or Denmark which are so organised that local authorities have a role in 'spatial planning' in order to protect water resource sources.

Before about 2000, the role of the voluntary sector was limited in water governance. It was largely recognised as having a role in general nature conservation (such as the county-wide Wildlife Trusts) or specifically concerned with recreational pastimes, notably reflecting angling interests. The focus was on state owned Regional Water Authorities (RWAs), or regulators such as the NRA and EA while private or - privatised - companies delivered water supply and other services (see Chapter 3). Into such an imperfect template, river basin NGOs are coming to represent a promising development, and these developments specifically involve delivering the complicated and varied requirements of WFD, as well as improving aspects of social inclusion.

10.7   Water and Economic Sectors

Across Britain all sectors are involved in water governance. In England and part of Wales, regulation is public, while delivery of utilities lies with private companies, while Welsh Water is a not-for profit company and in Scotland ownership is public. Yet, it is the rise of the 'third' or voluntary sector in environmental regulation and management that is in any real sense 'new' and its role is a real departure from the old private-public debate over ownership, regulation and operation. Water is today managed under heavy regulation that nonetheless leaves it largely treated as a commodity. Post-1989, this development remains contentious. Freed from state control, the newly private industry could hire and fire staff in the interest of efficiency and raise investment as it wished.

It is the consumer (and not the taxpayer) who has had to pay increased charges long term and importantly, regulation (economic, consumer and environmental) has remained essentially top-down and hence potentially closed to democratic inclusion. However, WFD from 2000 has demanded IWRM, social inclusion and operational and economic efficiency. Outcomes are to be measured, most notably in terms of the ecological status of water bodies; Britain is not yet 'there'.

The classic division of economic sectors involves 'primary' 'secondary' and 'tertiary' and it may be expanded to five. These are respectively:

a) Primary: production of raw materials (agriculture, forestry, mining, quarrying, fishing, etc).
b) The secondary sector of the economy manufactures finished goods such as industrial manufacturers, and construction.
c) The tertiary sector of the economy is the service industry including retail, transport, entertainment, professional services, and so on. Evidently this sector was too broad, and we now talk of.
d) A 'quaternary sector' consisting of intellectual activities including education, research, information technology.
e) The 'quinary sector' representing the 'highest form of decision making' including top executives from all sectors, but is this fine-tuning merited? (Rosenburg, 2015) The reader is invited to reflect on whether CEOs in a range of industries and in the universities, really merit their own sector.

The water industry actually has aspects of all the above. Specifically, one is dealing with abstraction (primary), production of a potable resource (secondary), distribution (tertiary) and information, research and public information (quaternary) and, naturally, policy and decision making. Then, throughout Britain there is separation of 'utility', that is the functions of water supply and sewage treatment from regulation (Chapter 3).

Less tidy in terms of functional definition, but growing in influence, is the 'voluntary', 'charity', 'third' and sometimes 'not-for profit' sector. Such terms are not always used interchangeably but they are taken to describe non-governmental organisations (NGOs) and these are real actors in environmental conservation and in modern water governance (Cook and Inman, 2012). Perceived 'virtues' of this sector include being value-driven (and hence more trusted by the public than other sectors) and purportedly free from bureaucratic interference. It can easily engage with direct and positive citizen participation, including engagement of groups considered to be marginalised achieving economic efficiency where public funds are involved and efficiently target their own funding to projects that may arise from voluntary donations. As such, voluntary bodies are well-placed to undertake activities such as stakeholder identification and stakeholder analysis.

The sector's role is complex with operations that cannot be separated from matters of governance, social inclusion or democratic accountability. NGOs are well-placed for flexibility and intermediary between stakeholders. The down-sides are complicated, but include under-resourcing, lack of continuity of staff, poor employment conditions, lack of ready technical information and expertise and lack of geographical and functional cover when compared with a fully funded state organisation. Generally speaking, the smaller the NGO the more problematic is its reliability of income and this is guaranteed to present these kinds of problem for both staff and functionality.

The UK Coalition Government (May 2010 to May 2015) promoted a flagship policy called the 'Big Society'; some also talked about a 'Big Green Society' the latest manifestation of a long-held desire to foster voluntarism and now public sector reform. Re-structuring in traditionally state organised sectors can press new functions on to the third sector, and as has been stated, it is value-driven and generally relatively free from bureaucratic control. There remains a strong possibility for negative interventions in labour markets, it lacks the resources required to function and also lacks a clear function and task allocation (Cook and Inman, 2012). Hence, it did not take long for questions about the 'Big Society' to be raised across the political spectrum (New Statesman, 2012). Water governance initiatives may yet prove a positive outcome of deliberations around NGO operations although most affected organisations pre-date the Big Society.

It remains a normative presumption that river basin management has to become more inclusive, that is incorporate multi-stakeholder engagement, and seeks to maximise democratic accountability in meeting complex challenges, including diffuse pollution (Smith and Porter, 2009). Economically, not only must there be full costing (eliminating negative externalities) but transaction costs must also be minimised. Implicit is the move away from 'top-down' and centralised control in water governance.

Governance is evolving and under new governance arrangements, an adaptable, decentralised, polycentric and multi-stakeholder approach might embrace:

- local leadership and decision making within the framework of higher regulation built on existing organisations and partnerships,
- adaptive management that exploits the benefits of stakeholder participation in programmatic design, implementation and monitoring,
- co-ordination and agreement on roles and responsibilities through vertical and horizontal integration of government agencies and other bodies and groups; that is more than mere delegation.

'NGOs' are distinct from public or private sector operations. The sector ranges in scale from small, local conservation charities to nationally important organisations (e.g., the National Trust or the Wildfowl and Wetlands Trust), and once largely task-oriented NGOs have campaigned and encapsulated ever deeper roles in planning and policy formulation so that increasingly community inclusion is at the core of its function. A working definition of NGOs over and above other (smaller) third sector organisations is their greater degree of environmental agency, something that may substitute for state agencies (Cook and Inman, 2012).

A positive view is that the third sector is value driven, issue focussed and considered economically efficient due to volunteer engagement and low administrative overheads in meeting conservation objectives. Furthermore, independence and flexibility make it an intermediary between stakeholders, government the private sector, and it is proving an effective vehicle for public engagement. NGOs are therefore emerging as a key player in environmental action, making them a partial replacement for 'big government action.' The UK NGO sector is poised for reasons of tradition, function, trust, non-suspect motives and political recognition to act as an honest broker or intermediary. Their advantage is being driven by feelings of commitment, belief and ideology and (importantly) it is motivated by altruism rather than economic gain. It is therefore useful in building trust with communities, a prerequisite for communities to adopt new sustainable management practices.

New governance has to incorporate the language characteristic of voluntary engagement. The British Trust for Conservation Volunteers (BTCV) is concerned with 'empowerment' for local people (nurturing 'ownership' of geographic areas and management problems), 'capacity building' (for practical skills, gaining confidence and building social networks), raising quality standards, providing training opportunities or paid employment while attracting funding and achieving judicious and efficient spending (Cook and Inman, 2012).

The requirement for public participation is manifest within the WFD, notably Article 14 (Danube Watch, n.d.). If this is to work at all, then not-for-profit, NGOs are value driven organisations that clearly have a role, at least where aligned with perceived objectives. Received wisdom tells us that the best rational geographical unit for such activity is the catchment.

Pertinent questions are therefore: to what extent do third sector organisations not only 'fit the bill' in a cultural and operational way? How might economic challenges be faced in an equable manner? How might real problems relating to the functionality of the voluntary sector be addressed, as well as delivering environmental goods? What is the function of the state and private sectors?

10.8   Analytical Frameworks

We may seek to analyse stakeholder interaction through theoretical frameworks that relate to roles and economic relationships. This is important, for the objective is not only clearly to demark task allocation, but also to engage with central issues of transaction costs and economic efficiency in achieving satisfactory outcomes. A helpful summary has been produced by Sabatier *et al.* (2005). Relevant examples include: Institutional based analyses frameworks such as Institutional Rational Choice, Transaction Costs Economics (TCE) and Institutional Analysis and Development (IAD). The Social Capital Framework is concerned with reciprocity and networks. The Advocacy Coalition Framework is actually well suited to 'wicked' problems for it concerns itself with goal conflict, technical disputes and multiple actors.

The Political Contracting Framework (PCF) is especially attractive (Lubell *et al.*, 2002) for it builds upon contracting property rights, yet it is concerned with transaction costs, as is Coase's Theorem. While mutually advantageous ('win-win') solutions are sought, it is the reduction of transaction costs that is favoured in application. Lubell (2004) finds that PCF may indeed optimise benefits for the stakeholders involved, and there may be unforeseen benefits in positive externalities. Otherwise, tangible benefits may include reduction in pollution, habit gain or some direct societal good proposed in a 'symbolic policy', but are not achieved because there is a disjuncture between promise and performance. We may imagine PCF analysis might not only expose such a situation but would help us understand why stakeholders might have failed to deliver an environmental or social good. PCF can be also be cast in terms of criteria for success or failure of a 'not-for profit' NGO other agency. Analysis of stakeholder relations may start with establishing a Political Contracting Framework (PCF) that is an assumed collective action 'agreement' between (for example) polluters and other stakeholders (Benson *et al.*, 2013).

In implementing PCF, the complex and diverse interests agree to a set of institutional rules to govern how the watershed resources are managed and used, they then co-operate to implement those rules (Sabatier *et al.*, 2005). Any 'contractual' outcomes can be then analysed in economic or political terms. PCF has been used by recent investigations in watershed partnerships basin studies (e.g., Benson *et al.*, 2013). In a very instrumental way, agreements between parties such as water companies, regulatory agencies, the voluntary sector and individual farmers are open to formation and analysis. Partnership formation, formal agreements and voluntary action are all implicated. The reduction of transaction costs is most efficient when collective action is addressing diffuse pollution, flood alleviation or dispersed habitat destruction across a catchment.

10.9   Ways, Means and Intermediaries

PCF therefore builds upon contracting property rights and is concerned with trans-action costs. Partnerships, including the identification of an 'Honest Broker' or neutral intermediary body that may be in the voluntary sector include the Sustainable Catchment Management Programme (SCaMP) funded by United Utilities and projects funded by Wessex Water and South West water described below. To put it another way, with roles defined and 'ownership' of land and activities understood, the transfer of financial and other resources may produce optimal outcomes if contractual roles are understood and flaws in a political process thus exposed. Here is an instrument to enact a process of 'adaptive catchment management'.

Regulation, as discussed, has largely been top-down in Britain. The system that evolved in the second half of the twentieth century was about Competent Authorities as regula-tors, separate from either state departments or from those who deliver water, sewage or other environmental services. Such is the EA, NE or their fellows in Scotland and Wales. The problem with these actors is that their main function is regulatory, or pejoratively *they act as 'enforcer'*. Even with good will on all sides, there remains a risk that they are perceived as 'official', for if the rules of some environmental scheme are transgressed, they have legal powers to prosecute (Cook and Inman, 2012). Mediation is therefore generally best organised by a person or persons that are outside either official or private interest; we have to conjure the role of an 'honest broker' intermediary.

The Compact Oxford Dictionary definition of an 'honest broker is: 'n. a mediator in inter-national, industrial, etc. disputes'. This implies significant conflict is threatened or underway; for example, at the international governance level (UNESCO, 2011). The role is therefore applicable at the highest governance level, and in potentially extreme conflict scenarios.

For present purposes, it is assumed there is no correct use of the term by a particular research or policy community and an honest broker intermediary conveys good inten-tion that all stakeholders are well represented and that proceedings progress to the benefit of all; that is 'without prejudice' and hence would be operating according to an agreed ethical standard converging on a 'win-win' outcome. A commonality might be that an individual or an organisation is operating as 'honest broker' intermediary implies no self-interest other than fulfilling a specified role, and that may be in keeping with a mission statement, or some agreement brought into a PCF process. It will be argued that the role is therefore, in river basin management context, suited to the voluntary sector for they operate 'without prejudice', are both value driven and 'not for profit'. Operations must also be open and accountable in order to foster and maintain trust. There are parallels with the openness required under the 'Chatham House Rule' for settling international disputes (Chatham House, 2014), where for example information

Figure 10.6 The honest broker model of the policy process (de Witt, 2005).

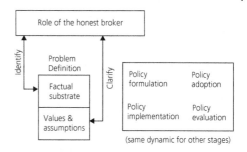

may be offered in negotiations, but not attributed. Yet there is a contradiction, for in open civil society the interests of transparency make it undesirable that stakeholder interests remain anonymous or that no record of business transactions are made!

Aside from mediation in ongoing disputes, the honest broker operates in the environmental policy formulation process itself. Within this is a need to reconcile positivist 'facts' with normative opinions. De Witt (2005) sees it imperative to include clarification of the scope of a problem, providing a balanced accounting of the pros and cons of various policy alternatives, explaining policy ideas and likewise the options in an objective fashion. At formulation stage, these policy processes are undertaken by 'parties-at-interest' who may indeed be contesting between facts and opinions.

De Witt's argument is one of *uber-demokratie* giving synergy with environmental politics. For the 'foundation principle of democratic theory is the assumption that a well-informed citizenry is the best possible basis for making good public policy.' Figure 10.6 shows one attempt to describe what is a potentially difficult process, for example, one that operates within a 'wicked problem' framed policy environment.

Pragmatic solutions should involve networks concerned with knowledge exchange and building trust. The presumption of most contemporary environmental political thinking is, indeed, to embrace a democratic process at (ideally) many levels. Even if there is little direct engagement with government, at some level (be it parish, town, rural district, county, etc.), there is engagement with many stakeholders. Analysis is required for both description and planning.

For convenience, characteristics required of an intermediary to delivering catchment management are:

- Political neutrality (it does not represent partisan interests).
- Distant from 'official' government and corporate policies.
- Able to build and sustain trust among stakeholders.
- 'Not for profit', although its own *raison d'etre* should be clear.
- Technically competent and able to call on outside expertise.
- Administratively efficient, transparent and financially competent.
- Has open access to data and to policy aspirations.
- Is capable of itself acting as consultee, for planning and policy-making purposes.
- Is a repository for local knowledge that, once gained, is sustained and increased.
- Is itself financially sustainable.
- Possesses flexibility and low operational overhead costs.
- The ability to report success and failure into other stakeholders and the policy formulation process.

The honest broker intermediary may be endowed with an economic function and Smith (2012) expands the instrumental roles of the honest broker:

- Set-up including research, stakeholder and partner consultation, legal agreements and monitoring arrangements
- Operational considerations, including:
- research and modelling to establish and target appropriate land management techniques and practices,
- identifying, recruiting, organising and advising ecosystem service sellers,
- negotiating agreements,
- demonstrating outcomes or practices,
- establishing protocols,
- arranging payments,
- monitoring of agreements,
- monitoring and evaluation of programme,
- Building and maintaining social capital and legitimacy,
- The characteristics of an honest broker are to be independent, credible, honest, transparent, accountable, trusted – especially locally accepted.

Challenges are minimising set-up, transaction and operational costs, maintaining equity and acceptance. A particular scheme, for example, may require that a scheme be based in 'additionality', that is providing for an outcome that is beyond those of existing, generally statutory, requirements.

10.10   Adopting the Catchment Approach

Modern river basin management adopts, largely of necessity, the catchment approach (Chapter 9). EA identified 10 pilot catchments in April 2011, with a further 15 identified in the following January (EA, 2014d) where new approaches in community engagement are being trialled in the light of WFD. Catchment-scale operations should ensure local knowledge is to effect change, through:

- identifying and understanding issues within a particular catchment,
- involving local groups in decision making,
- sharing evidence,
- identifying priorities for action,
- seeking to deliver integrated interventions in cost effective ways that protect local resources.

The new departure being that catchment vision is formed by stakeholder engagement, rather than all through expert opinion, albeit including a public consultation exercise. In addition there are Demonstration Test Catchments (DTC, n.d.) (the Wensum, Eden and Hampshire Avon) and the work do the Westcountry Rivers Trust is also being evaluated for impact of its participation methods (LWEC, n.d.). DTCs are closely monitored for diffuse pollution in the face of changing farming methods. In many cases, the EA defers evaluations to 'external hosts' including the Rivers Trusts.

The EA believes that to deliver the objectives outlined, successful adoption of the Catchment Based Approach will follow these principles (Defra, 2013d):

- That there is an environmentally focused planning and management process covering every catchment in England.
- That there is an opportunity for local engagement for every waterbody, irrespective of whether or not catchment partnerships exist.
- Formal catchment partnerships will be recognised by the EA. Leads in partnerships will be agreed with stakeholders in the catchment according to their ability to tackle the issues in the catchment in a collaborative way.
- Catchment partnerships look at the water environment in terms of all the ecosystems services connected to a healthy catchment and aim for better integration of planning and activities to deliver multiple benefits (for example, supporting the delivery of objectives for Water Framework Directive, Biodiversity 2020 (a governmental strategy for England's wildlife and ecosystem services) and flood risk management).
- Catchment partnerships inform the river basin district planning process and become integral to the way that WFD objectives are delivered providing a degree of flexibility to respond to emerging local evidence.
- Other groups in and across catchments continue to operate, particularly at a more local community scale or around a specific issue. They seek any formal recognition of their activities in River Basin Management Plans through the catchment partnership (where they exist) or the local Environment Agency catchment contact (where no partnership exists).

It will be recalled that 'localism' is a paradigmatic shift away from centralism and state planning. The EA finds, in the evaluation/review, a ringing endorsement of partnership formation. Scales of 'community' of sub catchment, catchment and river basin are recognised, and in its Appendix 3 there are four areas of competence, from the evaluation of the pilots, these are *Leadership, Co-ordination, Expert Facilitation* and *Technical Skills.* This hits the buttons of river basin governance, yet it is oddly lacking in prescriptions for improving water supply and the general water environment although flexibility and localism are implicit in tackling local issues.

Statutory groups emerged since the Environment Act 1996. Other groups and new institutions in partnerships included organisations engaged in 'collaborative governance' and these have included (Cook *et al.*, 2012): Regional Fisheries, Ecology and Recreation Advisory Committee (RFERAC), Regional Environmental Protection Advisory Committee (REPAC), and the extant Water Framework River Basin Liaison Panels (WFRBLPs) established in 2006, Catchment Sensitive Farming steering groups and Catchment Flood Management Plan Steering Groups. The first two were regionally (rather than catchment) based while all are thematic in nature. The exact relationships between vertical and horizontal political integration, between roles, task allocation and relations between sectors, remains fluid. Highly significant in WFD terms, WFRBLPs strongly relate to the six-year cyclic river basin planning process of the Environment Agency (the process is undertaken by SEPA in Scotland) in each RBD. RBDLPs see their role (IWA, 2013) as:

> 'RBMPs are prepared (to date) in consultation with 'stakeholders' in 'River Basin District Liaison Panels' (RBDLPs). These include some recreational/navigation

representatives but the vast majority of 'stakeholders' represent wildlife trusts/ projects, fishing and landowner bodies. Policies are developed by consensus in a series of workshops.'

Flood management has since been concentrated in Regional Flood and Coastal Committees (RFCCs) in England and Wales (Lorenzoni, *et al.*, 2015). RFCCs are regionally based, multi-actor and multi-functional and meant to provide a coherent mechanism for mediating government policy in respect of floods and coastal erosion risk and there remains a question over whether flood management has been concentrated in this process (Benson, pers. comm. 2015). Committees determine flood risk management decision-making in their regions and the management approach involves multiple actors making decisions in the Committees that generally speaking comprise a chairperson appointed by the Secretary of State from the local community, Lead Local Flood Authorities (LLFAs are county councils and unitary authorities) plus other actors appointed by the Environment Agency; primarily epistemic experts, NGOs and/or landowners. LLFAs are therefore in a strong position to engage stakeholders.

Representatives of the EA are also in attendance at meetings. Their functions are set out in the National Flood and Coastal Erosion Risk Management Strategy and include ensuring coherent plans are in place for identifying, communicating and managing flood and coastal erosion risks across catchments and shorelines, and they are concerned with deciding efficient investment in flood prevention. They provide a link between flood risk management authorities (Defra, 2014a). Money is raised from local authorities in order to undertake local priority flood defences and integrate regional and national strategic level planning, CFMPs and Shoreline Management Plans and local authority planning approaches within the region. The system, however, is complicated and liable to disjuncture due to confusion of remits and the question of effective communication between committees remains (Inman, 2015, pers. comm).

As in many countries, one important consequence for the UK is flood control and this has become increasingly politically, economically and socially significant; particularly in response to successive devastating floods since 2007. The 'Pitt Review' followed some calamitous flooding during the summer of 2007, yet this has proven useful in the extreme in UK policy evolution (Lorenzoni *et al.*, 2015; Cabinet Office, 2008). Four principles guided the Review:

1) There are needs of individuals and communities who have suffered flooding or are at risk.
2) Realisation that change will only happen with strong and more effective leadership across the board.
3) There is a need to be must be much clearer about who does what (people and organisations should be held to account).
4) There should be a willingness to work together and share information.

There was a serious re-occurrence of flooding in the winter 2013/2014. Conflicts have emerged over how flood defence investments are decided and the extent to which they reflect local preferences. The WFD requires 'the active involvement of all interested parties in its implementation' (Article 14). The UK government's Localism Act of 2011 required inter alia that a local authority, that is a lead local flood authority, must 'review and scrutinise the exercise by risk management authorities of both flood risk management functions and coastal erosion risk management functions which may affect the local authority's area' (Communities and Local Government, 2011). One of the UK government's responses has been to promote more local level participation in central flood protection investment decision-making in England and Wales via the Regional Flood and Coastal Committees (Lorenzoni *et al.*, 2015). Then there are the emergent NGOs.

Another suggestion for reform (HOC, 2016) is to establish a new National Floods Commissioner for England, to be accountable for delivery of strategic, long-term flood risk reduction outcomes agreed with Government. Delivery would be via new Regional Flood and Coastal Boards to take on current Lead Local Flood Authority and Regional Flood and Coastal Committee roles. Another proposal would be a new English Rivers and Coastal Authority, taking on current Environment Agency roles to focus on efficient delivery of national flood risk management plans.

Spatial planning, a concept growing internationally and especially favoured on the European mainland (Chapter 11) is more complex than simple land-use regulation and should also address problems such as tensions and contradictions among (sectoral) policies. For example, included is economic development, environmental and social cohesion policies. Outcomes promote the arrangement of activities while reconciling competing policy goals. Generally, there are identified long- or medium-term objectives and strategies for territories, dealing with land use and physical development. (Economic Commission for Europe, 2008). It is generally a tool of the public sector, concerned with government action; in governance terms, it is 'top-down'. There is currently no national spatial planning framework in England, something that contrasts with Scotland and Wales.

There emerges disconnect between the notions of governmental statutory planning and 'localism'. The latter is predicated on moving away from what is perceived as an over-centralised planning system to a system which is based on promoting a 'localist' agenda. This strengthens the role of local planning authorities and promotes neighbourhood-based planning and a greater role for existing residents in the determination of development proposals in their neighbourhood'. The dichotomy is addressed in concepts of 'spatial planning' (Bowie, 2010).

10.11 Voluntary Organisations and Public Participation

There has long been a normatively defined role for the voluntary sector (section 10.6) by UK governments of left and right, acknowledged by administrations from Atlee through Thatcher to New Labour and the Coalition 2010-2015. Complexity is acknowledged, so that an emerging aspirational governance structures may variously be described as 'polycentric', 'bottom-up' or 'socially inclusive' as opposed to 'top-down government knows best' (Cook *et al.*, 2012). The promotion of the Big Society and the ensuing UK Localism Act (2011) as a policy of the Coalition Government attempted to cash in on a noble tradition. Fostering multi-level governance and devolved power and action at the local level is both established and desirable, for while there have been problems with state intervention, so also there are with privatisation of national assets!

Voluntary sector bodies offering environmental management advice have been around for a long time dating from the 1960s, the Farming and Wildlife Advisory Group (FWAG) long had a pedigree that promoted biodiversity on farms, but not soil and water conservation. Both the Environment Agency and Natural England proffered advice, aside from any established sources of advice hitherto utilised by the farming community. Akin to the 'One-Stop-Shop' approach to agri-environmental advice for the agricultural sector, it was better to keep the regulatory functions (typified today by EA) and the linked grant-awarding role of Defra's Rural Payments Agency/Natural England at arm's length.

Into this Brave New World of ever-democratising water resource management emerged the third, or voluntary, sector culturally familiar in Britain and ready to assume a role. The third sector may become instrumental in meeting official agenda provided it is both competent and cost-effective (Cook and Inman, 2012). The sector may be answerable to state agencies (such as the Charity Commission and environmental agencies), it actually functions in a distributed way, being reliant on interpersonal relationships and on informal networks among the committed or linking communities with their local environment. It is *amateurish* in the sense that participation is 'for love' although often by non-experts. In many instances, particularly with long-established environmental NGOs, the outcome is actually a highly professional and adequately resourced organisation. But this takes time. Third sector organisations are in a good position to sell themselves as a neutral 'honest broker', an intermediary acting as a benign co-ordinator and engaging collaboratively with stakeholders. The sector is perceived as generally benign (being not for profit, value driven and trusted to engage stakeholders). The resulting trust gives rise to an opinion that the third sector is better serving the public good than statutory organisations. As NGOs increase in executive authority, there always remains the possibility of replacing one bureaucracy with another and they should always be careful to remain in contact with all stakeholders (A. Inman, 2015, pers. comm.).

This view does not does not imply that sufficient funds are available for successful fulfilment of mission by a given body (Cook *et al.*, 2012). While a neo-liberal 'rolling back of the state' might seem satisfied by such imperatives, reducing public expenditure and intervention in labour markets through job substitution (typically of public sector by voluntary sector employment) is open to questions around competence, continuity of function, funding and accountability. Voluntary bodies may well fill in the gaps in environmental action, but if they are to be effective and consistent they, like the public and private sectors, need the resources.

Instrumentally speaking, it is not just 'conservation organisations' such as the National Trust or Wildlife Trusts that are NGO actors. Engagement of the farming community is central to achieving wholesome catchment management. The honest broker seeks to engage landowners, farmers and others in a non-officious, non-judgemental and non-bureaucratic manner that may enhance the reputation, even increase the bottom-line profitability of third parties. The mechanism lies in building trust. Where investment in farm assets is required, externally sourced funds may come from industrial beneficiaries, including the water industry and insurance companies in PES. Where participation is required, there is already much social capital available in the voluntary sector and its growth ensures better networks and better social value.

There is an extensive literature on stakeholder participation and environmental conflict resolution long engaging 'citizen groups' (Crowfoot and Wondolleck, 1990; USEPA, n.d.). Famously, individual exploitation of open access, common property resources has led to the 'tragedy of the commons' (On The Commons, n.d.), thence to Ostrom's rules for common property resources. Environmental economists draw a distinction between 'rival' and 'excludable' goods (Fisher *et al.*, 2009), the former implies that, if one person benefits, then a second will potentially loose out; over-abstraction of water would be a good example. Excludable goods mean that if one person has proprietorial use, as with private use of land, then others will not benefit. If such actions in natural resource use lead to conflict, perhaps because property rights as well as responsibilities are not fully allocated, resulting in the exclusion of other parties by the generation of negative externalities, there is a potential for conflict.

Adaptive catchment strategies engaging agencies and communities, as practiced in the US, look to be promising (Smith and Porter, 2009). Yet, in the UK such a process has yet to reach a maturity, and a growing role for the voluntary sector is identified for environmental management and governance. There are also established private sector organisations who report that:

> 'Our work ranges from helping communities solve local challenges, academics share knowledge, through to international UN Conventions and international organisations.' (Dialogue Matters, 2011).

The development of both private consultants and voluntary sector initiatives in environmental governance and problem-solving reflects a political will to detach central government action, hoping that other sectors will jump into the breach, albeit the 'Big Society' has scarcely presented a systematic and funding-assured way forward (Cabinet Office, n.d.). On the other hand, mechanisms for the transfer of money destined for environmental improvement are certainly required.

10.12   PPP and PES

The polluter pays principle (PPP) and payment for ecosystem services (PES) are two economic instruments aimed to address market failure and to 'externalise' perceived negative externalities in activities that produce pollution or other ecosystem or environmental damage. Today, PPP is a recognized principle of international environmental law, and it is a fundamental principle of environmental policy of both the Organisation for Economic Co-operation and Development (OECD) and the European Community (Encyclopaedia of the Earth, 2008).

Apportion of blame is implicit in PPP and the principle has been favoured by Ofwat for England and Wales; in the UK and elsewhere PPP is achieved through charging for discharge licenses. In the case of water, this relates to licenses issued for volume, concentration and timing of discharges of potentially polluting substances. The revenue recovered supports the overall cost of regulation. In an ironic twist, the water industry has become its own (privatised) 'poacher and gamekeeper'. This is on account of it becoming both service provider (clean, potable water in appropriate volumes for the consumer), and polluter on account of its need to discharge treated effluent to the environment, largely through STWs. The switch to PES is, in reality to a 'Provider Pays Principle'. It is actually applied to interests external to the water industry.

Mobilisation of economic resources requires a trusted and efficient financial and regulatory system. The dynamic of river basins furthermore essentially links upstream users and down-stream recipients of the water and that is easier said than done. While the PPP implies a polluter pays for the 'benefit' of pollution and that seems to accord with natural justice, more controversially, a perceived beneficiary pays a presumed polluter in PES. Or simply stated, a party is paid **not** to pollute. To quote:

> 'PES schemes, effectively, provide incentives to address market failure by altering the economic incentives faced by land managers or owners. In this sense, PES can be argued to fit within the broad category of market-based (economic) instruments which include taxes and charges, subsidies and direct market creation. Given the diversity of reasons for market failure and the challenges for the natural environment, appropriate action will depend on a mix of policy instruments including regulation, economic instruments and other approaches including voluntary, behavioural and information tools' (Dunn, 2011).

Actually the quotation is clear that PES must function alongside more accepted means including regulation, but that would not be to the fore. The proverbial policeman is at arm's length. For the PES economic case, Fisher et al (2009) link PES directly to human welfare using cost-benefit analysis, and the process is explicitly concerned with minimising transaction costs. If PCF is evoked, PES schemes are voluntary, conditional between at least one 'seller' and one 'buyer' and relate to a defined environmental service (Wunder, 2005).

Figure 10.7 (Defra, 2011c) sets PES in a contemporary policy context. It stands apart from the long-established PPP because, rather than the 'offending party' paying, the beneficiary pays. In that sense, not only is it easy to define what the requirements (for improvement) on an ecosystem may be, but a price deemed to be proportional to that benefits is paid. Figure 10.7 shows PES contexts and contrasts the PPP and PES approaches.

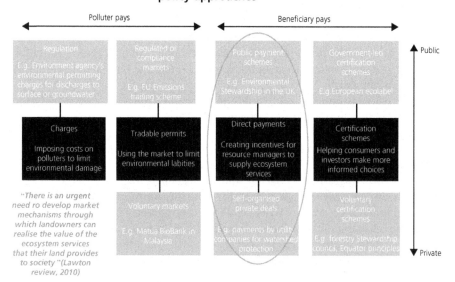

Figure 10.7 Policy context For PES (Defra, 2011c). Direct payments for ecosystem services as applied to the UK is ringed. The 'Environmental Stewardship' scheme is being replaced by 'Countryside.

In the catchment management context, payments to land managers for action (or non-action) ideally increase the quantity and/or quality of desired ecosystem services. *Conditionality* means there is an obligation on the recipient of the funds to undertake a course of beneficial action, but this should not be a 'reward for bad behaviour', something unlikely because of Cross-Compliance and Codes of Good Practice.

Inputs should therefore be additional, that is there should be aggregated benefits accruing from payment for a service, or a bundle of services, from the process that would likely not be realised without a PES agreement (Smith, 2012). Furthermore (and unlike for example the case of NVZs) the general consensus is that new services should be voluntary with no obligation to meet regulation or a mandatory cap (Wunder, 2005). This contrasts with a purely regulatory approach. Payment to land managers may be administered by government and sourced from the taxpayer (as a beneficiary) or the source of funding may be from a beneficiary of the ecosystem service that is provided. Actually, where the beneficiary is private, an NGO may be the intermediary, or trusted 'honest broker'. The basis for payment may be either output based (payment by results, such as reduced pollution) or else a specified (regarded as less damaging) land use is required. The adoption of outcomes is therefore in accordance with the 'achieving' good ecological status' required by WFD.

In Britain, PES remains very much in its infancy, although as will be seen the idea has been enthusiastically endorsement by South West Water (SWW) paid through the intermediary of the Westcountry Rivers Trust (WRT). According to Dunn (2011), 'There is growing evidence about the significant potential for long-term growth in emerging markets for biodiversity and ecosystem services'.

There inevitably are barriers and challenges that relate to a lack of awareness among potential beneficiaries to technical uncertainties to scaling of problems and set-up costs for PES schemes. However, in theory, the specificity of beneficiary requirements, the trust invested in honest brokers and the technical advice available to farmers or industry via government agencies or consultants should provide flexibility that goes beyond simple Codes of Good Practice or bespoke schemes such as NVZ.

The interest of the private sector in PES has also to relate to meeting notions of 'corporate social responsibility' or be simply instrumental in meeting real needs arising from consumer pressure, or from the statutory requirements arising from the regulatory environment (Dunn, 2011; Defra, 2011c). In this report for Defra, it is water quality, water supply and flood risk attenuation that are focussed, while it is noted that publicly funded agri-environmental schemes tend to be more ecologically focussed. PES may be additional to what is required under statute, or achieved as an outcome of an extant agri-environmental scheme. In short, it should not be adopted in preference to other instruments where good governance, economic and practical solutions may present themselves as appropriate in meeting positive water and environmental outcomes.

10.13 Down on the Farm

For context, the agricultural industry in UK has, since the late eighteenth century, and most notably post-1940, progressively undergone governmental intervention of one form or another in the national interest of food security, generally by protecting the domestic industry. Most notorious until the mid-nineteenth century were the 'Corn Laws'. The long-term result today, of dipping in and out of interventionist economic policy, is food security, then historically cheap food (Cook, 2010a). Long-term, this impacted on farm incomes and due to economics of scale, the most vulnerable holdings to losing out are small farm units. If equity is the issue, then PPP is neither relevant, nor fair on the small 'family' farms least able to survive fluctuations in the produce of agricultural produce for their profit margins are low and hypothetically they would have problems paying a 'pollution tax'. In England and Wales agriculture remains a major contributor to diffuse pollution to water and air and hence government funding for research has been substantial (Burke, 2011).

Yet, it is also recognised that farmers produce social and environmental goods – that is economic products other than food (Potter *et al.*, 1993). Economic instruments were once enacted to support the prices of agricultural produce, then to reduce grain surplus in UK. The policies that emerged since the 1990s support activities of farmers that are not directly related to food or energy crop production. These include not only 'sustainable practices', but also protect, enhance or extend conservation areas. Simply stated, the environmental case is made where there are significant benefits that directly accrue from payments. Critics evoking a requirement for additionality; that is, benefits delivered that go beyond what may be required by statutory requirement or Good Practice would point to extant schemes. These include Nitrate Vulnerable Zones (NVZ), Groundwater Source Protection zones (SPZs), Water Protection Zones (WPZs), HLS and England Catchment Sensitive Farming Delivery Initiative (ECSFDI). Importantly, these are financed using public funds (Chapter 9) and appear permanent fixtures (EA and Defra, 2014).

The water industry increasingly looks to farmers as environmental service providers (Ross, 2010, *pers. comm.*) rather than accusingly as 'reckless polluters'. This is largely because long-term food security policy and demands for cheap food make it difficult to square production with the PPP, without driving smaller enterprises out of business; hence many believe the position of small farmers rather lacks economic equity with other polluting enterprises. 'Family farms', after all remain part of the fabric of upland Britain and importantly *they maintain the landscape*. In the circumstances, we may reasonably ask the question, if not the tax payer or the farming industry, then who pays for pollution prevention and ecosystem services? PES actually does not rule out public funding and arguably aspects of current agri-environmental schemes represent just that where payments are delivered from central government funds. Because PES involves a voluntary transaction in which an environmental service is paid for by one or more buyers; that is beneficiaries pay for providers for ecosystem services. In catchments, this service may involve land use change to provide such a service (Couldrick, 2012, pers. comm.).

In policy terms, notions of 'fortress conservation' of environmental hotspots (typically wetlands, bogs, woodlands and grasslands) may have instrumental purposes protecting a specific species, habitat or landscape and are readily identified and understood by stakeholders. However, there has long been an argument that the approach has the potential for encouraging poor agricultural or conservation practice in the wider countryside. To succeed, there must be agreement to avoid aggravation of environmental problems in 'hotspots' and reduce overall degradation and pollution arising on farms.

Interest has grown in PES schemes due to the adoption of best practices in intensive systems and also a targeted reversion of land to more extensive management. This may deliver ecosystem services ranging from water quality improvements to flood attenuation and carbon capture. PES in essence is no different from other targeted agri-environmental benefit approaches, although such schemes are normally paid through statutory agencies. PES may assist in the delivery of other schemes such as ECSFDI (Chapter 9), although in general, PCF analysis suggests that the transaction costs of conventional agri-environmental solutions are high. This might include the cost of water treatment, flood defences, or the transaction costs of establishing and maintaining 'fortress conservation' areas.

Actually, the practical solutions put forward may involve investment, but they often save costs in the longer term that arise by the adoption of BMPs and following Codes of Good Practice for farmers. Examples include better maintenance of farm road surfaces that reduce veterinary bills arising from animal movement and efficient nutrient management that promotes the uptake of N and P by pasture or arable, thereby reducing loss of nutrients from artificial or organic fertiliser Defra (2009c).

10.14   Rivers Trusts: A Call to NGO Action?

A complementary approach to nature conservation in the form of 'fortress conservation' (typified by SSSI designation, Chapter 3) is that of 'Community Conservation' (WWF, 2010; The Wandle Trust, n.d.). By empowering local people to conserve their local resources, win-win situations are sought by which financial, social, environmental quality and habitat gain may be achieved. This is a step beyond the (nonetheless ever-useful) countryside conservation employed by the voluntary sector for decades. Key to this argument is the gaining of environmental goods from alleged perpetrators of problems, so that money is saved or gained by way of recompense for providing an Environmental Service.

As already discussed, NGOs are value driven rather than being for profit or driven by regulatory considerations. Ideally, stakeholders come to enjoy the fruits of 'win-win' situations, be that in environmentally beneficial outcomes or financial benefits or savings. Neither farms nor particular industries set to damage or pollute, such an outcome is in effect a by-products of market failure whereby *the true cost of producing a good is not factored in to costings*. In this instance, marginal costs involve not only balance-sheet changes but there is also a cost to the environment in the form of degradation increased if environmental factors such as contamination of air, water or land are not factored into ordinary calculations of profit returned from a particular production process.

An industrial funder may act in the public good and reduce its own economic margins and it may be seen as protecting its assets (clean water from wholesome catchments) and fulfilling legal obligations in terms of water supply and treatment. It too potentially provides a public good but through meeting its own obligations, as a 'for profit' organisation it is only answerable to consumers and to shareholders cannot be seen to be politically neutral, or without strong economic interest. None of this makes it a 'bad company'.

Ostrom's 'rules' bring us to the conclusion that collaborative stakeholder engagement processes must proceed through communication, including conflict resolution.

There should also be good understanding between outside bodies (including regulators). While an organisation in any sector might potentially link the supplier and benefactor of an ecosystem service (a role for and honest broker), the case for total neutrality is strong as it is 'value driven', yet hopefully economically and administratively efficient, not-for-profit – and politically neutral. The engagement of the private (profit motivated) and third (not for profit) sectors bring together beneficiaries (water industry) and alleged polluters (farmers, but also water service companies).

Trusts and similar organisations concerned with angling have been around for a long time. For example, the Salmon and Trout Association was formed 1903, and is today a company limited by guarantee. It was granted charitable status in March 2008 with a UK-wide membership of game anglers, fishery owners/managers, affiliated trades and members of the public with an interest in conserving the aquatic environment (Salmon and Trout Association, n.d.). Thus, not all organisations, from their inception, presumed a broad remit. The Wessex Salmon and Rivers Trust (formed in 1992) was a volunteer group of anglers, owners, river managers, scientists and environmentalists concerned about the decline of lowland river salmon (Wessex Salmon and Rivers Trust, 2015). Such organisations, while often campaigning, have been criticised for being 'sectarian' in that they represent the interests of angling above other stakeholders. We are not there yet. In practice what is emerging in UK are broad-based Rivers Trusts as NGO bodies, formally this was the nationally organised Association of Rivers Trusts, or 'ART'.

The normative view is there is an important role for the third sector in PES mediation, as honest broker, the concept being fairly new to the UK. The honest broker then has many characteristics of a researcher in requiring access to information. That role may not have such a strong role in policy formulation as understood at national (i.e., UK government) or supra-national (i.e., European Commission) levels, but must operate effectively within that policy environment. Furthermore, in the spirit of adaptive catchment strategies (Figure 10.4), there are numerous feedback loops which continually inform strategies in order to optimise outcomes. The honest broker cannot help but become instrumental in policy formulation over time.

ART (now the 'Rivers Trust') dates from 2001, a relative NGO newcomer. Given the long history of angling-focussed interests and recognition that solutions to catchment-wide problems are required to benefit not only fish populations, but wider ecological, societal and industrial interests, its formation has components of both reformation and revolution. Therefore, with the aim of presenting a national umbrella body for the development of the movement at the catchment and river basin level, The Rivers Trust helps secure substantial funds in pursuit of local initiatives. In the new governance environment, it builds solid bridges between statutory regulators and those on the ground including farmers, anglers and conservation groups it engages with the water industry and promotes development of 'social capital' through participation and more formal education. To quote (The Rivers Trust, n.d.a):

> 'Rivers trusts have been described as having "wet feet" because they have the reputation of being "doers", concentrating much of their effort on practical catchment, river and fishery improvement works on the ground. In the history of almost every trust there has been a key trigger leading to their formation. ...
>
> ....Most rivers trusts start out as riparian, fishing or river associations by combining the use of best available science and data drawn from the

Environment Agency etc. and their own resources, usually an energetic band of volunteers from angling clubs and riparian owners in each catchment or river basin. A river association covering the whole catchment provides an excellent means to identify problems and campaign for improvements.....'

In the North of England, The Eden River's Trust was established in 1996 and has run over 200 projects, identifying education as a major part of its activities as well as fish and other surveys and a great concern about adapting land use for climate change and flooding (Eden Rivers Trust, 2015). In order to counteract some of the threats to the river and to conserve, restore and protect the river environment, this Trust carries out the following work:

- Fencing; to prevent livestock damaging, and polluting, the river.
- Erosion Control; includes bank re-profiling and willow spiling to reinstate the river bank and prevent further erosion.
- Water Supplies; to provide an alternative water supply to livestock.
- Farm Advice; to help farmers work in a more environmentally friendly way.
- Barriers and Fish Passes; to allow spawning fish access to all parts of the river.
- Tree Management; to improve the habitat for wildlife in and around the river.
- Instream Works; to create gravel spawning grounds in canalised stretches of river.
- Water friendly farming; good practice guide.

10.15   The Westcountry Rivers Trust

WRT is an environmental charity (Charity no. 1135007, Company no. 06545646). It is a pioneer River Basin Organisation (RBO), established in 1995 with the aim:

> 'to secure the preservation, protection, development and improvement of the rivers, streams, watercourses and water impoundments in the Westcountry and to advance education of the public in the management of water' (WRT, n.d.a).

WRT has achieved non-hierarchical stakeholder participation, it became complementary with public policy and participatory means to deliver ECSFDI Natural England (2014), and all in the aspiration for IWRM. WRT experience is thus reported as an exemplar of 'good practice' (HOL, 2012). This Trust (Couldrick, 2010, pers. comm.) sees an intermediary honest broker instrumentally as agreeing target areas where there is some problem, for example a likely source of diffuse pollution or target for ecological restoration. It is important therefore to ensure multiple benefits for all stakeholders (e.g., Water Companies, Carbon Offsetting Businesses, Insurance Groups, Harbour Dredging Groups Tourism Groups, local authorities). Finally, there is a requirement to minimise administration (and hence transaction costs) for all parties

Affiliated with the Rivers Trust, the founders of WRT had identified that, despite the best efforts of statutory bodies and government departments, the integrity of the rivers was diminishing and that impacted on fish stocks. Discussions commenced in 1993 with Arlin Rickard as inaugural director and who, since 2004, became the National Director of ART/The Rivers Trusts). Charitable status was obtained because not only would this provide tax-effective but NGO status engenders trust among stakeholders.

The success of WRT is visible in the 'Cornwall River's Project' (2002-2006) that secured funds and engaged the community, especially farmers and the education sector in a multi-disciplinary project (Cornwall Rivers Project, n.d.). The Cornwall Rivers Project can be seen as the test-bed for the 'community conservation' approach in SW England, and indeed for the remainder of UK, largely via the emergent rivers trust movement. The genesis of the Trust in 1993 was focussed on perceived needs - hence there is a need to target both public and private funds. Achieving the protocols of 'Catchment Sensitive Farming', the current form of the programme of the ECSFDI, largely has its roots in the work of WRT.

Specific problems encountered prior to actions by WRT (Bright, 2010, pers comm) were many, partly based in a lack of local knowledge through nationally recruited advisors, and partly overburden of farmers bombarded by well-intentioned advice from several quarters. The WRT has no official membership scheme, and no core funding, leaving it 'political neutral' with funding largely project funded. In April 2004, Tamar Consulting was launched as the commercial arm of the charitable Trust and this donates money to WRT. The company was headed up by Alex Inman, and concentrated principally on ecological services, environmental market research, and running training courses to teach people how to write EU project bids. The late Dr Dylan Bright was WRT Director between 2007 and 2013. In 2010, the Trust employed around 18 people and delivered around £1 million worth of work every year. Since then, both its number of employees and turnover have increased and WRT comes into contact with thousands of people every year (Cook et al, 2014).

WRT subsequently entered the policy arena on account of its *modus operandi* on the ground and novel positioning in the new emerging polycentric, multi-level governance environment (Chapter 4). Operating within several sub-catchments of the Tamar system (Figure 10.8), WRTs operation as an NGO keeps it free from suspicion of being partisan (that is beyond fulfilling its stated mission) and its non-bureaucratic nature makes it cost-effective and flexible. The objection to the voluntary sector, that it is resource-scarce, when compared with a public sector agency would not seem to apply as the Trust has been successful in securing funds from EU and UK and achievements that has kept it free from the common assertion that NGOs are prone to under-resourcing.

WRT promotes the ' Ecosystem Approach' in spatial planning and also works to identify 'win-win' solutions to long term problems, notably as they affect farmers. This Approach seeks to combine 'political' considerations with economic and sound methodologies in environmental management and in spatial planning and WRT has evolved 12 principles (Box 10.3).

On account of its relative longevity, the WRT provides opportunity to scrutinise a partnership river basin organisation (Cook *et al.*, 2014), for WRT has been successful in mobilising resources, building collaborative relations and addressing environmental problems. Hence the WRT provides a model of partnership development for 'lesson-drawing' (Benson and Jordan, 2011). Indeed, WRT experience is reported in UK government documents (Cook et al 2015).

This Trust performs advisory and brokerage roles based around integrated catchment management principles. Primarily, the WRT delivers advice to farmers and riparian owners in order to implementing management change and change behaviour for the common good, ideally so as to improve profitability and diversification and to restore

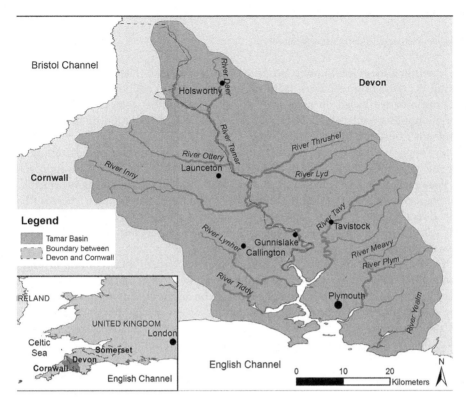

Figure 10.8 The river channel network in the Tamar catchment in the South West of England with main towns. (*Source:* Cook *et al.,* 2014).

Box 10.3 The ecosystem approach

The Ecosystem Approach serves as a tool in determining project scale, the targeting of effort, gaining engagement of stakeholders, empowering communities and to ensure a successful self-sustaining exit strategy.

In WRT context:

1) Objectives are a matter of societal choice, established at grass roots level.
2) Management should be undertaken at the lowest appropriate level.
3) Consider the effects on adjacent/other ecosystems to a river.
4) Understand and manage ecosystems in an economic context as a component of sustainability.
5) Conservation of ecosystem structure and functioning is a priority.
6) Manage within limits of functioning, this is pushed to the limit by agriculture.
7) Use appropriate spatial and temporal scales (catchment, sub catchment, farm, field, time of year, cycles of vulnerable species).
8) Objectives for ecosystem management should be long term (for land managers as well as policy makers).
9) Management must recognise that change is inevitable often reflecting economic considerations).
10) Keep an appropriate balance between integration of conservation use and use of biological diversity.
11) Consider all relevant information relevant to the 'water cycle'.
12) Involve all relevant sectors of society and science.

The above is adapted from: The Rivers Trust, n.d.b

ecosystem function in catchments in South West England. The WRT has proved successful in leveraging resources from multiple sources in support of an essentially bottom-up initiative based on consensus decision-making between diverse partners. A strong emphasis is placed on securing the voluntary collaboration of stakeholders alongside the utilisation of local knowledge.

Another approach increasingly utilised for achieving its aims for integrated catchment management is through facilitating payments for ecosystem services (PES). This approach is complementary to statutory regulation (Defra, 2011c). Under the polluter-pays principle, there is regulation or market compliance (permit trading being an example), whereas PES involves direct payments to the 'service supplier' from a beneficiary. The WRT, as we discuss below, is proving adept at broking such 'win–win' opportunities to support more effective catchment management. (Cook *et al.*, 2014).

Naturally, there are devils in the detail. There is concern that community and environmental imperatives can be overridden by market forces and hence there is no long-term guarantee that environmental goods can be delivered. Environmental drivers may be relatively easy to define, economic drivers are less straightforward. However, Ofwat, the economic regulator of the water industry, investigated assets over 30 years and has supported the idea of direct payment for ecosystem services to WRT as honest broker (Ross, 2010 pers. comm; Bright, 2012, pers. comm).

In essence, the cost-benefit of such actions are understood to favour PES while a model of mixed engagement emerged, having components of both advice and regulation. Stakeholders include the big players: WRT, EA and South West Water (SWW). The advice often goes beyond that which is required by law, so that farmers have to be engaged fully when asked to change practice in legally grey areas that nonetheless should prove effective for environmental improvement. The EA frequently refers farmers to WRT for advice.

To summarise from Couldrick (undated, accessed Nov 2010; Bright, 2010, 2012, pers comm.); PES as realised by WRT is founded in five principles (Box 10.4) distilled from the points listed in Box 10.3:

Box 10.4 PES as realised by the Westcountry Rivers Trust

- **Bioregional planning:** Taking a holistic approach to catchment management. Perhaps for the first time in UK environmental management, such an approach has a strong economic dimension that is incorporated within notions of economic development that are complementary with regulatory interventions.
- **Community conservation:** Nothing can progress without local leadership and support; achieving 'win-win' situations is central to the philosophy.
- **Payment is via the market:** A major source of funding for ecosystem services comes not through public expenditure (that is from European, UK or local revenues) but from the direct beneficiaries of environmental gain, provided the beneficiary perceives to gain in a strictly economic sense.
- **Working sequentially to engage the public:** Hitherto protected areas are restricted by boundaries and hence vulnerable to economic changes (especially in agriculture). Community conservation is a broader concept based in 'enlightened self-interest' such

---

that stakeholders are incentivised (not regulated) by the provider (South West Water) and it is important that the company is not seen as a regulator for it is a service provider. Naturally the process of economic mediation involves not only water resources and aquatic conservation, but also wider biodiversity in the catchment. The process has to accommodate 'free riders', provided their involvement is deemed positive for the environment. A single broker simplifies process and reduces transaction costs.

- **Seeking an 'honest broker' Intermediary:** While an organisation in any sector might potentially link the supplier and benefactor of an ecosystem service, the case for the voluntary sector is preferred as it is 'value driven', economically and administratively efficient, not-for-profit and politically neutral.

Bioregional planning requires further explanation. For example, any present analysis of sectoral interests would produce serious distortion towards the food production sector. The adoption of 'aspirational maps' produces a range of different land use and land management scenarios. These might focus on water quality (for habitat and consumption purposes), on habitat conservation and enhancement, on carbon sequestration, on recreation and so on. Maps utilising environmental information in 'intelligent catchment design' (WRT, n.d.b) are produced by weighting GIS overlays each on a scale 1 to 5 (theoretical maximum score of 35). These layers comprise: a clean and fresh water ecological layer, a clean and fresh drinking water layer, water regulation for flooding, water regulation for drought, a climate layer (greenhouse gas regulation and air quality) recreation and tourism, provision of habitat. These seven layers form a combined Environmental Services model that is compared with the intensity of agriculture. There is a 6% catchment area, where land use in intensive agriculture conflicts with conservation thereby requiring land targeting of potential 'hotspots'.

In order to achieve PES, whereby beneficiaries pay for ecosystem services, clear needs for investment are required. Identifying these needs is complicated, but as a generalisation, Ecosystem Services currently marketed are:

- Provisioning services of food and energy (hydropower, biomass fuels),
- Regulating services of carbon sequestration and climate regulation, flood and drought attenuation, nutrient dispersal and cycling,
- Supporting services such as the purification of water and air, pest control and biodiversity,
- Cultural services of cultural, intellectual and spiritual inspiration,
- Recreational experiences (including ecotourism).

Most of these are known to be influenced by the farmer, and where they benefit ecosystem service provision, they are provided free to other sectors. For example, goods with clear market significance include food or biomass provision and landscape. There are direct economic benefits to the readily identifiable markets from these. Other activities that might be seen presently as rather more regulatory include carbon sequestration (by wetlands, soils and actively growing biomass), nutrient utilisation and control (avoiding dispersal in air and water) or flood control and mitigation. Supporting services include purification of water, pest control and biodiversity in general. Recreational experiences include fisheries, tourism (both general, cultural and 'eco'). While the wider issues of regulation, laying largely with state institutions setting goals for air and water quality, carbon offsetting or identifying landscape and habitats for special attention, there remains the matter of *regulation and delivery of the PES process itself.*

The original focus may have been angling, but subsequently ART and its trustees not only encouraged broader ecosystem management, but promoted its members as being inclusive in terms of both statutory bodies and developing voluntarism. Tamar 2000 was the first major project that established protocols for dealing with diffuse pollution. The trigger was anglers' concern over decline in salmonid abundance in the Tamar catchment system and a perceived inability of public agencies to achieve any mitigation of the problem. The focus was to be land management and farm advisers were employed so that measures could be enacted at the landscape level: this was due to a lack of spatial planning. There are around 80 elements of farm advice, but at the core are minimising inputs and minimising wastes. Intended beneficiaries were to be habitats including those for birds and fish. Water quality and more even flow hydrographs were to be outcomes (Bright, 2012, pers comm.).

10.16   South West Water United Utilities and PES

WRT recognises the importance of economic analysis to enable improvement to profitability of the small farms that dominate the catchment. The locally identified 'win-win' situations provide positive economic outcomes in reducing costs and increasing profit. This was found far preferable to blunt regulatory advice (Bright, 2012, pers comm.). Here, the true cost of pollution could not be borne by a plethora of small farm enterprises who, in any case, represent the effective management and social infrastructure of the regional countryside. A solution has been sought in PES.

WRT extension workers easily identified systemic shortcomings, typically the overuse of fertilisers. This was in part due to ignorance of fertiliser indices for the soil in a particular field. There were also well-attested gains from river bank protection, from field-based soil conservation and farmyard measures associated with drainage and manure handling. A study of the economics of cattle farming in the area enabled the undertaking of a cost-benefit analysis. One issue is, once established, how incentives are weighted. Statutory bodies such as EA and Natural England are involved, and there are other candidate bodies, including the Wildlife Trusts, or local government bodies. Farming and Wildlife Advisory Group has long has a pedigree that promoted biodiversity on farms, but not soil and water conservation.

Technical matters were managed in concert with micro-economic studies. There remains a paucity of monitoring data, but even in this situation, the adoption of BMPs requires good technical knowledge. A common problem in hydrology and soil science is the scaling up of findings from agronomic plot experiments to a field, sub-catchment or even catchment scale. By analogy, where catchment restoration management deemed desirable through the restoration of peatlands, there are differing scales of information encountered introducing uncertainty at the catchment/headwater scale in landscape restoration (JNCC, 2011). Although the process introduces many uncertainties, generally associated with environmental variability as well as 'edge-effects' implicit in trials, the use of experimental data to parameterise computer-based process models is an essential process in description of physical behaviour. Actually in generating positive externalities the expectation is that benefits beyond those required by law are gained.

South West Water (SWW) is diverting funds to individual farmers and landowners in order to secure, specifically via WRT, multiple economic and environmental benefits, including to water supply and in corporate carbon reduction. Implementation of BMP is central to this process. This process, to date, is probably the most advanced and ambitious PES scheme in the UK. However, because it is operating in a changing governance environment, not only regulatory agencies but the WRT employed to facilitate the transfer of private sector funds, it is at the forefront of policy formulation.

Delivery of PES in holistic catchment management is via 'Upstream Thinking' (SWW, 2012) as a water company initiative. Stakeholders are: South West Water (SWW), EA, Devon and Cornwall Wildlife Trusts and the Exmoor and Dartmoor National Parks (Smith, 2013). WRT operates as 'honest broker', or intermediary, conveniently located between farmers and the water service company (Figure 10.9). Simply framed, this humorous and clear illustration represents the ideal yet contemporary evidence points to it being a fair representation. This evidence is borne out in economic terms by continued funding for both public engagement and for direct river improvement schemes (WRT, n.d.a). These have embraced species and habitat conservation, fisheries management, education and community and scientific research set that has developed bioregional planning and community conservation.

WRT became a major player in terms of the planning process, social capital and conservation. It may be still on a rising tide, supported for its activities from the public,

Figure 10.9 PES in action.

stakeholder groups and public and private sector organisation. As it builds itself as an institution, so it develops policy adopted by public (including government who seeks to promote Rivers Trusts and their role as honest brokers) and increasingly as agency within the private sector. Certainly, the role is suitable for a place within the PCF and is thus open for analysis. WRT had originally engaged with 'polluters' in order to reduce impact on fisheries.

Because the main cause of diffuse pollution has been agriculture, early success was (and had to be) achieved by contact with the farming community. Thus, once agreement was reached regarding 'win-win' situations regarding on farm operations and infrastructure, agreements typically led to improvements in farm balance sheet on the one hand, and reduction on pollution and sediment in runoff on the other. Positive externalities therefore became other benefits that may or may not have been predicted, and included animal welfare, increased biodiversity, and maybe positive changes in flood behaviour. Self-evidently negative externalities (mostly pollution) were reduced.

To summarise, the WRT approach has so far pioneered non-hierarchial stakeholder engagement and it is an operational (and economically efficient) model based in a workable contractual framework, poised for extension and adoption elsewhere in UK catchment management. The institutional design, strategic principles and democratic make-up of this partnership organisation may be considered key to its success. Underlying philosophies have also driven the partnership success. PES has opened up new opportunities to pursue management objectives. As a partnership organisation, the Trust has also provided avenues for democratic engagement by increasingly locating decision making close to stakeholders on the ground (Cook and Inman, 2012). In local authority participation, we may presume an electoral mandate.

Because the cost of water treatment is high and to reduce environmental loading of contaminants, Wessex Water (n.d.) aims to protect groundwater from nitrates and pesticides (metaldehyde in particular) by advising farmers directly through a catchment management team of catchment experts and agronomists to provide advice in 15 'at risk' catchments across the Wessex Water region that comprises 12 groundwater sources at risk from nitrate, one groundwater source threatened by pesticides. Apart from technical matters, advice is offered on financial arrangements associated with reducing agrochemical use. Indications are that early results are encouraging.

In the English uplands, the Sustainable Catchment Management Programme (SCaMP) was a partnership (Eftec, 2010) operated between 2005 and 2010 involving the water company United Utilities, the RSPB, local farmers and a wide range of other stakeholders. Its purpose is to invest in conservation activities in water catchment land with the aim of securing a wide range of environmental benefits, including water quality and conservation for 20,000 ha of land in the Trough of Bowland and the Peak District in Northwest England. That area includes 45 land holdings and 21 farms, the holders of which are incentivised to participate in sustainable land management of the area. Objectives were to: Deliver government targets for Sites of Special Scientific Interest (SSSIs); Enhance biodiversity, Ensure a sustainable future for the UU's agricultural tenants and Protect and improve water quality. With some caveats, ScAMP was regarded as successful and the scheme was extended (UU, 2016).

10.17   Success, Failure and Accountability

There next follows a sad story, for not all catchment management initiatives have or will prove successful. The Avon Valley Liaison Group (AVLG) existed between 1998 and 2007 collapsed due to lack of support. It was never a trust:

> 'The AVLG will focus on biodiversity and ecological issues, whilst having regard to land use, landscape, archaeological, historical recreational issues where they have a significant bearing on the river system's ecology' (AVLG, 2002).

The remit formally covered the 'five rivers' of Salisbury, including the New Forest streams that drain to the lower Avon that enters the sea at Christchurch.

It was a stakeholder forum, with little or no funding yet had laudable aims:

- To provide a forum for the exchange of views and information.
- To promote a collective view on issues having a bearing on the ecology of the river system.
- To coordinate survey and monitoring requirements.
- To raise public awareness and understanding of ecological issues in the river system.

Superficially, AVLG had all the makings of a successful stakeholder participatory group. It enjoyed wide technocratic representation and stakeholder interests included wildlife trusts and angling interests. Two newsletters were being produced by 2003, covering the upper and lower Avon respectively.

Tangibly there were (and still are) a range of important issues, within a highly prized river valley. There was a will to co-operate and apparently an *esprit de corps* among members attending. Discussion furthermore revealed the Group's breadth, permitted reporting back to other angling groups and showed potential for the transfer of experience between the upper and lower catchment. There was, however, also concern over the long-term ability to solve problems and the absence of farmers was causing concern towards the end of the group's existence. And there was no local authority representation, nor (later on) and farmer and landowner representation. Commentators remarked 'how busy certain people are'. Another angler commented:

> 'The AVLG has always appeared as a vehicle that has done its own thing and…for want of a better description and has failed to register on the radar in the actual day to day running of the catchment.'

Perhaps this view was originally against the notion of establishing a rivers trust on the model of the WRT, for 'this would probably add confusion'. He subsequently wrote: 'We are now some way down the road to changing the face of river conservation within the catchments of Hampshire, Wiltshire and Dorset. This was to become 'The Wessex Chalk Stream and Rivers Trust' (Wessex Salmon and Rivers Trust, 2008).

Staff from Wessex Water, the Environment Agencies and the voluntary sector, all regarded AVLG as a potential vehicle for information sharing and delivering catchment management, although there is regret that little was achieved on the ground. There was a call for AVLG to liaise with the objective of delivering 'favourable condition status' for a plethora of conservation designations. The loss of the Avon Valley Project Officer was lamented, no funding was found to replace her and it was implied that AVLG did not actually do anything as a result, or because there was confusion over a mission.

With mission statements apparently clear, AVLG failed due to lack of participation, communication, funding and there were differing views as to both its mission and achievements. There was no landowner representation and the angling lobby apparently just walked away - to form what was to become a Rivers Trust.

The view of Arlin Ricard (2012. pers comm.), Chief Executive Officer of the Rivers Trust, is simply that Rivers Trusts will fail where there is a failure to engage with stake-holders and where insufficient funding stifles development, notably at a difficult time for the economy. This, self-evidently will compromise political relationships or opera-tional frameworks, including the non-creation of informal understandings, of channels of communication including educative and transformative processes, of memoranda of Understandings (MoU) and, naturally, formal financial relationships are manifest in a contract. Likewise, engagement with state agencies and the public are central.

Out of an initiative at the end of the AVLG when the angling interests decided to estab-lish their own organisation, the Wessex and Chalk Stream and Rivers Trust was formed.

Its vision is one of healthy rivers within the Wessex region which are valued and nurtured by the community and which exhibit (Marshall, 2011):

- Sustainable and naturally abundant wildlife.
- High water quality and sustained natural flows.
- Fully functioning ecosystems which link the rivers with their valleys.
- Resilience to climate change and future stresses associated with social and economic development'.

Unlike AVLG, there is no explicit mention of information exchange, or of achieving collective views, or real community engagement, although partnerships are sought and the catchment approach is taken. Other potential issues are differences in farm structure and land tenure from WRT and Eden Rivers Trust, and the porous nature of the aquifers (mostly chalk) makes identification of pollution pathways to watercourses difficult. The WCSRT would appear to be behind the curve and apparently displays a less ambitious agenda than others, although it seeks to emulate them.

In the absence of direct intervention via Defra, the NGO model may not yet prove the only one for public participation and stakeholder engagement with the ultimate effort to meet WFD requirements. Ten pilot catchments have been selected by EA in order to test their participatory process that fell into criticism under judicial review relating to meeting WFD objectives (EA, 2014d). The EA had not been found successful in reaching stakehold-ers, and it ran into problems with the World Wildlife Fund when it attempted to stop the process on account of it being regarded as insufficient to meet WFD requirements (WWF, 2011). The aims and objectives are clear about the need for such as engagement, informa-tion sharing, co-ordination of action, assess action at a local level improve water and deliver multiple benefits; ensure early identification and involvement of relevant stakeholders, share understanding of the current problems in the catchment; ensure co-ordinated activity that will deliver multiple benefits assist delivery of WFD objectives EA (2011d).

The view from objectors was that the EA was considered high handed and disinclined to allow external agencies to engage stakeholders. The EA also wanted to control plans, rather than facilitate 'bottom-up' engagement (Rickard, 2012, pers comm.). While it may often be the case that EA is unresponsive to the views and needs of stakeholders, so the situation favours the development of a template that would enable planning; implicitly putting a PCF in place for the engagement of all relevant stakeholders, suggesting that success resides in the adoption of an iterative, adaptive process of the kind suggested by the USEPA (Smith and Porter, 2005). Looking forward, a PCF in place to reduce transaction costs would not only enable the scrutiny of who and what is involved, but also help analyse relationships as they develop in a fashion not unlike a long-established policy analysis and review process (Cook, 1998). This is especially important because in PES, all economic sectors are involved as are a range of stakeholder interests.

Rivers Trusts would seem to show a way forward meeting the requirements of an 'honest broker', not violating any Ostrom's rules, for they 'fit the bill' in both a cultural and operational way and how might PES overcome real problems that relate to the functionality of the voluntary sector as well as delivering environmental goods) are answerable in the affirmative. WRT has built, and also enjoys trust, as an NGO. Private sector funding through PES inevitably secures employment for professionally competent staff, and along with other (including public sector) funding the process builds corporate experience and wisdom.

This is not to rule out a more 'hands on approach' by other agencies, be they other environmental NGOs or regulators. The EA has been striving to test its participatory approach, and there may be situations where it is the appropriate 'honest broker'. These may be where voluntary sector bodies failed, or were never formed, or perhaps in difficult situations where a more 'hard-line' approach might be needed because of, for example, pollution situations that exceed our worse fears.

Notwithstanding all this, the apparent success in urban areas, particularly the Tyne Trust in Newcastle-upon-Tyne and that of the Wandle in south London do show adaptability of that model in delivering environmental goods. Perhaps the most striking outcome in PES is the lack of objections relating to blame. This is implicit in the (hitherto leading paradigm of PPP) that identifies, regulates and charges a polluter. This is a radical departure, for appeals to natural justice may lead to a conclusion that a minority of (often) marginally economic small farm enterprises apparently compromise the water supply of hundreds of thousands of water consumers. For example, there are more than 250,000 citizens in Plymouth alone. Raw economic appraisal would not do justice to the complexity of the problem. Here, the style of the contracting is one of participatory understanding and agreed practically defined outcomes, than blame culture.

The abandonment of production subsidies in agriculture has shifted to the agricultural sector to produce environmental goods (clean water, habitat provision and enhancement, flood control, soil and landscape conservation). This is the implicit logic from public funded agri-environmental schemes, presently manifest as Environmental Stewardship and since re-named 'Countryside Stewardship' (Countryside Stewardship, 2016). Where marginally profitable farm enterprises appear to lose out, win-win solutions, then formal PES agreements represent compensation. Two approaches run in tandem, one from public funds, the other supported by the private sector.

In contractual terms, 'regulation' has thus moved explicitly (but certainly not entirely) from state compensation and compulsion to the beneficiary (SWW) who needs to achieve in terms of CSR, customer satisfaction, reaching its statutory obligations in terms of water supply and meeting obligations in terms of carbon offsetting. Public agency regulation remains in the background, but it is a new kind of body that does the 'hands-on' work of public engagement and in doing so it askes communities to buy into and 'own' specific environmental issues. No opposition to this dualistic approach has been encountered during the research. There is evolving understanding and co-operation between state and NGO in this case study, perhaps implicit in the inability of the EA to engage stakeholders in demonstration catchments. The PCF approach can only report contractual nullity in such failure (Cook *et al.*, 2014).

10.18 Conclusions

Non-statutory bodies can fail and it can be argued this is a part of an evolutionary process in water governance, or a serious health warning about the retreat of official bodies. The voluntary sector is, however, set to remain a major player in environmental governance but it requires resources and support. Replacing regulatory approaches, Rivers Trusts have sought to engage with individual farms, no mean feat in a catchment with a fragmented farm structure comprising may family farms of limited area and often only marginally profitable. The relations between the farmers went from suspicion to 'buying-in' once the WRT engaged advisers were able to achieve benefits that improved farm profitability, solving on-farm problems and finally reduce deleterious impacts on water and habitat. The transaction costs on engagement and of physical modifications to land use and farm building infrastructure were reduced, resulting in improved profitability, in some instances. Reduction of risk of interference by state agencies (minor concern) to the outright fear of legal action paid no small part, making the farming community eager partners in catchment improvement. Indeed, such was the degree of mutualism that (at least) the opinion forming farmers came to lead discussion and demonstrate measures for their peers (Couldrick, 2010, pers comm.). The sense of contracting can be seen at this early stage of operations.

The cost-benefit of agri-environment schemes and the cost of engaging public agencies have been open to question (OECD, 2010). Pragmatically, with agreement of the farming community, the way becomes clear for PES. Transaction costs in terms of legal fees, policing, regulation and exploring new avenues for expansion of agri-environmental schemes (beyond ordinary compliance) seem to be avoided, or at least reduced. Finally, the idea of 'utility maximisation' failing and thus leading to non-participation recedes (Lubell *et al.*, 2002); or there is a perceived gain as stakeholder valuation occurs leads to the buying of partners (Sabatier *et al.*, 2005).

Yet there are always going to be questions raised. For one, with the official sanctioning of voluntary sector action, and an implied retreat of statutory regulatory bodies, there might emerge new asymmetries, those of funding, power and accountability. The perceived virtue that, while NGOs are independent, they provide environmental management 'on the cheap' is an ever-present concern and government may become concerned about their independence once it challenges the 'official' position from statutory government departments and organisations. We can only be sure that things will become ever more complicated. Furthermore, there are suggestions from UK Government that there should be reforms in flood governance with the creation of a National Floods Commissioner for England stripping responsibilities for flood management from the EA.

An even larger question, arising from the poor improvement in water quality in Britain to 2015 is simply there is a lack of effective governance, be it 'top-down' or 'bottom-up', or something in-between. This is manifest in elements of confusion over task allocation, lack of investment and the wielding of appropriate power by statutory authorities. This may be a manifestation of a lack of a working national Water Strategy that would seem desirable in addressing the wicked problems associated with water management.

11

The USA, Australia, Europe and Lessons to be Learned

Lessons to be Learned?

OK, Dr Cook, you have been doing agriculture in England for.......did you say 6,000 years? Here in the US it's around about 400 years, Australia was it only 200? And, Man, look at the mess we all made!

Summary delivered at the close of a seminar at Cornell University,
NY, USA (2002).

11.1 Them and Us?

The above quotation leads one to conclude that European people have moved around the world messing things up. Perhaps, to be accurate, it has been European-created technologies that created for us water pollution and associated problems, including more recently widely acknowledged global climate change. Industrialisation of manufactures, industrialised agriculture, urbanisation and poor environmental governance have been linked in a globalised mix, recently joined by newly industrialised countries including India and China that are re-running catchment problems otherwise experienced in Europe, the English-speaking world and elsewhere.

Out of context, that *ad hoc* group of Australian, British and American academics and watershed managers in a seminar room more than a decade ago could lead to a counsel of despair. Since then, the increase in greenhouse gases and impacts of climate change continue to grab the agenda of environmental politics such that it often seems that diffuse pollution of waters remains eclipsed by other, more global concerns. In actuality, this subset of planetary abuse may yet produce a quiet revolution in environmental governance that is in the long-term benefits of all humankind.

It is a truism that we have much to learn from each other about what works, be it from the United States and Australasia or nearer, from continental Europe. Implicitly, the theme of this book has been Europhilic, and for Britain it is debateable if the progress to date would have been achieved without the EU (a great subject for a student essay?). For, across Britain there have been well over 200 years of industrial development, urbanisation, afforestation and an unprecedented post-World-War-Two intensification of agriculture. All have left their mark, although it is diffuse pollution that appears to be the most intractable issue across Europe, bar only climate change. The landscape abounds with towns, reservoirs, forests, flood defences, drainage schemes, prairie-style agriculture, degraded natural, semi-natural and aquatic habitats. Less visible than

The Protection and Conservation of Water Resources, Second Edition. Hadrian F. Cook.
© 2017 John Wiley & Sons Ltd. Published 2017 by John Wiley & Sons Ltd.

the landscape, but just as real, are the changing mind-sets of water professionals, technocrats and stakeholders. Within one working lifetime the focus of resource management, and with it key attitudes have shifted from one of economically driven (albeit regulated) exploitation towards sustainable development. Never an easy concept, the shift is itself significant—and why?

Critics of UK water governance cite a 'top-down' and regulatory nature, incorporating legal and institutional frameworks dating from the nineteenth century (Cook, 1998, Ch. 3); their 'technocratic' nature is often cited as a cause of mission failure (Newson, 1992a, Ch. 7, Newson, 2008). Actually, the long-established requirement for 'integrated catchment management' directly addresses a need for joined-up thinking, at first in a purely technical sense. British water institutions were once responsible for the delivery of flood defence, wastewater treatment, drinking water supply, and had regard for ecology and resource planning. In the first instance, public institutions (Regional Water Authorities [RWAs]), National Rivers Authority (NRA) and the EA were criticised as having the perceived failings of nationalised industries, including operational and economic inefficiency, or were slammed by their detractors as 'quangos'.

Neither should we be unfair on previous generations of water professionals, for the industry of the past engendered a sense of public service. Even if RWAs were regarded as unwieldy and (with hindsight) remote from the democratic process following the exit of local government in water matters from 1974, alienation of stakeholders was mitigated through formal consultation in the statutory planning process or in drawing up Catchment Management Plans by the former NRA (Chapter 4). In recent decades, developments occurred in the manner by which engineers design channels, farmers manage the land and certain industrial concerns worry about how 'green' might be their corporate image. The apparent failure of multi-functional agencies post-1974 had a number of plausible reasons; not least under-investment by a government keen on privatisation. The goal posts had moved and there is little of the 'poacher and gamekeeper' problem in governance that was angled at the RWAs. There has also been a drive to improve the entire water environment.

Privatisation, once seen as the solution to the problems of the water industry—including raising money for investment—is also seen as a commodification of water services in England and with it remain issues of governance, tax paid by companies and the profit motive (Corporate Watch, 2013). In Scotland, the water industry incorporates catchment-based water management authorities while water and sewage utility and management remain in the public sector.

The degree to which the British experience can be exported is debatable because British water problems, when seen in a world context, are not very severe. Neither is there presently much to be proud about in a European setting. The relative prosperity of the UK as an EU member state allows an effective consenting system, monitoring and water planning in well-established statutory processes. The general affluence of the population is capable, in many instances, to provide a pro-active voluntary sector so flexibility is the word.

Frustrating though it is, a one-size-fits-all solution to water resource issues and even various 'templates' proposed, all in good faith, fall between the stools of being too prescriptive, or too generalised in their applications to particular problems, although those that build in adaption would have a head start (Chapter 10). One commentator, Mark Everard, proposes a framework for 'the deliberation and decisions for water management'. This reflects previous concerns of environmental, social (especially stakeholder involvement), economic, technical and legislative factors that synthesises many previous ideas, he summarises well:

> 'In short we remain far from achieving the goal so simply articulated by the three words, 'sustainable water management'. The harsh reality is that we have to adapt the principles exercised here to local situations. We cannot be divorced from solutions to sustainability problems in other areas of societal interest, as the management of water is central to so many other facets of development, from food and land security to redistribution of power and wealth' (Everard, 2013, p. 245).

The protection of British rivers, riparian habitats, groundwater and associated wetlands is central to wider water quality goals. The issue here is engagement of farmers, landowners and others directly concerned with land management and with an ability to protect the wider environment, including water resources. Recently, upland 'Water Protection Zones (WPZs)', that protect upland catchments from diffuse pollution (EA, 2013e, Chapter 9), have started to be defined for much of England and Wales. These zones are an implementing mechanism under the WFD. But there remains a matter of regulatory powers. The drive to establish Water Protection Zones WPZs is where voluntary initiatives fail (ENDS, 2010). A significant proportion of water sources in Britain arise in upland areas (CRC, 2010), and in approach, this is similar to the Nitrate Vulnerable Zone approach (EA and Defra, 2014) because would be imposed on technical criteria, with measures taken mandatory. It would, however, pave the way for meeting multiple objectives including peatland restoration, habitat conservation and protecting hill-farming systems.

It would seem that the issues around river basin management and the drive to achieve Integrated Water Resource Management (IWRM) potentially remain at 'high governance' level, leaving community-centred activities 'on the ground': a rich arena in which to develop progressive governance arrangements. However it remains to be seen how (and if) a top-down approach works in reality. In the United States, this approach has involved federal agencies, yet engaged with communities. For the UK, EU legislation has come to trump what arguably became a moribund legal and governance environment, while in Australia, successful public participation is actually driven by regional (i.e., state) government. These examples may all be from the English speaking world, but they serve to show different ways by which the 'problem' (if that is what it is) of democratically unaccountable 'technocratic' management is being usurped in favour of something more inclusive. However, there is no silver bullet.

This chapter will consider such issues as stakeholders/actors, 'joined-up thinking', payment, governance and the role of regulation. Six case examples are given for developed countries:

- U.S. case studies.
- New Zealand.
- SE Queensland, Australia.
- Netherlands.
- Aalborg, Jutland.
- Catchment groups in England.

Even casual observation suggests that significant differences are apparent in the processes of governance. The U.S. example of the upper Susquehanna basin finds local delivery essentially citizen-led. In the New York City watershed, local ownership of problems is achieved, but there has been significant investment from New York City itself in participation and in technical measures in its rural hinterland. The SE Queensland Healthy Waterways Partnership drives stakeholder and public participation, through by what amounts to regional government initiatives, as is frequently the case on the European continent. By contrast, in Britain, the emphasis has really been on the use of regulation, hence the struggle to adapt a more decentralized, regional scale approach; this is an inherent part of WFD.

It will be demonstrated that governance environments are seldom fully imported, yet there are lessons to be shared and, specifically for the UK, in investigating ways to achieve an appropriate level of de-centralisation in water governance. The problem, if that is what it is, is that there is poor connectivity between local government and—hence local democracy (perceived to be weak in Britain)—and water management issues at large. This is despite the point that, since 1996, there has been a strong local presence on the statutory groups that inform the Environment Agency (Chapter 10). For a long time, water management has been 'delegated' to competent authorities and to privatised water undertakings, at least in England and Wales. In Scotland (and Northern Ireland) water services remain public.

However, the potential for top-down is weak, for example, nowhere is one invited to elect a 'water commissioner', nor anywhere do water issues really feature in local political debates or in the manifestos of county, unitary authority, district, borough council, or parish council hopefuls, or from their parties. For comparison, the protection of water catchments supported at a high level is well-established in Singapore, including land-use planning (Leitmann, 1999).

True, flooding issues are a part of the statutory planning issue, as outlined in the National Planning and Policy Framework (DCLG, 2012). In terms of Sustainable Drainage Systems (SUDS) and in flood defence (Defra, 2011d), under the Flood and water Management Act 2010, local authorities enjoy a 'new role' in managing flood risk (Defra, 2010a), but for most instances, that is all. Resource planning, allocation, treatment, ecological considerations, supply issues, economic and finance arrangements and consultation, all lie elsewhere and in the last chapter the rise of RFCCs was mentioned. Resulting outcomes reflect not only a 'top down' political process, but also cultural concepts such as those around countryside, farmland and forest. While this is changing, there remain concerns over the role of 'competent authorities' in delivering public participation (as defined in WFD) and in general by the water industry to connect with its customers.

In different national jurisdictions such as the United States and Australia, the protection of river resources is being pursued by both top-down government agency approaches and more community-led initiatives predicated at the river basin scale (Benson *et al.*, 2013b). Common perceptions are that Britain promoted a rather top-down, centralised form of water management, although this is changing under the requirements of the WFD (EC, 2014a). The United States, meanwhile, outwardly exhibits strong degrees of more 'bottom-up' community engagement through 'watershed partnerships' that nonetheless have benefitted from agency support and regulation at local, state and, increasingly, federal level. The inference is that the UK could learn much from the U.S. approach as it struggles to implement the WFD. But how true is that perception? While lessons are not necessarily transferable inter-jurisdictionally between these two contexts, as conspicuous constraints often preclude effective transfer (Benson, 2009), differences in approaches could provide a basis for potential mutual learning and policy transfer (Benson and Jordan, 2011; Rose, 2005). Questions asked by analysts of water governance environment include:

- Who (in terms of the different actors engaged in environmental governance) governs? Top-down structures dominated by central government agencies may include only limited collaboration with non-state groups (consultees such as industry and 'the public'). More bottom-up approaches involve different and diverse community stakeholders.
- We may also ask how is environmental governance achieved in practice? To answer this question, both legal and policy instruments are engaged. It is usual to identify whether a top-down regulatory approach or more horizontal network steering that privileges the role of different stakeholders is prevalent.
- A third strand of enquiry is around funding arrangements (Cook *et al.*, 2014). Effective outcomes may not only involve public funded top-down initiatives, but also private sources (for example from industry) may be effective.
- Is governance sufficient to permit effective task allocation?

Discussed in this context are the New York City Watershed Programme (NYCWP); Upper Susquehanna Coalition (USC); the Upper Thurne Working Group (UTWG); and the Hampshire Water Partnership. Finally, we return to our initial questions to determine potential lessons on catchment management practice for the UK context.

11.2   U.S. Case Studies: An Introduction

First, we should consider scale. The English may think of the Thames as a large river. The Thames River Basin District covers an area of 16,133 km^2 from the source of the Thames (in Gloucestershire) through London to the North Sea (Defra and EA, 2009b). It is actually modest in size by continental European standards (Stanners and Bourdeau, 1995). The 'Tennessee valley', by comparison, occupies around 80 percent of the area of England and Wales with a discharge about 24 times that of the Thames (Newson, 1992b, p. 96), draining to the Ohio, then the Mississippi. The watershed was soon recognised as the basis for institutional and managerial arrangements. The Mississippi in turn drains around 41% of the contiguous 48 United States (US Army Corps of Engineers, n.d.).

The Tennessee Valley Authority (TVA) came into being in 1933 and there are numerous dam sites. It is a sound model providing a basis for large-scale plans for the development of river basins (Clarke, 1991). Integrated basin management has been achieved through appropriate institutional arrangements, and use of public funds, to control certain environmental problems while fostering industrial development. The TVA has provided a model for integrated river basin management throughout the world, either through example or via technology transfer by ex-TVA executives (McDonald and Kay, 1988), yet the model has been replicated nowhere else in its entirety. Indeed, attempts elsewhere have been disappointing. The effect of the TVA has been to produce a massive electricity generating utility, achieved through regional development.

The TVA influenced farmers (often through government agencies) to conserve soil, reclaim land from gulleying, limit the use of fertilisers, diversify crops and encourage marketing through co-operatives. Otherwise, navigability has been improved, industrial development made possible by huge electricity generation (now supplemented by coal-fired and nuclear plants), forests have been created on marginal land, and rural poverty alleviated through welfare and educational opportunities available to TVA employees. Malarial mosquitoes are controlled by maintaining water levels from April to June, and flooding has been controlled where appropriate. The authority has had its critics, at both the environmental and planning scale. Certain natural features of the valley, fisheries and habitats have been lost, and the TVA has drawn some harsh criticisms from conservationists; it was, after all, conceived in a 'pre-green' era. Other critics have a cultural problem with large-scale planning and public authorities which seem redolent of socialism, and have been prepared to criticise its financial arrangements (Newson, 1992b). In its modern form, it has also been seen by some as an agency of big business and the state, somewhat removed from its original objective of reclaiming land and alleviating grinding rural poverty (Clarke, 1991). In the present, economic development remains key to TVA operation (TVA, n.d.) for the organization aims to catalyse for sustainable economic development within its region. The TVA functions in four areas, namely recruiting major industrial operations, encouraging the location and expansion of companies to provide employment, helping communities develop assets that make them attractive to companies seeking a new location and encouraging entrepreneurs (including women and minorities) to start new businesses.

Returning to physical considerations, the U.S. Army Corps of Engineers has supplied skills in navigation and flood defence, and in planning since the mid-nineteenth century reflecting the importance of communication, settlement and agriculture during the pioneering phase. Today, the Corps has mostly civilian employees and concerns itself with issues other than water quality. Recently the cost of 'hard engineering' solutions such as embankment and straightening rivers, notably the Mississippi, has been brought into question. Another federal agency, the U.S. Bureau of Reclamation, is concerned with water conservation and transfer projects for irrigation and hydropower predominantly in the drier watersheds further west. This agency, along with the U.S. Army Corps of Engineers, has been active in the development of major irrigation schemes in California using diverted water from the Colorado River. Dams were built between 1910 and 1963, with the Hoover Dam being constructed in 1935 (Newson, 1992b); however, water is in short supply in the southwest of the United States and there are conflicts over its use between Arizona and California. There is also a clash of interests within California, between urban supply and irrigation.

The U.S. legal system is derived from English Common Law, and there is a comparable regulatory framework to the UK that operates at the federal level. Water supply institutions in the United States are, of necessity, complicated since it is a large and diverse country. In common with most societies, there is no ownership of running water. As boreholes and reservoirs were developed, a hierarchy of rights based upon a system of seniority of claim emerged. Later, claimants would have their supply curtailed first of all, and this then replaced a system of riparian rights (Newson, 1992b), although this approach remained operational within the eastern states. The politics are such that differences occur in the (federal) governmental system, and with the tradition of 'home rule' in many states, make local governments fiercely independent and focussed upon locally elected officials. Consequently, the United States has the structure for good participation at the community, stakeholder, municipality and county levels (Willet and Porter, 2001) that has led to all-important watershed community participation and local leadership. Yet, a political polarisation results between federalisation (making the United States potentially top-down) and democratisation at state level or below that drives community involvement.

Critics have blamed legislation for the enhancement of 'federalisation' (Kendall, 2004), with states averse to invading prerogatives of private land owners and municipal land use functions. There are two main interventionist influences. The regulatory United States Environmental Protection Agency (USEPA) is concerned with the contamination of the three media of land, water and air and the practice of land diversion out of agriculture for the benefits of soil and water, and regulation of forestry practices. This is manifest in the Conservation Reserve Programme or CRP (Potter *et al.*, 1991).

The main U.S. drivers in water quality have been pathogen control with chemical diffuse pollutants a close second, contrasting historically with European pre-occupations around nutrients. However, problems of nitrate in drinking water now cause concern as does concentrations in groundwater, surface water and from atmospheric deposition (USEPA, 2015a). Concern over Phosphorus in surface waters is not far behind. Furthermore, Pesticide Management Zones for the restricted use of pesticides have been defined in the United States, notably in California, although there are general Groundwater Protection Areas that include more preventative measures, whereas before designation followed detection in groundwater at levels deemed to be of concern (Bianchi, 2002).

In draft regulations, a Ground Water Protection Area is defined as "an area of land that has been determined by the Director [of the Department of Pesticide Regulation] to be sensitive to the movement of pesticides to ground water... The determination of a Ground Water Protection Area is based on factors such as soil type, climate, and depth to the ground water, that are characteristic of areas where legally applied pesticides or their breakdown products have been detected and verified in groundwater (Bianchi, 2002).

The federal USEPA is instrumental in responding to legislation, monitoring and setting water standards (Charnley and Englebert, 2005). At the Federal level, the Clean Water Act (Water Pollution Control Act) of 1972 (CWA) and the Safe Drinking Water Act of 1974 (SDWA) are legal instruments for improvement of US water quality. Under the SDWA § 300h-7(a), each state is required to establish its own protection programme by the Environmental Protection Agency (EPA, 1992). For groundwater, Well Head Protection (WHP) schemes are for the protection of groundwater supplies. Under the SDWA (1974) as amended 1986 § 300h-7(a), each state is required to establish its own protection program, the elements for each area are expected to specify the role of state agencies, delineate protection areas, identify contaminants, develop management approaches including technical, financial and education, identify alternative sources, and under § 300h-7(b), states shall maximise procedures that encourage public participation in WHP programme elements.

Both these acts stipulate watershed partnerships with states, through substantial delegation to them; the view that the two acts impose federal centralization is therefore unwarranted. The USEPA (1992) recognises stakeholders in three main ways:

- *Problem identification* of threats to human and ecosystem health within the watershed.
- *Stakeholder Involvement* with people most likely to be concerned or able to take action.
- *Integrated actions* taken in a comprehensive, integrated manner.

Despite the wellhead and source water protection provisions of the SDWA, it remains primarily a consumer protection act and is not designed to *protect water quality through environmental measures*. Stakeholder involvement and 'holism' are to the fore, but monitoring and evaluation is unfortunately not explicit—the CWA, by contrast, is the primary instrument of water quality protection and there are surveys pointing to limited improvement in U.S. water quality throughout the 1970s and 1980s (Smith *et al.*, 1994). Limited evidence shows that nitrate, phosphate and coliform bacteria were consistent or falling and pesticide contamination in fish may be falling. Serious deficiencies include inadequate monitoring of chemical, including toxic, substances.

The establishment of RWAs in England and Wales after 1974 caused these agencies to be responsible for the management of the whole hydrological cycle; however, there is no way that these were seen as an engine for economic development as is the TVA, even if it was hoped that they were proactive in water planning and development. With matters of flood control, navigation and land reclamation in the hands of long established agencies, environmental pollution regulation has been through the U.S. EPA. Its remit includes matters of air quality, water quality, disposal of hazardous waste and regulation of chemicals including pesticides and radioactive waste; in this way the pollution of the 'media' of air, water and land is covered, and the agency operates very much in the spirit of integrated pollution control. Established in 1970, the U.S. EPA has been the model for the Environment Agency in England and Natural Resources Wales, and for SEPA in Scotland. It constitutes and EPA is an independent agencies of the executive branch of the U.S. government. It was founded in 1970, works with state and local governments throughout the United States to control and abate pollution in the air and water, and to deal with the problems of solid waste, pesticides, radiation and toxic substances. EPA sets and enforces standards for air and water quality, evaluates the impact of pesticides and chemical substances, and manages the cleaning toxic waste sites (American History, n.d.).

The U.S. EPA is very much an anti-pollution agency, rather than an all-round environmental management organisation, and unlike its British counterparts it also tests car exhaust emissions. It is thus closer in spirit to SEPA than the EA. Most notably, flood defence rests elsewhere, with the U.S. Army Corps of Engineers, and abstraction licensing is notably absent. Voluntary compliance with standards is encouraged; however, under the terms of most statutes administered by the U.S. EPA, alleged polluters are notified of a violation of Agency standards and ordered to stop. If voluntary instructions are not obeyed, civil court proceedings may be taken to force compliance. The U.S. EPA also protects public drinking water by establishing standards and regulations setting maximum contaminant levels for any substance which may be hazardous to human health.

The state enforces these, but the agency can assume responsibility if the state fails (the responsibility for drinking water remains with DWI in England and Wales, and will remain outside the EA). Historically in the United States, the progress of standard setting has been slow. Pesticides are carefully controlled through registration with the U.S. EPA, which also provides training, monitoring and research. Stringent tests for new pesticides including metabolic tests on plants and animals, residues in soils, waters and living tissue, laboratory studies on the products of hydrolysis and ecosystem effects are required. In Britain, pesticides are the responsibility of Defra.

Major issues in U.S. watershed management are soil erosion and range management related to agriculture, and water yield and erosion relating to forestry. The U.S. Forest

Service is required to monitor diffuse pollution, including soil erosion, water temperature, dissolved oxygen, nutrients and pesticides. There is regard for the ecological wellbeing of water bodies, in that the aim is for 'water quality which provides for the protection and propagation of fish, shellfish and wildlife and provides for recreation in and on the water', and this is echoed in the responsibilities of the EA. Watershed management is based upon 76 'ecoregions', themselves based upon statistical considerations of natural water quality, and the U.S. EPA is instrumental in bringing considerations of water quality (especially from diffuse pollution) into watershed management (Newson, 1992b, p. 103). State quality control programmes monitor physical, chemical and biological quality in indicator watersheds for homogenous land uses, and from these set water quality objectives. Criticisms have been levelled at the U.S. EPA for the slow implementing of programmes. Efforts in controlling acid deposition during the 1980s were unsuccessful because the administration could not be persuaded, by the agency, to take strong action against emitters.

Consequently the U.S. EPA's work in the area was largely limited to research; however, it established that coal-burning plants in the Mid-West were linked to acid rain in the northeast United States and Canada, a cause of friction between the two countries. During the first Reagan administration of the 1980s the U.S. EPA was bedeviled by internal problems, scandal and personality clashes. President Ronald Regan had attempted to defund the EPA and cut its staff, arguably for partisan ideological reasons (Benson, 2015, pers. comm.). Despite political damage, the U.S. EPA survived, and has developed a market-based system of allowances by which a pollution-emitting utility gained credits by reducing emissions which were then sold to utilities which were unable to expand due to restrictions. A system of tradable permits is now established (USEPA, 2014; 2015b).

Priorities involve review of legislation, and an industry-by-industry regulatory approach to replace blanket emission regulations. At least in theory, the United States has been ahead of the UK in environmental pollution policy. The U.S. EPA predates its British counterparts by more than a quarter of a century. Its activity is directly concerned with the prevention of pesticide contamination, including the definition of Pesticide Management Zones. Although the idea of pesticide protection zones has been proposed, no policy has yet emerged in Britain. Groundwater supplies around half of the drinking water supplies in the United States, and is increasingly becoming the focus of the Agency. The legislation now available to the U.S. EPA is powerful and it enables charging retrospectively for a clean up of contaminated land.

Big questions may be asked about the role of agriculture, of advice to farmers, of public participation and the desirability of applying centralised polities. In the northeastern United States, (Smith and Porter, 2009) the EPA engages the water industry, farmers, business and local government in ways that successfully engage voluntary action. It prefers an 'adaptive management' strategy to one that presumes a 'one-size fits all' for watersheds. Even where problems faced are comparable, policy importation is difficult because driving factors and legislative and constitutional frameworks differ.

To be effective instrumentally, land-use planning employs GIS methodologies, especially for wellhead protection measures. These are also used in mapping the outcomes of dynamic models that characterise actual flows and contaminant loads. U.S. regulation requires establishment of Total Maximum Daily Loads (TMDL) of sediment, total phosphorus and other contaminants (USEPA, 2008). TMDLs are a tool for implementing state water quality standards and may be expressed in terms of mass per time, toxicity, or

Table 11.1 The Multiple Barrier Approach (MBA).

Barrier (with examples)	Characteristics of Barrier
Source: typically barn-yard concrete surfaces to divert washings, diversion of runoff from area and secure manure storage. Infiltration off 25-year floodplain, flow path around 30 m	• Contain materials, microorganism forms and potential pollutants. • Treat before release to soil or water. • Eliminate or reduce the presence of a potential pollutant on the farm.
Landscape: Farm fields where rotational grazing, manure spreading and nutrient planning programmes implemented. Strip cropping systems.	• Control where and when potential pollutants are released to soil • Reduce mobility
Stream Corridor: the margins of watercourses where livestock exclusion and filter strips including riparian vegetation are located.	• Capture runoff derived potential pollutants at the edge of a surface waterway, and protect stream edge from degradation

other appropriate measures that relate to a state's water quality standard (CWA, § 303). Process models may make predictions from land use planning scenarios (Ambrose *et al.*, 1996).

Monitoring across such a large country requires co-operation between federal government and the individual states as mandated by the CWA. However, high cost makes the monitoring over large areas uneconomic and it is pragmatic to employ as many means as is possible to reduce diffuse pollution. Best Management Practices (BMPs) are adopted using decision support material in consultation with farmers (Willett and Porter, 2001) ahead of any conclusive evidence of environmental improvement. Examples include the Multiple Barrier Approach', part of Whole Farm Planning (WFP) that aims to reduce sediment, pathogen and nutrient runoff through land-based measures (Marrison, 2007).

Table 11.1 summarises MBA for the protection of waters, ecosystems and public supply. The included WFP measures integrate the Cornell University Nutrient Management Planning System ('CuNMPS') that reduces agricultural inputs, with water quality gain achieved without reducing yields of crops and animal products (Tylutki et al, 2011).

Such tools are available to water professionals and stakeholders and provide objectives for change induced through public participation. US watershed groups assume a bottom-up approach (Sabatier *et al.*, 2005), a deeply rooted aspect of water governance (Leach and Pelkey, 2010). From the outset, issue definition, proposing solutions to problems and participation in the planning process are integral. They operate in a multi-level governance environment including federal legislative structures and statutory institutions.

11.3 New York City Wateshed Programme (NYCWP)

The NYCWP is publicly funded, engaging technical assistance from Cornell University and agricultural extension services. Disinfection and filtration methods relied upon since the early 1900s became viewed as insufficient and expensive, chemicals and their residuals increasingly pose problems and there is an ever threat from protozoan parasites (Porter and Smith, 2015). Funds that might otherwise be directed to water supply filtration are utilised to sustain the economic viability of farms, managed to minimise risk of pathogens (*Cryptosporidium* and *Giardia*) from entering watercourses and to reduce total phosphorous that encouraged algal blooms in reservoirs. Other pollutants that cause concern are pesticides, sediment and hydrocarbons and there is ongoing evaluation of BMPs (Bishop *et al.*, 2005).

Instead of investment in conventional water treatment alone to protect the drinking water for the New York City area, the policy decision was taken to protect water supplies at source, seen as the first barrier of defence against waterborne contamination. In the northeast United States, abandonment of lower quality land outside any CRP programme has led to an extensive cover by unmanaged secondary forest so that reduction in diffuse pollution to waters is accidental. In the New York City Watershed (NYCWP), 90% of the supply originates in the Catskill/Delaware Watershed. Where there is surviving agriculture, WFP is promoted via the Watershed Agricultural Council. Participation is voluntary, but is funded via the New York City Department of Environmental Conservation (NYC-DEC). Diffuse (non-point) pollution arises from largely agricultural contaminants such as nitrate, phosphorus, suspended sediment and pesticides (Tolson and Shoemaker, 2007; Watershed Agricultural Council, 2013).

This raises the economic cost of water treatment; it also causes loss of confidence in the total security of water treatment plants, forcing consideration of water source protection. This is regarded as economically efficient, but it cannot work without stakeholder and community action. In the Catskills, the Cannonsville Reservoir watershed is managed for water quality goals in the form of Total Maximum Daily Loads (TMDLs). A TMDL established in accord with the requirements of the CWA § 303 was established by the NYC-DEC equivalent to 50,000kg/year for phosphorus was discharged to the reservoir. The Delaware County Action Plan (that should not compromise economic growth) identifies three criteria for non-point pollution control: Low cost, adequate decision-making and scientific credibility. Delaware County, has instituted a comprehensive program to identify and manage all major non-point sources of phosphorus and pathogens in the Cannonsville Reservoir basin (Porter, 2006).

11.4 Upper Susquehanna Coalition (USC)

The Susquehanna reaches the Atlantic at Chesapeake Bay, some 444 miles (710 km) from its upper watershed at Otsego Lake. The Chesapeake Bay Programme specifically focuses on diffuse pollution, including nitrogen from atmospheric deposition and sediment loading (Porter, 2004). There is also ongoing evaluation of the effectiveness of BMPs through the US EPA's Chesapeake Bay Watershed Model (CEAM undated). The Chesapeake Bay watershed program engages professionals to develop strategies, partnerships and technical projects with the USC providing upstream strategic support to county members, local watershed organisations, town and county public works, planning officials, farmers and businesses. USC is a network of county-level resource professionals that form partnerships to represent citizen concern over flood and erosion protection issues. Funding is procured from US federal, state and local sources (USC, n.d.).

In a formalised response to the success of bottom-up management, US EPA actively seeks the 'adoption of watersheds' by communities (USEPA, 2012). Possible approaches to a problem are considered and implemented within constraints. These may be cost to and different degrees of landowner participation. USC works closely with the stakeholders and the public through outreach and education. 'Citizens for Catatonk Creek' is one such USC committee established following serious flooding during the 1990s, and it amasses technical and land use information (Curatolo, 2002, *pers comm*). USC is developing a database using GIS which will detail land use, agricultural information, wetlands (these are designated, ephemeral, potential and restored since 1992), and also topography, stream corridor information including riparian buffers and stormwater infrastructure. The committee are responsible for applying a Multiple Barrier Approach (MBA) approach to flood defence. Programmes include the re-instatement of drained wetlands as well as the construction of wetlands through small detention reservoir construction on the floodplain of the Hulbert Hollow Creek. More are planned along the nearby Sulphur Springs Creek at suitable locations where the landowner has co-operated. Of particular interest is the possibility of creating a wetland banking; the restoration and protection strategies will develop model codes by which local government will protect wetland through land use planning and regulation (Curatolo, 2009).

Wetlands trap sediment, nutrients and toxins. The Farm Act of 1996 allowed for CRP to fund a Wetland Reserve Programme through retirement of agricultural land, promote technical assistance and construction (Willett and Porter, 2001). The Farm Security and Rural Investment Act of 2002 (NRCS, 2002) or 'Farm Bill' placed strong emphasis on the conservation of (private) working lands, ensuring it remain both healthy and productive. As a result, conservation corridors, conservation grazing and haymaking, increased groundcover for wildlife and the 'Farmable Wetland Pilot Programme' was expanded. Increasing the wetland and lake area from zero in a watershed dramatically reduces flood peaks and 50% of this reduction may be achieved from the first 5% of area protected (Novitzki, 1985). Such ecosystem services make wetland re-instatement or establishment an attractive and cost-effective solution, delivering flood relief and conservation.

11.5 New Zealand

In New Zealand, comprehensive catchment-wide planning, including soil, water and land-use planning, has been emerging for some 50 years, and this has led the way in the development of the planning process in that country (Eriksen, 1990). The Resource Management Act 1991 requires regional councils to manage water resources in a sustainable manner (Encyclopaedia of New Zealand, n.d.). Meaning that the 12 Regional Council boundaries established in 1989, have a close co-incidence with drainage basins and hence Catchment Board responsibilities.

In New Zealand, a country generally well endowed with water, making catchment transfers is rare. Yet once more, it is that agriculture presents most problems. Demand is rising, and potentially produces stress during low rainfall periods. Specifically, non-point (diffuse) problems are caused by nutrient runoff caused particularly by the expanding dairy farming sector. Nitrogen is a particular problem in lakes and streams and their banks are damaged by cattle access, (a cause of degradation of lowland streams) and forestry. Urban development also presents problems (Cullen *et al.*, 2006). There is considerable public concern, with agriculture generally being blamed for pollution of streams, lakes and groundwater and for its consumption of water. There is furthermore a willingness to pay for improvement of lowland streams.

The goal is to achieve successful Integrated Water Resources Management (IWRM), and in New Zealand this is predicated on increased participation and locating decision-making set at an appropriate scale of management. In this respect, the New Zealand experience is proving most instructive (EPCR, 2013). IWRM operates at many scales, including that of the sub-catchment so that when dealing (for example) with hill country erosion, a 'clumping' process is called for. In governance terms, it may be that local government and the farming lobby are over-represented in provincial regional governments.

While the state operates a centralising tendency, local (including Maori rights) issues are better left to much smaller units for decision-making. Hence, the machinery is present for collaborative governance, but it may be that previous patterns of decision-making, that are not improving water quality in New Zealand, are allowed to persist. Statism is present, as in UK and the United States, to 'police' pollution and may prevent voluntary local collaborative initiatives in regulation.

This efficiency in resource management and governance would be difficult in Britain because, for one thing, local authority boundaries would be to re-drawn to coincide with catchment boundaries and hence improve water governance for the EA. Despite reorganisation of UK public bodies, this would be unlikely. Interestingly, political boundaries in the United States are similarly non-coincidental with watersheds, and in some instances rivers even define state boundaries!

There is also talk of a 'tyranny of the minority', where concentrated interests (such as farmers) facing high costs arising from a policy initiative, act against a relatively passive majority with little each to gain. This trumps open discourse. The presence of the state in ultimate regulation therefore remains central; this has helped to pull the dairy industry more into line, so that:

'If lessons are to be learnt from the New Zealand experience of IWRM, they must include that IWRM is not a silver bullet solution and that public engagement, of any type, is no guarantee of a better environment. Rather, as with any public institutions, IWRM processes and policies may only reify [i.e. to treat as if it had concrete existence] *existing power configurations* where concentrated interests can and apparently do capture process' (EPCR, 2013).

11.6 Murray-Darling Basin and SE Queensland

The Murray- Darling Basin (MDB) occupies over 1 million km^2 in SE Australia and has a population of 2.1 m people. The region supplies in the order of 40% of Australian agriculture (Ross and Connell, 2014). Like New Zealand, it is the legacy of European settlement and farming that is the major problem for water quality and it is the state (such as Queensland) that has the primary right to own or to control and use water so that allocation, planning and policy implementation that is at that level, although the federal government plays a major role in water policy development and financing.

As in many other places worldwide, there is an interesting history commencing with co-ordinated management of the MDB in 1915, two decades ahead of the TVA in the United States. Today, since a federal Water Act of 2007, there is a Murray-Darling Basin Authority that provides for a Murray-Darling Basin Plan. Since the 1950s, there was been a growth in water use, largely through irrigated agriculture (otherwise there is urban supply, domestic supply, industrial use and stock watering), although that fell in the decade from 1999 due to dry weather conditions and today water storage and regulating structures are being constructed to even out supply between years. Groundwater contributes about 16% of this supply across the MDB, but is higher in the north.

The issues facing the area include soil erosion and river flow diversion, pollution (especially from agriculture) resulting blue-green algal blooms, turbidity, low dissolved oxygen and salinity with acid sulphate toxicity in some soils (Murray-Darling Basin Authority, 2015). Today policy development centres around accountability by authorities, allocation and appropriate pricing through water markets. As in New Zealand, there is concern expressed about the allocation of powers between a high-level 'coercive' approach (Ross and Connell, 2014) and legitimacy at regional levels. The present tendency is to strengthen trust and hence cooperation between stakeholders.

Northeastwards across the great dividing range, in SE Queensland, it is a consortium of local authorities who are leading a not-for-profit, not federal government organisation. It aims to provide healthy ecosystems within watercourses and support livelihoods within catchment, managed through collaboration between community, government and industry'. For:

> 'Healthy Waterways works to understand and communicate the condition of our waterways to drive and influence future targets, policy and actions.'

And:

> 'Healthy Waterways is an independent, not-for-profit organisation working with our members from government, industry and the community to protect and improve our waterways.'

Evidently in these two important areas targeted for integrated water resource management, lies the tension between state-directed and hence 'top down' versus a more demonstrably democratic and stakeholder inclusive NGO approach. In Australia, as yet there is no ideal set of structures, strategies and procedures capable of driving integrated river basin management, albeit leadership and feedback are more important than the precise model adopted (Ross and Connell, 2014). This hints at a requirement for adaptive catchment management (Smith and Porter, 2009).

11.7   Prelude to Continental Europe: A Shared Experience?

Contentious though the notion of a collective idea of 'Europe' may be in certain political arenas, environmentally EU membership has proven extremely useful, in an instrumental sense, for British waters. Transnational river basins in the continent of Europe are regulated by many treaties, although issues relating to bulk water supply were not generally the subject of EU directives. So that before WFD 2000, the EC did not intervene in such matters as water quantity or flooding; it was concerned with water biological and chemical quality affecting the environment and human health related issues. Such were the concerns of the Drinking Water Directive of 1980 and the 'Nitrates Directive 1991', or the 'Bathing Water Directive 1976' concerned with protecting public health and the environment from faecal pollution from bathing waters. Since then, 'Framework Directives' (such as the WFD or Habitats Directive) have sought to consolidate older legislation, described in Chapters 3 and 10. The operation of the EU in an international context means that member states are prepared to accept transnational legislation on pollution which de facto limits their rights over resources (Clarke, 1991, p. 105).

Precipitation in Europe generally increased over the twentieth century, rising by 6 to 8% on average between 1901 and 2005, although there has been a reduction overall in eastern Europe and the Mediterranean (EEA, 2009). There may be variation from year to year and the south of Europe may be adversely affected, but it is often difficult to establish long-term trends in river discharge across the continent, although cyclical patterns have emerged (Pekarova *et al.*, 2006), but when periods of low rainfall occur they tend to affect large areas.

In 2009, the total abstraction of freshwater across Europe is around 288 km^3 yr^{-1} and represents, on average, 500 m^3 per capita. yr^{-1}. Overall, 44% of the total abstracted is for energy production, 24% for agriculture, 21% for the public water supply and 11% for industry with strong regional variation. In western member states, abstraction for electricity production predominates, contributing approximately 52% to total abstraction, followed by public water supply (29%) and industry (18%). In the south, it is abstraction of water is for agricultural purposes, specifically irrigation, which (at about 60%) predominates. Around 81% of freshwater abstracted is from surface water bodies, although groundwater dominates agricultural consumption and there is concern for surface water bodies for over-abstraction (EEA, 2009). Compared with Britain, the total abstraction of freshwater from groundwater from 2000 to 2012 fluctuated year on year, with a hike between 2011 and 2012, although this was largely attributed to an increase in water use for hydropower. Overall, most goes to public supply and electrical generation. The proportion from groundwater varies enormously by region depending on the presence of aquifers: in England and Wales this is around 35%, perhaps only 5% for Scotland (Chapter 1).

Considered overall, there is no water shortage in Europe, however, there are problems with water distribution across the continent with the highest amount of runoff generated in the north and west and the lowest in the south and east (Stanners and Bourdeau, 1995). Increasing shortages may change this situation. The lesson for the rest of the world may be one of nation states accepting international law aimed at marshalling resources, although issues of bulk water supply are potentially more explosive than pollution legislation.

Pulling it all together through coordinated governance approaches is something now well recognised on continental Europe. Continental member states, and indeed others, focus on the concept of 'spatial planning', concerned with problems of coordination and integration of the spatial dimension of sectoral policies (such as those of transport, agricultural and environment) within a territorially based strategy (Chapter 10).

11.8   The Netherlands

In a densely populated and heavily urbanised country, much of it reclaimed land on a river delta (Cook *et al.*, 2009) it is intuitive that the Netherlands might strongly gravitate towards spatial planning. Land drainage has been practiced for centuries, creating characteristic landscapes of and dikes and drainage ditches assisted by windmills (Figure 11.1). And that country is prone to high risks of flooding from both rivers and the North Sea. In the Netherlands, water governance is tiered from national through regional to local levels, embracing both water management in a practical sense and spatial planning in a strategic sense (National Committee of Water Assessment, n.d.). Water supply to households is a public sector function, that is water supply is municipalised, although private contracting is commonplace. Everywhere is the visitor made aware of the importance of water, and of its conservation (Figure 11.2).

Vink *et al.* (2014) identify differences in the Dutch and British approaches. In the Netherlands, deliberative governance (meaning governance that has been both discussed and planned) has been described as a neo-corporatist state tradition. This may have drawbacks, instilling apathy among politically elected decision-makers, but is arguably important given the immediacy of certain issues such as land drainage and flood protection. This contrasts with the pluralist state tradition of the UK where (clearly defined) rules and responsibilities yields negotiation and action in a flexible situation.

In the strongly centralised spatial planning environment of the Netherlands, since 2010, the ministry of Housing, Spatial Planning and the Environment has been merged with the Ministry of Transport, Public Works and Water Management. The new ministry,

Figure 11.1 Traditional Netherlands: Draining reclaimed land, Kinderdijk. (*Source:* H.F. Cook).

Figure 11.2 Modern Netherlands: Declaration of a Water Protection Zone in an urban area—Loosdrecht. (*Source:* H.F. Cook).

the Ministry of Infrastructure and the Environment, is broad in its remit: 'So the responsibility over water management and spatial planning is actually united within one ministry'. Within the ministry, the Directorate General for Spatial Planning is 'responsible for the national strategy on allocation of land and water resources for sustainable economic and social development.'

And:

> 'One of the key responsibilities of the Ministry of Infrastructure and Environment is preparing and coordinating national policy and strategies on spatial planning. This responsibility includes ensuring the implementation of EU legislation in national regulations.'

This also includes consultation, cooperation (between government bodies) and reaching consensus for decision-making (National Committee of Water Assessment, n.d.). Given the different attitudes towards both Big Government that emerged in Britain since the 1980s, and more recently poor attitudes by some to the whole EU project it is hard to imagine such language emerging, even in an official document, from a British governmental body! The more fluid attitude has been discussed in Chapter 10, including the creation of non-governmental river basin groups (Cook *et al.*, 2012). In the Netherlands, the expectation is that, effective, *adaptive catchment management can operate through statutory bodies*; especially important in the face of climate change and sea level rise. But the proof in the Dutch pudding is simple: does it work?

Swart et al. (2014) investigated climate change adaptation for 100 projects and water management for the Netherlands in a context of climate change, finding eight characteristics that would increase climate resilience, as well as the overall quality of a project. Investigating characteristics that would increase climate change resilience (including length of time needed for project completion, cost-effectiveness, breadth and integration of aspects leading to sustainability, breadth of scope, economic sectors including natural environmental and cultural heritage protection, stakeholder participation and entrepreneurial opportunities), it was concluded that spatial planning is working in that it aids implementation for climate change adaptation, but sound a note of caution about whether the Netherlands practice may be exported.

Work in the Waal river, the main distributary of the Rhine in the Netherlands, since the 1990s has taken things a long way, so that a 'Dutch School' of Landscape Architecture' has been declared (DSL, n.d.). This operates at an appropriate regional scale and has sought to 'soften the boundaries defined by dykes between the fluvial environment and conventional agriculture land'. Dyke profiles have been altered so they accommodate a broader base, gradually evening out, to accommodate reeds and other natural vegetation as well as new forest land, while maintaining sufficient channel capacity to constrain flooding and provide for recreational land.

Furthermore, it is a truism that the Netherlands are vulnerable to climate change and sea level change, and policy makers seek to improve the governance environment such that barriers to adaptation are lowered and risk-innovation becomes possible, notably learning from older mistakes and promoting novel approaches to adaptation (Biesbroek *et al.*, 2014).

As with 'Making space for water (2005)' in Britain, many in the Netherlands call for a transition to policies that actively manage flood risk to reduce flood impacts and

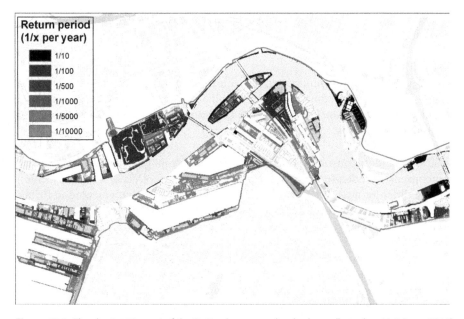

Return period
(1/x per year)
- 1/10
- 1/100
- 1/500
- 1/1000
- 1/5000
- 1/10000

Figure 11.3 Flood extent in part of the Rotterdam un-embanked area (based on Huizinga, 2010).

accommodate floods: 'living with water', rather than a mere focus on flood protection: 'fighting against water' (van Herk *et al.*,2013) and promote a transition using a multi-pattern approach (MPA) applied to understand the transition patterns to integrated flood risk management (IFRM). Policy makers can shape and monitor the outcomes that support a transition.

A surprising discovery is that the city of Rotterdam includes many un-embanked areas, and these are ripe for re-development. A current practise is to raise flood defences to a 1:4000 years flood level, which is not only costly, but it also causes problems for re-development where older building plots remain. The policy for Rotterdam now is to look for adaptive (or 'non-structural') measures to decrease flood damage areas that are ready for re-development (Figure 11.3). For example, better communication of flood risks is recommended to 'increase awareness and preparedness'; this should lead to a higher implementation of adaptive measures for example, embanking, elevating, dry-proofing (preventing flood water from entering a building by making reinforcing walls, windows and doors and making them impermeable) and wet-proofing (where flood water is allowed to flow in the building), construction materials not damaged by flood water (and hence easy to clean), temporal adaptation, floating, evacuation, regulations and communication. These are adopted for differing spatial scales (van Vliet *et al.*, 2014).

11.9   Aalborg in Denmark

Around 30% of the drinking water of the city of Aalborg in northwest Jutland, Denmark, is supplied by groundwater from the Drastrup well-field that penetrates the underlying (and largely unconfined) chalk aquifer, which has become contaminated by nitrates of agricultural origin (Municipality of Aalborg, 2002). Between 1990 and 2000, ground-water nitrate concentrations continued to rise well above the EU limit for nitrate in drinking water beneath the land surface, although below about 60 m it remained constant, leaving the possibility of it being blended to remain below the EU limit (Dhyr-Nielsen, 2009).

Land-use change involved a shift away from arable farming after 1994. Between 1998 and 2000 there was a dramatic decline in nitrate leaving the soil water in an area converted to grass, followed by a more gradual decline after 2000, with a downward trend in nitrate concentration in the shallow groundwater, which had reached up to 125 mg l^{-1} nitrate (Hiscock, 2005, p. 269-270). This was largely achieved by producing a recreational area on former arable land that comprised a mosaic of mixed age forest (230 ha, commenced in the late 1990s) and newly sown grassland, often serving as sports and picnic areas, although organic agriculture is encouraged. Despite a flush of nitrates during forest establishment, the peak soon fell off and it continues to be moni-tored. There remains concern over atmospheric inputs of nitrogen.

The groundwater capture zone became dominated by trees, grassland and low inten-sity grazing, in what has become a celebrated example of 'spatial planning'. The needs of the water company and its consumers have been balanced with those of agriculture. Farmers have received advise and have ultimately been compensated for lost productiv-ity, financially or in terms of land swaps outside the catchment area for the well-field (Figure 11.4). A further advantage is the reduction on pesticide loading on the former arable land. Groundwater modelling suggests the long-term outcome for the protection measures should be effective (Hørlück and Pedersen, 2012).

Figure 11.4 A picnic area near the city of Aalborg located on former agricultural land. (*Source:* H.F. Cook).

11.10   Example Catchment Groups in Britain

Voluntary sector environmental bodies have been operating in the UK for over a century despite chronic concerns over their economic sustainability (Cook and Inman, 2012). The last decade of the twentieth century saw the rise of directly instrumental Rivers Trusts (Chapter 10). The rise of watershed partnerships to complement top-down regulation is occurring both inside and outside the regulatory bodies (Benson *et al.*, 2013a). Some examples follow:

The Ribble Pilot Projects was used to test the EA's participative process. Various means of participation were tested, including a stakeholder forum that was considered "innovative and an effective means of involving stakeholders in the decision making process" (Davis and Rees, 2004; Louka, 2008, p. 237-239). Other changes in favour of greater participation occur through CAMS and formerly the CMP processes. The EA seeks consultation during its planning process, a significant break with the 'technocratic' past.

The Upper Thurne Working Group (UTWG) is noted for stakeholder participation. In the East Anglian Broadlands, water level management, water quality, recreation and visitor pressure are all long-term issues because agriculture, boating and septic tanks create diffuse pollution. The farming lobby enjoys strong representation through the National Farmer's Union, the Country Landowner and Business Association and Internal Drainage Boards. The EA seeks to improve water quality, and there is angling representation and a determination by Natural England and the voluntary sector to improve habitats. Participatory stakeholders are joined by elected stakeholders from local government. The meetings concentrate upon conflict resolution and in reaching agreements. The Upper Thurne Waterspace Management Plan provides the framework for partnership working in the area. It was written by the Upper Thurne Working Group covering between 2006 and 2009. UTWG management plans have been adopted by statutory bodies (Upper Thurne Water Space, 2012).

In southwest England, the River Tamar faces pressure from agriculture, wetland and other habitat loss pressure on fisheries, sewage effluent. In many ways, its problems parallel those of the New York City Watershed Agricultural Programme, because the agriculture comprises predominantly smaller, livestock farms. The Westcountry Rivers Trust works in partnership with farmers, communities and conservation bodies to reduce diffuse pollution, deliver CSF, reduce stream erosion, restore wetlands and other river corridor habitats and sustain local communities (Cook *et al.*, 2014). Similarities to the NYC Watershed program include the use of technical advisors, development of partnerships especially with the farming community, a voluntary approach rather than regulation and linking economic gains (including promoting BMP and implementing CSF) to environmental improvements (WRT, n.d.b, Chapter 10).

The Hampshire Water Partnership, formed was formed in 2001 and through the Water Champions initiative it sought to encourage local residents to provide information and support at the local community level in conserving rivers and wetlands (Hantsweb, 2015). Here was a direct effort on behalf of local government to engage the public in a participatory fashion linking regulatory imperatives directly into the community, much as the far more successful Rivers Trusts movement links stakeholders to 'take ownership' of local catchment issues including diffuse pollution and the wellbeing of fish populations. The lessons potentially learned require further elaboration, although the statutory basis for clear and direct involvement from local government is weak and limited to the statutory planning process.

In the light of the WFD, there are around 11 catchment management initiatives underway in Scotland and they work with river basin management plans. The Dee Catchment Partnership (DCP) is located in Aberdeenshire. The DCP is a voluntary association of stakeholders within the Dee catchment and its waters. Considered are issues including flood management, water pollution, river engineering and the impacts of recreation; specifically issues such as restoring urban watercourses, reducing pollution from septic tanks, reducing diffuse source pollution and managing flows that have been identified as priorities for action, delivered through specific projects (SNH, 2010).

The total length of the River Dee, including tributaries, is 1645 km and the system supplies over 300,000 people with drinking water. The Dee Catchment Partnership includes the following: Aberdeen Harbour Board, Aberdeen City Council, Aberdeenshire Council, Cairngorms National Park Authority, Dee District Salmon Fishery Board, Forestry Commission Scotland, James Hutton Institute (research), National Farmers Union Scotland, Royal Society for the Protection of Birds, Scottish Agricultural College, Scottish Environment Protection Agency, Scottish Government, Scottish Natural Heritage, and Scottish Water Areas designated for the protection of habitats or species.

The catchment is relatively unusual in Britain, in that it has predominantly upland and semi-natural land use. There are two geographically distinct regions, in the western and southern areas, semi-natural land covers types dominate, typically moorland, consisting of blanket bog and heather moorland on the upper and middle slopes and there is mountain alpine heath vegetation found on the highest summits. The lower slopes have been used for the establishment of managed forests (both coniferous and deciduous) and forests are mainly found in the river valleys. In the remaining areas of semi-natural vegetation and native 'Caledonian' pine woods.

Current and past projects include 'Aquarius' (a partnership between Aberdeenshire Council and the Macaulay Institute) that works with farmers and land managers to improve water quantity through land management under current and future scenarios of climate change. The focus is the Tarland Basin and the project will support development of the Tarland Flood Prevention Scheme. Other projects work with schools, creative workshops, buffer strip emplacement, work in priority sub-catchments for diffuse pollution, burn (stream) restoration (involving the river Dee Trust) as geomorphological restorations to improve flood management.

In the Dee catchment are sites designated under the Habitats Directive (92/43/EEC), known as Special Areas of Conservation (SACs), and Birds Directive (79/409/EEC), which are known as Special Protected Areas (SPAs). The river Dee and its tributaries support internationally important freshwater pearl mussel, Atlantic salmon and otter populations. For example, the Glen Tanar and Loch of Skene SPAs are in the Dee catchment; the headwaters of the Dee are part of the Cairngorms SPA. There are also Ramsar sites or designated wetlands of international importance. Most of the upper main river and its tributaries are Sites of Special Scientific Interest (SSSI). There are also NVZs in vulnerable areas (Dee Catchment Partnership, n.d.).

11.11 Britain: Lessons to be Learned?

If comparisons really are odious, then there is little advancement, especially as policy makers grapple with wicked water problems across continents. Transfer of policy options between legal systems, physical and biological environments and differing governance environments needs careful consideration, despite any difficulties that may arise. It needs to be done, not only to avoid re-invention, but to take advantage of a shared willingness among those with a real stake in river basin management to mutually improve matters.

The quotation at the start of this chapter is very real, for in that room were a group of American scientists and others influential in formulating watershed policy for the northeastern United States. The author had just returned from showing two Australian academics with comparable interests around the locale of Cornell University, home of the New York State Water Resources Institute (NYSWRI, 2015). Beyond an evident bonhomie in the room lurked mutual concern around the impact on agriculture on water resources at large. Geologists were trying to decide whether much of the sediment in the local rivers were an outcome of Pleistocene natural processes rather than soil erosion, water quality experts were concerned as to whether agriculture could be in some way managed to reduce nutrient (especially phosphorous) loading, or whether there was such a historic legacy from farming afflicting (even abandoned) land, now largely secondary forest, that other efforts may be fruitless. One man was talking to local towns about sampling and managing contamination from storm drains, students were asked to contemplate the impact of ill-managed septic tanks on otherwise beautiful lakes. The then Director operated as much as a diplomat as either manager or policy maker, successfully negotiating his way around the farming community, and inevitably the vested politics of agriculture, as elsewhere. Our Australian colleagues reflected on imported European agricultural practices on their fragile ecosystems. Erosion, salinization and diffuse pollution (later flooding) were massive problems. The author, far from playing the smug Englishman, had been concerned with adverse impacts on surface and groundwater in a temperate country that had not 'got it right' despite farming for millennia! Climate change, linked to flood risk, was everywhere climbing the agenda and, as ever, there was industry and sewage; naught for our comfort.

This chapter has looked at selected options from Australasia, parts of the United States and continental Europe. Clearly to sack all bar the latter into some notion of an 'Anglo-Saxon' cultural and political world is nonsense, due not only to physical differences in the environments mentioned, but to divergent governance arrangements. Nonetheless, as an outcome of such ruminations, the UK research councils (and other governmental money) instigated the Rural Economy and Land Use programme between 2004 and 2013, under which water issues and land use were to play an important part, and under which issues around governance and catchment communities were at least as important as technical measures aimed at improvements in flooding, habitat enhancement or diffuse pollution (RELU, 2015). The gloves were, in theory, off for tackling wicked problems; interdisciplinary research became almost respectable.

Official recognition of the search for means to achieve IWRM meant that scientists and policy makers looked to their peers around the world. While Britain, in the grip of a post neo-liberal environmental management crisis, looked to the voluntary sector to build partnerships, trust, social and environmental capital, politicians of left, right and centre are beginning to lament the weakness of local government, and while Scottish separatism looks like failing, devolution and, for England, growing regional autonomy may represent a swing back of the pendulum. Localism may indeed strengthen the statutory planning process, but broader environmental regulation remains with underfunded statutory bodies that have few formal links with elected local government bodies.

In Australia, state and local government arrangements are the very vehicles for partnership building and voluntary sector engagement while like New Zealand, there is some considerable angst around the allocation of powers between a high-level 'coercive' approach and legitimacy at regional levels; in New Zealand there is an expressed distaste for powerful minority voices 'the tyranny of a minority'. Task allocation, the power distribution between economic sectors and levers from 'top-down' to 'bottom-up' requires honing. In Denmark, local municipality powers are sufficient to permit a high degree of land use planning to reduce diffuse pollution from agriculture, a sector in that country has serious issues with nitrate contamination of groundwaters, manifest as the famed decision in pre-EU days to produce for the traditional English breakfast.

In these respects the all too brief discussion of 'spatial planning' in the water-locked Netherlands is especially interesting. It is not a great observation that the Dutch live in fear of water inundation, especially of their key cities and towns, nor that (like Denmark) there is an agrochemical pollution problem from intensive agriculture. Neither is it surprising that the authorities in the Netherlands will both take a strong grip on water management issues and also look for practical, innovative and cost-effective measures by way of problem resolution. Described as a 'neo-corporatist state tradition', the creation a super-ministry to drive IWRM and planning forward, pride in a professional responses (most noticeable in the profession of landscape architect) contrast with the disconnected manner by which water resource management has emerged into twenty-first-century Britain (Figure 10.5). To most continental minds, this is a mess. To neo-liberal opinion, this will resolve itself one way or another, possibly through spontaneous voluntary action away from the dead hand of statism. Funding as ever remains an issue, as does task allocation.

Most recently attention has been turned to upland environments, arguably at the expense of lowland farmed and densely populated areas. The justification is that these are a source of most of Britain's water supplies. The Commission for Rural Communities (CRC, 2010) adopts an integrated approach for upland English catchments. The recommendation is not to see upland areas in terms of agricultural disadvantage; rather there are opportunities for catchment protection, historic landscape, even carbon sequestration. This takes UK policy-making closer to the U.S. models. Hitherto, all-embracing catchment management planning in the UK differentiates it from arrangements in the United States where planning has tended to be much more 'issue oriented' as well as top-down. In theory, regulatory bodies act in the public interest by reflecting the needs and engaging the interests in respect of the environment (EA), drinking water quality (the Drinking Water Inspectorate) and pricing and consumer affairs, 'Ofwat' (Cook, 1998, Ch 3). Actually, democratic expression can have a hard time within a structure where un-elected power is exercised. Indeed, historically, the UK has been something of a closed information society. This caused a House of Lords Committee (HOL, 2000) to talk of a need for institutions to broaden dialogue with the public. Where statutory bodies call for public participation (typically via the statutory planning process) democratic input is welcome, *but is a response to the regulator's agenda rather than issue definition by stakeholders and communities themselves.*

As originally conceived, Catchment Management Planning set out a 'vision for the catchment', developed by the National Rivers Authority (1989-1996) and subsequently 'Local Environment Agency Plans' (LEAPs) by the Environment Agency for England and Wales (Cook, 1998, Ch 9). Post-LEAPs, in response to the WFD, the EA implemented Catchment Abstraction Management Strategies (CAMS) that aim to control volume abstraction (EA, 2009b). Catchment Flood Management Plans (CFMPs) provide for flood protection. For diffuse pollution, Nitrate Vulnerable Zones (NVZs) have been extensively implemented, but overall, trends across Britain for nitrate (and other pollutants) cause concern (Defra, 2013e). Despite this tendency in policy formulation to fragment task-allocation, the trend remains towards 'holism', 'sustainable development' and implementation of the precautionary principle. Legislation requires 'competent authorities' that are statutory agencies to provide for environmental goods and recent legislation reflect modern pre-occupations: the Water Act of 2003 is largely concerned with abstraction licensing, and the Floods and Water Management Act of 2010, covers flood management and coastal erosion.

The UK also has a strong ethos of countryside management that produces, through a desire to protect historic landscapes and biodiversity, spin-offs in terms of soil and water conservation. All-embracing Environmental Stewardship schemes require compliance, from basic environmental good practice to 'heritage farming' (Defra, 2014b). Riverine and floodplain environments are thus protected, and hence water quality improved and there is flood tolerance (Cook, 2010b). A similar approach to U.S. Whole Farm Planning in protecting soil and water has been taken under Catchment Sensitive Farming or CSF (Defra, 2008), although measures so far taken have caused its efficacy to be questioned, notably in terms of diffuse pollution reduction (Burn *et al.*, 2010). Initially within 50 priority catchments, technical advice will be given in the areas of: land use, soil management, manure management, protection of watercourses, livestock management, crop protection, nutrient management and voluntary advice for pesticide management. The promotion of 'strategic partnerships' in 10 more catchments where Catchment Steering Groups, incorporating the national and local voluntary sectors and nationally organised NGOs (including Rivers Trusts) and other stakeholder groups such as the National Farmers' Union and water companies.

If the United States seeks to create an ethos in public policy akin to UK in countryside management, England and Wales must seek an appropriate level of effective function for top-down catchment management. Regulation would be would be light touch with standards setting goals that are realised by adaptive management and local strategies; ideally these would be locally and community lead. Great uncertainty, however, remains around local government involvement, something that would at the very least inject representative democracy into the process. Community and stakeholder participation are central to issue definition, problem ownership and action. However, being cost-effective must never be seen as the 'cheap option', their strength lies in multi-level responsiveness to change. Table 11.2 summarises key differences in focus between the two regions in areas that protect water quality, habitat and fitness for public supply.

In the UK, top-down governance remains the driver, despite efforts by regulatory bodies to engage closer with the public. The CRC Report (2010) put its finger on some serious differences in governance within the UK, notably lack of participation, over-centralisation and fragmented in terms of sectors and agencies. Similarly, Cook (2010b) finds geographic fragmentation of floodplains at issue for policy makers. There are signs of change in UK water governance apparent through Rivers Trusts, local government and communities.

Table 11.2 Contrasting national approaches to watershed management.

UK focus	Northeast U.S. focus
Diffuse pollution control	Watershed regulation and management for human health (SDWA: New York City Watershed Program)
Countryside management, especially 'agri-environmental schemes'	Municipal priorities well defined (home rule)
Habitats (national and EC Habitats Directive)	Voluntary measures favoured
Rural economic diversification	Effective county-based professional services including co-operative extension in agriculture
Protection of 'riparian rights' (ancient)	Watershed protection promoted by the federal CWA in partnership with States
Protection of freshwater fisheries	Property rights fiercely defended
Flood protection (ancient)	Watershed approach is issue focussed
Abstraction controls (WFD)	Voluntary measures preferred
Local and central planning system (albeit poor on rural land use) achieving a new private-public balance	Leadership assumed by local communities
Catchment Management Planning broadly based, WFD and CAMS require participation	Decision making at all governmental levels is open and publicly accessible

One initiative to address such problems as fragmentation and uncertain governance arrangements in catchments is through catchment partnerships that arise from implementation of the catchment based approach to management (Defra, 2013d, Ch 10). Concerned as they are with engagement, use of information, delivery and monitoring these define research projects that demonstrate and empower partnerships with a catchment for both stakeholder engagement and the implementation of technical knowledge in order the co-ordinate planning and funding for the delivery of 'good ecological health'. There are over 100 such community partnerships and some 1500 organisations, including NGOs (including rivers trusts), cater companies, local authorities, government agencies, landowners, angling clubs, farmer representative bodies, academia and local businesses. These partnerships are seen by Defra as cost-effective (CaBA, n.d.).

While the United States retains a strong legislative and institutional framework, the tendency at state level towards 'home rule' has long caused involvement by local government and partnerships who marshal scientific and executive resources in self-organising groups; the United States is instinctively more 'bottom-up' in this respect. That country has been able to resolve tensions between the federal and state levels, in the implementation of water quality legislation, through local representation by elected officials and stakeholders. The USC was originally formed in 1992, quite independently of the concerns of the two federal Water Acts and it constitutes a true partnership (Leach and Pelkey, 2001) because it incorporates 'informal representation, including citizens and conservation group interests'. U.S. regulatory bodies still organise monitoring, undertake some modelling and remain backstop in case legal sanctions are required. The emergence of TMDLs supported by credible scientific support illustrates the successful protocols that emerge from linking community and technical expertise. Land-use measures, including riparian protection and wetland restoration, are central to TMDL success. This has been contingent on both appreciable funding (particularly in the case of the New York City watershed) and establishing scientific credentials.

Monitoring, however, is time consuming and expensive. To overcome unexpected change and uncertainty, adaptive management is a systematic process for improving management by learning from the outcomes of policies and practices (Smith and Porter, 2009). Hence, multi-level governance remains far stronger in the northeastern United States than in England and Wales, and is delivering joined up thinking throughout the water sector, largely due to self-organising networks. However, it should be noted that the New York state examples are relatively well-funded from public sources.

Statutory groups convened by statute, rather than possessing statutory powers, have emerged in the UK since the 1990s. Based in EA regions for England and Wales, these and are concerned *inter alia* with fisheries, ecological and water issues. Included is 10 Water Framework River Basin Liaison Panels (as well as the Catchment Steering groups) including stakeholders in all economic sectors, NGOs, local government, conservation bodies and the water industry. Since flooding became a major political issue, the Regional Flood and Coastal Committees (RFCCs) have been formed (Chapter 10).

Ownership of problems by communities (as is especially the case with USC) is essential to avoid a 'culture of blame', or worse, litigation, and this is cost effective. NYCWP and USC rely on the 'multiple barrier approach' to diffuse pollution including scientifically sound, economically viable but issue oriented monitoring protocols. DCAP and WFP contrasts with the 'vulnerable zone' approach in UK that requires a Statutory

Code or other statement of 'good practice'. Like BMPs, WFP and the MBA there is engagement of the farming community in watershed programmes, although long-term efficacy will need to be assessed and CSF in England and Wales is moving in the same direction. The TMDL process has further options for engaging the community and stakeholders both in specification of the standards, and in application of management measures. Evaluation must come to rely upon time and appropriate monitoring at all scales from the sub-catchment upwards. Here the UK—far smaller country—is ahead.

Yet, the United States has been slow to implement conservation as central to watershed management, arguably because a tradition of countryside management (as opposed to a 'deep green', wilderness conservation ethos) is lacking (Nash, 2001). Shifts in attitudes to wetlands and river corridors are now manifesting in UK policy and farmers have long been compensated against financial shortfall for managing the land in a sustainable manner including protection of historic landscapes (Cook, 2010b). This *de facto* shift—or modulation—in funding from agricultural price support towards environmental protection more-or-less leaves agricultural produce the market. Community involvement (as opposed to individual landowner or farmer participation) still remains lacking. The Ribble River basin vision was designed to widen participation and was driven by the EA, yet it is the voluntary sector in the Rivers Trust movement, local participation in the Upper Thurne Working Group and the attempt by Hampshire County Council to engage the public that are closer to the U.S. model. On the European mainland, there are models for strong but effective local government intervention in conservation of waters and in protection from flooding. At the time of writing, all UK agencies are facing severe financial constraints while the voluntary sector, as ever, suffers economic uncertainty (Cook and Inman, 2012).

Certainly there are no easy answers and the present governance arrangements need to be closely evaluated. Just as British water managers thought that they could see a way forward that effectively utilised all economic sectors (private, public and voluntary), first recession hit, then something akin to a constitutional crisis for the UK ensued and despite the vote in Scotland to retain the union. Then there is great uncertainty around the UK's future relationship with the European Union.

Still, we might ask if localism will lead to widespread devolution, but will there also be a strengthening of local democracy? This may or may not be relevant to water resources depending on the relationship with local governance and matters environmental, still largely a matter for statutory agencies. Should they be harder on polluters anyway? Do some farmers and other industries actually get away with too much? Should monitoring and 'policing' be tighter (and with it a greater associated cost to the public purse). Will a blend of voluntary action and PES become part of the warp and weft of British water governance? Should it all be brought together in a national Water Strategy that embraces professional and statutory bodies, academic and engineering concerns and embraces new and evolving governance structures? It would be so nice to end this tome on a stronger sense of optimism!

References

Abramitzky R. and Braggion, F. (n.d.). Malthusian and Neo-Malthusian Theories. Available at: http://web.stanford.edu/~ranabr/Malthusian%20and%20Neo%20Malthusian1%20for%20webpage%20040731.pdf [Accessed February 2015].

Adams, B. and Foster, S.S.D. (1992). Land surface zoning for groundwater protection. *J. Inst Water and Env Man.*, 6(3): 312–320.

ADAS. (2007). Nitrates in water – the current status in England (2006). Available at: http://archive.DEFRA.gov.uk/environment/quality/water/waterquality/diffuse/nitrate/documents/consultation-supportdocs/d1-nitrateswater.pdf [Accessed February 2015].

Agnew, C. and Anderson, E. (1992). Water Resources and the Arid Realm. Routledge, London, 329pp.

AIC. (2015). UK Fertiliser Consumption Trends & Statistic. Available at: https://www.agindustries.diazorg.uk/sectors/fertiliser/uk-fertiliser-consumption-trends-and-statistics/[Accessed January 2015].

Alchian, A.A. (2008). Property Rights. *The Concise Encyclopaedia of Economics*. Availablet at: http://www.econlib.org/library/Enc/PropertyRights.html [Accessed February 2015].

Ambrose R.B., Barnwell T.O., McCutcheon S.C., Williams, J.R. (1996). "Computer Models for Water-Quality Analysis", in Water Resources Handbook, Mays, Larry W., editor-in-chief. McGraw-Hill, New York.

American History (n.d). Available at: http://www.let.rug.nl/usa/outlines/government-1991/the-executive-branch-powers-of-the-presidency/the-independent-agencies.php [Accessed August 2016].

Anderson, H., M. Futter., I Oliver, J. Redshaw and A. Harper (2010).Trends in Scottish river water quality. Macaulay Institute and SEPA 178pp.

Andoh, R.Y.G. (1994). Urban runoff: Nature, characteristics and control. *J. Inst. Water and Env. Man.*, 8(4): 371–378.

Anglian Water. (n.d.). *Drinking water treatment*. Fact File 3. Available at: http://www.anglianwater.co.uk/_assets/media/Fact_File_3_-_Drinking_water_treatment.pdf [Accessed February 2015].

Anglian Water. (2010). *Water Resources Management Plan*. 139 pp. Available at: http://www.anglianwater.co.uk/_assets/media/AW_WRMP_2010_main_Report.pdf [Accessed February 2015].

Archer, J. and Lord, E. (1993). Results from the Pilot Nitrate Scheme. In: *Solving the Nitrate Problem*. Editor? MAFF, London, pp. 29–32.

Archer, J. and Thompson, R. (1993). Background to the nitrate problem in the UK. In: *Solving the Nitrate Problem*. Editor? MAFF, London, pp. 3–6.

The Protection and Conservation of Water Resources, Second Edition. Hadrian F. Cook.
© 2017 John Wiley & Sons Ltd. Published 2017 by John Wiley & Sons Ltd.

Armstrong Brown, S.M., Cook, H.F. and Lee, H.C. (2000). Topsoil physical and chemical characteristics from a paired farm survey of organic vs. conventional farms in Southern England *Biological Agriculture and Horticulture*, 18, 37–54.

Arnell, N.W., Jenkins, A., and George, D.G. (1994). *The Implications of Climate Change for the National Rivers Authority*. R&D Report 12 by the Institute of Hydrology. HMSO, London, 94pp.

Atkinson, B. (1979). Urban influences on precipitation in London. In: *Man's Impact on the Hydrological Cycle in the United Kingdom*, G.E. Hollis (ed.). Geo Abstracts, Norwich, pp. 123–134.

ASR. (n.d.). *ASR - Managing Water Resources*. Available at: http://www.asrforum.com/ [Accessed February 2015].

Aue, C. (2014). *Challenges for water suppliers and government regarding impact of agricultural landuse*. Report from Lower Saxony, Germany by OOWV. Available at: http://wef-conference.gwsp.org/uploads/media/A03_Aue.pdf [Accessed February 2015].

AVLG—Avon Valley Liaison Group. (2002). 'Summary of Aims and objectives'. Internal Document.

Baggs, E.M., Stevenson M., Pihlatie, M. Regar A., Cook H., and Cadisch, G. (2003). Nitrous oxide emissions following application of residues and fertiliser under zero and conventional tillage. *Plant & Soil*. 254: 361–370.

Bailey-Watts, A.E. (1994). Eutrophication. In: *The Freshwaters of Scotland*, P.S. Maitland, P.J. Boon and D.S. McLusky (eds). Wiley, Chichester, pp. 385–411.

Bannock, G., Baxter, R., and Davis, E. (2003). *Dictionary of Economics*, New York: John Wiley, 410pp.

Barton, B.M. and Perkins, M.A. (1994). Controlling the artesian boreholes of the South Lincolnshire Limestone. *J. Inst. Water and Env. Man.*, 8(2): pp. 183–196.

Batie, S. (1983). Resource policy in the future: Glimpses of the 1985 Farm Bill. In: *Farm and Food Policy - Critical Issues for Southern Agriculture*, M. Hammering and H. Haris (eds).

Battarbee, R.W. and Allot, T.E.H. (1994). Palaeolimnology. In: *The Freshwaters of Scotland*, P.S. Maitland, P.J. Boon and D.S. McLusky (eds). New Jersey: Wiley, Chichester, pp. 113–130.

Barrow, C. (1987). *Water Resources and Agricultural Development in the Tropics*. Longman, Harlow, 356pp.

Bayliss A.C. (1999) *The Flood Estimation Handbook*, vol 5: catchment descriptors. Institute of Hydrology, Wallingford, UK.

BBC. (2009). Call for widespread water meters. Available at: http://news.bbc.co.uk/1/hi/7971257.stm [Accessed February 2015].

BBC. (2012). *How much does your water company leak?* Available at: http://www.bbc.co.uk/news/uk-17622837 [Accessed February 2015].

BBC. (2015). What is fracking and why is it controversial? Available at: http://www.bbc.co.uk/news/uk-14432401 [Accessed January 2015].

Beckett, P.H.T., and Webster, R. (1971). Soil variability-A review. *Soils and Fertility*, 34: 1–15.

Benson, D. (2009) 'Review article: constraints to policy transfer'. CSERGE Working Paper Series. Norwich: CSERGE.

Benson D. and Jordan A.. (2011). What have we learned from policy transfer research? Dolowitz and Marsh Revisited. *Polit. Stud. Rev.* 9(3): 366e378. http://dx.doi.org/10.1111/j.1478-9302.2011.00240.x

Benson, D., Jordan A. Cook, H., Inman A., and Smith L. (2013a). Collaborative environmental governance: Are watershed partnerships swimming or are they sinking? *Land Use Policy* 30: 748–757. http://dx.doi.org/10.1016/j.landusepol.2012.05.016

Benson, D., Jordan, A.J., and Smith, L. (2013b). 'Is environmental management really more collaborative? A comparative analysis of putative 'paradigm shifts' in Europe, Australia and the USA.' *Environment and Planning A*. 30 (6). DOI: 10.1068/a45378

Best, G.A. (1994). Afforestation and forestry practice. In: *The Freshwaters of Scotland*, P.S. Maitland, P.J. Boon and D.S. McLusky (eds), Wiley, Chichester, pp. 413–433.

Beaumont, P. (1978). Man's impact upon river systems. *Area*, 10(1): 38–41.

BGS British Geological Survey. (2015). *Integrated surface-water groundwater modelling of the Thames catchment*. Available at: http://www.bgs.ac.uk/research/environmentalModelling/surfacewaterGroundwaterModelling.html [Accessed February 2015].

Bianchi M. (2002). Ground Water Protection Areas and Wellhead Protection Draft Regulations for California Agriculture FWQP reference sheet 8.3, Univ of California. Available at: http://groundwater.ucdavis.edu/files/136270.pdf [Accessed February 2105].

Biesbroek, G.R., Termeer, C.J., Klostermann, J.E., and Kabat, P. (2014). Rethinking barriers to adaptation: Mechanism-based explanation of impasses in the governance of an innovative adaptation measure. *Global Environmental Change*, 26: 108–118.

Binnie, C. (1995). The Past, Present and Future of Water. CIWEM Centenary Lecture, Brompton Study Centre, The Barracks, Chatham, 22 March 1995.

Binns, W.O. (1979). The hydrological impact of afforestation in Great Britain. In: *Man's Impact on the Hydrological Cycle in the United Kingdom*, G.E. Hollis (ed.), Geo Abstracts, Norwich, pp. 55–70.

Bishop, P.L., Hively, W.D., Stedinger, J.R., Rafferty, M.R., Lojpersberger, J.L., and Bloomfield J.A. (2005). Multivariate Analysis of Paired Watershed Data to Evaluate Agricultural Best Management Practice Effects on Stream Water Phosphorus. *J. Env. Qual*, 34: 1087–1101.

Bland, A. and Evans, D. (1984). Development of groundwater resources. In *Water Management-A Review of Current Issues.* Proceedings of IWES symposium held in London, 5/6 December 1984.

Bloomfield, C. (1957). The possible significance of polyphenols in soil formation. *J. Sci. Food Agric.*, 8: 389–392.

Boardman, J. (1986). The context of soil erosion. *SEESOIL*, 3: 2–13.

Boardman, J. (1988). Public policy and soil erosion in Britain. In: *Geomorphology in Environmental Planning*, J.M. Hooke (ed.). Wiley, Chichester, pp. 33–50.

Boardman, J. (1990). Soil erosion on the South Downs: A review. In: *Soil Erosion on Agricultural Land*, J. Boardman, I.D.L. Foster and J.A. Dearing (eds). Wiley, Chichester, pp. 87–105.

Boardman, J.B. (1992). 'Current erosion on the South Downs: implications for the past'. In: *Past and present soil erosion*, M. Bell and J. Boardman (eds.), Oxbow, Oxford, 9–19.

Boardman, J. (1995). Damage to property by runoff from agricultural land, South Downs, Southern England, 1976–93. *Geographical Journal*, 161(2): 177–191.

Boardman J. (2003). Soil erosion and flooding on the eastern South Downs, southern England, 1976–2001. *Transactions of the Institute of British Geographers*, 28(2): 176–196.

Boardman J. (2013). Soil Erosion in Britain: Updating the Record. *Agriculture*, 3: 418–442. doi:10.3390/agriculture3030418

Boardman, J. and Favis-Mortlock, D.F. (1992). Soil erosion and sediment loading of watercourses. *SEESOIL*, 7: 5–29.

Boon, P.J. (1992). Essential elements in the case for river conservation. In: *River Conservation and Management*. P.J. Boon, P.Calow and G.E. Petts (eds), Wiley, Chichester, pp. 11–33.

Boorman, D.B., Hollist, J.M., and Lilly, A. (1995). Hydrology of soil types: a hydrologically based classification of the soils of the United Kingdom Institute of Hydrology rept. No. 126, Wallingford. Available at: http://nora.nerc.ac.uk/7369/1/IH_126.pdf [Accessed August 2016].

Bowie D. (2010). Spatial Planning within a localism agenda: Developing new approaches to implementation. University of Westminster. Available at: Spatial-Planning-within-a-Localism-Agenda-September-2010.pdf [Accessed August 2016].

Bowers, J.K. (1988). Farm incomes and the benefit of environmental protection. In: *Economics, Growth and Sustainable Environments*. D. Collard, D. Pearce and D. Ulph (eds). Macmillan, London, pp. 161–171.

Bradley, R. (1978). *Prehistoric Settlement of Britain*. London: Routledge & Keegan Paul.

Braithwaite, F. (1861). On the Rise and Fall of the River Wandle; its Springs, Tributaries and Pollution, *Minutes and Proceedings of the Institution of Civil Engineers* 20: 192–196.

Briggs, A. (1968). Victorian Cities. Penguin, 414pp.

Broads Authority. (n.d.a). Issues Affecting Lakes and SSSI condition, Broads Lake Restoration Strategy. Appendix 1. Available at: http://www.broads-authority.gov.uk/__data/assets/pdf_file/0015/412521/Appendix_1_Issues_affecting_lakes-1.pdf [Accessed January 2015].

Broads Authority. (n.d.b). From Darkness to Light: the restoration of Barton Broad. Available at: http://www.broads-authority.gov.uk/__data/assets/pdf_file/0006/404268/Darkness_to_Light.pdf [Accessed January 2015].

Broads Authority. (1993). *No Easy Answers, Draft Broads Plan. The Authority*, Norwich, 158pp.

Broads Authority. (2001). *Broads Drained Marsh Strategy* 44p.

Broads Authority (2011). *Broads Plan 2011*. A strategic plan to manage the Norfolk and Suffolk Broads. Available at: http://www.broads-authority.gov.uk/__data/assets/pdf_file/0015/402045/Broads-Plan-2011.pdf [Accessed August 2016].

Brookes, A. (1994). *Environmental Impact Assessment in the NRA: Illustrated by Case Studies from South-Central England*. Institution of Water and Environmental Management South Eastern Branch Seminar.

Brooks, N.P. (1988). Romney Marsh in the Early Middle Ages. In J. Eddison and C. Green (eds), Romney Marsh, Evolution, Occupation, Reclamation. Oxford University Committee for Archaeology Monograph No. 24, pp. 90–104.

Brown, G., *et al.* (1984). *Block 4 Water Resources. The Earth's Resources S238*. Open University Press, 96pp.

Bureau of Land Management. (2015). *What are Best Management Practices (BMPs)?* Available at: http://www.blm.gov/wo/st/en/prog/energy/oil_and_gas/best_management_practices.html [Accessed February 2015].

Burgess, D.B. and Fletcher, S.W. (1988). Methods used to delineate groundwater protection zones in England and Wales In: Groundwater pollution, aquifer recharge and vulnerability Geological Society Special Publication, N.S. Robins (ed.). pp. 199–210.

Burke, S. (2011). A Synthesis of Diffuse Pollution Research in England and Wales funded by Defra and EA. Available at: http://www.demonstratingcatchmentmanagement.net/wp-content/uploads/2011/11/A-brief-review-of-diffuse-pollution-research-in-England-and-Wales.pdf [Accessed February 2015].

Burn, A.J., Erian, J.H., and Tytherleigh, A. (2010). Climate, Water and Soil: Science, Policy and Practice. Proceedings of the SAC and SEPA Biennial Conference, Edinburgh 31 March - 1 April 2010 (Eds.) K. Crighton and R. Audsley. The England Catchment Sensitive Farming Delivery Initiative and its Role in Tackling Diffuse Pollution Impacts on Sites of Special Scientific Interest, pp. 207–213.

Burnell, D. (ed.). (1994). *Waterfacts '94*. Water Services Association, London.

Business Dictionary Online. (2015). Available at: http://www.businessdictionary.com/ [Accessed February 2015].

Butcher, D.P., Claydon, J., Labadz, J.C., Pattinson, V.A., and Potter, A.W.R. (1992). Reservoir sedimentation and colour problems in Southern pennine reservoirs. *J Inst. Water and Env. Man.*, 6(4): 418–431.

Butler, J.J. (2005). Muddy Flooding on the South Downs, online papers archived by the Institute of Geography, School of Geosciences, University of Edinburgh. Available at: https://www.era.lib.ed.ac.uk/bitstream/1842/830/1/jbutler001.pdf [Accessed January 2015].

CaBA. (n.d.). *Catchment based Approach*. Available at: http://www. catchmentbasedapproach.org/ [Accessed March 2015].

Cabinet Office. (n.d.). Community and society. Available at: https://www.gov.uk/ government/topics/community-and-society [Accessed February 2015].

Cabinet Office. (2008). *The Pitt Review: learning lessons from the 2007 floods*. Available at: http://webarchive.nationalarchives.gov.uk/20100807034701/http://archive.cabinetoffice. gov.uk/pittreview/_/media/assets/www.cabinetoffice.gov.uk/flooding_review/ final_press_notice%20pdf.pdf [Accessed March 2015].

Calder, L.R. (1990). *Evaporation in the Uplands*. Wiley, Chichester, 148pp.

Calver, A. (1992). Some guidelines for the use of the Institute of Hydrology distributed model Available at: http://nora.nerc.ac.uk/14406/1/N014406CR.pdf [Accessed February 2015].

CAS (Centre for Agricultural Strategy). (1976). *Land for Agriculture*. CAS University of Reading, Reading.

Catt, J.A. (1985). Natural soil acidity. *Soil Use and Management*, 1(1): 8–10.

CCWater. (2016a). The Consumer Council for Water. Available at: http://www.ccwater.org. uk/ [Accessed January 2016].

CCWater. (2016b). CC Water challenges companies on rising leakage. Available at: http://www.ccwater.org.uk/ [Accessed February 2016].

Central Market Agency Scotland. (2016). About the Scottish Water Industry. Available at: http://www.cmascotland.com/ [Accessed January 2016].

Centre for Hydrology and Ecology. (n.d.a). *UK Hydrometric Register*. Available at: http://www.ceh.ac.uk/products/publications/documents/hydrometricregister_final_ withcovers.pdf [Accessed February 2015].

Centre for Hydrology and Ecology. (n.d.b). An overview of the 2010-12 drought and its dramatic termination. Available at: http://www.ceh.ac.uk/data/nrfa/nhmp/other_ reports/2012_Drought_Transformation.pdf [Accessed January 2015].

Chandler, T.J., Cooke, R.U., and Douglas, I. (1976). Physical problems of the urban environment. *Geographical J.*, 142: 57–80.

Charnley S, Engelbert, B. 2005. Evaluating public participation in environmental decision-making: EPA's superfund community involvement program. *J. Env. Man*, 77: 165–182.

Chatham House. (2014). *Chatham House Rule*. Available at: http://www.chathamhouse. org/about/chatham-house-rule [Accessed February 2015].

CIWEM. (n.d.). Reframing sustainable development: A critical analysis. Available at: http://www.ciwem.org/media/853658/Reframing%20Sustainability%20WS.pdf [Accessed January 2015].

CIWEM. (2011). *Integrated Water Management: A CIWEM briefing report*. Available at: http://www.ciwem.org/policy-and-international/current-topics/water-management/integrated-water-management.aspx [Accessed February 2015].

CIWEM. (2012a). *Planning Water Resources in England and Wales*. Available at: http://www.ciwem.org/knowledge-networks/panels/water-resources/planning-water-resources-in-england-and-wales-.aspx [Accessed January 2015].

CIWEM. (2015). Water efficiency in the home. Available at: http://www.ciwem.org/wp-content/uploads/2016/04/Water-efficiency-in-the-home.pdf [Accessed October 2016].

Ciria. (2016). Water Sensitive Urban Design in the UK. Available at: http://www.ciria.org/Resources/Free_publications/Water_Sensitive_Urban_Design.asp [Accessed February 2016].

Clapp, B. (1984). *An Environmental History of Britain since the Industrial Revolution*, London: Longmans, p. 87.

Coase, R.H. (1960). The Problem of Social Cost. *J Law and Economics*, 3: 1–44.

Coles, J.M. and Coles, B.J. (1989). *Prehistory of the Somerset Levels*. Somerset Levels Project, Thorverton, Devon, 64pp.

Committee for Climate Change. (2014). *Current spending plans may mean an extra £3 billion in future flood damages*. Available at: http://www.theccc.org.uk/blog/current-spending-plans-may-mean-an-extra-3-billion-in-future-flood-damages/ [Accessed January 2015].

Communities and Local Government. (2009). Comprehensive list of nationally defined consultees in the planning application process - information report. Draft for consultation. Available at: http://www.planningportal.gov.uk/uploads/kpr/Draft_list_of_stat_and_non_stat_consultees.pdf [Accessed January 2015].

Communities and Local Government, (2011). Available at: https://www.gov.uk/government/uploads/system/uploads/attachment_data/file/5959/1896534.pdf [Accessed August 2016].

Communities and Local Government. (2014a). *Planning practice guidance*. Available at: http://planningguidance.planningportal.gov.uk/blog/guidance/consultation-and-pre-decision-matters/other-organisations-non-statutory-consultees/ [Accessed January 2015].

Communities and Local Government. (2014b). *Planning practice guidance*. Available at: http://planningguidance.planningportal.gov.uk/blog/guidance/consultation-and-pre-decision-matters/is-it-possible-for-a-statutory-or-non-statutory-consultee-to-direct-refusal-of-an-application/ [Accessed January 2015].

Cook, H.F. (1986). *Assessment of drought resistance in soils*. PhD Thesis, University of East Anglia, Norwich.

Cook, H.F. (1987). Cluster analysis applied to grouping soil series according to draughtiness. *Soil Survey and Land Evaluation*, 7: 177–185.

Cook, H.F. (1991). Nitrate Protection Zones: Targeting and land use over an aquifer. *Land Use Policy*, 8(1): 16–28.

Cook, H.F. (1992). A spatial model for identifying nitrate protection zones over an aquifer. *SEESOIL* 7, 37–63.

Cook, H.F. (1993). Progress in water management in the lowlands. *Progress in Rural Policy and Planning*, 3: 91–103.

Cook, H.F. (1994). Field-scale water management in southern England to 1900. *Landscape History*, 16: 53–66.

Cook, H.F. (1998). *The Protection and Conservation of Water Resources: A British Perspective*, Wiley International, 336pp.

Cook, H.F. (1999a). Groundwater Development in England. *Env and Hist*, 5: 75–96.

Cook, H.F. (1999b). Hydrological management in reclaimed wetlands. in Cook and Williamson (eds). *Water Management in the English Landscape: Field, Marsh and Meadow*, Edinburgh University Press, pp. 84–100.

Cook, H.F. (2007). The Hydrology, Soils and Geology of the Wessex Water Meadows. In: *Water Meadows*, H.F. Cook and T. Williamson (eds). Windgather Press ch 8, pp. 94–106.

Cook, H.F. (2008a). A tale of two catchments: water management and quality in the Wandle and Tillingbourne, 1600 to 1990. *Southern History*. 2008: 78–103.

Cook, H.F. (2008b). Evolution of a floodplain landscape: A case study of the Harnham Water Meadows at Salisbury England. *Landscapes* 9(1): 50–73.

Cook, H.F. (2010a). Boom, slump and intervention: changing agricultural landscapes on Romney Marsh, 1790 to1990. In: *Romney Marsh: Persistence and Change in a Coastal Lowland*. M.P. Waller, E. Edwards, and L. Barber, (eds). Romney Marsh Research Trust, Sevenoaks, pp. 155–183.

Cook, H.F. (2010b). Floodplain agricultural systems: functionality, heritage and conservation *Journal of Flood Risk Management*, 3: 1–9. DOI:10.1111/j.1753-318X.2010.01069.x

Cook, H. F. (2015). 'An unimportant river in the neighbourhood of London': the use and abuse of the river Wandle (*The London Journal Special Issue*).

Cook, H.F., Benson, D., Inman, A., Jordan, A., and Smith, L. (2012). Community-based catchment management groups in England and Wales extent, roles and influences. *Water and Environment Journal*, 26(1): 47–54. DOI: 10.1111/j.1747-6593.2011.00262.x

Cook, H.F., Benson, D., and Inman, A. (2014). Partnering for success in the UK: The Westcountry Rivers Trust. In: *The Politics of River Basin Organizations: coalitions, institutional design choices and consequences*. D. Huitema, S. Meijerink, and S. Verduijnpp, (eds.). Cheltenham. Edward Elgar Pub, pp. 119–139.

Cook, H.F., Bonnett, S.A.F., and Pons, L.J. (2009). *Wetland and floodplain soils: Their characteristics, management and future*. In: The Wetlands Handbook, E. Maltby and T. Barker, (eds). Blackwell Scientific, Chapter 18, pp. 382–416.

Cook H.F., Cowan, M., and Tatton-Brown, T. (2008). *Harnham Water Meadows: History and Description*. Hobnob Press, Salisbury 44pp.

Cook, H.F., Cutting R.L., Buhler W., and Cummings, I.P.F. (2004). Productivity and soil nutrient relations of bedwork watermeadows in southern England. *Agriculture, Ecosystems and Environment* 102(1): 61–79.

Cook, H.F., Cutting, R.L., and Valsami-Jones, E. (2015). Flooding with constraints: water meadow irrigation impacts on temperature, oxygen, phosphorus and sediment in water returned to a river. J Flood Risk Management doi:10.1111/jfr3.12142

Cook, H.F. and Dent D.L. (1990). Modelling soil water supply to crops. *Catena*, 17: 25–39.

Cook, H. and Inman, A. (2012). The voluntary sector and conservation for England: Achievements, expanding roles and uncertain future. *J. Env. Man.* 112: 170–177. http://dx.doi.org/10.1016/j.jenvman.2012.07.013

Cook, H.F. and Lee, H.C. (eds.) (1995). Preface to *Soil Management in Sustainable Agriculture*. Wye College Press.

Cook, H.F. and Moorby, H. (1993). English marshlands reclaimed for grazing: a review of the physical environment. *Journal of Environmental Management*, 38: 55–72.

Cook, H. and Norman C. (1996). Targeting agricultural policy: an analysis relating to the use of Geographical Information Systems. *Land Use Policy*, 13(3): 217–228.

Cook, H.F., Valdes, G.S.B., and Lee, H.C. (2006). Effects of Mulch Material and Thickness on Rainfall Interception, Soil Physical Characteristics and Temperature Under Zea Mays L. *Soils and Tillage Research* 91: 227–235.

Cook H. and Williamson T. (eds.) (1999). Water Management in the English Landscape: Field, Marsh and Meadow, Edinburgh University Press, 274pp.

Clarke, R. (1991). *Water: The International Crisis*. Earthscan, London, 193pp.

Calling, J.H., Croll, B.T., Whincup, P.A.E., and Harward, C. (1992). Plumosolvency effects and control in hard waters. *J. Inst. Water and Env. Man.*, 6(3): 259–268.

Community Forests. (2005). *England's Community Forests*. Available at: http://www.communityforest.org.uk/aboutenglandsforests.htm [Accessed February 2015].

Community Forest Trust. (n.d.). *Community Forest Trust*. Available at: http://www.cf-trust.org/ [Accessed March 2015].

Consumer Council for Water. (2015). Drinking Water Quality. Available at: http://www.ccwater.org.uk/waterissues/currentkeywaterissues/drinkingwaterquality/ [Accessed January 2015].

Cooper, J.D. (1980). *Measurement of Moisture Fluxes in Unsaturated Soil in Thetford Forest*. Institute of Hydrology Report 66, Wallingford, Oxon.

Cornwall Rivers Project. (n.d). Available at: http://www.cornwallriversproject.org.uk/ [Accessed August 2016].

Corporate Watch. (2013). *Leaking away: The financial costs of water privatisation*. Available at: http://www.corporatewatch.org/news/2013/feb/14/leaking-away-financial-costs-water-privatisation [Accessed February 2016].

Correll, D.L. and Weller, D.E. (1989). *Factors limiting processes in freshwater wetlands: An agricultural primary stream riparian forest*. In: *Freshwater Wetlands and Wildlife*. R.R. Sharitz and J.W. Gibbons (eds.), USDOE Symposium Series No. 61, CONF-8603101.

Countryside Agency/Land Use Consultants. (2005). Evaluation of the Community Forest Programme. Final report 155 pages. Available at: http://www.communityforest.org.uk/resources/evaluation_report.pdf [Accessed January 2015].

Countryside Commission (CC). (1994). *Countryside Stewardship: Handbook and Application Form*. CCP No. 453, Cheltenham.

Countryside Stewardship. (2016). Available at: https://www.gov.uk/government/collections/countryside-stewardship-get-paid-for-environmental-land-management [Accessed January 2017].

Courtney, W.J.W. (1994). Groundwater- Managing a Scarce Resource. Keynote Address to Institution of Water and Environmental Management Symposium, 23rd March at the Ramada Hotel, Gatwick, East Sussex.

Covington, L. (2009). Appropriate Assessment for Somerset Authorities Core Strategies: Somerset Levels and Moors and Severn Estuary (Bridgwater Bay). Natura 2000 Sites Scoping Report Volume 1, Main Report. Royal Haskoning, 249pp, Available at: http://www.tauntondeane.gov.uk/irj/go/km/docs/CouncilDocuments/TDBC/Documents/Forward%20Planning/Evidence%20Base/Royal%20Haskoning%20Scoping%20Report.pdf [Accessed August 2016].

CPRE. (n.d.). Available at: http://protectkent.org.uk/canterbury/ [Accessed January 2015].

Commission for Rural Communities (CRC) (2010) High Ground, high potential: as future for England's upland communities. Available at: http://webarchive.nationalarchives.gov. uk/20110303145243/http:/ruralcommunities.gov.uk/wp-content/uploads/2010/06/CRC114_uplandsreport.pdf [Accessed February 2016].

Cummings, I. and Cook, H.F. (1992). Soil-water relations in an ancient coppiced woodland. In: *Ecology and Management of Coppice Woodland*, G.P. Buckley (ed.). Chapman & Hall, London, pp. 53–75.

Curatolo, J. (2009). "Integrating watershed-based wetland protection into the Upper Susquehanna Coalition Wetland Program." Proposal to US EPA Region 2 Wetland Development Grant EPA-R2-09WPDG.

Cutting R., Cook, H.F., and Cummings, I. (2003). Hydraulic conditions, oxygenation, temperature and sediment relationships of bedwork watermeadows. *Hydrol Process* 17: 1823–1843.

Croll, B.T. and Hayes, C.R. (1988). Nitrate in water supplies in the United Kingdom. *Environmental Pollution*, 50: 163–187.

Crook, C.E., Boar, R.R., and Moss, B. (1983). *The Decline of Reedswamp in the Norfolk Broadland: Causes, Consequences and Solutions*. BARS 6, Broads Authority, Norwich, 132pp.

Crowfoot J.E. and Wondolleck, J.M. (1990). Citizen organizations and environmental conflict. In *Environmental Disputes*, J.E. Crowfoot and J.M. Wondolleck, (Eds.), pp. 1–15. Island Press, Connecticut.

Cullen, R. Hughey, K., and Kerr, G. (2006). New Zealand freshwater management and agricultural impacts. *The Australian Journal of Agricultural and Resource Economics*, 50: 327–346.

Darteh, B., Sutherland, A., Chlebek, J., Denham, G., and Mackay, R. (n.d.) *Managing water risks in Birmingham, the city with the tastiest tap water in Britain*. Available at: http://www.switchurbanwater.eu/outputs/pdfs/w6-2_cbir_rpt_switch_city_paper_-_birmingham.pdf [Accessed August 2016].

Darby, H.C. (1940). *The Medieval Fen/and*. Cambridge Studies in Economic History. Cambridge University Press.

Darby, H.C. (1956). *The Draining of the Fens*, 2nd edition. Cambridge University Press.

Data 360. (n.d.). 'Average water use per person per day'. Available at: http://www.data360. org/dsg.aspx?Data_Set_Group_Id-757 [Accessed February 2016].

Danube Watch. (n.d.). *Water Framework Directive and public participation*. Available at: http://www.icpdr.org/icpdr/static/dw2003_2/dw0203p09.htm [Accessed February 2015].

Davies, G.P. and Barraclough, D. (1989). Nitrate leaching at Rushall Farm, Wiltshire, 1985–1988; a field monitoring study. *IFOAM Bulletin*, 7: 3–5.

Davis, M. and Rees, Y. (2004). "Public participation in the Ribble River Basin." Workpackage 5 of the Harmonicop Project, 32pp.

Davis, B. (1994). 'Introduction' and 'Conclusions' of *Meter Pilot Project*. IWEM Presentation, 25 April1994 at East Grinstead, Mid Kent Water.

Demars, B.O.L. and Harper, D.M. (2002). *Assessment of the Impact of Nutrient Removal on Eutrophic Rivers*. The Environment Agency, Bristol 139pp.

DTC- Demonstration Test Catchments. (n.d.). What is the programme about? Availablet at: http://www.lwec.org.uk/activities/demonstration-test-catchments [Accessed February 2015].

De Silva, S.H.S.A. and Cook, H.F. (2003). Soil physical conditions and performance of cowpea following organic matter amelioration of sand. *Communications in Soil Science and Plant Analysis*, 34(7&8): 1039–1058.

Deakin, N. (2001). *In Search of Civil Society*. Palgrave.

Dee Catchment Partnership. (n.d.). Available at: http://www.theriverdee.org/catchment-planning-in-scotland.asp [Accessed March 2015].

DETR. (2001). Sedimentation in storage reservoirs: final report. Available at: http://www.britishdams.org/reservoir_safety/DEFRA-reports/200102Sedimentation%20in%20storage%20reservoirs.pdf [Accessed January 2015].

DCLG - Department for Communities and Local Government. (2012). Technical Guidance to the National Planning Policy Framework. Available at: https://www.gov.uk/government/uploads/system/uploads/attachment_data/file/6000/2115548.pdf [Accessed February 2015].

DCLG. (2014.) Environmental Impact assessment. Available at: http://planningguidance.planningportal.gov.uk/blog/guidance/environmental-impact-assessment/the-purpose-of-environmental-impact-assessment/ [Accessed January 2015].

Defra. (n.d). *Land Use and Production*. Available at: https://www.gov.uk/government/organisations/department-for-environment-food-rural-affairs/about/statistics and http://www.ecifm.rdg.ac.uk/current_production.ht [Accessed January 2015].

Defra. (2001). Guidance note for the Control of Pollution (Oil Storage) (England) Regulations 2001. Available at: https://www.gov.uk/government/uploads/system/uploads/attachment_data/file/69255/pb5765-oil-storage-011101.pdf [Accessed February 2016].

Defra. (2004). Mapping the Problem: Risks of Diffuse Water Pollution from Agriculture. Available at: http://archive.DEFRA.gov.uk/foodfarm/landmanage/water/csf/documents/mapping-problem-lowres.pdf [Accessed January 2015].

Defra. (2005). *Making space for water*. Available at: http://archive.DEFRA.gov.uk/environment/flooding/documents/policy/strategy/strategy-response1.pdf [Accessed January 2015].

Defra. (2006a). Freshwater Quality: Pesticides. Available at: http://archive.DEFRA.gov.uk/evidence/statistics/environment/inlwater/iwpesticide.htm#iwtb13 [Accessed January 2015].

Defra. (2006b). Biological river quality surveys. Available at: http://archive.DEFRA.gov.uk/evidence/statistics/environment/inlwater/iwbiotest.htm [Accessed January 2015].

Defra. (2009a). *Adapting to climate change UK Climate projections*. Available at: https://www.gov.uk/government/uploads/system/uploads/attachment_data/file/69257/pb13274-uk-climate-projections-090617.pdf [Accessed February 2015].

Defra. (2009b). Safeguarding our Soils: A Strategy for England. Available at: https://www.gov.uk/government/uploads/system/uploads/attachment_data/file/69261/pb13297-soil-strategy-090910.pdf [Accessed February 2015].

Defra. (2009c). Protecting our Water, Soil and Air. Available at: https://www.gov.uk/government/uploads/system/uploads/attachment_data/file/268691/pb13558-cogap-131223.pdf [Accessed February 2015].

Defra. (2010a). Regional Flood defence Committees. Available at: http://archive.DEFRA.gov.uk/environment/flooding/who/rfdc.htm [Accessed January 2015].

Defra. (2010b). Environmental impact: Water Indicator DA3: Nitrate and phosphate levels in rivers. Available at: http://archive.DEFRA.gov.uk/evidence/statistics/foodfarm/enviro/observatory/indicators/d/da3_data.htm [Accessed January 2015].

Defra. (2010c) 'Sustainable Development'. Available at: http://webarchive.nationalarchives.gov.uk/20130402151656/http://archive.defra.gov.uk/sustainable/government/progress/regional/summaries/16.htm > [Accessed February 2015].

Defra (2010d). River water Quality Indicator for sustainable development – 2009 annual results. Available at: https://www.gov.uk/government/uploads/system/uploads/attachment_data/file/141697/rwq-ind-sus-2009-resultsv2.pdf [Accessed August, 2015].

Defra. (2011a). Water for Life Market reform proposals. Available at: https://www.gov.uk/government/uploads/system/uploads/attachment_data/file/69480/water-for-life-market-proposals.pdf [Accessed January 2015].

Defra. (2011b) *Buffer strip*. Available at: http://adlib.everysite.co.uk/adlib/DEFRA/content.aspx?doc=110582&id=119886 [Accessed March 2015].

Defra. (2011c). *Barriers and Opportunities to the Use of Payments for Ecosystem Services.* 214 pp. Defra Available at: PESFinalReport28September2011(FINAL)%20(1).pdf [Accessed August 2016].

Defra. (2011d). National Standards for sustainable drainage systems. Available at: https://www.gov.uk/government/uploads/system/uploads/attachment_data/file/82421/suds-consult-annexa-national-standards-111221.pdf [Accessed February 2015].

Defra Business Link. (2011). ADLib Glossary (B). Available at: http://adlib.everysite.co.uk/adlib/DEFRA/content.aspx?id=000HK277ZW.0A5QFJHNYYMI5Z [Accessed January 2015].

Defra. (2012). UK National Action Plan for the Sustainable Use of Pesticides (Plant Protection Products). Available at: https://www.gov.uk/government/uploads/system/uploads/attachment_data/file/82557/consult-nap-pesticides-document-20120730.pdf [Accessed January 2015].

Defra. (2013a). *Water Abstraction from Non-Tidal Surface Water and Groundwater in England and Wales, 2000 to 2012.* Available at: https://www.gov.uk/government/statistics/water-abstraction-estimates [Accessed January 2015].

Defra. (2013b). *Principles governing future water charges.* Available at: https://www.gov.uk/government/publications/principles-governing-future-water-charges [Accessed February 2015].

Defra. (2013c). Guidance on complying with the rules for Nitrate Vulnerable Zones in England for 2013 to 2016. Available at: https://www.gov.uk/government/uploads/system/uploads/attachment_data/file/261371/pb14050-nvz-guidance.pdf [Accessed February 2015].

Defra. (2013d) Catchment Based Approach: Improving the quality of our water environment. Available at: https://www.gov.uk/government/uploads/system/uploads/attachment_data/file/204231/pb13934-water-environment-catchment-based-approach.pdf [Accessed February 2015].

Defra. (2014a). *Flood risk management: information for flood risk management authorities, asset owners and local authorities.* Available at: https://www.gov.uk/flood-risk-management-information-for-flood-risk-management-authorities-asset-owners-and-local-authorities [Accessed January 2015].

Defra. (2014b). *Environmental Stewardship: funding to farmers for environmental land management.* Available at: https://www.gov.uk/environmental-stewardship [Accessed January 2015].

Defra. (2014c). Organic farming statistics 2013. Available at: https://www.gov.uk/government/statistics/organic-farming-statistics-2013 [Accessed February 2015].

Defra. (2014d). *Catchment Sensitive Farming: reduce agricultural water pollution.* Available at: https://www.gov.uk/catchment-sensitive-farming-reduce-agricultural-water-pollution [Accessed February 2015].

Defra. (2015a). *Protecting and enhancing our urban and natural environment to improve public health and wellbeing*. Available at: https://www.gov.uk/government/policies/protecting-and-enhancing-our-urban-and-natural-environment-to-improve-public-health-and-wellbeing [Accessed January 2015].

Defra. (2015b). *Improving water quality*. Available at: https://www.gov.uk/government/policies/improving-water-quality/supporting-pages/planning-for-better-water [Accessed January 2015].

Defra. (2015c). Water Abstraction from Non-Tidal Surface Water and Groundwater in England and Wales, 2000 to 2013. Available at: https://www.gov.uk/government/uploads/system/uploads/attachment_data/file/422246/Water_Abstractions_release_V1.pdf [Accessed January 2016].

Defra (2016) ENV15 - Water abstraction tables. Available at: https://www.gov.uk/government/uploads/system/uploads/attachment_data/file/493563/Water_Abstractions_2014.pdf [Accessed August 2016].

Defra and EA. (2004). Impact of climate change on flood flows in river catchments. Available at: https://www.gov.uk/government/uploads/system/uploads/attachment_data/file/290661/scho0305biwf-e-e.pdf [Accessed February 2015].

Defra and EA. (2009a). River Basin Management Plan, South West River Basin District. Available at: https://www.gov.uk/government/uploads/system/uploads/attachment_data/file/292791/gesw0910bstp-e-e.pdf [Accessed January 2015].

Defra and EA. (2009b). Water for Lives and Livelihoods: River Basin Management Plan Thames River Basin District. Available at: https://www.gov.uk/government/uploads/system/uploads/attachment_data/file/289937/geth0910bswa-e-e.pdf [Accessed January 2015].

Dent, D.L. (1984). An introduction to acid sulphate soils and their occurrence in East Anglia. *SEESOIL*, 2: 35–51.

Dent, D. (1999). Wetland Soils, in Cook and Williamson (eds). *Water Management in the English Landscape: Field, Marsh and Meadow*, Edinburgh University Press, pp. 73–83.

Dent, D.L. and Scammell, R.P. (1981). Assessment of long-term irrigation need by integration of data for soil and crop characteristics and climate. *Soil Survey and Land Evaluation*, 1(3): 51–57.

Dhyr-Nielsen, M. (2009). A Tale of Two Cities: Meeting Urban Water Demands through Sustainable Groundwater management. In: Integrated Water Resources Management in Practice, R. Lenton and M. Muller, (Eds.). Earthscan, pp. 29–44.

Dialogue matters. (2011). Available at: http://www.dialoguematters.co.uk [Accessed February 2015].

Diaz, G.E., Brown, T.C., and Sveinsson, O. (2015). Aquarius: A Modeling System for River Basin Water Allocation. USDA. Available at: http://www.fs.fed.us/rm/value/aquarius [Accessed January 2015].

Dillaha, T.A., Reneau, R.B., Mostaghimi, S., Shanholtz, V.O., and Magette, W.L. (1987). *Evaluating Nutrient and Sediment Losses from Agricultural Lands: Vegetated Filter Strips*. US Environmental Protection Agency Report No. CBP1fRS 2/87, Washington, DC.

DoE (Department of the Environment). (1971). *The Future Management of Water in England and Wales*. HMSO, London, 107pp.

DoE. (1973). *The New Water Industry*. HMSO, London, 84pp.

DoE. (1986). *Nitrate in Water. Pollution*. Paper No. 26, HMSO, London, 84pp.

DoE. (1988). *Privatisation of the Water Authorities in England and Wales*. Cmnd 9734, HMSO, London.

DoE. (1991). *The Potential Effects of Climate Change in the United Kingdom*. HMSO, London, 123pp.

DoE. (1992). *Digest of Environmental Protection and Water Statistics*. No. 15. HMSO, London.

Dodgson A. and Clarke, L. (2011). *Broadland flood alleviation project*. CIRIA briefing August 2011. Available at: http://www.broads-authority.gov.uk/news-and-publications/ publications-and-reports/conservation-publications-and-reports/water-conservation-reports/32.-Broadland-Flood-Alleviation-Project.pdf [Accessed January 2015].

Donaldson, T. (1999). Making stakeholder theory whole. *Academy of Management Review*, 24: 237–241.

Doneen, L.D. and Westcot, D.W. (1988). *Irrigation Practice and Water Management*. FAO Irrigation and Drainage Paper 1, Revision 1. FAO, Rome, 63pp.

Doorenbus, J. and Pruitt, W.O. (1976). *Guidelines for Predicting Crop Water Requirements*. FAO Irrigation and Drainage Paper No. 24, 2nd edition. FAO, Rome, 156pp.

DSL - Dutch school of landscape Architecture. (n.d.). Available at: http://www. dutchschooloflandscapearchitecture.nl/en/2012/07/hns-plan-ooievaar [Accessed February 2015].

Doornkamp, J.C., Gregory, K.J., and Brown, A.S. (Eds). (1980). *Atlas of Drought in Britain 1975-76.1BG*, London.

Duxbury, RM.C. and Morton, S.G.C. (1994). *Blackstone's Statutes on Environmental Law*, 2nd edition. Blackstone, 458pp.

DWI - Drinking Water Inspectorate. (2010). What are the drinking water standards? Available at: http://dwi.DEFRA.gov.uk/consumers/advice-leaflets/standards.pdf [Accessed January 2015].

DWI. (2013). DWI PR14 Guidance – Lead in Drinking Water. Available at: http://dwi. DEFRA.gov.uk/stakeholders/price-review-process/PR14-guidance-lead.pdf [Accessed February 2015].

DWI. (2014). The Drinking Water Inspectorate. Available at: http://dwi.DEFRA.gov. uk/[Accessed February 2015].

DWI. (2015). *About us*. Available at: http://dwi.DEFRA.gov.uk/about/index.htm [Accessed February 2015].

DWI. (n.d). Private water supplies in England and Wales. Available at: http://dwi.Defra.gov. uk/stakeholders/private-water-supplies [Accessed January 2016].

DTI. (2000). Flue Gas Desulphurisation (FGD) Technologies. Available at: http://webarchive. nationalarchives.gov.uk/+/http://www.dti.gov.uk/cct/pub/tsr012.pdf [Accessed August 2016].

Dunn H. (2011). Payments for ecosystem services. Defra Evidence and Analysis service paper 4.

Dymond, D. (1990). *The Norfolk Landscape*. The Alastair Press, Bury St Edmunds, 279pp.

EA. (n.d.a). *Underground, under threat*. Available at: http://webarchive.nationalarchives. gov.uk/20140328084622/http://cdn.environment-agency.gov.uk/geho0906bldb-e-e.pdf [Accessed January 2015].

EA. (n.d.b). Available at: http://webarchive.nationalarchives.gov.uk/20140328084622/ http://www.environment-agency.gov.uk/static/documents/Leisure/MIDS_Severn_Fact_ sheet_4_final.pdf [Accessed February 2015].

EA. (n.d.c). *Groundwater Source Protection Zone*. Available at: https://www.gov.uk/ government/uploads/system/uploads/attachment_data/file/290723/scho0199betq-e-e. pdf [Accessed March 2015].

EA. (1997a). Local Environment Agency Plans (LEAPS): The Future. Available at: http://www.dcrt.org.uk/wp-content/uploads/2011/09/900-years-of-the-RDon-fishery-5.pdf [Accessed January 2015].

EA. (1997b). *Demand Management Bulletin.* The Agency, Bristol, 26, December 1997.

EA. (2004). The State of England's Chalk Rivers, The Agency, Bristol. Available at: http://adlib.everysite.co.uk/resources/000/057/248/Summary_chalk_rivers.pdf [Accessed February 2015].

EA. (2005). Attenuation of nitrate in the sub-surface environment. Science Report SC030155/SR2. Available at: https://www.gov.uk/government/uploads/system/uploads/attachment_data/file/291473/scho0605bjcs-e-e.pdf [Accessed February 2015].

EA. (2008). Water Resources in England and Wales - current state and future pressures. Available at: http://webarchive.nationalarchives.gov.uk/20140328084622/http://cdn.environment-agency.gov.uk/geho1208bpas-e-e.pdf [Accessed August 2016].

EA. (2009a). Water for people and the environment: water resources strategy for England and Wales. Available at: http://webarchive.nationalarchives.gov.uk/20140328084622/http://cdn.environment-agency.gov.uk/geho0309bpkx-e-e.pdf > [Accessed August 2016].

EA. (2009b). *Catchment flood management plans.* Available at: https://www.gov.uk/government/collections/catchment-flood-management-plans [Accessed January 2015].

EA. (2009c). *Water for People and Environment.* Available at: http://webarchive.nationalarchives.gov.uk/20140328084622/http://cdn.environment-agency.gov.uk/geho0309bpkx-e-e.pdf [Accessed February 2015].

EA. (2009d). *Groundwater vulnerability national dataset user guide.* Available at: http://www.findmaps.co.uk/assets/pdf/Ground_Water_Vulnerability_user_guide_v2.0.0.pdf [Accessed February 2015].

EA (2009e) Groundwater Source Protection Zones – Review of Methods Integrated catchment science programme Science report: SC070004/SR1. Available at: https://www.gov.uk/government/uploads/system/uploads/attachment_data/file/290724/scho0309bpsf-e-e.pdf [Accessed August 2015].

EA. (2010a). About the Water Framework Directive. Available at: http://evidence.environment-agency.gov.uk/FCERM/en/SC060065/About.aspx [Accessed January 2015].

EA. (2011a). Managing drought in England and Wales. Available at: http://webarchive.nationalarchives.gov.uk/20140328084622/http://cdn.environment-agency.gov.uk/geho0911budj-e-e.pdf [Accessed February 2015].

EA. (2011b). River Itchen Augmentation scheme. Available at: http://webarchive.nationalarchives.gov.uk/20140328084622/http://www.environment-agency.gov.uk/static/documents/Business/Itchen_FAQs_July_final_PDF.pdf [Accessed February 2015].

EA. (2011c). *Large-scale water transfers Position statement.* Available at: https://www.gov.uk/government/uploads/system/uploads/attachment_data/file/297318/geho0811btvr-e-e.pdf [Accessed February 2015].

EA. (2011d). Extending the catchment-based approach. Available at: http://webarchive.nationalarchives.gov.uk/20140328084622/http://www.environment-agency.gov.uk/static/documents/Research/Guidance_for_hosts_v2.pdf [Accessed February 2015].

EA. (2012a). *Anglian Drought Plan.* Available at: http://webarchive.nationalarchives.gov.uk/20140328084622/http://cdn.environment-agency.gov.uk/gean0112bvyl-e-e.pdf [Accessed February 2015].

EA. (2012b). Kennet and Vale of White Horse Catchment Abstraction Licensing Strategy. Available at: https://www.gov.uk/government/uploads/system/uploads/attachment_data/file/289893/LIT_2517_39dc0f.pdf [Accessed February 2016].

EA. (2013a). River quality. Available at: http://apps.environment-agency.gov.uk/ wiyby/37813.aspx [Accessed January 2014].

EA. (2013b). Managing Water Abstraction. Available at: https://www.gov.uk/government/ uploads/system/uploads/attachment_data/file/297309/LIT_4892_20f775.pdf [Accessed January 2015].

EA. (2013c). CAMS: Essex abstraction licencing strategy. Available at: https://www.gov.uk/ government/publications/cams-essex-abstraction-licensing-strategy [Accessed February 2015].

EA. (2013d). *Options for maintaining water supply*. Available at: http://webarchive. nationalarchives.gov.uk/20140328084622/http://www.environment-agency.gov.uk/ homeandleisure/drought/138425.aspx [Accessed February 2015].

EA. (2013e). *Groundwater protection: Principles and practice (GP3)*. Available at: https:// www.gov.uk/government/uploads/system/uploads/attachment_data/file/297347/ LIT_7660_9a3742.pdf [Accessed February 2015].

EA. (2013f). Restrictions on the Kennet lifted but Environment Agency investigation continues. Available at: http://www.riverkennet.org/uploads/files/documents/Press% 20Releases/090713%20Restrictions_on_the_Kennet_lifted_but_Environment_Agency_ investigation_continues_9_July_2013.pdf [Accessed February 2016].

EA. (2013g). Test & Itchen Abstraction licensing strategy (March 2013). Available at: https://www.gov.uk/government/uploads/system/uploads/attachment_data/file/289879/ LIT_2494_0c58d2.pdf [Accessed February 2016].

EA. (2013h). Broadland Abstraction Licensing Strategy (Feb 2013). Available at: https:// www.gov.uk/government/uploads/system/uploads/attachment_data/file/289841/ LIT_7743_9e67bc.pdf [Accessed February 2016].

EA. (2014a). *River basin management plans*. Available at: https://www.gov.uk/government/ collections/river-basin-management-plans [Accessed January 2015].

EA. (2014b). *Briefing: Background to the Somerset Levels and Moors*. Available at: https:// www.gov.uk/government/uploads/system/uploads/attachment_data/file/286491/ Briefing_SoS_visit_to_Somerset__v2__140126.pdf [Accessed January 2015].

EA. (2014c). *Groundwater*. Available at: http://maps.environment-agency.gov.uk/wiyby/ wiybyController?x=357683.0&y=355134.0&scale=1&layerGroups=default&ep=map& textonly=off&lang=_e&topic=groundwater [Accessed February 2015]

EA. (2014d). *Catchment Based Approach for a healthier environment*. Available at: http:// webarchive.nationalarchives.gov.uk/20140328084622/http://www.environment-agency. gov.uk/research/planning/131506.aspx [Accessed February 2015].

EA. (2015). Water supply and resilience and infrastructure. Available at: https://www.gov. uk/government/uploads/system/uploads/attachment_data/file/504682/ea-analysis- water-sector.pdf [Accessed December 2016].

EA and Defra. (2014). Nitrate Vulnerable Zones. Available at: https://www.gov.uk/ nitrate-vulnerable-zones [Accessed January 2014].

Eagle, D.J. (1992). Leaching of pesticides to surface and groundwaters. *SEESOIL*, 7: 31–36.

Eastleigh Borough Council. (2012). *River Itchen Sustainability Study*. Available at: http:// www.eastleigh.gov.uk/meetings/mgAi.aspx?ID=4620 [Accessed January 2015].

EC Groundwater Directive (80/68/EEC) Appendix B. Available at: http://www.documents. hps.scot.nhs.uk/environmental/ppc/app-b-ppc-guide-v1.pdf [Accessed January 2015].

EC. (n.d.). Soil at the interface between Agriculture and Environment. Available at: http:// ec.europa.eu/agriculture/envir/report/en/inter_en/report.htm [Accessed March 2015].

EC. (2014a). *The EU Water Framework Directive - integrated river basin management for Europe*. Available at: http://ec.europa.eu/environment/water/water-framework/ [Accessed January 2015].

EC. (2014b). *EP supports ban of phosphates in consumer detergents*. Available at: http://europa.eu/rapid/press-release_IP-11-1542_en.htm [Accessed February 2015].

EC. (2015a). The EU Floods Directive. Available at: http://ec.europa.eu/environment/waer/flood_risk/ [Accessed January 2015].

EC. (2015b). Introduction to the new EU Water Framework. Available at: http://ec.europa.eu/environment/water/water-framework/info/intro_en.htm [Accessed January 2015].

EC. (2015c). Environmental Impact Assessment – EIA. Available at: http://ec.europa.eu/environment/eia/eia-legalcontext.htm [Accessed January 2015].

EC. (2015d). Strategic Environmental Assessment – SEA. Available at: http://ec.europa.eu/environment/eia/sea-legalcontext.htm [Accessed January 2015].

EC. (2015e). Ecological status and intercalibration. Available at: http://ec.europa.eu/environment/water/water-framework/objectives/status_en.htm [Accessed January 2015].

EC. (2015f). *The Aarhus Convention*. Available at: http://ec.europa.eu/environment/aarhus/ [Accessed January 2015].

Economic Commission for Europe. (2008). *Spatial Planning: Key Instrument for Development and Effective Governance with Special Reference to Countries in Transition United Nations*. Available at: http://www.unece.org/fileadmin/DAM/hlm/documents/Publications/spatial_planning.e.pdf [Accessed February 2015].

Eden Rivers Trust. (2015). Available at: http://trust.edenriverstrust.org.uk/ [Accessed February 2015].

Edie Net. (1999). *UK wetlands under threat*. Available at: http://www.edie.net/news/0/UK-wetlands-under-threat/1162/ [Accessed February 2015].

Edwards, A.C. and Dennis P. (2000). "The landscape ecology of water catchments: Integrated approaches to planning and management." *Landscape Research*, 25: 305–320.

Eftec. (2010). Valuing environmental impacts: Practical guidelines for the use of value transfer in policy and project appraisal. Available at: https://www.gov.uk/government/uploads/system/uploads/attachment_data/file/182370/vt-case-study2.pdf [Accessed February 2015].

Ellis, P.A. (2002). The impact of urban groundwater upon surface water quality: Birmingham – river Tame study. Unpub. PhD thesis Univ of Birmingham. Available at: http://core.kmi.open.ac.uk/download/pdf/76382.pdf [Accessed January 2015].

EPCR - European Consortium for Political Research. (2013). *The Legitimacy of IWRM Scale Politics: Lessons from New Zealand*. Available at: http://www.ecpr.eu/Events/PaperDetails.aspx?PaperID=3234&EventID=5 [Accessed February 2015].

EEA. (2009). *Water resources across Europe — confronting water scarcity and drought EEA Report*. No 2/2009. Available at: http://www.indiaenvironmentportal.org.in/files/Water-resources-final-low-res-17032009.pdf [Accessed February 2015].

EEA. (2012a). European Bathing water quality in 2012. Available at: http://www.eea.europa.eu/publications/european-bathing-water-quality-2012 [Accessed February 2015].

EEA. (2012b). *European waters — assessment of status and pressures. Nutrients in freshwater (CSI 020/WAT 003)*. Copenhagen. Orthophosphate. Available at: http://www.eea.europa.eu/data-and-maps/tags#c5=all&c0=5&b_start=0&c9=orthophosphate [Accessed January 2012].

EEA. (2012c). *European waters - assessment of status and pressures.* Available at: http://www.eea.europa.eu/themes/water/water-assessments-2012 [Accessed February 2015].

Elworthy, S. (1994). Farming and Drinking Water. Avebury Studies in Green Research, Aldershot, 123pp.

Encyclopaedia of the Earth. (2008). *Polluter Pays Principle.* Available at: http://www.eoearth.org/view/article/155292/ [Accessed February 2015].

Encyclopaedia of New Zealand. (n.d.). Managing Water Resources. Available at: http://www.teara.govt.nz/en/water-resources/page-6 [Accessed February 2015].

ENDS. (2010). Land & ecology, pollution & health, water: starting work on farming's water foul up. JULY, 425: 34–37.

Engineering Timelines. (n.d.). 'Roman Aqueduct, Dorchester. Available at: http://www.engineering-timelines.com/scripts/engineeringItem.asp?id=218 [Accessed January 2016].

English Nature. (1997). Somerset Levels and Moors Natural Area. A nature conservation profile. Available at: http://www.naturalareas.naturalengland.org.uk/science/natural/profiles%5CnaProfile85.pdf [Accessed August 2016].

Eriksen, N.J. (1990). New Zealand water planning and management; evolution or revolution? In: *Integrated Water Management*, B. Mitchell, (ed.). Belhaven Press, London, pp. 45–87.

Environment UK. (2012). Is the 'making space for water ideal already being diluted? Available at: http://www.enuk.net/home/news/features/299-featured-articles/2449-is-the-making-space-for-water-ideal-already-being-diluted [Accessed February 2015].

Ethics. (2013). *Normative Ethics.* Available at: http://www.ethicsmorals.com/ethicsnormative.html [Accessed February 2015].

Entec. (2008). Brighton Chalk groundwater body final report. Available at: http://webarchive.nationalarchives.gov.uk/20140328084622/http://cdn.environment-agency.gov.uk/geho0309bpsz-e-e.pdf [Accessed February 2016].

EU. (2006). Directive on the protection of groundwater against pollution and deterioration on 22 November 2006 on the Protection of Groundwater Against Pollution and Deterioration. Available at: http://www.groundwateruk.org/downloads/GWD_final.pdf [Accessed February 2015].

Europa. (2010). Water protection and management (Water Framework Directive). Available at: http://europa.eu/legislation_summaries/agriculture/environment/l28002b_en. [Accessed January 2015].

European Commission. (2015). Authorisation of Plant Protection Products. Available at: http://ec.europa.eu/food/plant/pesticides/authorisation_of_ppp/index_en.htm [Accessed February 2016].

Evans, R. (1990). Soils at risk of accelerated erosion in England and Wales. *Soil Use and Management*, 6(3): 125–131.

Evans, R. and Cook, S. (1986). Soil erosion in Britain. *SEESOIL*, 3: 28–58.

Evans, D., Moxon, I.R., and Thomas, J.H.C. (1993). Groundwater nitrate concentrations from the Old Chalford Nitrate Sensitive Area, West Oxfordshire. *J. Inst. Water and Env. Man.*, 7(5): 506–512.

Everard, M. (2013). The Hydro-politics of Dams: Engineering or Ecosystems? ZED Books. 301pp.

Environmental Protection Agency. (1992). "The watershed protection approach: Annual Report 1992". Act. EPA 840/6-S-93-001, July 1988. 44pp.

FAO. (2001). Major soils of the World. World Soil Resources report no. 94. Available at: http://www.isric.org/isric/webdocs/docs//major_soils_of_the_world/set9/pz/podzol.pdf [Accessed March 2015].

FAO (n.d.) Concepts and definitions. Available at: http://www.fao.org/docrep/005/y4473e/y4473e06.htm [Accessed January 2015].

Fairness On Tap. (2011). Making the case for metering. Available at: http://assets.wwf.org.uk/downloads/fairness_on_tap.pdf [Accessed February 2016].

Farming Scotland Magazine. (2012). Slurry spreading options in non-NVZ areas. Available at: http://www.nfus.org.uk/news/2014/june/nitrate-proposals-recognise-scottish-progress [Accessed March 2015].

Fisher, B., Turner, R.K., and Morling, P. (2009). Defining and classifying ecosystem services for decision making. *Ecological Economics*, 68, 643–653.

Flood and Water Management Act. (2010). Available at: http://www.legislation.gov.uk/ukpga/2010/29/contents [Accessed March 2015].

Forestry Authority. (1993). *Forests and Water Guidelines*, 3rd edition. Forestry Commission, 18 pp.

Forestry Commission. (n.d.a). Standing timber volume for coniferous trees in Britain. <http://www.forestry.gov.uk/pdf/FCNFI111.pdf/$FILE/FCNFI111.pdf > [Accessed January 2015].

Forestry Commission. (n.d.b). Legislation. Available at: http://www.forestry.gov.uk/forestry/INFD-8J3L28 [Accessed January 2015].

Forestry Commission. (2003). National Inventory of Woodland and Trees, Edinburgh, 68pp. <http://www.forestry.gov.uk/pdf/frnationalinventory0001.pdf/$FILE/frnationalinventory0001.pdf> Accessed August 2016.

Forestry Commission. (2005). Water Use by Trees. Available at: http://www.forestry.gov.uk/pdf/FCIN065.pdf/$file/FCIN065.pdf [Accessed March 2015].

Forestry Commission. (2011a). National Forest Inventory Woodland Area Statistics: England.

Forestry Commission. (2011b). National Forest Inventory Woodland Area Statistics: Scotland.

Forestry Commission. (2011c). National Forest Inventory Woodland Area Statistics: Wales.

Forestry Commission. (2011d). Forests and water, 5th ed. Available at: http://www.forestry.gov.uk/pdf/FCGL007.pdf/$FILE/FCGL007.pdf [Accessed January 2015].

Forest Research. (n.d.a). Effects of air pollution on soil sustainability. Available at: http://www.forestry.gov.uk/fr/infd-6269vz [Accessed January 2015].

Forest Research. (n.d.b). Managing riparian buffer areas. Available at: http://www.forestry.gov.uk/website/forestresearch.nsf/ByUnique/INFD-6MVK4U [Accessed February 2015].

Foster, S.S.D., Bridge, L.R., Geake, A.K., Lawrence, A.R. and Parker, J.M. (1986). *The Groundwater Nitrate Problem*. British Geological Survey Hydrogeological Report 86/2, NERC, Wallingford.

Foundation for Water Research FWR (2010). The Water Framework Directive (2000/60/EC). Available at: http://www.euwfd.com/html/wfd- a summary.html [Accessed January 2015].

Fowler, D., Cape, J.N. and Leith, I.D. (1985). Acid inputs from the atmosphere in the United Kingdom. *Soil Use and Management*, 1(1): 3–5.

Frontier Economics. (2013). *Investment in the Water sector: the role of financing. Prepared for Water UK*. Available at: https://dl.dropboxusercontent.com/u/299993612/News/Latest%20news/Industry/Price-setting/rep-investment-in-the-water-sector—the-role-of-financing-final-27-03-2013-stc-2.pdf [Accessed January 2015].

Frost, R.C. (1999). EU practice in setting wastewater emission limit values. Available at: http://www.wgw.org.ua/publications/ELV%20-%20EU%20practice.pdf [Accessed January 2015].

FWAG. (n.d.). Farming & Wildlife Advisory Group (FWAG). Available at: http://www.ukagriculture.com/conservation/fwag.cfm [Accessed February 2015].

FSA. (n.d). Pesticides. Available at: http://www.food.gov.uk/business-industry/farmingfood/pesticides/#.Un0QIstFDDc [Accessed January 2015].

Gardiner, J.L. (1994). Sustainable development for river catchments. *J. Inst. Water and Env. Man.*, 8(3): 308–319.

Gash, J.H.C. and Stewart, J.B. (1977). The evaporation from Thetford Forest during 1975. *J. Hydrol.*, 35: 385–396.

Geake, A.K. and Foster, S.S.D. (1989). Sequential isotope and solute profiling in the unsaturated zone of the British Chalk. *Hydrological Sciences Journal*, 34(1): 79–95.

George, M. (1992). *The Land Use, Ecology and Conservation of Broad/and*. Packard, 558pp.

Gibbs, G.M. and Lindley, M. (1990). Flood alleviation works for Ashford, Kent. *Proc. Inst. Water and Environmental Management Annual Symposium* held at the Scottish Exhibition Centre, Glasgow, 4-6 September 1990, paper 30, 10pp.

Gilvear, D.J. (1994). River flow regulation. In: *The Freshwaters of Scotland*, P.S. Maitland, P.J. Boon, and D.S. McLusky (eds). Wiley, Chichester, pp. 463–488.

Gilvear D.J., Heal, K.V., and Stephen, A. (2002). Hydrology and the ecological quality of Scottish river ecosystems. *The Science of the Total Environment*, 294: 131–159.

Global Water Partnership GWP. (n.d.). IWRM principles. Available at: http://www.gwp.org/en/The-Challenge/What-is-IWRM/IWRM-Principles [Accessed January 2015].

Griffiths, P., Heathwaite, A.L., and Parkinson, R.J. (1995). Transport and transportation of nutrients following applications of manure, slurry and inorganic fertiliser to sloping grassland. In: *Soil Management in Sustainable Agriculture*, H.F. Cook and H.C. Lee (eds.). Wye College Press, pp. 510–519.

Grimble, R., Aglioby, M-K.J., Quan, J., and the Natural Resources Forum. (1995). "Trees and Trade-offs: A Stakeholder Approach to Natural Resource Management," International Institute for Environment and Development, Gatekeeper Series No. 52, 19pp. Available at: http://pubs.iied.org/pdfs/6066IIED.pdf [Accessed February 2015].

Goddard N. (1996). A mine of wealth? The Victorians and the agricultural value of sewage, *J Hist Geog* 22(3): 274.

Goddard N. (2005). 'Sanitate Crescamus: Water Supply, Sewage Disposal and Environmental Values in a Victorian suburb'. In: *Resources of the City*, D. Schott, B. Luckin, and G. Massard-Guibaud (Eds.). Aldershot, pp. 132–148.

Goddard, N. (2007). The suburbanization of the English Landscape: Environmental Conflict in Victorian Croydon'. In: *Landscape History after Hoskins. Vol III, Post-Medieval Landscapes*, P.S. Barnwell and Marilyn Palmer (eds.). Macclesfield, pp.119–134.

Godwin, H. (1978). *Fenland: Its Ancient Past and Uncertain Future*. Cambridge University Press, 196pp.

Gomme, J.W., Shurvell, S., Hennings, S.M., and Clarke, L. (1991). Hydrology of pesticides in chalk catchments: surface waters. *J. Inst. Water and Env. Man.*, 5(5): 546–552.

Gosk, F., Vrba, J., and Zaporozec, A. (1994). Uses and limitations of groundwater vulnerability maps. In: *Guidebook on Mapping Groundwater Vulnerability*, Vol. 16, J. Vrba and A. Zaporozec (eds.). Heise, Hanover, pp. 75–82.

Gough, G.W. (1930). *The Mines of Mendip*. Clarendon Press, Oxford.

Gray, N.F. (1994). *Drinking Water Quality*. Wiley, Chichester, 315pp.

Green, F.H.W. (1979). Field under-drainage and the hydrological cycle. In: *Man's Impact on the Hydrological Cycle in the United Kingdom*, G.E. Hollis (ed.), Geo Abstracts, Norwich, pp. 9–18.

Grey, D.R.C., Kinniburgh, D.G., Barker, J.A., and Bloomfield, J.P. (1995). *Groundwater in the UK. A Strategic Study. Issues and Research Needs*. Groundwater Forum Report FR/GF 1.

Gustard, A., Bullock, A., and Dixon, J.M. (1992). Low flow estimation in the United Kingdom. Institute of Hydrology report no. 108 Wallingford. Available at: http://www.ceh.ac.uk/products/publications/documents/ih108_low_flow_estimation.pdf [Accessed January 2015].

Guy, S. and Marvin, S. (1995). *The commodification of water: new logics of water management in Britain*. Memo from the Centre for Urban Technology, University of Newcastle, Newcastle-upon-Tyne, January 1995.

Hall, C. (1989). *Running Water*. Robertson McCarta, 256pp.

Hall, R.L. and Roberts, J.M. (1989). Hydrological aspects of new broad-leaf plantations. *SEESOIL*, 6: 2–38.

Hanks, R.J. and Ashcroft, G.L. (1980). *Applied Soil Physics*. Springer-Verlag, New York, 159pp.

Hammerton, D. (1994). Domestic and industrial pollution. In: *The Fresh Waters of Scotland*, P.S. Maitland, P.J. Boon, and D.S. McLusky (eds.). Wiley, Chichester, pp. 347–364.

Hantsweb. (2015). *What our priorities are and how we are doing?* Available at: http://www3.hants.gov.uk/foi-publication-scheme/cx-foi-priorities.htm [Accessed March 2015].

Hassan, J. (1998). *A history of water in England and Wales*. Manchester University Press, Chapter 2.

Harvey, N. (1980). *The Industrial Archaeology of Farming in England and Wales*. Batsford, London.

Hardin, G. (1968). *The Tragedy of the Commons. Science (AAAS)*, 162(3859): 1243–1248. doi:10.1126/science.162.3859.1243

Harding, R.J., Hall, R.L., Neil, C., Roberts, J.M., Rosier, P.T.W., and Kinniburgh D.G. (1992). *Hydrological Impacts of Broadleaved Woodlands: Implications for water use and water quality*. Projet Report 115/03/ST. National Rivers Authority, Bristol UK.

Hawkins, D. (1982). *Avalon and Sedgemoor*. Alan Sutton, Gloucester, 192pp.

Hazelden, J. (1990). *Soils in Norfolk VIII* Soil Survey Record No. 115. Silsoe.

Hazelden, J. Loveland, P.J., and Sturdy, R.G. (1986). *Saline Soils in North Kent*. Soil Survey of England and Wales, Special Survey No. 14. Harpenden.

Haycock, N.E. and Burt, T.P. (1993). The sensitivity of rivers to nitrate leaching: The effectiveness of near-stream land as a nutrient retention zone. In: *Landscape Sensitivity*, D.S.G. Thomas and R.J. Allison (eds.). Wiley, Chichester, pp. 261–272.

Hayes, J. (1977). Prior Wibert's Waterworks. *Friends of Canterbury Cathedral Chronicle*, 1977: 17–26.

Headworth, H.G. (1994). *The Groundwater Schemes of Southern Water, 1970–1990. Recollections of a Golden Age*. Southern Science, Crawley, Sussex.

Headworth, H.G. (2004). Recollections of a golden age: the groundwater schemes of Southern Water 1970 to 1990. In: *200 Years of British Hydrogeology*, J.D. Mather (ed.). Geological Society, London, Special Publications, pp. 339–362.

Headworth, H.G. and Fox, G.B. (1986). The South Downs chalk aquifer: its development and management. *J. Inst. Water Eng. and Sci.*, 40: 345–361.

Healthy Waterways. (2014). Available at: http://healthywaterways.org [Accessed February 2015].

Herbertson, P.W. (1994). Restoring flows to the River Darent. In: *Groundwater - Managing a Scarce Resource*. Proceedings of Institution of Water and Environmental Management Symposium, 23rd March at the Ramada Hotel, Gatwick, East Sussex.

Herrera-Pantoja, M. and Hiscock, K.M. (2008). The effects of climate change on potential groundwater recharge in Great Britain. *Hydrological Processes*, 22(1): 73–86. DOI: 10.1002/hyp.6620

Hey, R.D. (1994). Environmentally sensitive river engineering. In: *The Rivers Handbook*, Volume 2, P. Callow and G.E. Petts (eds). Blackwell, Oxford, pp. 337–362.

HGCA. (2005). *Field margins – guidelines for Entry Level Stewardship in England*. Available at: http://archive.hgca.com/publications/documents/cropresearch/FieldMargins_complete.pdf [Accessed February 2015].

Hill, A.R. (1976). The environmental impact of land drainage. *J. Env. Man.*, 4: 251–274.

Hillary, R. (1994). *The New EU Approach to Achieving Environmental Improvements in Products and Processes*. Centre for Environmental Technology, Imperial College of Science and Technology, University of London.

Hiscock, K.M., Lloyd, J.W., and Lemer, D.N. (1991). Review of natural and artificial denitrification of groundwater. *Water Research*, 25(9): 1099–1111.

Hiscock, K. (1995). Groundwater pollution and protection. In: *Environmental Science for Environmental Management*, T. O'Riordan (ed.). Longman, Harlow, pp. 246–262.

Hiscock, K. (2005). *Hydrogeology: Principles and Practice*. Blackwell, Oxford 390pp.

Hogan, C.M. (2012). Thermal Pollution. Available at: http://www.eoearth.org/view/article/156599/ [Accessed February 2016].

Hogwood, B.W. and Gunn, L.A. (1984). *Policy Analysis for the Real World*. Oxford University Press, 289pp.

Hollis, G.E. (1974). The effect of urbanisation on floods in the Cannon's Brook, Harlow, Essex. In: *Fluvial Processes in Instrumented Watersheds*, K.J. Gregory and D.E. Walling (eds.). Institute of British Geographers, Special Publication 6, pp. 123–139.

Hørlück, M.D and Pedersen, C.W. (2012). *Modelling of scale dependent solute transport in fractured limestone: Groundwater protection against nitrate pollution in Drastrup*. Master's dissertation Aalborg University.

Hornung, M. (1985). Natural soil acidity. *Soil Use and Management*, 1(1): 24–28.

HOL - Authority of the House of Lords. (2006). Water Management. Volume I: Report. House of Lords Science and Technology Committee. 8th Report of Session 2005-06, p. 69. The Stationery Office Limited, London, 6 June 2006.

HOL - House of Lords Select Committees. (2012). *Ch 2: Implementation of EU water legislation*. Available at: http://www.publications.parliament.uk/pa/ld201012/ldselect/ldeucom/296/29605.htm [Accessed February 2015].

HOC - House of Commons Environment, Food and Rural Affairs Committee report on future of flood prevention. October 2016 Available at: http://www.publications.parliament.uk/pa/cm201617/cmselect/cmenvfru/115/115.pdf [Accessed November 2016].

Hove, H. (2004). Critiquing Sustainable Development: A Meaningful Way of Mediating the Development Impasse? *Undercurrent*, I, 1.

Howarth, W. (1992). Regulation of the acquatic environment. In O. Lomas (ed.), *Managing the Environmental Challenge*. IBC Legal Studies and Services Ltd, Module IV.

Howell, D.L. (1994). Role of the environmental agencies. In: *The Fresh Waters of Scotland*, P.S. Maitland, P.J. Boon, and D.S. McLusky (eds.). Wiley, Chichester, pp. 577–612.

HMIP - Her Majesty's Inspectorate of Pollution. (1994). *Pollution Prevention, Our Common Concern*. DoE Pamphlet, 93, EP 314, March 1994.

Howells, G., Dalziel, T.R.K., and Turnpenny, A.W.H. (1992). Loch Fleet: liming to restore a brown trout fishery. *Environmental Pollution*, 78: 131–139.

Huizinga, J. (2010). HSRR02: Water safety: Flood depth and extent. Available at: http://knowledgeforclimate.climateresearch netherlands.nl/nl/25222818-Knowledge_transfer.html [Accessed August 2016].

Huntingford, C. *et al.* (2014). Potential influences on the United Kingdom's floods of winter 2013/14.*Nature Climate Change Nature Climate Change* (4), 769–777. Available at: http://www.nature.com/nclimate/journal/v4/n9/full/nclimate2314.ht [Accessed January 2014].

HSE. (n.d). Approvals for Pesticides in the UK. Available at: http://www.pesticides.gov.uk/guidance/industries/pesticides/topics/pesticide-approvals [Accessed January 2015].

Huby, M. and J. Bradshaw (2012). *Water Poverty in England and Wales*. Available at: http://www.york.ac.uk/inst/spru/presentations/JRBWaterPovFRSUM2012.pdf [Accessed February 2015].

Hurst, P., Hay, A., and Dudley, N. (1991). *The Pesticide Handbook*. Journeyman, 358pp.

Huscheke, R.E. (ed.) (1959). *Glossary of Meteorological Terms*. American Meteorological Society.

IGS (Institute of Geological Sciences). (1970). *Hydrogeological Map of the Chalk and Lower Greens and of Kent*, Sheet 1 Chalk, Regional Hydrogeological Characteristics and Explanatory Notes. Scale 1:126 720. London.

IGS. (1976). *Hydrogeological Map of Northern East Anglia, Sheet 1, Regional Hydrogeological Chacteristics and Explanatory Notes*. Scale 1:125 000. London.

IGS/Anglian Water Authority. (1981). *Hydrogeological Map of Southern East Anglia, Sheet 2, Chalk, Crag and Lower Greensand: Geological Structure*. Scale 1:125 000. London.

IGS/Southern Water Authority (SWA). (1978). *Hydrogeological Map of the South Downs and Adjacent Parts of the Weald*. Scale 1:100 000. London.

IGS/SouthernWater Authority (SWA).. (1979). *Hydrogeological Map of Hampshire and the Isle of Wight*. Scale 1:100 000. London.

Ineson, J. (1963). Applications and limitations of pumping tests: (b) hydrogeological significance. *Journal of the Institution of Water Engineers*, 17: 200–215.

Institute of Hydrology. (1976). *Water Balance of the Head water Catchments of the Wye and Severn 1970–75*. Report no. 33, NERC.

Institute of Hydrology. (1980). Low flows study report, Institute of Hydrology, Wallingford

Inman, A. (2006). Soil erosion in England and Wales: causes, consequences and policy options for dealing with the problem World Wildlife Fund 30 pp. Available at: http://www.wwf.org.uk/filelibrary/pdf/soilerosionengwales.pdf [Accessed March 2015].

Inman, M. (2014). The fracking fallacy. *Nature* 16 pp pp28–30.

Institute of Hydrology. (1980). Low Flow Studies report, Institute of Hydrology, Wallingford.

Irvine, K.A., Moss, B. and Balls, H.R. (1989). The loss of submerged plants with eutrophication Il. Relationships between fish and zooplankton in a set of experimental ponds, and conclusions. *Freshwater Biology*, 22: 89–107.

IWA- Inland waterways Association (2013). The Water Framework Directive. Available at: https://www.waterways.org.uk/waterways/restoration/restoration_resources/considerations/water_framework_directive [Accessed February 2015].

Jarman, R (1999). The National Trust policy on soil erosion. Available at: http://eusoils.jrc.ec.europa.eu/projects/scape/uploads/19/Jarman.pdf [Accessed February 2016].

Jenkins, G. J., Murphy, J. M., Sexton, D. M. H., Lowe, J. A., Jones, P. and Kilsby, C. G. (2009). UK Climate Projections: Briefing report. Met Office Hadley Centre, Exeter, UK.

JNCC (2010). Managing upland catchments: Priorities for water and habitat conservation. Available at: http://jncc.DEFRA.gov.uk/page-2098 [Accessed January 2015].

JNCC (2013). Special Areas of Conservation (SAC). Available at: http://jncc.DEFRA.gov.uk/page-23 [Accessed February 2015].

Johnes, P.J. and Burt, T.P. (1991). Water quality trends in the Windrush catchment: Nitrogen speciation and sediment interaction. In: *Sediment and Stream Water Quality in a Changing Environment: Trends and Explanation*, N.E. Peters and D.E. Walling (eds). IAHS Publication, No. 203, IAHS Press, Wallingford, pp. 349–357.

Johnson, RC. (ed.) (1995). *Effects of Upland Afforestation on Water Resources. The Balquhidder Experiment 1981–1991*. Institute of Hydrology Report No. 116, 2nd edition, Wallingford.

Johnson, D.B. and Hallberg, K.B. (2005). Acid mine drainage remediation options: a review *Science of the Total Environment* 338: 3–14.

Johnson, R.C. and Law, J.T (1992). *The Physical Nature of the Balquhidder Catchments*. Institute of Hydrology, Wallingford, 16pp.

Joint Nature Conservation Committee (JNCC). (2011). Towards an assessment of the state of UK Peatlands. Available at: http://jncc.defra.gov.uk/pdf/jncc445_web.pdf [Accessed February 2016].

Jones H.K. and Robins, N.S. (eds. 1999). *The Chalk Aquifer of the South Downs*. British Geological Survey Keyworth. 124 pp Available at: http://nora.nerc.ac.uk/12713/1/SD99001.pdf [Accessed February 2015].

Jones, P.D., Ogilvie, A.E.J., and Wigley, T.M.L. (1984a). *Riverflow Data for the United Kingdom: Reconstruction Data Back to 1844 and Historical Data Back to 1556*. Climatic Research Unit, University of East Anglia, Norwich.

Jones, P.D., Briffa, K.R., and Pilcher, J.R. (1984b). Riverflow reconstruction from tree rings in southern Britain. *J. Climatology*, 4: 461–472.

Jordan, T.E., Correll, D.L., and Weller, D.E. (1993). Nutrient interception by a riparian forest receiving inputs from adjacent cropland. *J. Environmental Quality*, 22: 467–472.

Kelly, A. (2008). Review of Sediment Removal. Broads Lake Restoration Strategy Appendix 3. The Broads Authority. Available at: http://www.broads-authority.gov.uk/__data/assets/pdf_file/0016/412504/Appendix_3_Review_of_Sediment_Removal.pdf [Accessed January 2015].

Kendall, D.T. (2004). *Redefining Federalism: Listening to the States in shaping Our Federalism*. Environmental Law Institute, Washington, DC, 176pp.

Kinnersley, D. (1988). *Troubled Water*. Chapter 3. Hilary Shipman, London.

Kinnersley, D. (1994). Coming Clean. Penguin, 229pp.

Kirby, C. (1984). *Water in Great Britain*. Pelican Books, 160pp.

Kjeldsen T.R. (2007). Flood Estimation Handbook Supplementary Report No. 1. Centre for Ecology and Hydrology. Available at: http://www.ceh.ac.uk/feh2/documents/fehsr1finalreportx.pdf [Accessed August 2016].

Kjeldsen, T., Jones, D., Bayliss, A., Spencer P., Surendran, S., Laeger S., Webster, P., and McDonald, D. (2008). Improving the FEH statistical Model. In: *Flood & Coastal Management Conference 2008*, University of Manchester, 1–3 July 2008.

Lapworth, D.J., Gooddy, D., Harrison, I., and Hookey, J. (2005). *Pesticides and their metabolites in groundwater - diuron in the Isle of Thanet chalk aquifer of southeast England*. British Geological Survey Internal Report, IR/05/049. 23pp. Available at: http://nora.nerc.ac.uk/4491/1/IR05049.pdf [Accessed February 2015].

Lapworth, D., Stuart, M., Hart, A., Crane, E., and Baran, N. (2011). Emerging contaminants in groundwater. UK Groundwater Forum, 24th May 2011, London. Available at: http://www.groundwateruk.org/downloads/7_D_Lapworth.pdf [Accessed February 2015].

Law, F. (1956). The effects of afforestation upon the yield of water catchment areas. *Journal of the British Waterworks Association*, 38: 489–494.

LEAF. (n.d.). *Example Farm Landscape and Nature Conservation and Enhancement Plan.* Available at: http://www.leafuk.org/resources/001/028/196/FINAL_LEAF_Landscape_and_Nature_Conservation_and_Enhancement_Plan_Report_template_for_v._12.0.pdf [Accessed February 2015].

Leach, W.D., and Pelkey, N.W. (2001). "Making watershed partnerships work: A review of the Empirical Literature." *J Water Resou. Plan and Man.* Nov/Dec 2001: 378–385.

Leitmann, J. (1999). *Integrating the Environment in Urban Development: Singapore as a Model of Good Practice Urban Development Division.* The World Bank, Washington, DC, Working Paper no 7.

Leeson, J.D. (1995). *Environmental Law.* Pitman, London, 479pp.

Lenntech. (n.d.). *Nitrates removal.* Available at: http://www.lenntech.com/processes/nitrates/nitrate-removal.htm [Accessed February 2015].

Lerner, D.N. and Tellam J.H. (1993). The protection of urban groundwater from pollution. In: *Water and the Environments*, J.C. Currie and A.T. Pepper (eds.). Ellis Horwood, London, pp. 322–335.

Life. (2008). River Basin Planning in England and Wales. Available at: http://www.wagrico.org/publishor/system/component_view.asp?LogDocId=157&PhyDocId=184 [Accessed February 2015].

Lilly, A. (2010). A hydrological classification of UK soils based on soil morphological data. 19th World Congress of Soil Science, Soil Solutions for a Changing World – 6th August 2010, Brisbane, Australia.

LWEC - Living with Environmental Change. (n.d.). Test Catchments. Available at: http://www.lwec.org.uk/activities/demonstration-test-catchments [Accessed February 2015].

Lloyd, J.W., Williams, G.M., Foster, S.S.D., Ashley, R.P., and Lawrence, A.R. (1991). Urban and industrial groundwater pollution. In: *Applied Groundwater Hydrology*, R.A. Downing and W.B. Wilkinson (eds.). Clarendon Press, Oxford, pp. 134–148.

Locke, G.M. (1987). *Census of Woodlands and Trees 1979-82.* Forestry Commission Bulletin No. 63, HMSO, London.

Londoncouncils. (n.d.). 'London key facts'. Available at: http://www.londoncouncils.gov.uk/londonfacts/default.htm?category=8 [Accessed February 2015].

Longstaff, L.L., Aldous, P.J., Clark, L. Flavin, R.J., and Partington, J. (1992). Contamination of the chalk aquifer by chlorinated solvents. *Water and Environment Journal*, 6(6): 541–550. Available at: http://onlinelibrary.wiley.com/doi/10.1111/j.1747-6593.1992.tb00789.x/

Lorenzoni, I., Benson, D., and Cook, H. (2015). 'Regional rescaling in UK climate adaptation governance: from agency to collaborative control?' In: *Climate Adaptation Governance: Theory, Concepts and Praxis in Cities and Regions*, J. Knieling and K. Klindworth (eds.). Chichester: Wiley.

Louka, E. (2008). *Water Law and Policy: Governance Without Frontiers*, Oxford, Ch.15. 446pp.

Lubell, M. (2004). Collaborative environmental institutions: all talk and no action? *Journal of Policy Analysis and Management* 23(3): 549–573.

Lubell, M., Schneider, M., Scholz, J.T., and Mete, M. (2002).Watershed Partnerships and the Emergence of Collective Action Institutions. *American Journal of Political Science*, 46(1): 148–163.

Ludwig, D. (2001). The era of management is over. *Ecosystems*, 4: 758–764.

Macdonald, T.D. (1994). Water supply. In: *The Fresh Waters of Scotland*, P.S. Maitland, P.J. Boon, and D.S. McLusky (eds.). Wiley, Chichester, pp. 279–296.

McDonald, A.T. and Kay, D. (1988). *Water Resources Issues and Strategies*. Longman, Harlow (with Wiley, New York).

McHugh M., Harrod, T., and Morgan, R. (2002). The extent of soil erosion in upland England and Wales. *Earth Surface Process and Landforms*, 27(1): 99–107. DOI: 10.1002/esp.308.

Mackay, D.W. (1994). Pollution control. In: *The Freshwaters of Scotland*, P.S. Maitland, P.J. Boon, and D.S. McLusky (eds.). Wiley, Chichester, pp. 517–529.

MAFF. (1993). *Our Living Heritage, Environmentally Sensitive Areas*. MAFF, London.

MAFF. (1996). *Nitrate Sensitive Areas Scheme: Monitoring of Nitrate Concentrations in Water Draining from Subsoil, Results Winter 1995-96*. Internal Report 6pp. + tables. MAFF, London.

MAFF/ADAS. (1982). *Irrigation*, 5th edition. Reference Book 138, HMSO, London, 122pp.

MAFF/STW (Severn Trent Water)/DoE. (1988). Severn Trent Water, Birmingham. *The Hatton Catchment Nitrate Study*. Severn Trent Water, Birmingham.

MAFF/WOAD. (1994). *Designation of Vulnerable Zones in England and Wales under the EC Nitrate Directive (91/676): A Consultation Document*. PB 1715, The Ministry, London.

Mannion, A.M. (1991). *Global Environmental Change*. Longman, Harlow, 404pp.

Maplecroft. (2012). 'Water stress Index 2012'. Maplecroft Business Management Consultancy. Available at: http://reliefweb.int/sites/reliefweb.int/files/resources/map_2103.pdf [Accessed January 2015].

Marrison, D.L. (2007). "Whole Farm Planniing Model Building for the Successful Transition of Your Agriculture Business." Fact Sheet Ser. Ohio State Univ. Extension Jefferson, OH.

Marrs, R.H. and Gough, M.W. (1989). Soil fertility, a potential problem for habitat restoration. In: *Biological Habitat Restoration*, G.P. Buckley (ed.). Belhaven Press, pp. 29–44.

Marsh, T., Cole, G., and Wilby, R. (2007). Major droughts in England and Wales, 1800–2006. *Weather*, 62(4): 87–93. DOI: 10.1002/wea.67.

Marsh T.J. and Lees, M.L. (2003). Hydrological Data United Kingdom, *Hydrometric register and statistics 1996-2000*. Natural Environment Research Council, Centre for Ecology and Ecology, Wallingford, Oxford.

Martinez-Espineira R. (2007). An estimation of residual water demand using co-integration and error correction. Vol X. *Journal of Applied Economics*, 10(1): 161–184.

Marshall B. (2011). *Wessex Chalk Stream & Rivers Trust*. Available at: http://www.gethooked.co.uk/game/wessex_chalk_stream_and_rivers_trust [Accessed February 2015].

Martin S.A., Blackwell, D., Hogan, D.V., Pinay, G., and Maltby, E. (2009). The role of buffer zones for Agricultural Runoff. In: *The Wetlands Handbook*, E. Maltby and T. Barker (eds.). Blackwell Scientific ch. 19, pp. 417–439.

Matthews, P.J. (1992). Sewage sludge disposal in the UK: a new challange for the next twenty years. *J. Inst. Water and Env. Man.*, 6(5): 551–559.

Mason, C.F. (1991). *The Biology of Freshwater Pollution*, 2nd edition. Longman, Harlow, 333pp.

Meat Info. (2008). 70% of England designated as Nitrate Vulnerable Zone. Available at: http://www.meatinfo.co.uk/news/archivestory.php/aid/1166/70_25_of_England_designated_as_Nitrate_Vulnerable_Zone.html [Accessed March 2015].

Medway Council. (2007). *Provision of water for Medway's future needs*. Available at: http://www.medway.gov.uk/pdf/06key_facts_and_issues.pdf [Accessed February 2015].

Met Office. (n.d.a). General weather and atmospheric data. Available at: http://www.metoffice.gov.uk/services/industry/data/general [Accessed February 2015].

Met Office. (n.d.b). Climate data. Available at: http://www.metoffice.gov.uk/public/weather/climate/gcjfgescd [Accessed February 2015].

Met Office. (n.d.c). *The UK's wet summer, the jet stream and climate change*. Met Office News Blog. Available at: https://metofficenews.wordpress.com/2012/07/12/the-uks-wet-summer-the-jet-stream-and-climate-change/[Accessed February 2015].

Met Office. (2013). England and Wales drought 2010 to 2012. Available at: http://www.metoffice.gov.uk/climate/uk/interesting/2012-drought [Accessed January 2015].

Miles R. (1993). Maintaining Groundwater Supplies during Drought Conditions in the Brighton Area. *J Inst. Water and Env. Man.*, 7(4): 382–393.

Minford, P. (1991). *The Supply Side Revolution in Britain*. Edward Elgar, Aldershot, 262pp.

Mitchell R.K., Agle, R.A., and Wood, D.J. (1997). Towards a theory of stakeholder identification and and salience: defining the principle of who really matters and what really counts. *Academy of Management Review*, 22(4): 853–866.

Mitsch, W.J. and Gosselink, J.G. (1993). Wetlands, 2nd edition. Van Nostrand Reinhold, New York, 722pp.

Monkhouse, R.A. and Richards, H.J. (1982). *Groundwater Resources of the United Kingdom*. Th. Schafer Druckerei GmbH, Hannover.

Moorby, H. and Cook, H.F. (1992). Use of fertiliser-free grass strips to protect dyke water from nitrate pollution. *Aspects of Applied Biology*, 30: 231–234.

Moorby, H. and Cook, H.F. (1993). *The Use of Fertiliser-Free Buffer Strips to Protect Dyke Flora from Nitrate Pollution, on Romney Marsh SSSI*. Final Report to the National Rivers Authority NRA R&D Project F1(90)16F, March 1993, 58pp.

Morgan, R.P.C. (1992). Soil conservation options in the UK. *Soil Use and Management*, 8: 176–180.

Morgan, R.P.C., Morgan, D.D.V., and Finney, H.J. (1986). A simple model for assessing annual soil erosion on hillslopes. In: *Agricultural Nonpoint Source Pollution: Selection and Application*, A. Giorgini and F. Zingales (eds.). Elsevier, Oxford.

Morris, J. (1989). *Land Drainage: Agricultural Benefits and Impacts*. Technical papers from the Annual Symposium of IWEM. 21-22 March 1989, York.

Morrison, B.R.S. (1994). Acidification. In: *The Freshwaters of Scotland*, P.S. Maitland, P.J. Boon, and D.S. McLusky (eds.). Wiley, Chichester, pp. 435–463.

Moss, B. (1988). *Ecology of Fresh Waters*, 2nd edition. Blackwell Scientific, Oxford, 417pp.

Movahedi Naeni, S.A.R. and Cook, H.F. (2000). Influence of compost on temperature, water, nutrient status and the yield of maize in a temperate soil. *Soil Use and Man.* 16: 215–221.

Municipality of Aalborg. (2002). *The Drastrup Project*. Available at: http://ec.europa.eu/ environment/life/project/Projects/index.cfm?fuseaction=home.showFile&rep=file&fil= Sustainable_Landuse_in_Groundwater_Areas.pdf [Accessed February 2015].

Muñoz-Carpena, R. Shukla, S., and Morgan, K. (n.d). Field Devices for Monitoring Soil Water Content *Southern regional water programme*. Available at: https://www.bae.ncsu. edu/topic/go_irrigation/docs/Field-devices-monitoring-.pdf [Accessed August 2016].

Murray-Darling Basin Authority. (2015). Environmental challenges and issues in the Basin. Available at: http://www.mdba.gov.au/about-basin/basin-environment/challenges-issues [Accessed February 2015].

Muscutt, A.D., Harris, G.L., Bailey, S.W., and Davies, D.B. (1993). Bufferzones to improve water quality: a review of their potential use in UK agriculture. *Agriculture, Ecosystems and Environment* 45: 59–77.

Nace, R.L. (1974), Pierre Perrault: the man and his contribution to modern hydrology. *J. American Water Resources Association*, 10: 633–647. doi: 10.1111/j.1752-1688.1974. tb05623.x

Nash, R.F. (2001). "Wilderness and the American Mind." Yale University Press, New Haven, 413pp.

Natural England. (2015). Environmental Stewardship: guidance and forms for existing agreement holders. Available at: http://www.naturalengland.org.uk/ourwork/farming/ funding/es/default.aspx [Accessed August 2016].

National Academy of Sciences NAS. (2007). Fossil Water in Libya. Available at: http:// www.koshland-science-museum.org/water/html/en/Sources/Fossil-Water-in-Libya.html [Accessed January 2015].

Natural England. (2012). Valuing Ecosystem Services: Case Studies from Lowland England. Natural England Commissioned Report NECR101. 18pp. Available at: http://publications. naturalengland.org.uk/publication/2319433> Available at: NECR101_edition_1.pdf [Accessed February 2012].

Natural England. (2013). Natural Character Area Profile. No *142: Somerset Levels and Moors.* Available at: http://publications.naturalengland.org.uk/publication/12320274? category=587130 [Accessed August 2016].

Natural England. (2016). Catchment Sensitive Farming: reduce agricultural water pollution. Available at: https://www.gov.uk/catchment-sensitive-farming-reduce-agricultural-water-pollution [Accessed August 2016.

Natural England and Defra. (2013). *Sites of special scientific interest: what you're allowed to do.* Available at: https://www.gov.uk/guidance/protected-areas-sites-of-special-scientific-interest [Accessed August 2016].

Natural Resources Wales (NRW). (2014). 'Our purpose'. Available at: http:// naturalresourceswales.gov.uk/about-us/our-reports-and-priorities/our-corporate-plan/ our-purpose/?lang=en [Accessed January 2015].

National Audit Office. (2007). Environment Agency: Building and maintaining river and coastal flood defences in England. Available at: http://www.nao.org.uk/report/building-

and-maintaining-river-and-coastal-flood-defences-in-england/ [Accessed January 2015].

National Committee of Water Assessment. (n.d.). Water Management and Spatial Planning in the Netherlands. Available at: http://www.restorerivers.eu/Portals/27/Publications/Water%20Management%20and%20Spatial%20Planning%20in%20the%20Netherlands.pdf [Accessed February 2015].

National Parks. (n.d.) Moor restoration in the Peak District. Available at: http://www.nationalparks.gov.uk/lookingafter/climate-change/moor-restoration-in-peak-district [Accessed January 2015].

National Trust. (n.d.). Buttermere and Crummock Water. Available at: http://www.nationaltrust.org.uk/buttermere-valley/features/buttermere-and-crummock-water [Accessed February 2015].

NERC. (n.d.). Groundwater flow and quality *UK Groundwater Forum.* Available at: http://www.groundwateruk.org/downloads/groundwater_flow_and_quality.pdf [Accessed March 2015].

NERC. (1975). Regional storm and flood risk assessment. Available at: http://www.ceh.ac.uk/products/publications/documents/ih102floodandstormassessment.pdf [Accessed January 2014].

NERC. (n.d.). Small world. Available at: http://www.nerc.ac.uk/research/funded/programmes/nanoscience/eni-brochure/ [Accessed February 2016].

NERPB (North East River Purification Board). (1994). River Ythan Catchment Review. 79pp.

New Civil Engineer NCE. (2011). *Environment Agency reveals 10 most improved rivers.* Available at: http://www.nce.co.uk/news/water/environment-agency-reveals-10-most-improved-rivers/8619124.article [Accessed February 2015].

Newson, M.D. (1979). The results of ten years' experimental study on Plynlimon, Mid-Wales, and their importance for the water industry. *J. Inst. Water Engineers and Scientists*, 33: 321–333.

Newson, M. (1992a). Land, Water and Development. Routledge, London, 351pp.

Newson, M. (1992b). Land and water: convergence, divergence and progress in U.K. policy. *Land Use Policy*, 9(2): 111–121.

Newson, M. (1994). *Hydrology and the River Environment.* Clarendon Press, Oxford.

Newson, M. (2008). Land, Water and Development: Sustainable and Adaptive Management of Rivers 3rd edition Routledge.

New Statesman. (2012). *Where art thou, "big society"?* Available at: http://www.newstatesman.com/blogs/the-staggers/2012/03/society-cameron-state-sofa [Accessed February 2015].

Nicholson, B. (2001). *Nature Conservation Management Plan for the for the lower Wandle local nature reserve* London Conservation services ltd., 39pp.

NMTWG (National Metering Trials Working Group). (1993). *Water Metering Trials, Final Report-Summary.* Published by Water Services Association, Water Companies' Association, Office of Water Services, WRC and DoE.

Norman, C., Potter, C., and Cook, H. (1994). Using GIS to target agri-environmental policy. In: *Innovations in GIS, no. I*, M.F. Worboys (ed.). Taylor & Francis, London.

Novitzki R.P. (1985). "The effects of lakes and wetlands on flood flows and base flows in selected northern and eastern states." In: *Proceedings of a Wetland conference of the Chesapeake, Env Law Institute*, H.A. Groman *et al.* (eds.). Washington D.C., pp. 377–388.

NRA (National Rivers Authority). (1991a). *Proposals for Statutory Water Quality Objectives.* Report of the National Rivers Authority, Water Quality Series no. 5, The Authority, Bristol.

NRA. (1991b). *Somerset Levels and Moors Water Level Management and Nature Conservation Strategy.* NRA, Wessex Region.

NRA. (1991c). *River Medway Catchment Management Plan Phase 1.* The Authority, Worthing.

NRA. (1991d). *Summary Report of the 1989 Test Pumping of the Alre Scheme, no. 91/R/160.* The Authority, Worthing.

NRA. (1992a). *Drought in the South.* The Authority, Worthing.

NRA. (1992b). *Ely Ouse Catchment Management Plan Consultation Report Summary.* The Authority, Peterborough, P/7617/92.

NRA. (1992c). *The Influence of Agriculture on the Quality of Natural Waters in England and Wales.* Report of the National Rivers Authority, Water Quality Series no. 6, The Authority, Bristol.

NRA. (1992d). *Policy and Practice for the Protection of Groundwater.* The Authority, Bristol, 52pp.

NRA. (1992e). *River Itchen Catchment Management Plan.* The Authority, Worthing.

NRA. (1992f). *Flood Defence Law.* The Authority, Bristol, 54pp.

NRA. (1992g). *Water Resources Development Strategy.* The Authority, Bristol.

NRA. (1993a). *River Medway Catchment Management Plan Final Report.* The Authority, Worthing.

NRA. (1993b). *Water Resources Development Strategy.* The Authority, Bristol.

NRA. (1993c). *Low Flows and Water Resources.* The Authority, Bristol.

NRA. (1993d). *Regional Water Resources Strategy, Severn Trent Region.* The Authority, Solihull.

NRA. (1993e). The river Kennet Catchment Management Plan Consultation report. The Authority Reading.

NRA. (1994a). *Discharge Consents and Compliance.* Report of the NRA, Water Quality Series no. 17, HMSO, 76pp.

NRA. (1994b). *The Drought of 1988–1992 in Anglia: A review.* The Authority, Peterborough, P70/1/94.

NRA. (1994c). *Hampshire Avon Catchment Management Plan, Final Report.* The Authority, Blandford Forum.

NRA. (1994d). *Broadland: Flood Alleviation Strategy.* The Authority, Peterborough, P94/6/94.

NRA. (1994e). *Water Resources in Anglia.* The Authority, Peterborough, P81/9/94, 115pp.

NRA. (1994f). *Guidance Notes for Local Planning Authorities on the Methods of Protecting the Water Environment through Development Plans.* The Authority, Bristol.

NRA. (1994g). *Water: Nature's Precious Resource. An Environmentally Sustainable Water Resources Development Strategy for England and Wales.* The Authority, March 1994, HMSO, London, 93pp.

NRA. (1994h). *The Quality of Rivers and Canals in England and Wales (1990 to 1992).* NRA Water Quality Series no. 19. HMSO, London, 39pp.

NRA. (1994i). *East Sussex Rather Catchment Management Plan Consultation Report.* The Authority, Worthing.

NRA. (1994j). *Shropshire Groundwater Scheme, Third Phase: Leaton Area.* The Authority,

NRA. (1994k). *Guidance Notes for Local Planning Authorities on the Methods of Protecting the Water Environment through Development Plans.* The Authority, Bristol.

NRA. (1994l). *The Ely Ouse Catchment Management Plan*. The Authority, Peterborough, P67/l/94. Solihull.

NRA. (1994m). *Groundwater Vulnerability 1. · 100 000 Map Series Sheet 47. East Kent*. HMSO, London.

NRA. (1995a). *Bathing Water Quality: Report to the NRA*. The Authority, Bristol, 80pp.

NRA. (1995b). *Catchment Management Planning Guidelines*. HMSO, London.

NRA. (1995c). *The Drought of 1995*. The Authority, Bristol, 52pp.

NRA. (1995d). *Saving Water: The NRA's Approach to Water Conservation and Demand Managemet*. The Authority, Bristol, 70pp.

NRA. (1995e). *Upper Thames Catchment Management Plan Consultation Report*. The Authority, Wallingford.

NRA. (1995f). *Somerset Levels and Moors Water Level Management and Nature Conservation Strategy Summary*. NRA South West Region, Exeter.

NRA/Binnie and Partners. (1991). *A Flood Alleviation Strategy for Broad/and, Initial Consultation Document*. In association with Oakwood Environmental.

NRC (National Research Centre). (1993). *Ground Water Vulnerability Assessment: Predicting Relative Contamination Potential under Conditions of Uncertainty*. National Academy Press, Washington, DC, 28pp.

NRCS - Natural Resources Conservation Service. (2002). "Farm Bill 2002, fact sheet." The Natural Resources Conservation Service, May 2002, 5pp.

NRFA. (n.d.). National River Flow Archive Hydrological report for 2002. Available at: http://www.nerc-wallingford.ac.uk/ih/nrfa/yb/yb2002/evaporation.html [Accessed January 2015].

NRFA. (2014). Hydrometry in the UK, Hydrometric Areas. Available at: http://www.ceh.ac.uk/data/nrfa/hydrometry/has.html [Accessed January 2015].

NYSWRI - New York State Water Resources Institute. (2015). *About us*. Available at: http://wri.cals.cornell.edu/about/ [Accessed February 2015].

OECD. (2010). Guidelines for Cost-effective Agri-environmental Policy Measures. OECD Publishing.

OMF - Observatory Monitoring Framework. (2012). Environmental Framework: Water. Available at: https://www.gov.uk/government/uploads/system/uploads/attachment_data/file/160780/DEFRA-stats-observatory-indicators-da1-120224.pdf [Accessed January 2015].

OMF - Observatory Monitoring Framework. (2014). Environmental Framework: Water abstraction for agriculture. Available at: https://www.gov.uk/government/uploads/system/uploads/attachment_data/file/270336/agindicator-da5-09jan14.pdf [Accessed February 2015].

Office for National Statistics. (2012). *2011 Census - Population and Household Estimates for England and Wales, March 2011*. Available at: http://www.ons.gov.uk/ons/dcp171778_270487.pdf [Accessed February 2015].

Ofwat (Office of Water Services). (1991). *Paying for Water: The Way Ahead*. Ofwat pamphlet 12/91. 30pp.

Ofwat. (1993a). *Paying for Growth: A Consultation Paper on the Framework for Reflecting the Costs of Providing for Growth in Charges*. February 1993, 24pp.

Ofwat. (1994). 1993-94 Report on the Cost of Water Delivered and Sewage Collected. Ofwat, Birmingham, 43pp.

Ofwat. (2001). Water and sewerage bills - how they are controlled. Available at: http://www.ofwat.gov.uk/consumerissues/chargesbills/prs_inf_wandsbills [Accessed February 2015].

Ofwat/EA. (n.d.). The case for change – reforming water abstraction management in England. Available at: http://webarchive.nationalarchives.gov.uk/20140328084622/http://cdn.environment-agency.gov.uk/geho1111bveq-e-e.pdf [Accessed January 2015].

Ofwat/Defra. (2006). The development of the water industry in England and Wales. Available at: http://www.ofwat.gov.uk/publications/commissioned/rpt_com_devwatindust270106.pdf [Accessed January 2015].

Ofwat. (n.d.). Glossary of Terms. Available at: http://www.ofwat.gov.uk/aboutofwat/gud_pro_ofwatglossary.pdf [Accessed January 2015].

Ofwat. (2011a). Water supply licensing –guidance on eligibility. Available at: http://www.ofwat.gov.uk/competition/wsl/gud_pro_wslelig.pdf [Accessed February 2015].

Ofwat. (2011b). Your water and sewerage bill 2011-12. Available at: http://www.ofwat.gov.uk/consumerissues/chargesbills/prs_lft_charges2011-12.pdf [Accessed February 2015].

Ofwat. (2015a). Board leadership, transparency and governance – updated assessment of monopoly water companies' governance arrangements. Available at: http://www.ofwat.gov.uk/wp-content/uploads/2015/11/pap_pos20150615boardleadership.pdf [Accessed February 2016].

Ofwat. (2015b). The economic regulator of the water and sewerage sectors in England and Wales. Available at: http://www.ofwat.gov.uk/ [Accessed January 2016].

Ofwat. (2015c). IB 08/11: Ofwat investigating leakage failures. Available at: http://www.ofwat.gov.uk/ib-0811-ofwat-investigating-leakage-failures/ [Accessed February 2016].

Ofwat. (2015d). Ofwat investigating leakage failures http://www.ofwat.gov.uk/ib-0811-ofwat-investigating-leakage-failures/11leakage [Accessed February 2015].

On The Commons. (n.d.). Elinor Ostrom's 8 Principles for Managing A Commons. Available at: http://www.onthecommons.org/magazine/elinor-ostroms-8-principles-managing-commmons [Accessed February 2015].

Oklahoma Water Resources Board. (2015). Surface water studies. Available at: http://www.owrb.ok.gov/studies/surface/investigations.php [Accessed January 2015].

Osborn, S. and Cook, H.F. (1997). Nitrate Vulnerable Zones and Nitrate Sensitive Areas: A policy and technical analysis of groundwater source protection in England and Wales. *J. Env. Planning and Management*, 40(2): 217–233.

Osborn T. and D. Maraun (2008). Changing intensity of rainfall over Britain. *Climatic Research Unit*, UEA. Available at: http://www.cru.uea.ac.uk/documents/421974/1295957/Info+sheet+%2315.pdf/8b8457b7-7bd2-49fc-888a-9b3f6785a40e [Accessed January 2015].

Ospar Commission. (2009). Impacts of microbiological contamination on the marine environment of the North-East Atlantic. Available at: http://qsr2010.ospar.org/media/assessments/p00466_Microbiological_contamination_09-10-22.pdf [Accessed January 2015].

O'Shea, M.J. (1984). Borehole recharge of the Folkestone Beds at Hardham, Sussex. *J. Inst. Water Eng. and Scientists*, 38(1): 9–24.

O'Shea, M.J (1995). The hydrogeology of the Enfield-Haringey artificial recharge scheme, north London. *Quarterly Journal of Engineering Geology and Hydrogeology*, 28: S115–S129. doi:10.1144/GSL.QJEGH.1995.028.S2.03).

Owen, M., Headworth, H.G., and Morgan-Jones, M. (1991). Groundwater in basin management. In R.A. Downing and Wilkinson, W.B. (eds), *Applied Groundwater Hydrology: A British Perspective*. Clarendon Press, Oxford, pp. 16–33.

Owen, M. (1991). Groundwater abstraction and river flows. *J. Inst. Water and Env. Man.*, 5(6): 697–702.

Oxfordshire County Council. (2013). Available at: http://www.oxfordshire.gov.uk/cms/content/proposed-reservoir-oxfordshire [Accessed January 2015].

Packman, J.C. (1979). The effect of urbanisation on flood magnitude and frequency. In: *Man's Impact on the Hydrological Cycle in the United Kingdom*, G.E. Hollis (ed.). Geo Abstracts, Norwich, pp.154–172.

Palmer, R.C. (1993). *1:625 000 Scale Map: Risk of Soil Erosion in England and Wales by Water on Land under Winter Cereal Cropping*. Soil Survey and Land Research Centre, Silsoe, Bedfordshire.

Parker, D.J. and Penning-Rowsell, E.C. (1980). *Water Planning in Britain*. George Alien & Unwin, London, 277pp.

Parker, D.E., Legg, T.P., and Folland, C.K. (1992). A New Daily Central England Temperature Series, 1772-1991. *Int. J. Clim.*, 12: 317–342.

Parrett Internal Drainage Board. (2010). *North Moor and Salt Moor Water Level Management Plan*. Available at: http://www.somersetdrainageboards.gov.uk/media/North-Moor-WLMP-Parr-approved-Jul-10.pdf [Accessed March 2015].

Pekarova P., Miklenek, P., and Pekar, J. (2006). Long-term trends and runoff fluctuations of European rivers *Climate Variability and Change—Hydrological Impacts* (Proceedings of the Fifth FRIEND World Conference held at Havana, Cuba, November 2006), IAHS Publ. 308, 2006. Pp520–525. Available at: http://147.213.145.2/pekarova/WEBClanky/ADEA01.pdf [Accessed February 2015].

Pesticide News. (2006). Factsheet 20-21. Pesticides News 56 June 2002. Available at: http://www.pan-uk.org/pestnews/Actives/atrazine.htm [Accessed January 2015].

Petersen, R.C., Petersen, L.B.M., and Lacoursiere, J. (1992). A building block model for stream restoration. In P.J. Boon, P. Callow and G.E. Petts (eds), *River Conservation and Management*. Wiley, Chichester, pp. 293–309.

Petts, G.E. (1988). Regulated rivers in the United Kingdom. *Regulated Rivers Research and Management*, 2:201–220.

Petts, G.E. and Foster, I. (1985). *Rivers and Landscape*. Edward Arnold, London, 274pp.

Petts, G.E. and Lewin, J. (1979). Physical effects of reservoirs on river systems. In: *Man's Impact on the Hydrological Cycle in the United Kingdom*, G.E. Hollis (ed.). Geo Abstracts, Norwich, pp. 79–91.

Pierre, J. and Peters, B.G., (2000). Governance, Politics and the State, Ch.1. Macmillan, Basingstoke.

Pinsent Masons. (2012). Available at: http://www.pinsentmasons.com/en/media/dco-news/september-2012/26-september-2012/government-to-underwrite-thames-tideway-tunnel/ [Accessed January 2015].

Pike, T. (2004). Dry fly off Wandsworth High Street (*first published in Fly Fishing & Fly Tying magazine*, February 2004. Available at: http://www.wandlepiscators.net/?page_id=2 [Accessed January 2015].

Politics. (2015). Water regulation and Ofwat. Available at: http://www.politics.co.uk/reference/water-regulation-and-ofwat [Accessed February 2015].

Porter K.S. (2004). "Does the CAP fit? Rectifying Eutrophication in the Chesapeake Bay." *J Water Law.* 15(5): 188–192.

Porter K.S. (2006). "Fixing our Drinking Water: From Field and Forest to Faucet." *Pace University School of Law, Environmental Law Review*, 23(2): 389–422.

POST (Parliamentary Office of Science and Technology). (1993). *Dealing with Drought*, 86pp.

POST. (2007). Urban flooding. Available at: http://www.managingclimaterisk.org/document/urban%20flooding%20UK.pdf [Accessed February 2007].

POST. (2008). River Basin Management Plans. Available at: http://www.parliament.uk/documents/post/postpn320.pdf [Accessed January 2015].

Potter, C., Cook, H., and Norman, C. (1993). The targeting of rural environmental policies: an assessment of agri-environmental schemes in the UK. *J. Environmental Planning and Management*, 36(2): 199–215.

Potter, C., Burnham, P., Edwards, A., Gasson, R., and Green, B. (1991). *The Diversion of Land: Conservation in a Period of Farming Contraction*. Routledge, London, 130pp.

Price, M. (1996). *Introducing Groundwater*. Second ed. George Allien & Unwin, Abingdon, 278pp.

Prudhomme C. and Williamson, J. (2003). Derivation of RCM-driven potential evapotranspiration for hydrological climate change impact analysis in Great Britain: a comparison of methods and associated uncertainty in future projections. *Hydrol. Earth Syst. Sci.*, 17: 1365–1377. doi:10.5194/hess-17-1365-2013.

Purseglove, J. (1988). *Taming the Flood*. Oxford University Press, 307pp.

Purseglove, J. (2015). Taming the Flood. 2nd ed. William Collins, 2015, 3980pp.

Rackham, O. 1986. *The History of the Countryside*. Dent, p. 131.

Rabesandratana, T. (2013). Pesticidemakers Challenge E.U. Neonicotinoid Ban in Court. *Science Insider*, 28th Aug, 2013. Available at: http://www.sciencemag.org/news/2013/08/pesticidemakers-challenge-eu-neonicotinoid-ban-court [Accessed August 2016].

Rafferty, L.J. (n.d.). 'The River Wandle: Pollution' unpub. notes, The Wandle Industrial Museum.

Rahtz, P. (1993). *The English Heritage Book of Glastonbury*. Batsford/English Heritage, London.

Ramsar. (n.d.). What are wetlands? *Ramsar Information Paper no.1*. Available at: http://ramsar.rgis.ch/pdf/about/Info2007-01-e.pdf [Accessed January 2015].

Reckon. (2007). Water Act 2003. Available at: http://www.reckon.co.uk/open/Water_Act_2003 [Accessed January 2015].

Redclift, M. (1994). Reflections on the 'sustainable development' debate. *Int. J. Sustainable Development and World Ecology*, 1: 3–21.

Refsgaard, J.C., Storm, B., and Clausen, T. (2010). Système Hydrologique Europeén (SHE): review and perspectives after 30 years development in distributed physically-based hydrological modelling. *Hydrology Research*, 41(5): 355–377.

Reganold, J.P. (1995). Soil quality and farm profitability studies of biodynamic and conventional farming systems. In: *Soil Management in Sustainable Agriculture*, H.F. Cook and H.C. Lee (eds.). Wye College Press, pp. 1–17.

Regis, E. (n.d.). *The Doomslayer*. Wired. Available at: http://archive.wired.com/wired/archive/5.02/ffsimon_pr.html [Accessed February 2015].

Rees, J. and Williams, S. (1993). *Water for Life*. Council for the Protection of Rural England.

RELU – Rural Economy and Land Use Programme. (n.d.). Fourth Call Adapting Rural Living and Land Use to Environmental Change. UK Research Councils.

RELU. (2009). Catchment management for the protection of water resources: The Ecosystem Health Report Card. Rural Economy and Land Use Programme Policy and practice note 7. Available at: http://www.relu.ac.uk/news/policy%20and%20practice%20notes/Smith/Smith%20PP7%20final.pdf [Accessed August 2016].

RELU - Rural Economy and Land Use Programme. (2015). Available at: http://www.relu.ac.uk/ [Accessed January 2015].

Richards, K. (1982). *Rivers: Form and Process in Alluvial Channels*. Methuen, London, 361pp.

Rittel, H.W. and Webber, M.W. (1973). Dilemmas in a General Theory of Planning. *Policy Sciences*, 4(1973): 155–169.

The Rivers Trust. (n.d.a). *The Rivers Trust Movement*. Available at: http://www.theriverstrust.org/riverstrusts/trust_movement.html [Accessed February 2015].

The Rivers Trust. (n.d.b). The Ecosystem Approach & Rivers Trusts. Available at: http://www.theriverstrust.org/environment/ecosystem/eco_approach.htm. [Accessed February 2015].

Robson, A. and Reed, D. (1999). *Flood estimation handbook: statistical procedures for flood estimation, vol 3*. IoH, Wallingford.

Rodda, J.C. (1976). Basin studies. In: *Facets of Hydrology* J.C. Rodda (ed.). Wiley, Chichester, pp. 257–297.

Rodriguez S.I., M. S. Roman, S. C. Sturhahn and E H. Terry (2002). Sustainability Assessment and reporting for University of Michigan, Ann Arbour Campus. Center for Sustainable Systems, Report No. CSS02-04, Available at: http://www.css.snre.umich.edu/css_doc/CSS02-04.pdf [Accessed December 2016].

Rose, R. (2005). *Learning from comparative public policy: a practical guide*. London: Routledge.

Rose, S.C. and Armstrong, A.C. (1992). Agricultural field drainage and the changing environment. *SEESOIL*, 8: 53–68.

Rosenburg M. (2015). Sectors of the Economy. *About Education.* Available at: http://geography.about.com/od/urbaneconomicgeography/a/sectorseconomy.htm [Accessed February 2015].

Rosier, P.T.W., Harding, R.J., and Neal, C. (1990). *The Hydrological Impacts of Broadleaved Woodland in Lowland Britain*. Institute of Hydrology Report to the National Rivers Authority no. 2, October 1990, NERC.

Ross, A. and Connell, D. (2014). The evolution of river basin management in the Murray-Darling Basin. In: The Politics of River Basin Organisations, D Huitema and S Meijerink (eds.). Edward Elgar Cheltenham pp. 326–355.

Rowan, J.S, Bradley, S.B., and Walling, D.E. (1992). Fluvial restribution for Chernobyl fallout: reservoir evidence in the Severn Basin. *J. Inst. of Water and Env. Man.*, 6(6): 659–666.

RSPB. (n.d.). *The Uplands*. Available at: http://www.rspb.org.uk/Images/uplands_tcm9-166286.pdf [Accessed March 2015].

Rushton, K.R. (2003). *Groundwater Hydrology: Conceptual and Computational Models*. Wiley 430pp.

Ryden, R.C., Bull, P.R., and Garwood, E.A. (1984). Nitrate leaching from grassland. *Nature*, 2311: 50–53.

Sabatier, P.A., Leach, W.D., Lubell, M., and Pelkey, N.W., (2005). Theoretical frameworks explaining partnership success. In: Swimming Upstream: Collaborative Approaches To

Watershed Management, P.A. Sabatier, W. Focht, M. Lubell, Z. Trachtenberg, A. Vedlitz, M. Matlock (Eds.). MIT Press, Cambridge, MA, pp. 173–199.

Salmon and Trout Association. (n.d.). *About Us*. Available at: http://www.salmon-trout. org/c/about/ [Accessed February 2015].

Saxon Water. (n.d.). Available at: http://www.saxonwater.com/images/UK_Water_ Industry_Map.jpg [Accessed January 2015].

Schiff, A.L. (1962). *Fire and Water: Scientific Heresy in the Forest Service*. Harvard University Press, pp. 116–163.

Scientific Software Group. (2015). FLOWPATH II Detailed Description. Available at: http://www.scisoftware.com/environmental_software/detailed_description. php?products_id=195 [Accessed February 2015].

Schumm, S.A. (1977) *The Fluvial System*. Wiley, New York.

Scotland's Environment. (n.d.). 'Groundwater'. Available at: http://www.environment. scotland.gov.uk/media/54815/Water-Groundwater.pdf [Accessed January 2016].

Scottish Environment Link. (2015). Available at: http://www.macaulay.ac.uk/LINK/ link_part3_25_intro.html [Accessed February 2015].

Selby, K.H. and Skinner, A. C. (1979). Aquifer protection in the Severn-Trent Region: Policy and practice. *Water Pollution Control*, 2: 254–269.

SEPA. (n.d.a). *Trends in Scottish river water quality*. Available at: http://www.sepa.org.uk/ science_and_research/data_and_reports/water/scottish_river_water_quality.aspx [Accessed January 2015].

SEPA. (n.d.b). *Abstraction regime*. Available at: http://www.sepa.org.uk/water/water_ regulation/regimes/abstraction.aspx [Accessed January 2015].

SEPA. (n.d.c). *About us*. Available at: http://www.sepa.org.uk/about_us.aspx [Accessed February 2015].

SEPA. (2003). *Catchment Management Plans*. Available at: http://www.sepa.org.uk/water/ water_publications/catchment_plans.aspx [Accessed February 2015].

SEPA. (n.d.d). *Flood Risk Management*. Available at: http://www.sepa.org.uk/flooding/ flood_risk_management/who_is_responsible.aspx [Accessed February 2015].

SEPA. (n.d.e). *River Basin Planning*. Available at: http://www.sepa.org.uk/water/river_ basin_planning.aspx [Accessed February 2015].

SEPA. (n.d.f). *Climate change homepage*. Available at: http://www.sepa.org.uk/climate_ change.aspx [Accessed February 2015].

SEPA. (n.d.g). *Pollution control*. Available at: http://www.sepa.org.uk/regulations/water/ pollution-control/ [Accessed February 2016].

Scottish Executive. (2002). Climate Change Flooding Occurrences Review. Available at: http://www.scotland.gov.uk/Resource/Doc/156664/0042098.pdf [Accessed February 2015].

Scottish Government. (n.d.a). Key Scottish Environmental Statistics 2012. Available at: http://www.gov.scot/Publications/2012/08/2023/26 [Accessed August 2016].

Scottish Government. (n.d.b). Responsibilities for managing flood risk. Available at: www. scotland.gov.uk/Resource/Doc/921/0052798.doc [Accessed March 2015].

Scottish Government. (n.d.c). 'Building standards, water efficiency'. Available at: http:// www.scotland.gov.uk/resource/buildingstandards/2013Domestic/chunks/ch04s28.html [Accessed February 2015].

Scottish Government. (n.d.d). Water Resource Management. Available at: http://www. scotland.gov.uk/Publications/2009/12/08131259/1 [Accessed February 2015].

Scottish Government. (2008). The Future of Flood Risk Management in Scotland: A Consultation Document. Available at: http://www.scotland.gov.uk/Publications/2008/02/13095729/8 [Accessed February 2015].

Scottish Government and Natural Scotland. (n.d.). The river basin management plan for the Scotland river basin district 2009–2015. Available at: www.sepa.org.uk/water/idoc.ashx?docid=fbcdf339-4d78-4ccb [Accessed February 2015].

Scottish Government. (2010). Public Water Supplies - Water Abstracted and Supplied: 2002/03-2009/10. Available at: http://www.scotland.gov.uk/Publications/2010/09/08094058/24 [Accessed February 2015].

Scottish Government. (2012). Flood Risk Management Act 2009. Available at: http://www.scotland.gov.uk/Topics/Environment/Water/Flooding/FRMAct [Accessed February 2015].

Scottish Government. (2014a). Scottish Water Industry Background. Available at: http://www.scotland.gov.uk/Topics/Business-Industry/waterindustryscot [Accessed January 2015].

Scottish Government. (2014b). Drinking Water Quality Regulator Annual Summary Report. Available at: http://www.scotland.gov.uk/Publications/2014/08/1070/1 [Accessed January 2015].

Scottish Government. (2015). Review of NVZ Designated Areas. Available at: http://www.gov.scot/Topics/farmingrural/Agriculture/Environment/NVZintro/ArchiveNVZNewsGuidance/NVZRevisions2015 [Accessed January 2016].

Scottish Natural Heritage. (2006). Natural Heritage Trends: riparian woodlands in Scotland – 2006 Available at: http://www.snh.org.uk/pdfs/publications/commissioned_reports/Report%20No204.pdf [Accessed February 2015].

Scottish Water. (n.d.). Public Water Supplies - Water abstracted and supplied (Ml/d)(1): 2002/03 to 2013/14. Available at: http://www.scotland.gov.uk/SESO/Resources/StatFiles/WATERpublicwatersupply.xls [Accessed February 2015].

Sheail, J. (1982). Underground water abstraction: indirect effects of urbanisation on the countryside. *J Hist Geog.* 8: 395–408.

Sheail, J. (1988). Regulated rivers in the United Kingdom. *Regulated Rivers, Research and Management*, 2: 201–220.

Sheail, J. (1992). The South Downs and Brighton's water supplies - an inter-war study of resource management. *Southern History*, 14: 93–111.

Sheail, J. (2002). *An Environmental History of the Twentieth Century*. Palgrave.

Shirazi, S.M., Imran, H.M., and Shatirah, A. (2012). GIS-based DRASTIC method for groundwater vulnerability assessment: a review. *Journal of Risk Research* 15(8): 991–1011.

Simmons, E.A. (1995). A concensus on sustainability? In: *Soil Management in Sustainable Agriculture*, H.F. Cook and H.C. Lee (eds.). Wye College Press, pp. 588–590.

Skinner, A.C. (1991). Groundwater -legal controls and organizational aspects. In: *Applied Groundwater Hydrology: A British Perspective*, R.A. Downing and W.B. Wilkinson (eds.). Clarendon Press, Oxford, pp. 8–15.

Skinner, A.C. (1994). *The NRA's groundwater protection policy. Proceedings of Symposium on Groundwater- Managing a Scarce Resource* held 23 March at the Ramada Hotel, Gatwick.

Skinner, K. (2012). 'Test & Itchen River Restoration Strategy Technical Report to the Environment Agency', Atkins, 31st March 2012. Available at: https://consult.environment-agency.gov.uk/file/2444733 [Accessed March 2015].

Skinner, A. and Foster, S. (1995). Managing land to protect water; the British experience in groundwater protection. Paper XXVI *Congress of the International Association of Hydrogeologists,* Edmonton, Canada, June1995.

Smith, K. (1972). *Water in Britain.* Macmillan, London.

Smith, M. (1992). *Expert Consultation on Revision of FAO Methodologies for Crop Water Requirements.* Land and Drainage Division of the Food and Agricultural Organisation of the United Nations, Rome, 60pp.

Smith R.A., Alexander R.B., and Lanfear, K.J. (1994). "Stream Water Quality in the Conterminous United States – Status and Trends of Selected Indicators During the 1980's." National Water Summary 1990-91 -- Stream Water Quality, U.S. Geological Survey Water-Supply Paper 2400 Available at: http://water.usgs.gov/nwsum/sal/summary.html [Accessed August 2016].

Smith, K. and Bennett, AM. (1994). Recently increased river discharge in Scotland: effects on flow hydrology and some implications for water management. *Applied Geography,* 14: 123–133.

Smith, L.E.D., and Porter, K.S., (2009). Management of catchments for the protection of water resources: drawing on the New York City watershed experience. *Reg. Environ. Change.* 10(4):311–326. http://dx.doi.org/10.1007/s10113-009-0102- z.

Smith, L. (2012). Paying for ecosystem services (PES) and the water industry Protecting Water Catchments from Diffuse Pollution - the Emerging Evidence. Royal Society of Chemistry Birmingham, 21 February 2012.

Smith, L.E.D. (2013). The United Kingdom case study: payments for ecosystem services (PES) and collective action – 'Upstream Thinking in the South West of England', in OECD (2013), Providing Agri-environmental Public Goods through Collective Action, OECD Publishing.

Porter, K. and Smith, L. (2015). New York City Watershed program: A national paradigm? In: *Catchment and River Basin Management Integrating Science and Governance,* L. Smith, K. Porter, K. Hiscock, M.J. Porter, and D. Benson. Routledge. 292pp.

SNH – Scottish Natural Heritage. (2010). Catchment management. Available at: http://www.snh.gov.uk/land-and-sea/managing-freshwater/catchment-management/ [Accessed March 2015].

SNIFFER. (2006). An online handbook of climate trends across Scotland. Available at: http://www.climatetrendshandbook.adaptationscotland.org.uk/ [Accessed February 2015].

SOED. (1994). *Public Water Supplies 1992-3, Water Resources Survey.* Engineering, Water and Waste Directorate, Edinburgh, February 1994.

The Somerset Levels and Moors Flood Action Plan. (2014). Available at: http://www.mendip.gov.uk/CHttpHandler.ashx?id=8813&p=0 [Accessed August 2015].

South Downs Joint Committee. (2008). The South Downs Management Plan 2008-2013. Available at: http://www.southdowns.gov.uk/__data/assets/pdf_file/0003/123294/SD_Mgt_Plan_Intro_part_A.pdf [Accessed March 2015].

South Downs National Park Authority. (2011). Official website. Available at: http://www.southdowns.gov.uk/ [Accessed January 2015].

South Downs National Park Authority. (2014). South Downs take action against flooding and soil erosion. Available at: http://www.southdowns.gov.uk/about-us/news/press-notices/south-downs-take-action-against-flooding-and-soil-erosion [Accessed February 2015].

Speed, R. (2013). Basin Water Allocation Planning: Principles, Procedures and Approaches for Basin Allocation Planning. UNESCO. Available at: http://unesdoc.unesco.org/images/0022/002208/220875e.pdf [Accessed January 2014].

Spencer, J.F. and Fleming, J.M. (1987). A comparison of the cooling water screens catches of Dungeness 'A' and 'B' Power Station in relation to their differing cooling water inlet end capping arrangements, CEGB.

Staddon C. (n.d). Do Water Meters Reduce Domestic Consumption?: a summary of available literature. University of the West of England, Bristol. Available at: http://www.heednet.org/metering-DEFRAHEEDnet.pdf [Accessed February 2015].

Stanier P. (2000). *Discover Dorset: Mills*. The Dovecote Press. Wimborne. 80pp.

Stanners, D. and Bourdeau, P. (1995). *Europe's Environment: The Dobris Assessment*. European Environment Agency, Copenhagen, 676pp.

Stansfield, S., Moss, B., and Irvine, K. (1989). The loss of submerged plants with eutrophication HI. Potential role of organochlorine pesticides: a palaeoecological study. *Freshwater Biology*, 22: 109–132.

Stearne, K. and Cook, H. (2015). Water Meadow Management in Wessex: Dynamics of Change from 1800 to the Present Day, *Landscape Research*, 40 (3), 377–395. DOI: 10.1080/01426397.2013.818109.

Stedinger J.R. (2000). Flood Frequency Analysis and statistical estimation of Flood Risk In: *Inland Flood Hazards*, E.E. Wohl (ed.). Cambridge pp. 334–358.

Swart, R., Sedee, a.G.J., de Pater, F., *et al.* (2014). Climate-Proofing Spatial Planning and Water Management Projects: An Analysis of 100 Local and Regional Projects in the Netherlands. *Journal of Environmental Policy & Planning*, 16(1): 55–74. DOI:10.1080/15 23908X.2013.817947

Stokes, S.N. (2001). Modeling Groundwater Vulnerability Using GIS and DRASTIC. University of Texas. Available at: http://www.crwr.utexas.edu/gis/gishydro02/Classroom/trmproj/Stokes/Stokes.htm [Accessed February 2015].

Stopes, C. and Philipps, L. (1992). Organic farming and nitrate leaching. *SEESOIL*, 7: 37–48.

Swann, D. (1988). *The Retreat of the State- Deregulation and Privatisation in the UK and US*. Harvester Wheatsheaf, 344pp.

Sylvester R. (1999). Medieval reclamation of marsh and fen. In: *Water Management in the English Landscape: Field, Marsh and Meadow*, Cook and Williamson (eds.). Edinburgh University Press, pp. 122–140.

Sylvester-Bradley, R. (1993). Fertiliser nitrogen for arable crops. In: *Solving the Nitrate Problem*, MAFF PB, 1092, London, ch. 2.

Taylor C. (1999). Post-medieval drainage of marsh and fen. In: *Water Management in the English Landscape: Field, Marsh and Meadow*, Cook and Williamson (eds.). Edinburgh University Press, pp. 141–156.

Taylor, R. *et al.* (2012). Groundwater and global hydrological change – current challenges and new insight. Xth Kovacs Colloquium held in Paris, 2-3 July 2010, 2010, Paris, France. 338, pp. 51–61. Available at: https://hal.archives-ouvertes.fr/file/index/docid/708163/filename/2010_taylor_etal_kovacs_2010.pdf [Accessed February 2015].

Test and Itchen Catchment Partnership quarterly newsletter no 1. (2014). Available at: http://adlib.everysite.co.uk/resources/000/057/248/Summary_chalk_rivers.pdf [Accessed January 2015].

Tester, D.J. (1994). Statutory Water Quality Objectives on the River Cam. *J. Inst. Water and Env. Man.*, 8(3): 246–255.

Thames Water. (n.d.). *1989-1995 - Investment in infrastructure.* Available at: http://www.thameswater.co.uk/about-us/850_2613.htm [Accessed February 2015].

Thames Water. (2013). *Water resource planning and the statutory process.* Available at: http://www.thameswater.co.uk/about-us/5387.htm [Accessed January 2015].

Thames Water. (2014). Beddington. Available at: http://www.thameswater.co.uk/about-us/17939.htm [Accessed January 2015].

Thames Water. (2015). *Our five-year plan for 2010 to 2015.* Available at: http://www.thameswater.co.uk/about-us/6759.htm [Accessed February 2015].

Thorsen D.E. and Lie, A. (2007, English version, n.d). What is Neoliberalism? Department of Political Science University of Oslo pub online. Available at: http://folk.uio.no/daget/What%20is%20Neo-Liberalism%20FINAL.pdf [Accessed January 2015].

Tiede, K. *et al.* (2011). Review of the risks posed to drinking water by man-made nano-particles. DWI 70/2/246. University of York. Available at: http://dwi.defra.gov.uk/research/completed-research/reports/dwi70_2_246.pdf [Accessed February 2016].

Tinson A. and P. Kenway. (n.d.). The water industry: a case to answer. A report by the New Policy Institute. Unison. Available at: https://www.unison.org.uk/upload/sharepoint/On%20line%20Catalogue/21 [Accessed January 2015].

Todd, D.K. (1980). *Groundwater Hydrology.* Wiley, Chichester, 535pp.

Tolson B. and Shoemaker, C.A. (2007). "Cannonsville Reservoir Watershed SWAT2000 model development, calibration and validation." *Journal of Hydrology*, 337(1-2): 68–89. doi:10.1016/j.jhydrol.2007.01.017

Toplis, F. (1879). Suggestions for dividing England and Wales into watershed Districts. *Journal of the Royal Society of Arts*, 27: 696–709.

Tubbs, C.R. (2001). New Forest, the history, ecology and conservation. New Forest Ninth Centenary Trust, Lyndhurst.

Tutor2u.(n.d). What are externalities? Available at: http://tutor2u.net/economics/content/topics/externalities/what_are_externalities.htm [Accessed February 2015].

TVA - Tenessee valley Authority. (n.d). *Economic Development.* Available at: http://www.tva.com/econdev/index.htm [Accessed February 2015].

T. P. Tylutki, D.G. Fox, and M. McMahon (2011). Implementation of the CuNMPS· Development and evaluation of Alternatives. Available at: https://www.inmpwt.cce.cornell.edu/documents/Handout%2011_15_02%20Tylutki.pdf [Accessed August 2016].

Ullman, D. (1979). *Hydrobiology.* Wiley Interscience, Chichester.

UK Agriculture. (n.d.) *Set-Aside.* Available at: http://www.ukagriculture.com/crops/setaside.cfm [Accessed February 2015].

UK Groundwater Forum. (n.d.). *Industrial and urban pollution of groundwater.* Available at: http://www.groundwateruk.org/downloads/industrial_and_urban_pollution_of_groundwater.pdf [Accessed August 2016].

UK Groundwater Forum. (2011a). UK Thermal and Mineral Springs. Available at: http://www.groundwateruk.org/UK_thermal_springs.aspx [Accessed February 2015].

UK Groundwater Forum. (2011b). Groundwater Projects. Available at: http://www.groundwateruk.org/Shropshire-Groundwater-Scheme.aspx [Accessed February 2015].

UK Parliament. (2005). Memorandum submitted by Water UK (Z34). Available at: http://www.publications.parliament.uk/pa/cm200405/cmselect/cmenvfru/258/258we25.htm [Accessed February 2015].

UK Parliament. (2012). The Water White Paper 2 Managing water resources. Available at: http://www.publications.parliament.uk/pa/cm201213/cmselect/cmenvfru/374/37405.htm [Accessed February 2015].

UK Parliament. (2014). Water Act 2014. Available at: http://www.legislation.gov.uk/ukpga/2014/21/contents/enacted [Accessed February 2016].

UK Technical Advisory Group UKTAG. (2008). UK Environmental standards and conditions (Phase 1). Available at: http://www.wfduk.org/sites/default/files/Media/Environmental%20standards/Environmental%20standards%20phase%201_Finalv2_010408.pdf [Accessed January 2015].

UN. (2014). *Framework Convention on Climate Change*. Available at: http://unfccc.int/adaptation/nairobi_work_programme/knowledge_resources_and_publications/items/5497.php [Accessed February 2015].

UN Documents. (n.d.). Our Common Future, Chapter 2: Towards Sustainable Development. Available at: http://www.un-documents.net/ocf-02.htm [Accessed January 2015].

UNESCO. (2011). Earth Sciences: Geoengineering. Available at: http://www.unesco.org/new/en/natural-sciences/environment/earth-sciences/emerging-issues/geo-engineering/ [Accessed February 2015].

United Utilities. (2015). *About Your Water*. Available at: http://www.unitedutilities.com/About-your-water.aspx [Accessed February 2015].

United Utilities. (n.d). United Utilities Water Business Plan 2010-2015. Available at: http://corporate.unitedutilities.com/documents/B3_-_Section_3_Water_Business_Cases.pdf [Accessed February 2016].

UK Agriculture. (2007). *Life after set-aside*. Available at: http://www.ukagriculture.com/pdfs/Life%20after%20set-aside.pdf [Accessed August 2016].

United States Army Corps of Engineers. (n.d.). The Mississippi Drainage Basin. Available at: http://www.mvn.usace.army.mil/Missions/MississippiRiverFloodControl/MississippiRiverTributaries/MississippiDrainageBasin.aspx [Accessed February 2015].

Upper Susquehanna Coalition. (USC). About the USC. Available at: http://www.u-s-c.org/html/Aboutus.htm [Accessed February 2015].

Upper Thurne Waterspace. (2012). *Management Plan 2012*. Available at: http://www.broads-authority.gov.uk/__data/assets/pdf_file/0005/404258/Upper-Thurne-management-plan-summary.pdf [Accessed March 2015].

USEPA - United States Environmental Protection Agency. (n.d.). *Resource Guide: Resolving Environmental Conflicts in Communities*. Available at: http://www.epa.gov/adr/Resguide.pdf [Accessed February 2015].

USEPA. (2001). *Protecting and Restoring America's Watersheds*. Available at: http://www.epa.gov/owow/protecting/restore725.pdf [Accessed January 2015].

USEPA. (2008). *Handbook for Developing Watershed Plans to Restore and Protect Our Waters*. United States Environmental Protection Agency. EPA 841-B-08-002. Available at: http://water.epa.gov/polwaste/nps/upload/2008_04_18_NPS_watershed_handbook_handbook-2.pdf [Accessed February 2015].

USEPA. (2012). *Adopt Your Watershed*. Available at: http://water.epa.gov/action/adopt/index.cfm [Accessed February 2015].

USEPA. (2014). *Views from the Former Administrators.* Available at: http://www2.epa.gov/aboutepa/views-former-administrators [Accessed February 2015].

USEPA. (2015a). *The Problem.* Available at: http://www2.epa.gov/nutrientpollution/problem [Accessed February 2015].

USEPA. (2015b). *Trading and Marketable Permits.* National Center for Environmental Economics. Available at: http://yosemite.epa.gov/ee/epa/eed.nsf/dcee735e22c76aef8f4116/f2834a11a3177603a09!OpenDocument [Accessed February 2015].

United Utilities (UU). (2016). SCaMP 1 & 2. Available at: http://corporate.unitedutilities.com/cr-scamp.aspx [Accessed February 2016].

USGS. (2014). MODPATH: A Particle-Tracking Model for MODFLOW. Available at: http://water.usgs.gov/ogw/modpath/index.html [Accessed February 2015].

Uson, A. and Cook, H.F. (1995). Water relations in a soil amended with composed organic waste. In: *Soil Management in Sustainable Agriculture,* H.F. Cook and H.C. Lee (eds.). Wye College Press, 453–460.

Utility Week. (2015). *Clear, secure property rights needed for successful water trading.* Available at: http://www.utilityweek.co.uk/news/clear-secure-property-rights-needed-for-successful-water-trading/766232#.VNZN20-zXIV [Accessed February 2015].

Vaguely Interesting. (2012). *Water, water everywhere – but not enough to drink?* http://www.vaguelyinteresting.co.uk/?p=1271 [Accessed February 2015].

Van Herk, C., Van Herk, C.N., Zevenbergen A, and Ashley, R. (2013). Understanding the transition to integrated floodrisk management in the Netherlands. *Environmental Innovation and Societal Transitions,* 15: 84–100.

Van Vliet, M. and Aerts, J.C.J.H. (2014). Adaptation to climate change in urban water management – flood management in the Rotterdam Rijnmond Area. In: Understanding and Managing Urban Water in Transition, R.Q. Grafton, M.B. Ward, K.A. Daniell, C. Nauges, J.D. Rinaudo, and W.W. Chan (eds.). Springer.

Vickers, J. and Wright, V. (1989). The politics of privatisation in Western europe: an overview. In: *The Politics of Privatisation in Western Europe,* J. Vickers and V. Wright (eds.). Frank Cass, London, 145pp.

Vink, M.J., Benson, D., Boezeman, D., Cook, H., Dewulf, A., and Termeer, C. (2014). Do state traditions matter? Comparing deliberative governance initiatives for climate change adaptation in Dutch corporatism and British pluralism. *Water and Climate Change,* 6(1): 71–88.

Walling, D.E. (1979). The hydrological impact of building activity, a study near Exeter. In: *Man's Impact on the Hydrological Cycle in the United Kingdom,* G.E. Hollis (ed.). Geo Abstracts, Norwich, pp. 135–153.

Wastebook. (n.d). The role and function of the Environment Agency. Available at: http://www.wastebook.org/append2.htm [Accessed February 2015].

Walling, D.E. (1990). Linking the field to the river: Sediment delivery from agricultural land. In: *Soil Erosion on Agricultural Land,* J. Boardman, I.D.L. Foster, and J.A. Dearing (eds.). Wiley, Chichester, pp. 129–152.

Water Management Alliance. (n.d.). *Water Level Management Plans.* Available at; http://www.broads-authority.gov.uk/news-and-publications/publications-and-reports/conservation-publications-and-reports/water-conservation-reports/31.-WLMPs.pdf [Accessed February 2015].

The Wandle Trust. (n.d.). South Rivers East Trust. Available at: http://www.wandletrust.org/?page_id=4289 [Accessed February 2015].

Wandle Trust. (2014). The river Wandle: Catchment Plan. Available at: http://www.wandletrust.org/?page_id=193 [Accessed March 2015].

Ward R.C. and Robinson M. *Principles of Hydrology*, (3rd ed) McGraw-Hill, London 366pp

Ward R.C. and Robinson M. (2000). *Principles of Hydrology*, (4th ed) McGraw-Hill, London 450pp.

The Wastebook Appendix. (n.d.). Available at: http://www.wastebook.org/append2.htm [Accessed August 2016].

Waterwise. (2016). Save water. Available at: http://www.waterwise.org.uk/ [Accessed February 2016].

Water Briefing. (2013). Thames Water uses £2m Canadian technology to produce phosphorous-based fertiliser. Available at: http://waterbriefing.org/home/technology-focus/item/8360-thames-water-uses-%C2%A32m-canadian-technology-to-produce-phosphorous-%C2%ADbased-fertiliser [Accessed February 2016].

Water UK. (n.d) *Climate Change adaptation*. Available at: http://www.water.org.uk/policy/environment/climate-change [Accessed January 2014].

Water UK. (2016a). About water UK. Available at: http://www.water.org.uk/ [Accessed February 2016].

Water UK. (2016b). Financing the Water Industry. Available at: http://www.water.org.uk/policy/financing-industry [Accessed February 2016].

Water Act. (2003). Available at: http://www.legislation.gov.uk/ukpga/2003/37/notes/division [Accessed January 2003].

Water Industry Act. (1999). Available at: http://www.legislation.gov.uk/ukpga/1999/9/contents [Accessed January 2015].

Watershed Agricultural Council. (2013). Available at: http://www.nycwatershed.org/ [Accessed February 2015].

Watson, N., Mitchell, B., and Mulamoottil, G. (1996). Integrated resource management: Institutional arrangements regarding nitrate pollution in England. *J. of Env. Planning and Management*, 39(1): 45–64.

Wensum Alliance. (2010). Demonstration Test Catchment Project. Available at: http://www.wensumalliance.org.uk/wensum.html [Accessed February 2015].

Wessex Salmon and Rivers Trust. (2008). Archive for December, 2008. Available at: https://silverrunpublishing.wordpress.com/2008/12/ [Accessed February 2015].

Wessex Trout and Salmon Association. (2015). *Who We Are*. Available at: http://www.wcsrt.org.uk/Who_We_Are [Accessed February 2015].

Wessex Water. (n.d.). Catchment Management. Available at: https://www.wessexwater.co.uk/uploadedFiles/Corporate_Site/Catchment%20management.pdf [Accessed February 2016].

Wessex Water Authority WWA. (1979). *Somerset Local Land Drainage District Land Drainage Survey Report*. Bridgwater, Somerset.

Wellings, S.R. and Bell, J.P. (1980). Movement of water and nitrate in the unsaturated zone of upper chalk near Winchester, Rants, *England. J. Hydrol.*, 48: 119–136.

Wellings, S.R. and Cooper, J.D. (1983). The variability of recharge of the English Chalk aquifer. *Agricultural Water Management*, 6: 243–253.

Welsh Water. (2013). Available at: http://www.dwrcymru.com/en/Company-Information/Glas-Cymru.aspx [Accessed February 2016].

Wheeler, B.D. and Shaw, S.C. (1992). *Valley Fens in East Anglia*. Environmental Consultancy, University of Sheffield, Contract F27-13-09, Report to English Nature, June 1992.

White, R.E. (1979). *Introduction to the Principles and Practice of Soil Science*. Blackwell Scientific, Oxford, 198pp.

Whitaker, W. (1908). The Water Supply of Kent. Memoir of the Geological Survey of England and Wales. HMSO, London, 400pp.

White, S.K., Cook, H.F. and Garraway, J.L. (1996). Nitrate attenuation under fertiliser-free grass strips. In O. Van Cleemput et al. (eds), *Progress in Nitrogen Cycling Studies*, Kluwer Academic Publishers, Netherlands, 695–701.

White, S.K., Cook, H.F., and Garraway, J.L. (1998). Use of fertiliser-free grass buffer strips to attenuate nitrate input to marshland dykes. *Water & Env Man J.*, 12(1): 54–59.

White, P.J. and Hammond, J.P. (2006). *Updating the estimate of the sources of Phosphorus in UK Waters*, Defra funded project, WT0701CSF 57pp.

Wildlife and Countryside Link. (2013). 'Blueprint for water briefing on the abstraction reform consultation'. Available at: http://www.wcl.org.uk/docs/blueprint_for_water_abstraction_reform_consultation_briefing.pdf. [Accessed February 2015].

Willet, I.R. and Porter, K.S. (2001). "Watershed management for water quality improvement: the role of agricultural research." Australian Centre for International Agricultural Research Working Paper No. 52, September 2001 Canberra, Australia.

Williams, M. (1970). *The Draining of the Somerset Levels*. Cambridge University Press.

Williams, R.J., Brooke, D.N., Matthiessen, P., Miles, M., Turnbull, A., and Harrison, R.M. (1995). Pesticide transport to surface waters within an agricultural catchment. *J. Inst. Water and Env. Man.*, 9(1): 72–81.

Williams, G.M., Young, C.P., and Robinson, H.D. (1991). Landfill disposal of wastes. In: *Applied Groundwater Hydrology*, R.A. Downing and W.B. Wilkinson (eds.). Clarendon Press, Oxford, pp. 114–133.

Williamson, T. (1999). Post-medieval field drainage. In: *Water Management in the English Landscape: Field, Marsh and Meadow*, Cook and Williamson, (eds.). Edinburgh University Press, pp. 41–52.

Wilkinson, W.B. and Brassington, F.C. (1991). Rising groundwater levels - an international problem. In: *Applied Groundwater Hydrology*, R.A. Downing and W.B. Wilkinson (eds.). Clarendon Press, Oxford, pp. 35–53.

Wilson, E.M. (1983). *Engineering Hydrology*, 3rd edition. Macmillan Education, London, 309pp.

Withers, B. and Vipond, S. (1974). *Irrigation: Design and Practice*. Batsford, London, 306pp.

Withers, P.J.A. and Sharpley, A.N. (1995). Phosphorus management in sustainable agriculture. In: *Soil Management in sustainable agriculture*, H.F. Cook and H.C. Lee (eds.). Wye College Press, pp. 201–207.

WISE. (n.d.). Intercalibration: A common scale for Europe's waters. Water Note no. 7. Available at: http://ec.europa.eu/environment/water/participation/pdf/waternotes/water_note7_intercalibration.pdf [Accessed March 2015].

Withers, P.J.A. and Sharpley, A.N. (1995). Phosphorus management in sustainable agriculture. In: *Soil Management in sustainable agriculture*, H.F. Cook and H.C. Lee (eds.). Wye College Press, pp. 201–207.

de Witt L. (2005). The Honest Broker Stage of Policymaking: Supplying the Missing Piece to Current Policy Models. Available at: http://www.larrydewitt.org/Essays/HonestBroker.htm [Accessed February 2005].

Worldatlas. (n.d.). 'Countries of the world'. Available at: http://www.worldatlas.com/aatlas/populations/ctypopls.htm [Accessed February 2015].

World Bank. (n.d.) What is sustainability? Available at: http://www.worldbank.org/depweb/english/sd.html [Accessed January 2015].

World Press. (2104). *The Somerset Levels and Moors Flood Action Plan.* Available at: https://somersetnewsroom.files.wordpress.com/2014/03/20yearactionplanfull3.pdf [Accessed January 2015].

World Resources Institute. (WRI). (2003). 'World Resources eAtlas. 16. Environmental Water Scarcity'. Index by Basin. Available at: http://pdf.wri.org/watersheds_2003/gm16.pdf [Accessed February 2015].

WWF - World Wildlife Fund (2010). *Riverside Tales.* Available at: http://assets.wwf.org.uk/downloads/riverside_tales.pdf?_ga=1.107890369.232473216.140915635 [Accessed February 2015].

WWF. (2011). Legal action pays off on water. Available at: http://www.wwf.org.uk/wwf_articles.cfm?unewsid=4881 [Accessed February 2015].

WWT. (2015). Project Focus: Drone imagery helps map flood risk. Available at: http://wwtonline.co.uk/features/project-focus-drone-imagery-helps-map-flood-risk#.VrtB51SLTIU [Accessed February 2016].

Worthing, C.R. and Ranee, R.J. (eds). (1991). *The Pesticide Manual*, 9th edition. The British Crop Protection Council, 1141pp.

Wright, P. (1995). Water resources management in Scotland. *J. Inst. Water and Env. Man.*, 9(2): 153–163.

WRT - Westcountry Rivers Trust. (n.d.a). Available at: http://wrt.org.uk/Westcountry Rivers [Accessed February 2015].

WRT. (n.d.b). Intelligent Catchment Design. Available at: http://www.groundwateruk.org/model/Downloads/Paling_N%28WRT%29.pdf [Accessed February 2015].

Wunder, S. (2005). Payments for Environmental Services: Some Nuts and Bolts. CIFOR, Occasional Paper No.42.

Zaporozec, A. and Vrba, J. (1994). Classification and review of groundwater vulnerability maps. In: *Guidebook on Mapping Groundwater Vulnerability*, J. Vrba and A. Zaporozec (eds.). International Association of Hydrogeologists. Verlag Heinz Heise GmbH & Co., Hanover, pp. 21–29.

Index

The Protection and Conservation of Water Resources, Second Edition. Hadrian F. Cook.
© 2017 John Wiley & Sons Ltd. Published 2017 by John Wiley & Sons Ltd.